John N. Mordeson, Kiran R. Bhutani, Azriel Rosenfeld

Fuzzy Group Theory

Studies in Fuzziness and Soft Computing, Volume 182

Editor-in-chief
Prof. Janusz Kacprzyk
Systems Research Institute
Polish Academy of Sciences
ul. Newelska 6
01-447 Warsaw
Poland
E-mail: kacprzyk@ibspan.waw.pl

John N. Mordeson
Kiran R. Bhutani
Azriel Rosenfeld

Fuzzy
Group Theory

 Springer

John N. Mordeson
Creighton University
Center for Research in
Fuzzy Mathematics and
Computer Science
Omaha, NE 68178
USA
E-mail: mordes@creighton.edu

Kiran R. Bhutani
Department of Mathematics
The Catholic University of America
620 Michigan Avenue
Washington DC 20064
USA
E-mail: bhutani@cua.edu

Azriel Rosenfeld †
Center for Automation Research
University of Maryland
College Park, MD 20742
USA

Library of Congress Control Number: 2005925382

ISSN print edition: 1434-9922
ISSN electronic edition: 1860-0808
ISBN-10 3-540-25072-7 Springer Berlin Heidelberg New York
ISBN-13 978-3-540-25072-2 Springer Berlin Heidelberg New York

Springer is a part of Springer Science+Business Media
springeronline.com
© Springer-Verlag Berlin Heidelberg 2005
Printed in The Netherlands

Typesetting: by the authors and TechBooks using a Springer LATEX macro package
Cover design: E. Kirchner, Springer Heidelberg

Printed on acid-free paper   SPIN: 10936443   89/TechBooks   5 4 3 2 1 0

Preface

Lotfi A. Zadeh introduced the notion of a fuzzy subset of a set in his paper published in 1965. Zadeh's ideas marked a new direction and stirred the interest of researchers worldwide. It provided tools and an approach to model imprecision and uncertainty present in phenomena that do not have sharp boundaries. Rapid theoretical developments and practical applications based on the concept of a fuzzy subset were seen to emerge soon after that.

In 1971, Azriel Rosenfeld used the notion of a fuzzy subset of a set to introduce the notion of a fuzzy subgroup of a group. Rosenfeld's paper inspired the development of fuzzy abstract algebra. He also introduced fuzzy graphs, an area which has been growing actively since then. This is the first book dedicated entirely to the rapidly growing field of fuzzy group theory. It is not easy to present in a single 300 page book all that has been done in fuzzy group theory up-to-date. However, the authors have made a sincere effort to present in a systematic way some results that have appeared in papers and conference proceedings (including some work by the authors themselves). We thank the researchers worldwide for their contributions to this growing field and special thanks to those whose work is referenced in our book. We hope that the reader will find this book crisp and not fuzzy in presentation, as well as rewarding and motivating for developing further results and applications of fuzzy group theory. The material presented in this book has been selected so as to make this a good reference for graduate students and researchers working in fuzzy group theory. The end of each chapter lists numerous references. While some of those have contributed to the material in the chapters, others are directly related to the material presented and so have been listed there.

In Chapter 1, we first present some basic material concerning fuzzy subsets of a set. We then introduce the notion of a fuzzy subgroup of a group and develop some concepts such as normal fuzzy subgroups and complete and weak direct products of fuzzy subgroups. We also present the notion of the fuzzy order of an element of a group.

In Chapter 2, we present several fuzzy versions of Lagrange's Theorem and Caley's Theorem. We consider fuzzy quotient groups, characteristic fuzzy subgroups and conjugate fuzzy subgroups.

The notion of the ascending central series of a fuzzy subgroup is presented in Chapter 3 and used to define nilpotency of a fuzzy subgroup. The notion of the descending central series of a fuzzy subgroup is also presented and used to define the nilpotency of a fuzzy subgroup. It is shown that these two definitions are not equivalent. The notion of commutators to generate the derived chain of a fuzzy subgroup is introduced and is used to define a solvable fuzzy subgroup. Fuzzy versions of well-known crisp results are presented in this chapter.

Fuzzy subgroups of Hamiltonian, solvable, P-Hall, and nilpotent groups are examined in Chapter 4. The notions of generalized characteristic fuzzy subgroups, fully invariant fuzzy subgroups, and characteristic fuzzy subgroups are introduced. It is shown that if G is a finite group all of whose Sylow subgroups are cyclic, then a fuzzy subgroup of a group G is normal if and only if it is a generalized fuzzy subgroup of G. Normal fuzzy subgroups, quasi-normal fuzzy subgroups, (p, q)-subgroups, fuzzy cosets, fuzzy conjugates and $SL(p, q)$-subgroups are also considered in this chapter.

In Chapter 5, we present two approaches to show the existence of free fuzzy subgroups. One features the approach by Garzon and Muganda.

In Chapter 6, we study fuzzy subgroups of Abelian groups. We develop the notions of independent generators, primary fuzzy subgroups, divisible fuzzy subgroups, and pure fuzzy subgroups. We determine a complete system of invariants for those fuzzy subgroups which are direct sums of fuzzy subgroups whose supports are cyclic. We also develop the notions of basic fuzzy subgroups and p-basic fuzzy subgroups.

In Chapter 7, we introduce the notion of the fuzzy direct product of fuzzy subgroups defined over subgroups of a group. These ideas are applied to the problem in group theory of obtaining conditions under which a group G can be expressed as the direct product of its normal subgroups.

The number of fuzzy subgroups of certain finite Abelian groups with respect to a suitable equivalence relation as determined by Murali and Makamba are considered in Chapter 8. The Abelian groups under consideration are those which are direct sums of cyclic groups of prime order and those of order $p^n q^m$ for distinct primes p and q and nonnegative integers n and m.

In Chapter 9, we present the work of Tom Head concerning methods for deriving fuzzy theorems from crisp ones and embedding lattices of fuzzy subgroups into lattices of crisp subgroups. We also present the work of Ajmaal and Thomas concerning properties of lattices of fuzzy subgroups.

The first part of Chapter 10 is concerned with deriving membership functions from similarity relations. An algebraic approach for the construction of fuzzy subgroups is also considered. The chapter closes with some applications of fuzzy subgroups to a generalized recognition problem.

We have done our best to provide the reader with a complete bibliography used for writing this book. We welcome comments and suggestions by the readers and apologize in advance if we inadvertently missed any source of reference in our bibliography.

The authors are grateful to the staffs of Springer-Verlag, especially Frank Holzwarth, Gabriele Maas, Heather King, Janusz Kacprzyk and Dr. Thomas Ditzinger. We are indebted to Dr. Timothy Austin, Dean, Creighton College of Arts and Sciences and to Dr. and Mrs. George Haddix for their support of our work. We also wish to thank Professor Paul Wang of Duke University for his strong support of fuzzy mathematics. The first author dedicates the book to his grandchildren, Emily and twins Emma and Jenna. The second author dedicates the book to her supportive husband Ravi, and loving sons Navin and Manoj. Together, we dedicate this book to the family of Professor Azriel Rosenfeld.

John Mordeson
Creighton University

Kiran R. Bhutani
The Catholic University of America

Azriel Rosenfeld
University of Maryland

We were deeply saddened by the passing away of Professor Rosenfeld in February 2004. His presence would have added further strength to our book. We have lost a great collaborator, an outstanding scientist and a wonderful mentor. He will be missed sorely.

John Mordeson
Kiran R. Bhutani

Azriel Rosenfeld (1931-2004)

Azriel Rosenfeld was a tenured research Professor, a Distinguished University Professor (since 1995), and founder (1983) and Director of the Center for Automation Research, a department level unit of the College of Computer, Mathematical and Physical Sciences at the University of Maryland in College Park. He directed the Center until his retirement in June 2001. Upon his retirement he was designated a Distinguished University Professor Emeritus in the University's Institute for Advanced Studies. He also held affiliate professorships in the Departments of Computer Science and Psychology and in the College of Engineering. He held a Ph. D. in mathematics from Columbia University (1957), rabbinic ordination (1952) and a Doctor of Hebrew Literature degree (1955) from Yeshiva University and honorary Doctor of Technology degrees from Linkoping University, Sweden (1980) and Oulu University, Finland (1994), and an honorary Doctor of Humane Letters degree from Yeshiva University (2000).

Dr. Rosenfeld was widely regarded as the leading researcher in the world in the field of computer image analysis. Over a period of more than 35 years he made many fundamental and pioneering contributions to nearly every area of that field. He wrote the first textbook in the field (1969): was founding editor of its first journal (1972); and was co-chairman of its first international conference (1987). He published 30 books and over 600 book chapters and journal articles, and directed over 50 Ph. D. dissertations.

He was a Fellow of the Institute of Electrical and Electronics Engineers (1971), and won its Emanuel Piore Award in 1985; he was a founding Fellow of the American Association for Artificial Intelligence (1990) and of the Association for Computing Machinery (1993); he was a Fellow of the Washington Academy of Sciences (1998), and won its Mathematics and Computer Science Award in 1988; he was a founding Director of the Machine Vision Association of the Society of Manufacturing Engineers (1985-8), won its President's Award in 1987, and was a certified Manufacturing Engineer (1988); he was a founding member of the IEEE Computer Society's Technical Committee on Pattern Analysis and Machine Intelligence (1965), served as its Chairman (1985-7), and received the Society's Meritorious Service Award in 1986; and its Harry Goode Memorial Award in 1944 and became a Golden Core member of the Society in 1996; he received the IEEE Systems, Man, Cybernetics Norbert Wiener Award in 1995; he received an IEEE Standards Medallion in 1990, and the Electronic Imaging International Imager of the Year Award in 1991; he was a founding member of the Governing Board of the International Association for Pattern Recognition (1978-85), served as its President (1980-2), won its first K. S. Fu Award in 1988, and became one of its founding Fellows in 1994; he was a Foreign Member of the Academy of Science of the German Democratic Republic (1988-92), and was a Corresponding Member of the National Academy of Engineering of Mexico (1982).

In 1985, he served as chairman of a panel appointed by the National Research Council to brief the President's Science Advisor on the subject of computer vision; he has also served (1985-8) as a member of the Vision Committee of the National Research Council. In 1982-3 and 1986-8 he served as a member of a task force appointed by the Defense Science Board to review the state of the art in automatic target recognition, and in 1993-4 he chaired a panel that conducted an assessment of foreign pattern recognition and image understanding research and development.

We were deeply saddened by his death on Sunday, February 22, 2004. His love of knowledge, his passion and dedication to research, his generosity and kindness will always be remembered.

Contents

1

Fuzzy Subsets and Fuzzy Subgroups

The pioneering work of Zadeh on fuzzy subsets of a set in [53] and Rosenfeld on fuzzy subgroups of a group in [43] led to the fuzzification of algebraic structures. In this chapter, we begin the study of fuzzy subgroups of a group.

We write $\wedge$ for minimum or infimum and $\vee$ for maximum or supremum. Throughout this book, we assume that the least upper bound of the empty set is 0 and the greatest lower bound of the empty set is 1. We let X, Y, and Z denote nonempty sets. We let $\mathbb{N}$ denote the set of positive integers, $\mathbb{Z}$ the set of all integers, $\mathbb{Q}$ the set of rational numbers, $\mathbb{R}$ the set of real numbers, and $\emptyset$ the empty set. Unless otherwise stated I denotes an arbitrary nonempty index set. Many of the results of Sections 1.1-1.5 are taken from Yu, Mordeson and Cheng [52]. An extensive list of references of the material used in the development of [52] can be found there, for example, [13], [30], [48] and [51]. Many related papers can be found in the references.

1.1 Fuzzy Subsets

In this section, we present some basic concepts of fuzzy set theory. In particular, we present the important principle called the extension principle.

Definition 1.1.1. *A **fuzzy subset** of X is a function from X into $[0,1]$. The set of all fuzzy subsets of X is called the **fuzzy power set** of X and is denoted by $\mathcal{FP}(X)$.*

Definition 1.1.2. *Let $\mu \in \mathcal{FP}(X)$. Then the set $\{\mu(x) \mid x \in X\}$ is called the **image** of μ and is denoted by $\mu(X)$ or $Im(\mu)$. The set $\{x \mid x \in X, \mu(x) > 0\}$, is called the **support** of μ and is denoted by μ^*. In particular, μ is called a **finite fuzzy subset** if μ^* is a finite set, and an **infinite fuzzy subset** otherwise.*

If $\mu \in \mathcal{FP}(X)$, then μ is said to have the **sup property** if every subset of $\mu(X)$ has a maximal element.

Definition 1.1.3. *Let* $\dot{Y} \subseteq X$ *and* $a \in [0,1]$. *We define* $a_Y \in \mathcal{FP}(X)$ *as follows:*

$$a_Y(x) = \begin{cases} a \text{ for } x \in Y \\ 0 \text{ for } x \in X \backslash Y. \end{cases}$$

In particular, if Y *is a singleton, say* $\{y\}$, *then* $a_{\{y\}}$ *is called a **fuzzy point** (or **fuzzy singleton**), and is sometimes denoted by* y_a. *Let* 1_Y *denote the* **characteristic function** *of* Y. *If* $\mathcal{S}$ *is a set of fuzzy singletons, then we let* $\text{foot}(\mathcal{S}) = \{y \in X \mid y_a \in \mathcal{S}\}$.

Definition 1.1.4. *Let* $\mu, \nu \in \mathcal{FP}(X)$. *If* $\mu(x) \leq \nu(x) \; \forall \; x \in X$, *then* μ *is said to be **contained** in* ν *(or* ν **contains** μ*), and we write* $\mu \subseteq \nu$ *(or* $\nu \supseteq \mu$*). If* $\mu \subseteq \nu$ *and* $\mu \neq \nu$, *then* μ *is said to be **properly contained** in* ν *(or* ν **properly contains** μ*) and we write* $\mu \subset \nu$ *(or* $\nu \supset \mu$*).*

Clearly, the inclusion relation $\subseteq$ is a partial ordering on $\mathcal{FP}(X)$.

Definition 1.1.5. *Let* $\mu, \nu \in \mathcal{FP}(X)$. *Define* $\mu \cup \nu$ *and* $\mu \cap \nu \in \mathcal{FP}(X)$ *as follows:* $\forall \; x \in X$,

$$(\mu \cup \nu)(x) = \mu(x) \vee \nu(x),$$
$$(\mu \cap \nu)(x) = \mu(x) \wedge \nu(x).$$

Then $\mu \cup \nu$ *and* $\mu \cap \nu$ *are called the **union** and **intersection** of* μ *and* ν, *respectively.*

For any collection, $\{\mu_i \mid i \in I\}$, of fuzzy subsets of X, where I is a nonempty index set, the least upper bound $\cup_{i \in I} \mu_i$ and the greatest lower bound $\cap_{i \in I} \mu_i$ of the μ_i's are given by $\forall x \in X$,

$$(\cup_{i \in I} \mu_i)(x) = \vee_{i \in I} \mu_i(x)$$
$$(\cap_{i \in I} \mu_i)(x) = \wedge_{i \in I} \mu_i(x),$$

respectively. We write

$$\cap_{i \in I} \mu_i = \cap_{i=1}^n \mu_i = \mu_1 \cap \mu_2 \cap ... \cap \mu_n$$

and

$$\cup_{i \in I} \mu_i = \cup_{i=1}^n \mu_i = \mu_1 \cup \mu_2 \cup ... \cup \mu_n$$

if $I = \{1, 2, ..., n\}$.

Definition 1.1.6. *Let* $\mu \in \mathcal{FP}(X)$. *For* $a \in [0,1]$, *define* μ_a *as follows:*

$$\mu_a = \{x \mid x \in X, \mu(x) \geq a\}.$$

μ_a *is called the* a-**cut** *(or* a-**level set** *) of* μ.

It follows easily that for all $\mu, \nu \in \mathcal{FP}(X)$,
  (1) $\mu \subseteq \nu$, $a \in [0,1] \Rightarrow \mu_a \subseteq \nu_a$,
  (2) $a \leq b$, $a, b \in [0,1] \Rightarrow \mu_b \subseteq \mu_a$,
  (3) $\mu = \nu \Leftrightarrow \mu_a = \nu_a \; \forall a \in [0,1]$.
  The next two theorems state some basic properties of cuts. Their proofs are straightforward.

Theorem 1.1.7. *Suppose that $\{\mu_i \mid i \in I\} \subseteq \mathcal{FP}(X)$. Then for any $a \in [0,1]$,*
(1) $\cup_{i \in I}(\mu_i)_a \subseteq (\cup_{i \in I}\mu_i)_a$,
(2) $\cap_{i \in I}(\mu_i)_a = (\cap_{i \in I}\mu_i)_a$,
Moreover, if I is finite, then we have equality in (1).

Theorem 1.1.8. *Let $\mu \in \mathcal{FP}(X)$ and $\{a_i \mid i \in I\}$ be a non-empty subset of $[0,1]$. Let $b = \wedge_{i \in I} a_i$ and $c = \vee_{i \in I} a_i$. Then the following assertions hold:*
(1) $\cup_{i \in I} \mu_{a_i} \subseteq \mu_b$,
(2) $\cap_{i \in I} \mu_{a_i} = \mu_c$.

Theorem 1.1.9. *Let $\mu \in \mathcal{FP}(X)$. Then $\mu = \cup_{a \in [0,1]} a_{\mu_a} = \cup_{a \in \mu(X)} a_{\mu_a}$.*

Proof. Let $x \in X$. Then $(\cup_{a \in [0,1]} a_{\mu_a})(x) = \vee_{a \in [0,1]} a_{\mu_a}(x) = \vee\{a \in [0,1] \mid a \leq \mu(x)\} = \mu(x)$. Thus $\mu = \cup_{a \in [0,1]} a_{\mu_a}$. Similarly, $\mu = \cup_{a \in \mu(X)} a_{\mu_a}$. $\square$

Definition 1.1.10. *Let I be a nonempty index set and let $\{X_i \mid i \in I\}$ be a collection of nonempty sets. Let X denote the Cartesian product of the X_i's, namely,*

$$X = \prod_{i \in I} X_i = \{(x_i)_{i \in I} \mid x_i \in X_i, i \in I\}.$$

Let $\mu_i \in \mathcal{FP}(X_i)$ for all $i \in I$. Define the fuzzy subset μ of X by

$$\mu(x) = \wedge_{i \in I} \mu_i(x_i)$$

$\forall x = (x_i)_{i \in I} \in X$. Then μ is called the **complete direct product** of the μ_i's and is denoted by

$$\mu = \widetilde{\prod_{i \in I}} \mu_i.$$

If $I = \{1, 2, \ldots, n\}$, then

$$X = \prod_{i \in I} X_i$$
$$= X_1 \times X_2 \times \ldots \times X_n$$
$$= \{(x_1, x_2, \ldots, x_n) \mid x_i \in X_i, i = 1, 2, \ldots, n\},$$

and we write

$$\widetilde{\prod_{i \in I}} \mu_i = \mu_1 \widetilde{\otimes} \mu_2 \widetilde{\otimes} \ldots \widetilde{\otimes} \mu_n.$$

Clearly, if $\mu_i, \nu_i \in \mathcal{FP}(X_i)$ with $\mu_i \subseteq \nu_i$ for all $i \in I$, then

$$\widetilde{\prod_{i \in I}} \mu_i \subseteq \widetilde{\prod_{i \in I}} \nu_i.$$

Definition 1.1.11 (Extension Principle). *Let f be a function from X into Y, and let $\mu \in \mathcal{FP}(X)$ and $\nu \in \mathcal{FP}(Y)$. Define the fuzzy subsets $f(\mu) \in \mathcal{FP}(Y)$ and $f^{-1}(\nu) \in \mathcal{FP}(X)$ by $\forall y \in Y$,*

$$f(\mu)(y) = \begin{cases} \vee\{\mu(x) \mid x \in X, f(x) = y\} & \text{if } f^{-1}(y) \neq \emptyset, \\ 0 & \text{otherwise} \end{cases}$$

and $\forall x \in X$,

$$f^{-1}(\nu)(x) = \nu(f(x)).$$

*Then $f(\mu)$ is called the **image** of μ under f and $f^{-1}(\nu)$ is called the **preimage** (or **inverse image**) of ν under f.*

Recall that in Definition 1.1.11, the least upper bound of the empty set is the zero element.

Theorem 1.1.12. *Let f be a function from X into Y and g a function from Y into Z. Then the following assertions hold.*

(1) *For all $\mu_i \in \mathcal{FP}(X)$, $i \in I$, $f(\cup_{i \in I}\mu_i) = \cup_{i \in I}f(\mu_i)$ and so*

$$\mu_1 \subseteq \mu_2 \Rightarrow f(\mu_1) \subseteq f(\mu_2) \ \forall \mu_1, \mu_2 \in \mathcal{FP}(X).$$

(2) *For all $\nu_j \in \mathcal{FP}(Y)$, $j \in J$, where J is a nonempty index set,*

$$f^{-1}(\cup_{j \in J}\nu_j) = \cup_{j \in J}f^{-1}(\nu_j),$$
$$f^{-1}(\cap_{j \in J}\nu_j) = \cap_{j \in J}f^{-1}(\nu_j),$$

and therefore

$$\nu_1 \subseteq \nu_2 \Rightarrow f^{-1}(\nu_1) \subseteq f^{-1}(\nu_2) \ \forall \nu_1, \nu_2 \in \mathcal{FP}(Y).$$

(3) *$f^{-1}(f(\mu)) \supseteq \mu \ \forall \mu \in \mathcal{FP}(X)$. In particular, if f is an injection, then*

$$f^{-1}(f(\mu)) = \mu \ \forall \mu \in \mathcal{FP}(X).$$

This means $\mu \mapsto f(\mu)$ is an injection from $\mathcal{FP}(X)$ into $\mathcal{FP}(Y)$ and $\nu \mapsto f^{-1}(\nu)$ is a surjection from $\mathcal{FP}(Y)$ onto $\mathcal{FP}(X)$.

(4) *$f(f^{-1}(\nu)) \subseteq \nu \ \forall \nu \in \mathcal{FP}(Y)$. In particular, if f is a surjection, then*

$$f(f^{-1}(\nu)) = \nu \ \forall \nu \in \mathcal{FP}(Y),$$

and therefore $\mu \mapsto f(\mu)$ is a surjection from $\mathcal{FP}(X)$ onto $\mathcal{FP}(Y)$ and $\nu \mapsto f^{-1}(\nu)$ is an injection from $\mathcal{FP}(Y)$ into $\mathcal{FP}(X)$.

(5) *$f(\mu) \subseteq \nu \Leftrightarrow \mu \subseteq f^{-1}(\nu) \ \forall \mu \in \mathcal{FP}(X)$ and $\forall \nu \in \mathcal{FP}(Y)$.*

(6) *$g(f(\mu)) = (g \circ f)(\mu) \ \forall \mu \in \mathcal{FP}(X)$ and*

$$f^{-1}(g^{-1}(\xi)) = (g \circ f)^{-1}(\xi) \ \forall \xi \in \mathcal{FP}(Z).$$

Proof. Using the extension principle Definition 1.1.11 and the definitions following Definition 1.1.5, one can show that (1) and (2) hold. To prove (3), consider any $\mu \in \mathcal{FP}(X)$. Then

$$
\begin{aligned}
f^{-1}(f(\mu))(x) &= f(\mu)(f(x)) \\
&= \vee\{\mu(x') \mid x' \in X, f(x') = f(x)\} \\
&\supseteq \mu(x)
\end{aligned}
$$

$\forall x \in X$. In particular, if f is an injection, then

$$
\begin{aligned}
f^{-1}(f(\mu))(x) &= \vee\{\mu(x') \mid x' \in X, f(x') = f(x)\} \\
&= \mu(x)
\end{aligned}
$$

$\forall\, x \in X$. Hence (3) is true.

To prove (4), we consider any $\nu \in \mathcal{FP}(Y)$. Then

$$
\begin{aligned}
f(f^{-1}(\nu))(y) &= \vee\{f^{-1}(\nu)(x) \mid x \in X, f(x) = y\} \\
&= \vee\{\nu(f(x)) \mid x \in X, f(x) = y\} \\
&= \begin{cases} \nu(y) \text{ for } y \in f(X) \\ 0 \quad \text{ otherwise} \end{cases} \\
&\leq \nu(y)
\end{aligned}
$$

$\forall\, y \in Y$. Thus $f(f^{-1}(\nu))(y) = \nu(y) \,\forall\, y \in Y$ if f is a surjection. Hence assertion (4) holds.

Assertion (5) is an immediate consequence of assertions (1) through (4).

To prove assertion (6), we consider any $\mu \in \mathcal{FP}(X)$ and any $z \in Z$. Then

$$
\begin{aligned}
g(f(\mu))(z) &= \vee\{f(\mu)(y) \mid y \in Y, g(y) = z\} \\
&= \vee\{\vee\{\mu(x) \mid x \in X, f(x) = y\} \mid y \in Y, g(y) = z\} \\
&= \vee\{\mu(x) \mid x \in X, (g \circ f)(x) = z\} \\
&= (g \circ f)(\mu)(z)
\end{aligned}
$$

$\forall\, z \in Z$. Further, for all $\xi \in \mathcal{FP}(Z)$ and $\forall\, x \in X$,

$$
\begin{aligned}
((g \circ f)^{-1}(\xi))(x) &= \xi(g(f(x))) \\
&= g^{-1}(\xi)(f(x)) \\
&= f^{-1}(g^{-1}(\xi))(x).
\end{aligned}
$$

$\square$

1.2 Fuzzy Subgroups

For the remainder of this chapter, G denotes an arbitrary group with a multiplicative binary operation and identity e. In order to define the notion of a fuzzy subgroup and to examine its properties, we introduce some operations on a fuzzy subset of a group G in terms of the group operation.

Definition 1.2.1. *We define the binary operation "∘" on $\mathcal{FP}(G)$ and the unary operation $^{-1}$ on $\mathcal{FP}(G)$ as follows: $\forall \mu, \nu \in \mathcal{FP}(G)$ and $\forall x \in G$,*

$$(\mu \circ \nu)(x) = \vee\{\mu(y) \wedge \nu(z) \mid y, z \in G, yz = x\} \text{ and } \mu^{-1}(x) = \mu(x^{-1}).$$

*We call $\mu \circ \nu$ the **product** of μ and ν, and μ^{-1} the **inverse** of μ.*

It is easy to verify that the binary operation ∘ in Definition 1.2.1 is associative.

The next theorem can be proved easily using previously defined notions and thus we omit its proof.

Theorem 1.2.2. *Let $\mu, \nu, \mu_i \in \mathcal{FP}(G)$, $i \in I$. Let $a = \vee\{\mu(x) \mid x \in G\}$. Then the following assertions hold:*

(1) $(\mu \circ \nu)(x)$
$$= \vee_{y \in G}(\mu(y) \wedge \nu(y^{-1}x))$$
$$= \vee_{y \in G}(\mu(xy^{-1}) \wedge v(y))$$

$\forall x \in G$;

(2) $(a_y \circ \mu)(x) = \mu(y^{-1}x) \ \forall x, y \in G$;
(3) $(\mu \circ a_y)(x) = \mu(xy^{-1}) \ \forall x, y \in G$;
(4) $(\mu^{-1})^{-1} = \mu$;
(5) $\mu \subseteq \mu^{-1}$

$$\Leftrightarrow \mu^{-1} \subseteq \mu$$
$$\Leftrightarrow \mu = \mu^{-1}$$
$$\Leftrightarrow \mu(x) \leq \mu(x^{-1}) \ \forall x \in G$$
$$\Leftrightarrow \mu(x^{-1}) \leq \mu(x) \ \forall x \in G$$
$$\Leftrightarrow \mu(x) = \mu(x^{-1}) \ \forall x \in G;$$

(6) $\mu \subseteq \nu \Leftrightarrow \mu^{-1} \subseteq \nu^{-1}$;
(7) $(\cup_{i \in I} \mu_i)^{-1} = \cup_{i \in I} \mu_i^{-1}$;
(8) $(\cap_{i \in I} \mu_i)^{-1} = \cap_{i \in I} \mu_i^{-1}$;
(9) $(\mu \circ \nu)^{-1} = \nu^{-1} \circ \mu^{-1}$.

Definition 1.2.3. *Let $\mu \in \mathcal{FP}(G)$. Then μ is called a **fuzzy subgroup** of G if*

(1) $\mu(xy) \geq \mu(x) \wedge \mu(y) \ \forall x, y \in G$ and
(2) $\mu(x^{-1}) \geq \mu(x) \ \forall x \in G$.

Denote by $\mathcal{F}(G)$, the set of all fuzzy subgroups of G. If $\mu \in \mathcal{F}(G)$, we let

$$\mu_* = \{x \in G \mid \mu(x) = \mu(e)\}$$

and recall from Definition 1.1.2 that μ^* denotes the support of μ. If $\mu \in \mathcal{FP}(G)$ satisfies condition (1) of Definition 1.2.3, then

$$\mu(x^n) \geq \mu(x) \ \forall x \in G,$$

where $n \in \mathbb{N}$. Also, μ satisfies conditions (1) and (2) of Definition 1.2.3 if and only if $\mu(xy^{-1}) \geq \mu(x) \wedge \mu(y) \ \forall x, y \in G$.

Comment 1.2.4. If $\mu \in \mathcal{F}(G)$ and H is a subgroup of G, then $\mu|_H \in \mathcal{F}(H)$.

Lemma 1.2.5. *Let $\mu \in \mathcal{F}(G)$. Then $\forall\ x \in G$,*
   (1) $\mu(e) \geq \mu(x)$;
   (2) $\mu(x) = \mu(x^{-1})$.

Proof. Let $x \in G$.
   (1) $\mu(e) = \mu(xx^{-1}) \geq \mu(x) \wedge \mu(x^{-1}) \geq \mu(x) \wedge \mu(x) = \mu(x)$.
   (2) $\mu(x) = \mu((x^{-1})^{-1}) \geq \mu(x^{-1}) \geq \mu(x)$. Hence $\mu(x) = \mu(x^{-1})$.   □

We note that if μ is a fuzzy subgroup of a group G and if $x, y \in G$ with $\mu(x) \neq \mu(y)$, then $\mu(xy) = \mu(x) \wedge \mu(y)$: Suppose $\mu(x) > \mu(y)$. Then $\mu(y) = \mu(x^{-1}xy) \geq \mu(x^{-1}) \wedge \mu(xy) = \mu(x) \wedge \mu(xy)$. Thus $\mu(y) \geq \mu(x) \wedge \mu(xy)$ and since $\mu(x) \geq \mu(y)$, it follows that $\mu(y) \geq \mu(xy) \geq \mu(x) \wedge \mu(y) = \mu(y)$. Thus $\mu(xy) = \mu(x) \wedge \mu(y)$. A similar argument can be used for the case $\mu(y) > \mu(x)$.

Lemma 1.2.6. *Let $\mu \in \mathcal{FP}(G)$. Then μ is a fuzzy subgroup of G if and only if μ_a is a subgroup of $G\ \forall\ a \in \mu(G) \cup \{b \in [0,1] \mid b \leq \mu(e)\}$.*

Proof. Suppose μ is a fuzzy subgroup of G and let $a \in \mu(G)$. Since $\mu(e) \geq \mu(x)$ $\forall\ x \in G$, $e \in \mu_a$. Thus $\mu_a \neq \emptyset$. Let $x, y \in \mu_a$. Then $\mu(x) \geq a$ and $\mu(y) \geq a$. Since μ is a fuzzy subgroup, $\mu(xy^{-1}) \geq \mu(x) \wedge \mu(y) \geq a \wedge a = a$. Hence $xy^{-1} \in \mu_a$ and so μ_a is a subgroup of G . Similarly, if $a \leq \mu(e)$, then one can show that μ_a is a subgroup of G. Conversely, suppose μ_a is a subgroup of G $\forall\ a \in \mu(G) \cup \{b \in [0,1] \mid b \leq \mu(e)\}$. Then $\forall a \in \mu(G)$ we must have $e \in \mu_a$ and so it follows that $\mu(e) \geq a$. Let $x, y \in G$ and let $\mu(x) = a$ and $\mu(y) = b$. Let $c = a \wedge b$. Then $x, y \in \mu_c$ and $c \leq \mu(e)$. By hypothesis, μ_c is a subgroup of G and so $xy^{-1} \in \mu_c$. Hence $\mu(xy^{-1}) \geq c = a \wedge b = \mu(x) \wedge \mu(y)$. Thus μ is a fuzzy subgroup of G.   □

The next three results can be proved using previous notions.

Corollary 1.2.7. *If $\mu \in \mathcal{F}(G)$, then μ_* is a subgroup of G.*

Theorem 1.2.8. *If $\mu \in \mathcal{F}(G)$, then μ^* is a subgroup of G.*

Theorem 1.2.9. *Let $\mu \in \mathcal{FP}(G)$. Then $\mu \in \mathcal{F}(G)$ if and only if μ satisfies the following conditions:*
(1) $\mu \circ \mu \subseteq \mu$;
(2) $\mu^{-1} \subseteq \mu$ *(or $\mu^{-1} \supseteq \mu$, or $\mu^{-1} = \mu$).*

Theorem 1.2.10. *Let $\mu, \nu \in \mathcal{F}(G)$. Then $\mu \circ \nu \in \mathcal{F}(G)$ if and only if $\mu \circ \nu = \nu \circ \mu$.*

Proof. Suppose $\mu \circ \nu \in \mathcal{F}(G)$. Then $\mu \circ \nu = \mu^{-1} \circ \nu^{-1} = (\nu \circ \mu)^{-1} = \nu \circ \mu$.
   Conversely, suppose that $\mu \circ \nu = \nu \circ \mu$. Then $(\mu \circ \nu)^{-1} = (\nu \circ \mu)^{-1} = \mu^{-1} \circ \nu^{-1} = \mu \circ \nu$ and

$$(\mu \circ \nu) \circ (\mu \circ \nu) = \mu \circ (\nu \circ \mu) \circ \nu$$
$$= \mu \circ (\mu \circ \nu) \circ \nu$$
$$= (\mu \circ \mu) \circ (\nu \circ \nu)$$
$$\subseteq \mu \circ \nu.$$

Consequently, by Theorem 1.2.9, $\mu \circ \nu \in \mathcal{F}(G)$.   □

Theorem 1.2.11. *Let $\mu \in \mathcal{F}(G)$ and H be a group. Suppose that f is a homomorphism of G into H. Then $f(\mu) \in \mathcal{F}(H)$.*

Proof. Let $u, v \in H$. Suppose either $u \notin f(G)$ or $v \notin f(G)$. Then $f(\mu)(u) \wedge f(\mu)(v) = 0 \leq f(\mu)(uv)$. Now assume $u \notin f(G)$. Then $u^{-1} \notin f(G)$. Thus $f(\mu)(u) = 0 = f(\mu)(u^{-1})$. Now suppose $u = f(x)$ and $v = f(y)$ for some $x, y \in G$. Then

$$(f(\mu))(uv) = \vee\{\mu(z) \mid z \in G, \ f(z) = uv\}$$
$$\geq \vee\{\mu(xy) \mid x, y \in G, \ f(x) = u, \ f(y) = v\}$$
$$\geq \vee\{\mu(x) \wedge \mu(y) \mid x, y \in G, \ f(x) = u, \ f(y) = v\}$$
$$= (\vee\{\mu(x) \mid x \in G, \ f(x) = u\}) \wedge (\vee\{\mu(y) \mid y \in G, \ f(y) = v\})$$
$$= (f(\mu))(u) \wedge (f(\mu))(u).$$

Also $(f(\mu))(u^{-1}) = \vee\{\mu(z) \mid z \in H, \ f(z) = u^{-1}\} = \vee\{\mu(z^{-1}) \mid z \in H, \ f(z^{-1}) = u\} = (f(\mu))(u)$. Hence $f(\mu) \in \mathcal{F}(H)$.   □

Theorem 1.2.12. *Let H be a group and $\nu \in \mathcal{F}(H)$. Let f be a homomorphism of G into H. Then $f^{-1}(\nu) \in \mathcal{F}(G)$.*

Proof. Let $x, y \in G$. Then $f^{-1}(\nu)(xy) = \nu(f(xy)) = \nu(f(x)f(y)) \geq \nu(f(x)) \wedge \nu(f(y)) = f^{-1}(\nu)(x) \wedge f^{-1}(\nu)(y)$. Further, $f^{-1}(\nu)(x^{-1}) = \nu(f(x^{-1})) = \nu(f(x)^{-1}) = \nu(f(x)) = f^{-1}(\nu)(x)$. Hence $f^{-1}(\nu) \in \mathcal{F}(G)$.   □

Theorem 1.2.13. *Let $\{\mu_i \mid i \in I\} \subseteq \mathcal{F}(G)$. Then $\cap_{i \in I} \mu_i \in \mathcal{F}(G)$.*

Proof. Let $x, y \in G$. Then
$$(\cap_{i \in I} \mu_i)(xy^{-1}) = \wedge\{\mu_i(xy^{-1}) \mid i \in I\} \geq \{\mu_i(x) \wedge \mu_i(y) \mid i \in I\}$$
$$= (\wedge\{\mu_i(x) \mid i \in I\}) \wedge (\wedge\{\mu_i(y) \mid i \in I\})$$
$$= (\cap_{i \in I} \mu_i)(x) \wedge (\cap_{i \in I} \mu_i)(y).$$ □

Definition 1.2.14. *Let $\mu \in \mathcal{FP}(G)$. Let*

$$\langle \mu \rangle = \cap\{\nu \mid \mu \subseteq \nu, \nu \in \mathcal{F}(G)\}.$$

*Then $\langle \mu \rangle$ is called the fuzzy subgroup of G **generated** by μ.*

Clearly, $\langle \mu \rangle$ is the smallest fuzzy subgroup of G which contains μ. We now present another procedure of constructing $\langle \mu \rangle$. Define $\mu^1 = \mu$ and $\mu^n = \mu^{n-1} \circ \mu \ \forall n \in \mathbb{N}, \ n > 1$.

Theorem 1.2.15. *Let $\mu \in \mathcal{FP}(G)$ and let $a = \wedge\{\eta(e) \mid \mu \subseteq \eta, \eta \in \mathcal{F}(G)\}$.*
Then

$$\langle \mu \rangle = e_a \cup \left(\cup_{n=1}^{\infty}(\mu \cup \mu^{-1})^n\right) = \cup_{n=1}^{\infty}(e_a \cup \mu \cup \mu^{-1})^n.$$

Proof. Let $\nu = e_a \cup \left(\cup_{n=1}^{\infty}(\mu \cup \mu^{-1})^n\right)$.

For all $x \in G$, $\nu(x^{-1}) = \left(e_a \cup (\cup_{n=1}^{\infty}(\mu \cup \mu^{-1})^n)\right)(x^{-1})$
$= e_a(x^{-1}) \vee \left(\cup_{n=1}^{\infty}(\mu \cup \mu^{-1})^n\right)(x^{-1}) = e_a(x^{-1}) \vee \left(\vee_{n=1}^{\infty}(\mu \cup \mu^{-1})^n(x^{-1})\right)$
$= e_a(x) \vee \left(\vee_{n=1}^{\infty}(\mu \cup \mu^{-1})^n(x)\right) = \left(e_a \cup (\cup_{n=1}^{\infty}(\mu \cup \mu^{-1})^n)\right)(x) = \nu(x)$.

Let $x, y \in G$ and let $n \in \mathbb{N}$ be such that $n \geq 2$.

Then $(\mu \cup \mu^{-1})^n(xy) = \vee\{(\mu \cup \mu^{-1})(\nu_n) \wedge \cdots \wedge (\mu \cup \mu^{-1})(\nu_1) | xy = \nu_n \cdots \nu_1, \nu_i \in G, i = 1, \ldots, n\} \geq \vee\{(\mu \cup \mu^{-1})(x_1) \wedge \cdots \wedge (\mu \cup \mu^{-1})(x_k) \wedge (\mu \cup \mu^{-1})(y_{k+1}) \wedge \cdots \wedge (\mu \cup \mu^{-1})(y_n)\} | x = x_1 \ldots x_k, \ y = y_{k+1} \ldots y_n, \ x_i, y_j \in G, i = 1, \ldots, k, j = k+1, \ldots, n; \ k \in \{1, 2, \ldots, n-1\}\} \geq \vee\{(\mu \cup \mu^{-1})(x_1) \wedge \cdots \wedge (\mu \cup \mu)^{-1}(x_k) | x = x_1 \ldots x_k, \ x_i \in G, \ i = 1, \ldots, k\} \wedge \vee\{(\mu \cup \mu^{-1})(y_{k+1}) \wedge \cdots \wedge (\mu \cup \mu^{-1})(y_n) | y = y_{k+1} \ldots y_n, y_j \in G, j : k+1, \ldots, n\} = (\mu \cup \mu^{-1})^k(x) \wedge (\mu \cup \mu^{-1})^{n-k}(y)$ for $k \in \{1, \ldots, n-1\}$. Hence $(\mu \cup \mu^{-1})^n(xy) \geq (\mu \cup \mu^{-1})^k(x) \wedge (\mu \cup \mu^{-1})^{n-k}(y), \ k = 1, \ldots, n-1$.

Thus $\left(\cup_{n=1}^{\infty}(\mu \cup \mu^{-1})^n\right)(xy) \geq (\mu \cup \mu^{-1})^k(x) \wedge (\mu \cup \mu^{-1})^{n-k}(y), \forall k = 1, \ldots, n-1$. Hence

$$\left(\cup_{n=1}^{\infty}(\mu \cup \mu^{-1})^n\right)(xy) \geq \left(\vee_{n=1}^{\infty}(\mu \cup \mu^{-1})^n\right)(x) \wedge \left(\vee_{n=1}^{\infty}(\mu \cup \mu^{-1})^n\right)(y).$$

Thus it follows that ν is a fuzzy subgroup of G. Clearly, $\mu \subseteq \nu$. Hence $\langle \mu \rangle \subseteq \nu$. Let ξ be a fuzzy subgroup of G such that $\mu \subseteq \xi$. Then $e_a \subseteq \xi$ and $(\mu \cup \mu^{-1})^n \subseteq \xi$. Thus $\nu \subseteq \xi$. Hence $\nu \subseteq \langle \mu \rangle$. □

1.3 Normal Fuzzy Subgroups

The notion of a normal subgroup is one of the central concepts of classical group theory. It plays an important role in the study of the general structure of groups. Just as a normal subgroup plays an important role in the classical group theory, a normal fuzzy subgroup plays a similar role in the theory of fuzzy subgroups.

Theorem 1.3.1. *Let $\mu \in \mathcal{FP}(G)$. Then the following assertions are equivalent:*

(1) $\mu(yx) = \mu(xy) \ \forall \ x, y \in G$; *in this case, μ is called an* **Abelian** *fuzzy subset of G.*

(2) $\mu(xyx^{-1}) = \mu(y) \ \forall \ x, y \in G$.

(3) $\mu(xyx^{-1}) \geq \mu(y) \ \forall \ x, y \in G$.

(4) $\mu(xyx^{-1}) \leq \mu(y) \ \forall \ x, y \in G$.

(5) $\mu \circ \nu = \nu \circ \mu \ \forall \ \nu \in \mathcal{FP}(G)$.

Proof. (1) $\Rightarrow$ (2): Let $x, y \in G$. Then $\mu(xyx^{-1}) = \mu(x^{-1} \cdot xy) = \mu(y)$.

(2) $\Rightarrow$ (3): Immediate.

(3) $\Rightarrow$ (4): $\mu(xyx^{-1}) \leq \mu(x^{-1} \cdot xyx^{-1} \cdot (x^{-1})^{-1}) = \mu(y) \ \forall \ x, y \in G$.

$(4) \Rightarrow (1)$: Let $x, y \in G$. Then

$$\begin{aligned}
\mu(xy) &= \mu(x \cdot yx \cdot x^{-1}) \\
&\leq \mu(yx) \\
&= \mu(y \cdot xy \cdot y^{-1}) \\
&\leq \mu(xy).
\end{aligned}$$

Hence $\mu(xy) = \mu(yx)$.

$(1) \Rightarrow (5)$: Let $x \in G$. Then

$$\begin{aligned}
(\mu \circ \nu)(x) &= \vee_{y \in G}\{\mu(xy^{-1}) \wedge \nu(y)\} \\
&= \vee_{y \in G}\{\mu(y^{-1}x) \wedge \nu(y)\} \\
&= (\nu \circ \mu)(x).
\end{aligned}$$

Hence $\mu \circ \nu = \nu \circ \mu$.

$(5) \Rightarrow (1)$: Now $1_{\{y^{-1}\}} \circ \mu = \mu \circ 1_{\{y^{-1}\}} \, \forall y \in G$. Thus $(1_{\{y^{-1}\}} \circ \mu)(x) = (\mu \circ 1_{\{y^{-1}\}})(x) \, \forall x, y \in G$. Hence $\mu(yx) = \mu(xy) \, \forall \, x, \, y \in G$. $\qquad \square$

Definition 1.3.2. *Let $\mu \in \mathcal{F}(G)$. Then μ is called a **normal fuzzy subgroup** of G if it is an Abelian fuzzy subset of G. Let $\mathcal{NF}(G)$ denote the set of all normal fuzzy subgroups of G.*

If $\mu, \nu \in \mathcal{F}(G)$ and there exists $u \in G$ such that $\mu(x) = \nu(uxu^{-1}) \, \forall x \in G$, then μ and ν are called **conjugate fuzzy subgroups** (with respect to u) and we write, $\mu = \nu^u$, where $\nu^u(x) = \nu(uxu^{-1})$ for all $x \in G$.

Clearly, 1_G and $1_{\{e\}}$ are normal fuzzy subgroups of G. If G is a commutative group, every fuzzy subgroup of G is normal. A fuzzy subgroup μ of G is normal if and only if $\mu = \mu^z \, \forall z \in G$.

Theorem 1.3.3. *Let $\mu \in \mathcal{FP}(G)$. Then $\mu \in \mathcal{NF}(G)$ if and only if μ_a is a normal subgroup of $G \, \forall a \in \mu(G) \cup \{b \in [0,1] \mid b \leq \mu(e)\}$.*

Proof. Suppose that $\mu \in \mathcal{NF}(G)$. Let $a \in \mu(G) \cup \{b \in [0,1] \mid b \leq \mu(e)\}$. Since $\mu \in \mathcal{F}(G)$, μ_a is a subgroup of G. If $x \in G$ and $y \in \mu_a$, it follows from Theorem 1.3.1 that $\mu(xyx^{-1}) = \mu(y) \geq a$. Thus $xyx^{-1} \in \mu_a$. Hence μ_a is a normal subgroup of G.

Conversely, assume that μ_a is a normal subgroup of $G \, \forall a \in \mu(G) \cup \{b \in L \mid b \leq \mu(e)\}$. By Lemma 1.2.6, we have that $\mu \in \mathcal{F}(G)$. Let $x, \, y \in G$ and $a = \mu(y)$. Then $y \in \mu_a$ and so $xyx^{-1} \in \mu_a$. Hence $\mu(xyx^{-1}) \geq a = \mu(y)$. That is, μ satisfies condition (3) of Theorem 1.3.1. Consequently, it follows from Theorem 1.3.1 that $\mu \in \mathcal{NF}(G)$. $\qquad \square$

Theorem 1.3.4. *Let $\mu \in \mathcal{NF}(G)$. Then μ_* and μ^* are normal subgroups of G.*

Proof. Since $\mu \in \mathcal{F}(G)$, it follows from Lemma 1.2.6 that μ_* and μ^* are subgroups of G. Let $x \in G$ and $y \in \mu_*$. Since μ satisfies condition (2) of Theorem 1.3.1, we have $\mu(xyx^{-1}) = \mu(y) = \mu(e)$ and thus $xyx^{-1} \in \mu_*$. Hence

μ_* is a normal subgroup of G. Let $x \in G$ and $y \in \mu^*$. Since μ satisfies condition (2) of Theorem 1.3.1, it follows that $\mu(xyx^{-1}) = \mu(y) > 0$ and so $xyx^{-1} \in \mu^*$. Therefore, μ^* is a normal subgroup of G. □

Example 1.3.5. *The converse of Theorem 1.3.4 is not true as can be seen by the following example: Let G be a group and H be a subgroup of G which is not normal. Define the fuzzy subset μ of G by $\mu(e) = 1, \mu(x) = \frac{1}{2}$ if $x \in H \setminus \{e\}$, and $\mu(x) = \frac{1}{4}$ if $x \in G \setminus H$. Then μ is a fuzzy subgroup of G since its level sets are subgroups of G. Now $\mu_{\frac{1}{2}} = H$ is not normal in G. Hence μ is not a normal fuzzy subgroup of G. However $\mu_* = \{e\}$ and $\mu^* = G$ are normal in G.*

The proof of the following result can be found in [[52], Theorem 5.2.6, p. 122] and Chapter 9. We state it here for information purposes.

Theorem 1.3.6. *Let $a \in (0, 1]$ and let $\mathcal{NF}_a(G) = \{\mu \in \mathcal{NF}(G) \mid \mu(e) = a\}$. The set $\mathcal{NF}_a(G)$ together with $\subseteq$ constitutes a complete lattice whose meet is fuzzy subset intersection $\cap$ and whose join is the product $\circ$. Moreover, $\mathcal{NF}_a(G)$ is closed under intersection.*

Theorem 1.3.7. *Suppose $\mu \in \mathcal{F}(G)$. Let $N(\mu) = \{x \mid x \in G, \mu(xy) = \mu(yx)$ $\forall \ y \in G\}$. Then $N(\mu)$ is a subgroup of G and the restriction of μ to $N(\mu)$, $\mu|_{N(\mu)}$, is a normal fuzzy subgroup of $N(\mu)$.*

Proof. Clearly, $e \in N(\mu)$. Let $x, y \in N(\mu)$. For any $z \in G$, we see that

$$
\begin{aligned}
\mu(xy^{-1} \cdot z) &= \mu(x \cdot y^{-1}z) \\
&= \mu(y^{-1}z \cdot x) \\
&= \mu(x^{-1}z^{-1} \cdot y) \\
&= \mu(y \cdot x^{-1}z^{-1}) \\
&= \mu(z \cdot xy^{-1}).
\end{aligned}
$$

Thus $xy^{-1} \in N(\mu)$. Hence $N(\mu)$ is a subgroup of G.

By Comment 1.2.4, it follows that $\mu|_{N(\mu)} \in \mathcal{F}(N(\mu))$ and $\mu|_{N(\mu)}(xy) = \mu|_{N(\mu)}(yx) \ \forall \ x, y \in N(\mu)$. Therefore, $\mu|_{N(\mu)} \in \mathcal{NF}(N(\mu))$. □

The subgroup $N(\mu)$ of G defined in Theorem 1.3.7 is called the **normalizer** of μ in G.

Theorem 1.3.8. *Let $\nu \in \mathcal{F}(G)$. Then the cardinal number of the set $\{\nu^u \mid u \in G\}$ is equal to the index $[G : N(\nu)]$ of the normalizer $N(\nu)$ in G.*

Proof. Let $u, v \in G$. Then $\nu^u = \nu^v \Leftrightarrow \nu(uxu^{-1}) = \nu(vxv^{-1}) \ \forall \ x \in G \Leftrightarrow \nu(uv^{-1} \cdot x) = \nu(x \cdot uv^{-1}) \ \forall \ x \in G \Leftrightarrow uv^{-1} \in N(\nu) \Leftrightarrow u^{-1}N(\nu) = v^{-1}N(\nu)$. Thus $\nu^u \mapsto u^{-1}N(\nu)$ is a bijection from $\{\nu^u \mid u \in G\}$ onto $\{uN(\nu) \mid u \in G\}$. □

Theorem 1.3.9. *Let $\nu \in \mathcal{F}(G)$. Then $\cap_{u \in G} \nu^u \in \mathcal{NF}(G)$ and is the largest normal fuzzy subgroup of G that is contained in ν.*

Proof. Since $\nu^u \in \mathcal{F}(G) \, \forall \, u \in G$, $\cap_{u \in G} \nu^u \in \mathcal{F}(G)$. For all $x \in G$, observe that $\{\nu^u \mid u \in G\} = \{\nu^{ux} \mid u \in G\}$. Thus

$$\wedge_{u \in G} \nu^u(xyx^{-1}) = \wedge_{u \in G} \nu(uxy(ux)^{-1})$$
$$= \wedge_{u \in G} \nu^{ux}(y)$$
$$= \wedge_{u \in G} \nu^u(y)$$

$\forall \, x, \, y \in G$. Hence by Theorem 1.3.1, $\cap_{u \in G} \nu^u \in \mathcal{NF}(G)$.

Now let $\mu \in \mathcal{NF}(G)$ with $\mu \subseteq \nu$. Then $\mu = \mu^u \subseteq \nu^u \, \forall \, u \in G$. Thus $\mu \subseteq \cap_{u \in G} \nu^u$. Therefore, $\cap_{u \in G} \nu^u$ is the largest normal fuzzy subgroup of G that is contained in ν. $\qquad \square$

Let $\mu \in \mathcal{F}(G)$ and $x \in G$. The fuzzy subsets $\mu(e)_{\{x\}} \circ \mu$ and $\mu \circ \mu(e)_{\{x\}}$ are referred to as the **left coset** and **right coset** of μ with respect to x, and written as $x\mu$ and μx, respectively. From Theorem 1.3.1, we know that if $\mu \in \mathcal{NF}(G)$, then the left coset $x\mu$ is just the right coset μx. Thus in this case, we call $x\mu$ a **coset** for short.

Theorem 1.3.10. *Let $\mu \in \mathcal{F}(G)$. Then for all $x, \, y \in G$,*
  (1) $x\mu = y\mu \Leftrightarrow x\mu_* = y\mu_*$;
  (2) $\mu x = \mu y \Leftrightarrow \mu_* x = \mu_* y$.

Proof. Suppose that $x\mu = y\mu$. Then $\mu(e)_{\{x\}} \circ \mu = \mu(e)_{\{y\}} \circ \mu$ which means that $\mu(x^{-1}z) = \mu(y^{-1}z) \, \forall \, z \in G$. Choosing $z = y$ yields $\mu(x^{-1}y) = \mu(y^{-1}y) = \mu(e)$ and thus $x^{-1}y \in \mu_*$. Hence $x\mu_* = y\mu_*$. Conversely, suppose that $x\mu_* = y\mu_*$. Then $x^{-1}y \in \mu_*$ and $y^{-1}x \in \mu_*$. Hence

$$\mu(x^{-1}z) = \mu(x^{-1}y \cdot y^{-1}z)$$
$$\geq \mu(x^{-1}y) \wedge \mu(y^{-1}z)$$
$$= \mu(e) \wedge \mu(y^{-1}z)$$
$$= \mu(y^{-1}z)$$

$\forall \, z \in G$. Similarly, $\mu(y^{-1}z) \geq \mu(x^{-1}z) \, \forall z \in G$. Therefore, $\mu(x^{-1}z) = \mu(y^{-1}z)$ $\forall z \in G$ which shows that $x\mu = y\mu$. Similar reasoning shows that (2) holds. $\quad \square$

Theorem 1.3.11. *Let $\mu \in \mathcal{NF}(G)$ and $x, \, y \in G$. If $x\mu = y\mu$, then $\mu(x) = \mu(y)$.*

Proof. Suppose that $x\mu = y\mu$. By Theorem 1.3.10, $x^{-1}y \in \mu_*$ and $y^{-1}x \in \mu_*$. Since $\mu \in \mathcal{NF}(G)$ it follows that $\mu(x) = \mu(y^{-1}xy) \geq \mu(y^{-1}x) \wedge \mu(y) = \mu(e) \wedge \mu(y) = \mu(y)$. Similarly, $\mu(y) \geq \mu(x)$ and therefore $\mu(x) = \mu(y)$. $\quad \square$

Theorem 1.3.12. *Let $\mu \in \mathcal{NF}(G)$. Set $G/\mu = \{x\mu \mid x \in G\}$. Then the following assertions hold:*
  (1) $(x\mu) \circ (y\mu) = (xy)\mu \, \forall \, x, \, y \in G$.
  (2) $(G/\mu, \circ)$ is a group.
  (3) $G/\mu \cong G/\mu_*$.
  (4) Let $\mu^{(*)} \in \mathcal{FP}(G/\mu)$ be defined by $\mu^{(*)}(x\mu) = \mu(x) \, \forall \, x \in G$. Then $\mu^{(*)} \in \mathcal{NF}(G/\mu)$.

Proof. (1) For all $x, y \in G$, we have

$$
\begin{aligned}
(x\mu) \circ (y\mu) &= (\mu(e)_{\{x\}} \circ \mu) \circ (\mu(e)_{\{y\}} \circ \mu) \\
&= \mu(e)_{\{x\}} \circ (\mu \circ \mu(e)_{\{y\}}) \circ \mu \\
&= \mu(e)_{\{x\}} \circ (\mu \circ \mu) \circ \mu(e)_{\{y\}} \\
&= \mu(e)_{\{x\}} \circ \mu \circ \mu(e)_{\{y\}} \\
&= \mu(e)_{\{x\}} \circ (\mu \circ \mu(e)_{\{y\}}) \\
&= \mu(e)_{\{x\}} \circ (\mu(e)_{\{y\}} \circ \mu) \\
&= (\mu(e)_{\{x\}} \circ \mu(e)_{\{y\}}) \circ \mu \\
&= (xy)\mu
\end{aligned}
$$

by Theorems 1.2.2 and 1.3.1.

(2) By (1), G/μ is closed under the operation $\circ$. Also, $\circ$ satisfies the associative law. Now

$$\mu \circ (x\mu) = (e\mu) \circ (x\mu) = (ex)\mu = x\mu \ \forall x \in G$$

and

$$(x^{-1}\mu) \circ (x\mu) = (x^{-1}x)\mu = e\mu = \mu \ \forall x \in G.$$

Hence $(G/\mu, \circ)$ is a group.

(3) Since $\mu \in \mathcal{NF}(G)$, it follows by Theorem 1.3.4 that μ_* is a normal subgroup of G. Hence (G/μ_*) is a group and the function $f : (G/\mu) \to (G/\mu_*)$ given by $x\mu \to x\mu_*$ is an isomorphism by Theorem 1.3.10 and the fact that $x\mu \circ y\mu = (xy)\mu$ and $x\mu_*y\mu_* = (xy)\mu_*$.

(4) By Theorem 1.3.11, $x\mu = y\mu$ implies that $\mu(x) = \mu(y)$. Thus $\mu^{(*)}$ is single-valued. Since

$$\mu^{(*)}((x\mu)^{-1}) = \mu^{(*)}(x^{-1}\mu) = \mu(x^{-1}) = \mu(x) = \mu^{(*)}(x\mu)$$

$\forall \ x \in G$ and

$$
\begin{aligned}
\mu^{(*)}((x\mu) \circ (y\mu)) &= \mu^{(*)}(xy\mu) \\
&= \mu(xy) \\
&\geq \mu(x) \wedge \mu(y) \\
&= \mu^{(*)}(x\mu) \wedge \mu^{(*)}(y\mu)
\end{aligned}
$$

$\forall \ x, y \in G$, it follows that $\mu^{(*)} \in \mathcal{F}(G/\mu)$. Moreover, since

$$
\begin{aligned}
\mu^{(*)}((x\mu) \circ (y\mu)) &= \mu^{(*)}(xy\mu) \\
&= \mu(xy) \\
&= \mu(yx) \\
&= \mu^{(*)}(yx\mu) \\
&= \mu^{(*)}((y\mu) \circ (x\mu))
\end{aligned}
$$

$\forall \ x, y \in G$, we have that $\mu^{(*)} \in \mathcal{NF}(G/\mu)$. $\qquad\square$

The group G/μ defined in Theorem 1.3.12 is called the **quotient group** (or **factor group**) of G relative to the normal fuzzy subgroup μ.

Theorem 1.3.13. *Let $\nu \in \mathcal{F}(G)$ and let N be a normal subgroup of G. Define $\xi \in \mathcal{F}P(G/N)$ as follows:*

$$\xi(xN) = \vee\{\nu(z) \mid z \in xN\} \ \forall x \in G.$$

Then $\xi \in \mathcal{F}(G/N)$.

Proof. Now

$$\begin{aligned}
\xi((xN)^{-1}) &= \xi(x^{-1}N) \\
&= \vee\{\nu(z) \mid z \in x^{-1}N\} \\
&= \vee\{\nu(w^{-1}) \mid w^{-1} \in x^{-1}N\} \\
&= \vee\{\nu(w) \mid w \in xN\} \\
&= \xi(xN)
\end{aligned}$$

$\forall\, x \in G$;

$$\begin{aligned}
\xi(xNyN) &= \vee\{\nu(z) \mid z \in xyN\} \\
&= \vee\{\nu(uv) \mid u \in xN, v \in yN\} \\
&\geq \vee\{\nu(u) \wedge \nu(v) \mid u \in xN, v \in yN\} \\
&= (\vee\{\nu(u) \mid u \in xN\}) \wedge (\vee\{\nu(v) \mid v \in yN\}) \\
&= \xi(xN) \wedge \xi(yN)
\end{aligned}$$

$\forall\, x,\, y \in G$. Hence $\xi \in \mathcal{F}(G/N)$. □

The fuzzy subgroup ξ defined in Theorem 1.3.13 is called the **quotient fuzzy subgroup** (or **factor fuzzy subgroup**) of the fuzzy subgroup ν of G relative to the normal subgroup N of G and is denoted by ν/N.

Theorem 1.3.14. *Let $\mu \in \mathcal{N}\mathcal{F}(G)$ and H be a group. Suppose that f is an epimorphism of G onto H. Then $f(\mu) \in \mathcal{N}\mathcal{F}(H)$.*

Proof. By Theorem 1.2.11, $f(\mu) \in \mathcal{F}(H)$. Now let x, $y \in H$. Since f is a surjection, $f(u) = x$ for some $u \in G$. Thus

$$\begin{aligned}
f(\mu)(xyx^{-1}) &= \vee\{\mu(w) \mid w \in G, f(w) = xyx^{-1}\} \\
&= \vee\{\mu(u^{-1}wu) \mid w \in G, f(u^{-1}wu) = y\} \\
&= \vee\{\mu(w) \mid uwu^{-1} \in G, f(w) = y\} \\
&= \vee\{\mu(w) \mid w \in G, f(w) = y\} \\
&= f(\mu)(y).
\end{aligned}$$

Therefore, it follows from Theorem 1.3.1 that $f(\mu) \in \mathcal{N}\mathcal{F}(H)$. □

Theorem 1.3.15. *Let H be a group and $\nu \in \mathcal{N}\mathcal{F}(H)$. If f is a homomorphism from G into H, then $f^{-1}(\nu) \in \mathcal{N}\mathcal{F}(G)$.*

Proof. By Theorem 1.2.12, $f^{-1}(\nu) \in \mathcal{F}(G)$. Now for any x, $y \in G$, we have $f^{-1}(\nu)(xy) = \nu(f(xy)) = \nu(f(x)f(y)) = \nu(f(y)f(x)) = \nu(f(yx)) = f^{-1}(\nu)(yx)$. Hence $f^{-1}(\nu) \in \mathcal{N}\mathcal{F}(G)$. □

1.4 Homomorphisms and Isomorphisms

This section consists of two parts. The first part is mainly concerned with a generalization of the concept of a normal fuzzy subgroup. We introduce the concept of a normal fuzzy subgroup of a fuzzy subgroup and examine its basic properties. In the second part, we introduce the concepts of homomorphisms and isomorphisms of fuzzy subgroups and use them to develop results concerning fuzzy subgroups of a fuzzy subgroup.

Definition 1.4.1. *Let μ, $\nu \in \mathcal{F}(G)$ and $\mu \subseteq \nu$. Then μ is called a **normal fuzzy subgroup** of the fuzzy subgroup ν, written $\mu \lhd \nu$, if*

$$\mu(xyx^{-1}) \geq \mu(y) \wedge \nu(x) \ \forall x, y \in G.$$

The following statements are immediate from Definition 1.4.1.

(1) If G_1 and G_2 are subgroups of G, then G_1 is a normal subgroup of G_2 if and only if 1_{G_1} is a normal fuzzy subgroup of 1_{G_2}.

(2) If $\mu \in \mathcal{NF}(G)$, $\nu \in \mathcal{F}(G)$, and $\mu \subseteq \nu$, then μ is a normal fuzzy subgroup of ν.

(3) Every fuzzy subgroup is a normal fuzzy subgroup of itself.

(4) $\mu \in \mathcal{FP}(G)$ is a normal fuzzy subgroup of G if and only if μ is a normal fuzzy subgroup of the fuzzy subgroup 1_G.

Theorem 1.4.2. *Let μ, $\nu \in \mathcal{F}(G)$ and $\mu \subseteq \nu$. Then the following assertions are equivalent:*

(1) μ *is a normal fuzzy subgroup of* ν.

(2) $\mu(yx) \geq \mu(xy) \wedge \nu(y) \ \forall \ x, \ y \in G$.

(3) $\mu(e)_{\{x\}} \circ \mu \supseteq (\mu \circ \mu(e)_{\{x\}}) \cap \nu \ \forall \ x \in G$.

Proof. (1) $\Rightarrow$ (2): Since μ is a normal fuzzy subgroup of ν, it follows that $\mu(yx) = \mu(yxyy^{-1}) \geq \mu(xy) \wedge \nu(y) \ \forall x, y \in G$.

(2) $\Rightarrow$ (3): For any $z \in G$,

$$\begin{aligned}
(\mu(e)_{\{x\}} \circ \mu)(z) &= \vee\{[\mu(e)_{\{x\}}(p) \wedge \mu(q)] | pq = z\} \\
&= \vee\{[\mu(e)_{\{x\}}(x) \wedge \mu(q)] | xq = z\} \\
&= \vee[\mu(e) \wedge \mu(x^{-1}z)] \\
&= \vee[\mu(x^{-1}z)] \\
&= \vee\mu(z^{-1}x) \\
&\geq \vee\mu(xz^{-1}) \wedge \nu(z^{-1}) \\
&= \vee\mu(zx^{-1}) \wedge \nu(z^{-1}) \\
&= (\mu \circ \mu(e)_{\{x\}})(z) \wedge \nu(z) \\
&= ((\mu \circ \mu(e)_{\{x\}}) \cap \nu)(z).
\end{aligned}$$

(2) $\Rightarrow$ (1):

$$\begin{aligned}
\mu(xyx^{-1}) &\geq \mu(yx^{-1}x) \wedge \nu(x) \\
&= \mu(y) \wedge \nu(x).
\end{aligned}$$

(3) $\Rightarrow$ (2): $\forall$ $x, y \in G$,

$$
\begin{aligned}
\mu(yx) &= \mu(x^{-1}y^{-1}) \\
&= (\mu(e)_{\{x\}} \circ \mu)(y^{-1}) \\
&\geq ((\mu \circ \mu(e)_{\{x\}}) \cap \nu)(y^{-1}) \\
&= \mu(y^{-1}x^{-1}) \wedge \nu(y^{-1}) \\
&= \mu(xy) \wedge \nu(y).
\end{aligned}
$$

$\square$

Theorem 1.4.3. *Let* μ, $\nu \in \mathcal{F}(G)$. *Then* μ *is a normal fuzzy subgroup of* ν *if and only if* μ_a *is a normal subgroup of* ν_a $\forall a \in \{b \in [0,1] \mid b \leq \mu(e)\}$.

Proof. Suppose μ is a normal fuzzy subgroup of ν. Let $a \in \{b \in [0,1] | b \leq \mu(e)\}$. Then μ_a is a subgroup of ν_a. Let $y \in \mu_a$ and $x \in \nu_a$. Then $\mu(y) \geq a$ and $\nu(x) \geq a$. By hypothesis, $\mu(xyx^{-1}) \geq \mu(y) \wedge \nu(x) \geq a \wedge a = a$. Hence $xyx^{-1} \in \mu_a$. That is, μ_a is a normal subgroup of ν_a.

Conversely suppose μ_a is a normal subgroup of ν_a for all $a \in \{b \in [0,1] | b \leq \mu(e)\}$. Let $\mu(y) = a$, $\nu(x) = b$ and suppose that $b \geq a$. Then $x \in \nu_a$ and so $xyx^{-1} \in \mu_a$ by hypothesis. Thus $\mu(xyx^{-1}) \geq a = a \wedge b = \mu(y) \wedge \nu(x)$. Suppose $b < a$. Then $y \in \mu_b$. Since μ_b is a normal subgroup of ν_b, this implies $xyx^{-1} \in \mu_b$. Thus, $\mu(xyx^{-1}) \geq b = b \wedge a = \nu(x) \wedge \mu(y)$. $\square$

Theorem 1.4.4. *Let* μ, $\nu \in \mathcal{F}(G)$ *and* μ *be a normal fuzzy subgroup of* ν. *Then* μ_* *is a normal subgroup of* ν_* *and* μ^* *is a normal subgroup of* ν^*.

Proof. If $x \in \mu_*$ (respectively $x \in \mu^*$) and $y \in \nu_*$ (respectively $y \in \nu^*$), then μ a normal fuzzy subgroup of ν implies that $\mu(y^{-1}xy) \geq (\mu(x) \wedge \nu(y)) = \mu(e) \wedge \nu(e) = \mu(e)$ (respectively $\mu(y^{-1}xy) > 0$). This shows μ_* is a normal subgroup of ν_* (respectively μ^* is a normal subgroup of ν^*). $\square$

Theorem 1.4.5. *If* $\mu \in \mathcal{NF}(G)$ *and* $\nu \in \mathcal{F}(G)$, *then* $\mu \cap \nu$ *is a normal fuzzy subgroup of* ν.

Proof. Clearly, $\mu \cap \nu \in \mathcal{F}(G)$ and $\mu \cap \nu \subseteq \nu$. Now

$$
\begin{aligned}
(\mu \cap \nu)(xyx^{-1}) &= \mu(xyx^{-1}) \wedge \nu(xyx^{-1}) \\
&= \mu(y) \wedge \nu(xyx^{-1}) \\
&\geq \mu(y) \wedge \nu(x) \wedge \nu(y) \wedge \nu(x^{-1}) \\
&= (\mu \cap \nu)(y) \wedge \nu(x)
\end{aligned}
$$

$\forall$ $x, y \in G$. Hence $\mu \cap \nu$ is a normal fuzzy subgroup of ν. $\square$

Theorem 1.4.6. *Let* μ, v, $\xi \in \mathcal{F}(G)$ *be such that* μ *and* ν *are normal fuzzy subgroups of* ξ. *Then* $\mu \cap \nu$ *is a normal fuzzy subgroup of* ξ.

Proof. Observe that $\mu \cap \nu \in \mathcal{F}(G)$ and $\mu \cap \nu \subseteq \xi$. Now

$$
\begin{aligned}
(\mu \cap \nu)(xyx^{-1}) &= \mu(xyx^{-1}) \wedge \nu(xyx^{-1}) \\
&\geq (\mu(y) \wedge \xi(x)) \wedge (\nu(y) \wedge \xi(x)) \\
&\geq (\mu \cap \nu)(y) \wedge \xi(x).
\end{aligned}
$$

Therefore, $\mu \cap \nu$ is a normal fuzzy subgroup of ξ. $\square$

Theorem 1.4.7. *Let* μ, $\nu \in \mathcal{F}(G)$ *and* μ *be a normal fuzzy subgroup of* ν. *Let* H *be a group and* f *a homomorphism from* G *into* H. *Then* $f(\mu)$ *is a normal fuzzy subgroup of* $f(\nu)$.

Proof. Note, $f(\mu)$, $f(\nu) \in \mathcal{F}(H)$ and $f(\mu) \subseteq f(\nu)$. Now

$$
\begin{aligned}
(f(\mu))(xyx^{-1}) &= \vee\{\mu(z) \mid z \in G, f(z) = xyx^{-1}\} \\
&\geq \vee\{\mu(uvu^{-1}) \mid u, v \in G, f(u) = x, f(v) = y\} \\
&\geq \vee\{\mu(v) \wedge \nu(u) \mid u, v \in G, f(u) = x, f(v) = y\} \\
&= (\vee\{\mu(v) \mid v \in G, f(v) = y\}) \wedge (\vee\{\nu(u) \mid u \in G, f(u) = x\}) \\
&= (f(\mu))(y) \wedge (f(\nu))(x)
\end{aligned}
$$

$\forall\, x, y \in H$. Hence $f(\mu)$ is a normal fuzzy subgroup of $f(\nu)$. $\qquad\square$

Theorem 1.4.8. *Let* H *be a group. Let* μ, $\nu \in \mathcal{F}(H)$ *and* μ *be a normal fuzzy subgroup of* ν. *Let* f *be a homomorphism from* G *into* H. *Then* $f^{-1}(\mu)$ *is a normal fuzzy subgroup of* $f^{-1}(\nu)$.

Proof. Clearly, $f^{-1}(\mu)$, $f^{-1}(\nu) \in \mathcal{F}(G)$. It follows easily that $f^{-1}(\mu) \subseteq f^{-1}(\nu)$. Now

$$
\begin{aligned}
f^{-1}(\mu)(xyx^{-1}) &= \mu(f(xyx^{-1})) \\
&= \mu(f(x)f(y)(f(x))^{-1}) \\
&\geq \mu(f(y)) \wedge \nu(f(x)) \\
&= f^{-1}(\mu)(y) \wedge (f^{-1}(\nu))(x)
\end{aligned}
$$

$\forall\, x, y \in G$. Hence $f^{-1}(\mu)$ is a normal fuzzy subgroup of $f^{-1}(\nu)$. $\qquad\square$

Definition 1.4.9. *Let* G *and* H *be groups and let* $\mu \in \mathcal{F}(G)$ *and* $\nu \in \mathcal{F}(H)$.

(1) *A homomorphism* f *of* G *onto* H *is called a* **weak homomorphism** *of* μ *into* ν *if* $f(\mu) \subseteq \nu$. *If* f *is a weak homomorphism of* μ *into* ν, *then we say that* μ *is* **weakly homomorphic** *to* ν *and we write* $\mu \overset{f}{\sim} \nu$, *or simply* $\mu \sim \nu$.

(2) *An isomorphism* f *of* G *onto* H *is called a* **weak isomorphism** *of* μ *into* ν *if* $f(\mu) \subseteq \nu$. *If* f *is a weak isomorphism of* μ *into* ν, *then we say that* μ *is* **weakly isomorphic** *to* ν *and we write* $\mu \overset{f}{\simeq} \nu$, *or simply* $\mu \simeq \nu$.

(3) *A homomorphism* f *of* G *onto* H *is called a* **homomorphism** *of* μ *onto* ν *if* $f(\mu) = \nu$. *If* f *is a homomorphism of* μ *onto* ν, *then we say that* μ *is* **homomorphic** *to* ν *and we write* $\mu \overset{f}{\approx} \nu$, *or simply* $\mu \approx \nu$.

(4) *An isomorphism* f *of* G *onto* H *is called an* **isomorphism** *of* μ *onto* ν *if* $f(\mu) = \nu$. *If* f *is an isomorphism of* μ *onto* ν, *then we say that* μ *is* **isomorphic** *to* ν *and we write* $\mu \overset{f}{\cong} \nu$, *or simply* $\mu \cong \nu$.

Let μ, $\nu \in \mathcal{F}(G)$. Suppose that μ is a normal fuzzy subgroup of ν. Then μ^* is a normal subgroup of ν^* by Theorem 1.4.4. Clearly, $\nu|_{\nu^*}$ is a fuzzy subgroup of ν^*. Thus by Theorem 1.3.13, the factor fuzzy subgroup of $\nu|_{\nu^*}$ relative to μ^* exists. For convenience sake, we denote this factor fuzzy subgroup by ν/μ and call it the **quotient subgroup** (or **factor subgroup** of ν relative to μ.

Theorem 1.4.10. *Let* μ, $\nu \in \mathcal{F}(G)$ *and* μ *be a normal fuzzy subgroup of* ν. *Then* $\nu|_{\nu^*} \approx \nu/\mu$.

Proof. Let f be the natural homomorphism from ν^* onto ν^*/μ^*. Then

$$f(\nu|_{\nu^*})(x\mu^*) = \vee\{\nu|_{\nu^*}(z) \mid z \in \nu^*, f(z) = x\mu^*\}$$
$$= \vee\{\nu(y) \mid y \in x\mu^*\}$$
$$= (\nu/\mu)(x\mu^*)$$

$\forall\, x \in \nu^*$. Therefore, $\nu|_{\nu^*} \overset{f}{\approx} \nu/\mu$. $\qquad\square$

Lemma 1.4.11. *If* $f : X \to Y$ *and* $\mu \in \mathcal{FP}(G)$, *then* $(f(\mu))^* = f((\mu)^*)$

Theorem 1.4.12. *Let* $\nu \in \mathcal{F}(G)$. *Suppose that* H *is a group and* $\xi \in \mathcal{F}(H)$ *is such that* $\nu \approx \xi$. *Then there exists a normal fuzzy subgroup* μ *of* ν *such that* $\nu/\mu \cong \xi|_{\xi^*}$.

Proof. Since $\nu \approx \xi$, there exists an epimorphism f of G onto H such that $f(\nu) = \xi$. Define $\mu \in \mathcal{FP}(G)$ as follows: $\forall\, x \in G$,

$$\mu(x) = \begin{cases} \nu(x) & \text{if } x \in \text{Ker } f \\ 0 & \text{otherwise.} \end{cases}$$

Clearly, $\mu \in \mathcal{F}(G)$ and $\mu \subseteq \nu$. If $x \in \text{Ker } f$, then $yxy^{-1} \in \text{Ker } f \,\forall\, y \in G$, and so

$$\mu(yxy^{-1}) = \nu(yxy^{-1}) \geq \nu(x) \wedge \nu(y) = \mu(x) \wedge \nu(y) \,\forall y \in G.$$

If $x \in G\backslash Ker\ f$, then $\mu(x) = 0$ and so

$$\mu(yxy^{-1}) \geq \mu(x) \wedge \nu(y) \,\forall y \in G.$$

Hence μ is a normal fuzzy subgroup of ν. Also, $\nu \overset{f}{\approx} \xi$ implies $f(\nu) = \xi$ which further implies $(f(\nu))^* = \xi^*$. By Lemma 1.4.11 it follows that $f(\nu^*) = \xi^*$. Let $g = f|_{\nu^*}$. Then g is a homomorphism of ν^* onto ξ^* and Ker $g = \mu^*$ by the definition of μ. Thus there exists an isomorphism h of ν^*/μ^* onto ξ^* such that $h(x\mu^*) = g(x) = f(x) \,\forall\, x \in \nu^*$. For such an h, we have

$$h(\nu/\mu)(z) = \vee\{(\nu/\mu)(x\mu^*) \mid x \in \nu^*, h(x\mu^*) = z\}$$
$$= \vee\{\vee\{\nu(y) \mid y \in x\mu^*\} \mid x \in \nu^*,\ g(x) = z\}$$
$$= \vee\{\nu(y) \mid y \in \nu^*, g(y) = z\}$$
$$= \vee\{\nu(y) \mid y \in G, f(y) = z\}$$
$$= \xi(z)$$

$\forall\, z \in \xi^*$. Therefore, $\nu/\mu \overset{h}{\cong} \xi|_{\xi^*}$. $\qquad\square$

Theorem 1.4.13. *Let* $\mu \in \mathcal{NF}(G)$ *and* $\nu \in \mathcal{F}(G)$ *be such that* $\mu(e) = \nu(e)$. *Then*

$$\nu/(\mu \cap \nu) \simeq (\mu \circ \nu)/\mu.$$

Proof. From Theorem 1.3.4, we have that μ^* is a normal subgroup of G. By the Second Isomorphism Theorem for Groups,

$$\nu^*/(\mu^* \cap \nu^*) \cong (\mu^* \nu^*)/\mu^*.$$

One can verify that

$$(\mu \cap \nu)^* = \mu^* \cap \nu^*,$$
$$(\mu \circ v)^* = \mu^* \nu^*.$$

Consequently, we have

$$\nu^*/(\mu \cap \nu)^* \overset{f}{\cong} (\mu \circ \nu)^*/\mu^*,$$

where f is given by

$$f(x(\mu \cap \nu)^*) = x\mu^*$$

$\forall\, x \in \nu^*$. Thus

$$
\begin{aligned}
f(\nu/(\mu \cap \nu))(y\mu^*) &= (\nu/(\mu \cap \nu))(y(\mu \cap \nu)^*) \quad &\text{(since } f \text{ is one-to-one)}\\
&= \vee\{\nu(z) \mid z \in y(\mu \cap \nu)^*\}\\
&\leq \vee\{(\mu \circ \nu)(z) \mid z \in y(\mu^* \cap \nu^*)\}\\
&\leq \vee\{(\mu \circ \nu)(z) \mid z \in y\mu^*\} \quad &\text{(since } \mu(e) = \nu(e))\\
&= ((\mu \circ \nu)/\mu)(y\mu^*)
\end{aligned}
$$

$\forall\, y \in \nu^*$. Hence $f(\nu/(\mu \cap \nu)) \subseteq (\mu \circ \nu)/\mu$. Therefore, $\nu/(\mu \cap \nu) \overset{f}{\simeq} (\mu \circ \nu)/\mu$. $\square$

The next lemma can be proved using Theorem 1.4.10.

Lemma 1.4.14. *Let μ, ν, $\xi \in \mathcal{F}(G)$ be such that μ and ν are normal fuzzy subgroups of ξ and $\mu \subseteq \nu$, then (ν/μ) is a normal fuzzy subgroup of (ξ/ν).*

Theorem 1.4.15. *Let μ, ν, $\xi \in \mathcal{F}(G)$ be such that $\mu \subseteq \nu$ and μ and ν be normal fuzzy subgroups of ξ. Then*

$$(\xi/\mu)/(\nu/\mu) \cong \xi/\nu.$$

Proof. By Theorem 1.4.4, μ^* is a normal subgroup of ν^*, and both μ^* and ν^* are normal subgroups of ξ^*. By the Third Isomorphism Theorem for Groups,

$$(\xi^*/\mu^*)/(\nu^*/\mu^*) \overset{f}{\cong} \xi^*/\nu^*,$$

where f is given by

$$f(x\mu^* \cdot (\nu^*/\mu^*)) = x\nu^* \; \forall x \in \xi^*.$$

$$\begin{aligned}
f((\xi/\mu)/(\nu/\mu))(x\nu^*) &= ((\xi/\mu)/(\nu/\mu))(x\mu^* \cdot (\nu^*/\mu^*)) \\
&= \vee\{(\xi/\mu)(y\mu^*) \mid y \in \xi^*, y\mu^* \in x\mu^* \cdot (\nu^*/\mu^*)\} \\
&= \vee\{\vee\{\xi(z) \mid z \in y\mu^*\} \mid y \in \xi^*, y\mu^* \in x\mu^* \cdot (\nu^*/\mu^*)\} \\
&= \vee\{\xi(z) \mid z \in \xi^*, z\mu^* \in x\mu^* \cdot (\nu^*/\mu^*)\} \\
&= \vee\{\xi(z) \mid z \in x\mu^* \cdot (\nu^*/\mu^*)\} \\
&= \vee\{\xi(z) \mid z \in \xi^*, f(z) \in x\nu^*\} \\
&= (\xi/\nu)(x\nu^*)
\end{aligned}$$

$\forall\, x \in \xi^*$, where the first equality holds since f is one-to-one. Hence

$$(\xi/\mu)/(\nu/\mu) \overset{f}{\cong} \xi/\nu.$$

$\square$

We now define a concept that is used later.

Definition 1.4.16. *Let f be a homomorphism of a group G into a group H. Then a fuzzy subgroup μ of G is called f-invariant if for all $x, y \in G$, $f(x) = f(y)$ implies $\mu(x) = \mu(y)$.*

Remark 1.4.17. We note that μ being f-invariant is equivalent to the statement that $\forall x \in G, f(\mu)(f(x)) = \mu(x)$: Suppose this latter condition holds. Suppose $f(x) = f(y) \,\forall x, y \in G$. Then $f(\mu)(f(x)) = f(\mu)(f(y))$ and so $\mu(x) = \mu(y)$. Conversely suppose μ is f-invariant. Let $x \in G$. Then $f(\mu)(f(x)) \geq \mu(x)$. Suppose $f(\mu)(f(x)) > \mu(x)$. Then there exists a $y \in G$ such that $\mu(y) > \mu(x)$ and $f(x) = f(y)$, a contradiction. Thus $f(\mu)(f(x)) = \mu(x)$.

1.5 Complete and Weak Direct Products

In order to define the complete direct product of fuzzy subgroups, we first review the concept of the complete direct product of groups.

Let $\{G_i \mid i \in I\}$ be a collection of groups and let e_i denote the identity of G_i for all $i \in I$. Multiplication on the Cartesian product $\prod_{i \in I} G_i$ is defined by

$$(x_i)_{i \in I}(y_i)_{i \in I} = (x_i y_i)_{i \in I},$$

$\forall (x_i)_{i \in I}$, $(y_i)_{i \in I} \in \prod_{i \in I} G_i$. Then the Cartesian product $\prod_{i \in I} G_i$ together with the multiplication just defined form a group with identity $(e_i)_{i \in I}$. This group is called the **complete direct product** of the G_i's and is denoted by $\overset{\sim}{\prod}_{i \in I} G_i$. In particular, when $I = \{1, 2, \ldots, n\}$ (where $n \in \mathbb{N} \setminus \{1\}$), we regard the Cartesian product $\prod_{i \in I} G_i$ as the set

$$G_1 \times G_2 \times \ldots \times G_n = \{(x_1, x_2, \ldots, x_n) \mid x_i \in G_i, \; i \in I\}$$

and denote the complete direct product $\overset{\sim}{\prod\limits_{i \in I}} G_i$ by $G_1 \overset{\sim}{\otimes} G_2 \overset{\sim}{\otimes} \ldots \overset{\sim}{\otimes} G_n$.

It follows that if $G = \overset{\sim}{\prod\limits_{i \in I}} G_i$ and $\mu_i \in \mathcal{F}(G_i)$ for all $i \in I$, then $\overset{\sim}{\prod\limits_{i \in I}} \mu_i \in \mathcal{F}(G)$, where $\forall (x_i)_{i \in I}$

$$(\overset{\sim}{\prod_{i \in I}} \mu_i)((x_i)_{i \in I}) = \wedge_{i \in I} \mu_i(x_i).$$

In this case, we call the fuzzy subgroup $\overset{\sim}{\prod\limits_{i \in I}} \mu_i$ of G the **complete direct product** of the μ_i's.

Unless otherwise specified, it is assumed that G_i is a group with identity e_i for all $i \in I$, $G = \overset{\sim}{\prod\limits_{i \in I}} G_i$, and so $e = (e_i)_{i \in I}$.

Theorem 1.5.1. *Let $\mu_i \in \mathcal{NF}(G_i) \ \forall \ i \in I$. Then $\mu = \overset{\sim}{\prod\limits_{i \in I}} \mu_i \in \mathcal{NF}(G)$.*

Proof. Now $\mu \in \mathcal{F}(G)$. For all $x = (x_i)_{i \in I}$ and $y = (y_i)_{i \in I}$ in G, we have $\mu(xy) = \wedge_{i \in I} \mu_i(x_i y_i) = \wedge_{i \in I} \mu_i(y_i x_i) = \mu(yx)$. Hence $\mu \in \mathcal{NF}(G)$. $\square$

We introduce the notion of a weak direct product of fuzzy subgroups.

Let $x_i \in G$ for $i \in I$, where $|I| > 1$. Suppose that there are at most finitely many x_i not equal to e. We define $\prod_{i \in I} x_i$ as follows: if x_i is equal to e for every $i \in I$, then we interpret $\prod_{i \in I} x_i$ as e; and if $x_i \neq e$ for some $i \in I$, then $\prod_{i \in I} x_i$ is merely the product of those x_i that are not equal to e.

Note that if there are infinitely many x_i's not equal to e, then $\prod_{i \in I} x_i$ is undefined. For the sake of convenience, we assume that $\prod_{i \in I} x_i$ is meaningful whenever it appears in the ensuing discussion.

In the remainder of this section $|I| > 1$ and we write $\prod x_i$ for $\prod_{i \in I} x_i$.

Clearly, every element x of G can be expressed in the form $x = \prod x_i$.

Now let G_i be a submonoid of G for all $i \in I$ and assume that

$$x_i \in G_i \text{ and } x_j \in G_j \text{ for } i, j \in I \text{ with } i \neq j \Rightarrow x_i x_j = x_j x_i.$$

Then

$$x = \prod x_i \text{ and } y = \prod y_i, \text{ where } x_i, y_i \in G_i \ \forall \ i \in I \Rightarrow xy = \prod (x_i y_i).$$

For subsets A_i $(i \in I)$ of G, the set $\{\prod x_i \mid x_i \in A_i, \ i \in I\}$ is called the **weak product** of the A_i's and written as $\overset{*}{\prod\limits_{i \in I}} A_i$.

If $e \in A_i \ \forall \ i \in I$, then $A_j \subseteq \overset{*}{\prod\limits_{i \in I}} A_i \ \forall \ j \in I$.

For subgroups G_i of G $(i \in I)$, the group G is called the **weak direct product** of the G_i's, and written as $G = \prod_{i \in I} \dot{G_i}$, if the following conditions are satisfied:

(1) $G = \prod_{i \in I}^{*} G_i$,

(2) $x_i \in G_i$ and $x_j \in G_j$ for $i, j \in I$ with $i \neq j \Rightarrow x_i x_j = x_j x_i$,

(3) $\prod x_i = \prod y_i$, where $x_i, y_i \in G_i \Rightarrow x_i = y_i \ \forall \ i \in I$.

It can be shown that condition (2) is equivalent to the following condition:

(2)$'$ every G_i is a normal subgroup of G;

and condition (3) is equivalent to either of the following two conditions:

(3)$'$ $e = \prod x_i$, where $x_i \in G_i \Rightarrow x_i = e \ \forall \ i \in I$;

(3)$''$ $G_j \cap (\prod_{i \in I \setminus \{j\}}^{*} G_i) = \{e\} \ \forall \ j \in I$.

(If $I \setminus \{j\}$ is a singleton $\{i_0\}$, we write $\prod_{i \in I \setminus \{j\}}^{*} G_i = G_{i_0}$.)

Recall from group theory that if I is finite, then the complete direct product and the weak direct product coincide. If G is the weak direct product $\prod_{i \in I} \dot{G_i}$ and I is finite, say $I = \{1, 2, \ldots, n\}$, then we sometimes use the notation $G_1 \dot{\otimes} G_2 \dot{\otimes} \ldots \dot{\otimes} G_n$ or the more standard notation $G_1 \otimes G_2 \otimes \ldots \otimes G_n$ to denote the weak direct product of the G_i's, and call G the **direct product** of the G_i's.

We now define the weak product of fuzzy subsets as follows.

Definition 1.5.2. *For all $i \in I$, let $\mu_i \in \mathcal{FP}(G)$. Define $\mu \in \mathcal{FP}(G)$ as follows:*

$$\mu(x) = \vee\{(\wedge_{i \in I}\mu_i(x_i)) \mid x_i \in G, i \in I, \prod x_i = x\}$$

$\forall \ x \in G$. *Then μ is called the **weak product** of the μ_i's and is denoted by*

$$\mu = \prod_{i \in I}^{*} \mu_i.$$

Let $I = \{1, 2, \ldots, n\}$, where $n \in \mathbb{N} \setminus \{1\}$. If $\mu_i(e) \geq \mu_j(x)$ and $\mu_i \subseteq \mu \ \forall \ i, j \in I$ and $\forall x \in G$, then

$$\prod_{i \in I}^{*} \mu_i \subseteq \mu_1 \circ \mu_2 \circ \ldots \circ \mu_n.$$

However, if the μ_i's also satisfy the following condition:

$$x_i \in \mu_i^* \text{ and } x_j \in \mu_j^* \text{ for } i, j \in I \text{ with } i \neq j \Rightarrow x_i x_j = x_j x_i,$$

then

$$\prod_{i\in I}^{*} \mu_i = \mu_1 \circ \mu_2 \circ \ldots \circ \mu_n.$$

Therefore, the product of finitely many fuzzy subsets of a commutative group can be treated as a special case of the weak product.

Theorem 1.5.3. *For all* $i \in I$, *let* $\mu_i \in \mathcal{F}(G)$ *and let* $\mu = \prod_{i\in I}^{*} \mu_i$. *Then the following assertions hold:*

(1) $\mu_* \supseteq \prod_{i\in I}^{*}(\mu_i)_*.$

(2) *If* $\vee((\cup_{i\in I}\mu_i(G)) \setminus \{\mu(e)\}) < \mu(e)$, *then* $\mu_* = \prod_{i\in I}^{*}(\mu_i)_*.$

Theorem 1.5.4. *For all* $i \in I$, *let* $\mu_i \in \mathcal{FP}(G)$ *and let* $\mu = \prod_{i\in I}^{*} \mu_i$. *Then*

$$\mu^* = \prod_{i\in I}^{*} \mu_i^*.$$

Let $\mu \in \mathcal{FP}(G)$. Then μ is called a **fuzzy subsemigroup** of G if $\mu(xy) \geq \mu(x) \wedge \mu(y) \ \forall x, y \in G$ and a **fuzzy submonoid** of G if in addition to being a fuzzy subsemigroup of G, $\mu(e) \geq \mu(x) \ \forall x \in G$.

Theorem 1.5.5. *For all* $i \in I$, *let* $\mu_i \in \mathcal{FP}(G)$ *and let* $\mu = \prod_{i\in I}^{*} \mu_i$. *Suppose that* G *is a commutative group. Then the following assertions hold:*

(1) *If every* μ_i *is a fuzzy subsemigroup of* G, *then so is* μ.

(2) *If every* μ_i *is an fuzzy submonoid of* G *and* $\mu_i(e) = \mu_j(e) \ \forall i, j \in I$, *then* μ *is the smallest fuzzy submonoid of* G *that contains all* μ_i's.

(3) *If every* μ_i *is an fuzzy subgroup of* G *and* $\mu_i(e) = \mu_j(e) \ \forall i, j \in I$, *then* μ *is the smallest fuzzy subgroup of* G *that contains all* μ_i's, *that is,* $\mu = \langle \cup_{i\in I}\mu_i \rangle$.

Proof. (1) If every μ_i is a fuzzy subsemigroup of G, then

$$\begin{aligned}
\mu(xy) &= \vee\{\wedge_{i\in I}\mu_i(z_i) \mid z_i \in G, i \in I, \textstyle\prod z_i = xy\} \\
&\geq \vee\{\wedge_{i\in I}\mu_i(x_iy_i) \mid x_i, y_i \in G, i \in I, \textstyle\prod x_i = x, \prod y_i = y\} \\
&\geq \vee\{\wedge_{i\in I}(\mu_i(x_i) \wedge \mu_i(y_i)) \mid x_i, y_i \in G, i \in I, \textstyle\prod x_i = x, \prod y_i = y\} \\
&\geq \vee\{(\wedge_{i\in I}\mu_i(x_i)) \wedge (\wedge_{i\in I}\mu_i(y_i)) \mid x_i, y_i \in G, i \in I, \\
&\qquad \textstyle\prod x_i = x, \prod y_i = y\} \\
&= (\vee\{\wedge_{i\in I}\mu_i(x_i) \mid x_i \in G, i \in I, \textstyle\prod x_i = x\}) \\
&\qquad \wedge(\vee\{\wedge_{i\in I}\mu_i(y_i) \mid y_i \in G, i \in I, \textstyle\prod y_i = y\}) \\
&= \mu(x) \wedge \mu(y)
\end{aligned}$$

$\forall \ x, y \in G$. Hence μ is a fuzzy subsemigroup of G.

(2) If every μ_i is a fuzzy submonoid of G, then μ is also a fuzzy subsemigroup G, and in addition we have

$$\mu(e) = \wedge_{i \in I} \mu_i(e).$$

Thus μ is a fuzzy submonoid of G. Now since

$$\mu_i \subseteq \mu \; \forall i \in I$$

and for any fuzzy submonoid ν of G that contains all μ_i's,

$$x = \prod x_i, \text{ where } x, x_i \in G \text{ for } i \in I \Rightarrow \wedge_{i \in I} \mu_i(x_i) \leq \wedge_{i \in I} \nu(x_i) \leq \nu(x),$$

it follows that $\mu \subseteq \nu$. Therefore, μ is the smallest fuzzy submonoid of G that contains all μ_i's.

(3) If every μ_i is a fuzzy subgroup of G, then μ is a fuzzy submonoid of G. Also,

$$\begin{aligned}
\mu(x^{-1}) &= \vee\{\wedge_{i \in I} \mu_i(y_i) \mid y_i \in G, i \in I, \prod y_i = x^{-1}\} \\
&= \vee\{\wedge_{i \in I} \mu_i(y_i^{-1}) \mid y_i \in G, i \in I, \prod y_i^{-1} = x\} \\
&= \mu(x)
\end{aligned}$$

$\forall \, x \in G$. Thus μ is also a fuzzy subgroup of G. It has been shown that μ, being a fuzzy submonoid of G, is the smallest fuzzy submonoid of G that contains all μ_i's. Naturally, μ is also the smallest fuzzy subgroup of G that contains all μ_i's, that is, $\mu = \langle \cup_{i \in I} \mu_i \rangle$. □

In the preceding theorem, G is assumed to be a commutative group. For an arbitrary group, we have the following theorem.

Theorem 1.5.6. *For all $i \in I$, let $\mu_i \in \mathcal{FP}(G)$ and $\mu = \overset{*}{\underset{i \in I}{\prod}} \mu_i$. Suppose that the μ_i's satisfy the following condition:*

$$x_i \in \mu_i^* \text{ and } x_j \in \mu_j^* \text{ for } i \neq j \Rightarrow x_i x_j = x_j x_i.$$

If every μ_i is a fuzzy subsemigroup (fuzzy submonoid, fuzzy subgroup, normal fuzzy subgroup) of G with $\mu_i(e) = \mu_j(e) \; \forall i, j \in I$, then so is μ.

Proof. The proofs of the first three parts of the theorem are similar to those of Theorem 1.5.5 and thus are omitted. We now prove the last part. Assume that every μ_i is a normal fuzzy subgroup of G. By the third part, we conclude that μ is a fuzzy subgroup of G. Hence

$$\begin{aligned}
\mu(xyx^{-1}) &= \vee\{\wedge_{i \in I} \mu_i(z_i) \mid z_i \in G, i \in I, \prod z_i = xyx^{-1}\} \\
&= \vee\{\wedge_{i \in I} \mu_i(xy_ix^{-1}) \mid y_i \in G, i \in I, \prod xy_ix^{-1} = xyx^{-1}\} \\
&= \vee\{\wedge_{i \in I} \mu_i(y_i) \mid y_i \in G, i \in I, \prod y_i = y\} \\
&= \mu(y)
\end{aligned}$$

$\forall \, x, y \in G$. Thus μ is a normal fuzzy subgroup of G. □

Theorem 1.5.7. *For all $i \in I$, let $\mu_i \in \mathcal{F}(G)$ be such that $\mu_i(e) = \mu_j(e)$ $\forall i, j \in I$. Let $\mu = \prod_{i \in I}^{*} \mu_i$. Suppose that H and H_i $(i \in I)$ are subgroups of G such that $\mu_i^{*} \subseteq H_i \ \forall \ i \in I$ and $H = \prod_{i \in I} H_i$, i.e., H is the weak direct product of the H_i's. Then the following assertions hold.*

(1) $\mu \in \mathcal{F}(G)$.

(2) Every μ_i is a normal fuzzy subgroup of the fuzzy subgroup μ.

(3) $\mu_j \cap (\prod_{i \in I \setminus \{j\}}^{*} \mu_i) = \mu(e)_{\{e\}} \ \forall \ j \in I$.

(When $I \setminus \{j\}$ is a singleton $\{i_0\}$, we write $\prod_{i \in I \setminus \{j\}}^{*} \mu_i = \mu_{i_0}$).

(4) If $\mu_i|_H \in \mathcal{NF}(H) \ \forall \ i \in I$, then $\mu|_H \in \mathcal{NF}(H)$.

Proof. Let $x, y \in G$ and consider $\mu(xy)$. For either $x \notin H$ or $y \notin H$, clearly $\mu(xy) \geq 0 = \mu(x) \wedge \mu(y)$. If $x, y \in H$, we have $xy \in H$ and the proof that (1) holds follows from Theorem 1.5.5.

Now we show assertion (2) holds. Let $j \in I$. Since $\mu_j \in \mathcal{F}(G)$ and $\mu_j \subseteq \mu$, in view of part (2) of Theorem 1.4.2, it needs only to be shown that

$$\mu_j(yx) \geq \mu_j(xy) \wedge \mu(y) \ \forall x, y \in G.$$

If either $xy \notin H_j$ or $y \notin H$, it is clear that

$$\mu_j(yx) \geq 0 = \mu_j(xy) \wedge \mu(y).$$

Now assume that $xy \in H_j$ and $y \in H$. Since $H_j \subseteq H$, we have $x \in H$. Also, from the fact that H_j is a normal subgroup of H, it follows that $yx = x^{-1} \cdot xy \cdot x \in H_j$. Thus using the fact that $xy, yx \in H_j$, we have that

$$\mu(xy) = \mu_j(xy) \text{ and } \mu(yx) = \mu_j(yx).$$

Hence

$$\begin{aligned}
\mu_j(yx) &= \mu(yx) \\
&= \mu(y \cdot xy \cdot y^{-1}) \\
&\geq \mu(xy) \wedge \mu(y) \\
&= \mu_j(xy) \wedge \mu(y).
\end{aligned}$$

Therefore, μ_j is a normal fuzzy subgroup of μ, that is, assertion (2) holds.

It now follows that (3) holds.

To show (4), we suppose that $\mu_i|_H \in \mathcal{NF}(H) \ \forall \ i \in I$. Clearly, $\mu|_H \in \mathcal{F}(H)$. Next, for any $x, y \in H$, x and y have a unique expression as $x = \prod x_i$ and $y = \prod y_i$, where $x_i, y_i \in H_i \ \forall \ i \in I$. Hence $xy = \prod(x_i y_i)$ and $yx = \prod(y_i x_i)$. Thus

$$\mu(xy) = \wedge_{i \in I} \mu_i(x_i y_i) = \wedge_{i \in I} \mu_i(y_i x_i) = \mu(yx).$$

Therefore, $\mu|_H \in \mathcal{NF}(H)$. That is, assertion (4) holds. $\qquad\square$

Further, it is easy to show that $\mu|_H$ in Theorem 1.5.7 is isomorphic to some fuzzy subgroup of the group $\widetilde{\prod\limits_{i \in I}} H_i$.

Definition 1.5.8. *Let $\mu \in \mathcal{F}(G)$ and $\mu_i \in \mathcal{F}(G)$ for all $i \in I$. Suppose that $\mu_i(e) = \mu_j(e) \ \forall i, j \in I$. Then μ is called the **weak direct product** of the μ_i, written $\mu = \prod\limits_{i \in I}^{\cdot} \mu_i$, if the following conditions are satisfied:*

(1) $\mu = \prod\limits_{i \in I}^{*} \mu_i$.

(2) *Every μ_i is a normal fuzzy subgroup of μ.*

(3) $\mu_j \cap (\prod\limits_{i \in I \backslash \{j\}}^{*} \mu_i) = \mu(e)_{\{e\}} \ \forall \ j \in I$.

In addition, when $I = \{1, 2, \ldots, n\}$ $(n \in \mathbb{N} \backslash \{1\})$, we sometimes write $\mu_1 \otimes \mu_2 \otimes \ldots \otimes \mu_n$ for $\prod\limits_{i \in I}^{\cdot} \mu_i$ and call it the **direct product** of the μ_i's.

Theorem 1.5.9. *Let $\mu \in \mathcal{F}(G)$ and $\mu_i \in \mathcal{F}(G)$ for all $i \in I$. Suppose that $\mu_i(e) = \mu_j(e) \ \forall i, j \in I$. Then a necessary and sufficient condition for $\mu = \prod\limits_{i \in I}^{\cdot} \mu_i$ is that*

$$\mu^* = \prod\limits_{i \in I}^{\cdot} \mu_i^* \ and \ \mu = \prod\limits_{i \in I}^{*} \mu_i.$$

Proof. The sufficiency follows directly from Theorem 1.5.7 with $H_i = \mu_i^*$ and $H = \mu^*$.

For the necessity, suppose that $\mu = \prod\limits_{i \in I}^{\cdot} \mu_i$. Then by Theorem 1.5.4, we obtain that $\mu^* = \prod\limits_{i \in I}^{*} \mu_i^*$ and by Theorem 1.4.4, it follows that for all $j \in I$, μ_j^* is a normal subgroup of μ^*. Again, by Theorem 1.5.4,

$$x \in \mu_j^* \cap (\prod\limits_{i \in I \backslash \{j\}}^{*} \mu_i^*) \Leftrightarrow x \in \mu_j^* \cap (\prod\limits_{i \in I \backslash \{j\}}^{*} \mu_i)^*$$

$$\Leftrightarrow \mu_j(x) > 0 \ and \ (\prod\limits_{i \in I \backslash \{j\}}^{*} \mu_i)(x) > 0$$

$$\Leftrightarrow (\mu_j \cap (\prod\limits_{i \in I \backslash \{j\}}^{*} \mu_i))(x) > 0$$

$$\Leftrightarrow x = e.$$

for all $j \in I$. Therefore, the desired result holds. $\quad\square$

Theorem 1.5.10. *Let $\mu \in \mathcal{F}(G)$ and $\{\mu_i \mid i \in I\}$ be a collection of fuzzy subgroups of G such that $\mu_i \subseteq \mu \ \forall \ i \in I$. Suppose that either*

(1) $\bigcup_{i \in I} |\mu_i(G)|$ *is finite or*

(2) $\mu^* = \prod_{i \in I} \mu_i^*.$

Then $\mu = \prod_{i \in I \setminus \{j\}}^* \mu_i$ *if and only if* $\mu_a = \prod_{i \in I \setminus \{j\}}^* (\mu_i)_a \ \forall \ a \in \{b \in [0,1] \mid b \leq \mu(e)\}.$

Proof. We first show that $\mu_i(e) = \mu(e) \ \forall \ i \in I$ in either direction of the proof. Suppose that $\mu = \prod_{i \in I}^* \mu_i$. Then $\mu(e) = \vee\{\wedge\{\mu_i(x_i) \mid i \in I\} \mid e = \prod_{i \in I} x_i\} \leq \vee\{\wedge\{\mu_i(e) \mid i \in I\} \mid e = \prod_{i \in I} x_i\}$ (since $\mu_i(x_i) \leq \mu(e)$) $= \wedge\{\mu_i(e) \mid i \in I\} \leq \mu(e)$ (since $\mu_i \subseteq \mu$). Thus $\mu(e) = \mu_i(e) \ \forall i \in I$. Suppose $\mu_a = \prod_{i \in I}(\mu_i)_a \ \forall$ $a \in \{b \in [0,1] \mid b \leq \mu(e)\}$. Now $e \in \mu_{\mu(e)}$ and so $e \in \prod_{i \in I}^* (\mu_i)_{\mu(e)}$. Thus $(\mu_i)_{\mu(e)} \neq \emptyset$. Hence $\mu_i(e) \geq \mu(e) \ \forall i \in I$.

Suppose that $\mu = \prod_{i \in I}^* \mu_i$. Then

$$x \in (\prod_{i \in I}^* \mu_i)_a \Leftrightarrow (\prod_{i \in I}^* \mu_i)(x) \geq a$$
$$\Leftrightarrow \vee\{\wedge\{\mu_i(x_i) \mid i \in I\} \mid x = \prod_{i \in I} x_i\} \geq a$$
$$\Leftarrow \mu_i(x_i) \geq a \text{ for some representation of } x = \prod_{i \in I} x_i \text{ and } \forall i \in I$$
$$\Leftrightarrow x \in \prod_{i \in I}^* (\mu_i)_a.$$

If condition (2) holds, then x has a unique representation. Clearly then, either condition (1) or condition (2) implies that the $\Leftarrow$ becomes $\Leftrightarrow$ in the above sequence of implications. Conversely, suppose that $\mu_a = \prod_{i \in I}^* (\mu_i)_a \ \forall \ a \in \{b \in [0,1] \mid b \leq \mu(e)\}$. Let $x \in G$ and $\mu(x) = a$. Then $x \in \mu_a$ and so there exist $y_i \in (\mu_i)_a$ such that $x = \prod_{i \in I} y_i$. Thus

$$(\prod_{i \in I}^* \mu_i)(x) = \vee\{\wedge\{\mu_i(x_i) \mid i \in I\} \mid x = \prod_{i \in I} x_i\} \geq a = \mu(x).$$

Hence it follows that $\mu = \prod_{i \in I}^* \mu_i$. $\square$

Theorem 1.5.11. *Let $\mu \in \mathcal{F}(G)$ and $\{\mu_i \mid i \in I\}$ be a collection of fuzzy subgroups of G such that $\mu_i \subseteq \mu \ \forall \ i \in I$. Suppose that $\mu_i(e) = \mu_j(e) \ \forall i, j \in I$. Then*

$$\mu = \prod_{i \in I}^{\cdot} \mu_i \text{ if and only if } \mu_a = \prod_{i \in I}^{\cdot} (\mu_i)_a$$

$\forall \ a \in \{b \in [0,1] \mid b \leq \mu(e)\} \setminus \{0\}.$

Proof. Suppose that $\mu = \dot{\prod}_{i \in I} \mu_i$. Then $\mu^* = \dot{\prod}_{i \in I} \mu_i^*$ by Theorem 1.5.9. Thus $\forall$ $a \in \{b \in [0,1] \mid b \leq \mu(e)\} \backslash \{0\}$, $\mu_a = \overset{*}{\prod}_{i \in I} (\mu_i)_a$ by Theorem 1.5.10 and

$$(\mu_i)_a \cap (\mu_j)_a \subseteq \mu_i^* \cap \mu_j^* = \{e\}$$

if $i \neq j$. Thus

$$\mu_a = \dot{\prod}_{i \in I} (\mu_i)_a$$

$\forall a \in \{b \in [0,1] \mid b \leq \mu(e)\} \backslash \{0\}$ by Theorem 1.4.3. Conversely, suppose that $\mu_a = \dot{\prod}_{i \in I} (\mu_i)_a \forall a \in \{b \in [0,1] \mid b \leq \mu(e)\} \backslash \{0\}$. Then by Theorem 1.5.10, $\mu = \overset{*}{\prod}_{i \in I} \mu_i$. Let $x \in \mu_i^* \cap \mu_j^*$. Then for $a = \mu_i(x) \wedge \mu_j(x)$, we have that $a > 0$ and $x \in (\mu_i)_a \cap (\mu_j)_a = \{e\}$ for $i \neq j$. Hence $x = e$ and so $\mu_i^* \cap \mu_j^* = \{e\}$ for $i \neq j$. Thus $\mu^* = \dot{\prod}_{i \in I} \mu_i^*$ by Theorem 1.4.4 and hence $\mu = \dot{\prod}_{i \in I} \mu_i$ by Theorem 1.5.9. $\square$

Theorem 1.5.12. *Let $\mu \in \mathcal{F}(G)$ and $\{\mu_i \mid i \in I\}$ be a collection of fuzzy subgroups of G such that $\mu_i \subseteq \mu \forall i \in I$. Suppose that $\mu = \dot{\prod}_{i \in I} \mu_i$ and $\mu_* = \overset{*}{\prod}_{i \in I} (\mu_i)_*$. Then $\mu_* = \dot{\prod}_{i \in I} (\mu_i)_*$.*

Proof. By Corollary 1.2.7, μ_* is a subgroup of G. Since by definition, every μ_i is a normal fuzzy subgroup of μ, it follows from Theorem 1.4.4 that every $(\mu_i)_*$ is a normal subgroup of μ_*. Let $j \in I$ and $x \in (\mu_j)_* \cap \overset{*}{\prod}_{i \in I \backslash \{j\}} (\mu_i)_*$. Then $x \in \mu_j^* \cap \overset{*}{\prod}_{i \in I \backslash \{j\}} \mu_i^*$ and so $(\mu_j \cap (\prod_{i \in I \backslash \{j\}}^* \mu_i))(x) > 0$. Hence $x = e$. Therefore, $\mu_* = \dot{\prod}_{i \in I} (\mu_i)_*$. $\square$

The following two theorems are direct consequences of Theorems 1.4.5 and 1.4.6, respectively.

Theorem 1.5.13. *Let $\mu \in \mathcal{F}(G)$, $\{\mu_i \mid i \in I\}$ be a collection of fuzzy subgroups of G such that $\mu_i \subseteq \mu \forall i \in I$. Let $\nu_i \in \mathcal{NF}(G)$. Suppose that $\mu = \dot{\prod}_{i \in I} \mu_i$. Let $\xi = \overset{*}{\prod}_{i \in I} (\mu_i \cap \nu_i)$. Then $\xi = \dot{\prod}_{i \in I} (\mu_i \cap \nu_i)$.*

Theorem 1.5.14. *Let $\mu \in \mathcal{F}(G)$ and $\{\mu_i \mid i \in I\}$ be a collection of fuzzy subgroups of G such that $\mu_i \subseteq \mu \forall i \in I$. Suppose that for all $i \in I$, ν_i is* *a normal fuzzy subgroup of μ and $\mu = \prod_{i \in I}^{\cdot} \mu_i$. Let $\xi = \prod_{i \in I}^{*}(\mu_i \cap \nu_i)$. Then*

$$\xi = \prod_{i \in I}^{\cdot}(\mu_i \cap \nu_i).$$

We now consider some results from [3] and [12]. Let $I = \{1, 2, ..., n\}$. Let G_i be a group, $i = 1, 2, ..., n$, and let μ a fuzzy subgroup of $G_1 \overset{\sim}{\otimes} G_2 \overset{\sim}{\otimes} ... \overset{\sim}{\otimes} G_n$. We show there exist fuzzy subgroups $\mu_1, \mu_2, ..., \mu_n$ of $G_1, G_2, ..., G_n$, respectively, such that the complete direct product $\prod_{i \in I}^{\sim} \mu_i \subseteq \mu$. We find necessary and sufficient conditions for $\mu = \prod_{i \in I}^{\sim} \mu_i$.

For all $i = 1, 2, ..., n$, let μ_i be a fuzzy subgroup of a group G_i. Let e_i denote the identity of G_i, $i = 1, 2, ..., n$. Recall that the complete direct product of the μ_i $(i = 1, 2, ..., n)$ is the fuzzy subset $\prod_{i \in I}^{\sim} \mu_i$ of $G_1 \overset{\sim}{\otimes} G_2 \overset{\sim}{\otimes} ... \overset{\sim}{\otimes} G_n$ defined by $(\prod_{i \in I}^{\sim} \mu_i)(x_1, x_2, ..., x_n) = \mu_1(x_1) \wedge \mu_2(x_2) \wedge ... \wedge \mu_n(x_n)$ for all $x_i \in G_i$, $i = 1, 2, ..., n$.

Theorem 1.5.15. *Let $G_1, G_2, .., G_n$ be groups and let μ be a fuzzy subgroup of $G_1 \overset{\sim}{\otimes} G_2 \overset{\sim}{\otimes} ... \overset{\sim}{\otimes} G_n$. Let μ_i be the fuzzy subset of G_i defined by $\mu_i(x) = \mu(e_1, ..., e_{i-1}, x, e_{i+1}, ..., e_n) \forall x \in G_i$, $i = 1, 2, ..., n$. Then μ_i is a fuzzy subgroup of G_i, $i = 1, 2, ..., n$, and $\prod_{i \in I}^{\sim} \mu_i \subseteq \mu$.*

Proof. Let $K_i = \{e_1\} \overset{\sim}{\otimes} ... \overset{\sim}{\otimes} \{e_{i-1}\} \overset{\sim}{\otimes} G_i \overset{\sim}{\otimes} \{e_{i+1}\} \overset{\sim}{\otimes} ... \overset{\sim}{\otimes} \{e_n\}$ for $i = 1, 2, ..., n$. Then K_i is a subgroup of $G_1 \overset{\sim}{\otimes} G_2 \overset{\sim}{\otimes} ... \overset{\sim}{\otimes} G_n$. Clearly, $\mu|_{K_i}$ is a fuzzy subgroup of K_i. Let $\phi : G_i \to K_i$ be the function defined by $\phi(x) = (e_1, ..., e_{i-1}, x, e_{i+1}, ..., e_n) \forall x \in G_i$. Then ϕ is an isomorphism and $\mu_i = \mu|_{K_i} \circ \phi$. Hence μ_i is a fuzzy subgroup of G_i. Now

$\mu(x_1, x_2, ..., x_n) = \mu((x_1, e_2, ..., e_n)(e_1, x_2, ..., x_n)) \geq$
$\mu(x_1, e_2, ..., e_n) \wedge \mu(e_1, x_2, ..., x_n) =$
$\mu_1(x_1) \wedge \mu(e_1, x_2, x_3, ..., x_n) =$
$\mu_1(x_1) \wedge ((\mu e_1, x_2, e_3, ..., e_n)(e_1, e_2, x_3, ..., x_n)) \geq$
$\mu_1(x_1) \wedge \mu((e_1, x_2, e_3, ..., e_n) \wedge \mu(e_1, e_2, x_3, ..., x_n)) =$
$\mu_1(x_1) \wedge (\mu_2(x_2) \wedge \mu(e_1, e_2, x_3, ..., x_n)) \geq$
$\mu_1(x_1) \wedge (\mu_2(x_2) \wedge (\mu_3(x_3) \wedge \mu(e_1, e_2, e_3, x_4, ..., x_n))) \geq ... \geq$
$\mu_1(x_1) \wedge (\mu_2(x_2) \wedge ... \wedge (\mu_{n-2}(x_{n-2}) \wedge (\mu_{n-1}(x_{n-1}) \wedge \mu_n(x_n))) =$
$\mu_1(x_1) \wedge \mu_2(x_2) \wedge ... \wedge \mu_{n-2}(x_{n-2}) \wedge \mu_{n-1}(x_{n-1}) \wedge \mu_n(x_n).$

Thus $\mu(x_1, x_2, x_3, ..., x_n) \geq \mu_1(x_1) \wedge \mu_2(x_2) \wedge ... \wedge \mu_n(x_n) = (\prod_{i \in I}^{\sim} \mu_i)(x_1, x_2, ..., x_n).$ $\quad\square$

Lemma 1.5.16. *Let* $G_1, G_2, ..., G_n$ *be groups and let* μ *be a fuzzy subgroup of* $G_1 \tilde{\otimes} G_2 \tilde{\otimes} ... \tilde{\otimes} G_n$. *If* $\mu(e_1, e_2, ..., e_{i-1}, x_i, e_{i+1}, ..., e_n) \geq \mu(x_1, x_2, ..., x_n)$ *for* $i = 1, 2, ..., k-1, k+1, ..., n$, *then* $\mu(e_1, e_2, ..., e_{k-1}, x_k, e_{k+1}, ..., e_n) \geq \mu(x_1, x_2, ..., x_n)$.

Proof. $\mu(e_1, ..., e_{k-1}, x_k, e_{k+1}, ..., e_n) =$
$\quad \mu((x_1, x_2, ..., x_n)(x_1^{-1}, ..., x_{k-1}^{-1}, e_k, x_{k+1}^{-1}, ..., x_n^{-1})) \geq$
$\quad \mu(x_1, x_2, ..., x_n) \wedge (\mu(x_1^{-1}, x_2^{-1}, ..., x_{k-1}^{-1}, e_k, x_{k+1}^{-1}, ..., x_n^{-1})) =$
$\quad \mu(x_1, x_2, ..., x_n) \wedge \mu(x_1, x_2, ..., x_{k-1}, e_k, x_{k+1}, ..., x_n) \geq$
$\quad \mu(x_1, x_2, ..., x_n) \wedge (\mu(e_1, e_2, ..., e_{k-2}, x_{k-1}, e_k, e_{k+1}, ..., e_n) \wedge$
$\quad \mu(x_1, x_2, ..., x_{k-2}, e_{k-1}, e_k, x_{k+1}, ..., x_n)) =$
$\quad \mu(x_1, x_2, ..., x_n) \wedge \mu(x_1, ..., x_{k-2}, e_{k-1}, e_k, x_{k+1}, ..., x_n) \geq$
$\quad \mu(x_1, x_2, ..., x_n) \wedge (\mu(e_1, ..., e_{k-3}, x_{k-2}, e_{k-1}, e_k, e_{k+1}, ..., e_n) \wedge$
$\quad \mu(x_1, ..., x_{k-3}, e_{k-2}, e_{k-1}, e_k, x_{k+1}, ..., x_n)) =$
$\quad \mu(x_1, x_2, ..., x_n) \wedge \mu(x_1, ..., x_{k-3}, e_{k-2}, e_{k-1}, e_k, x_{k+1}, ..., x_n) \geq ... \geq$
$\quad \mu(x_1, x_2, ..., x_n) \wedge \mu(e_1, ..., e_{k-2}, e_{k-1}, e_k, e_{k+1}, ..., e_{n-1}, x_n) =$
$\quad \mu(x_1, x_2, ..., x_n)$.
Thus $\mu(e_1, e_2, ..., e_{k-1}, x_k, e_{k+1}, ..., e_n) \geq \mu(x_1, x_2, ..., x_n)$.   □

Theorem 1.5.17. *Let* $G_1, G_2, ..., G_n$ *be groups and let* μ *be a fuzzy subgroup of* $G_1 \tilde{\otimes} G_2 \tilde{\otimes} ... \tilde{\otimes} G_n$. *Then for all* $x_i \in G_i$, $\mu(e_1, e_2, ..., e_{i-1}, x_i, e_{i+1}, ..., e_n) \geq \mu(x_1, x_2, ..., x_n)$ *for* $i = 1, 2, ..., k-1, k+1, ..., n$, *if and only if* $\mu = \prod_{i \in I}^{\sim} \mu_i$, *where* $\mu_1, \mu_2, ..., \mu_n$ *are fuzzy subgroups of* $G_1, G_2, ..., G_n$, *respectively.*

Proof. Define $\mu_j : G_j \to [0,1]$ by $\mu_j(x_j) = \mu(e_1, ..., x_j, ..., e_n)$ for $j = 1, 2, ..., n$. Let $K_j = \{e_1\} \tilde{\otimes} \{e_2\} \tilde{\otimes} ... \tilde{\otimes} G_j \tilde{\otimes} ... \tilde{\otimes} \{e_n\}$. Then each $\mu|_{K_j}$ is a fuzzy subgroup of K_j for $j = 1, 2, ..., n$. By Theorem 1.5.15, $\prod_{i \in I}^{\sim} \mu_i \subseteq \mu$. By Lemma 1.5.16, $\mu(e_1, e_2, ..., e_{i-1}, x_i, e_{i+1}, ..., e_n) \geq \mu(x_1, x_2, ..., x_n)$ for $i = 1, 2, ..., n$. $\mu_j(x_j) = \mu(e_1, ..., x_j, ..., e_n) \geq \mu(x_1, x_2, ..., x_n)$ for $j = 1, 2, ..., n$. Thus $(\prod_{i \in I}^{\sim} \mu_i)(x_1, x_2, ..., x_n) = \mu_1(x_1) \wedge \mu_2(x_2) \wedge ... \wedge \mu_n(x_n) \geq \mu(x_1, x_2, ..., x_n)$. Hence $\prod_{i \in I}^{\sim} \mu_i = \mu$. The other direction is straightforward.   □

Lemma 1.5.18. *Let* G *be a group and let* μ *be a fuzzy subgroup of* G. *Let* $x \in G$ *be of finite order* k. *If* $r \in \mathbb{N}$ *and* k *are relatively prime, then* $\mu(x) = \mu(x^r)$.

Proof. There exist integers s and t such that $1 = sr + kt$. Thus $\mu(x) = \mu(x^{sr+tk}) = \mu(x^{sr} x^{tk}) = \mu(x^{sr}) \geq \mu(x^r) \geq \mu(x)$.   □

Lemma 1.5.19. *Let* $G_1, G_2, ..., G_n$ *be finite groups and let* μ *be a fuzzy subgroup of* $G_1 \tilde{\otimes} G_2 \tilde{\otimes} ... \tilde{\otimes} G_n$. *If the orders of* G_i *and* G_j *are relatively prime for all* $i, j \in I, i \neq j$, *then* $\mu(x_1, ..., x_i, ..., x_n) \leq \mu(e_1, ...e_{i-1}, x_i, e_{i+1}, ..., e_n)$ *for all* $x_i \in G_i, i = 1, 2, ..., n$.

Proof. Let $(x_1, ..., x_i, ..., x_n) \in G_1 \overset{\sim}{\otimes} G_2 \overset{\sim}{\otimes} ... \overset{\sim}{\otimes} G_n$. Let $i \in I$ and $r_i = |G_1|...|G_{i-1}||G_{i+1}|...|G_n|$. Then r_i and $|G_i|$ are relatively prime as are r_i and the order of x_i for all $i \in I$. Clearly, $(x_j)^{r_i} = e_j$ for all $i, j \in I, j \neq i$. Hence

$$\mu(x_1, ..., x_i, ..., x_n) \leq \mu((x_1, ..., x_i, ..., x_n)^{r_i}) =$$
$$\mu((x_1, e_i, ..., e_n)^{r_i}...(e_1, ..., e_{i-1}, x_i, e_{i+1}, ..., e_n)^{r_i}...(e_1, ..., e_{n-1}, x_n)^{r_i}) =$$
$$\mu(e_1, ..., e_{i-1}, x_i^{r_i}, e_{i+1}, ..., e_n) = \mu(e_1, ..., e_{i-1}, x_i, e_{i+1}, ..., e_n), \text{ where the}$$

last equality holds by Lemma 1.5.18.   □

Theorem 1.5.20. *Let $G_1, G_2, ..., G_n$ be finite groups and let μ be a fuzzy subgroup of $G_1 \overset{\sim}{\otimes} G_2 \overset{\sim}{\otimes} ... \overset{\sim}{\otimes} G_n$. If the orders of G_i and G_j are relatively prime for all $i, j \in I, i \neq j$, then $\mu = \prod\limits_{i \in I} \mu_i$, where $\mu_1, \mu_2, ..., \mu_n$ are fuzzy subgroups of $G_1, G_2, ..., G_n$, respectively.*

Proof. The result follows from Theorem 1.5.17 and Lemma 1.5.19.   □

1.6 Fuzzy Order Relative to Fuzzy Subgroups

We present the notion of the fuzzy order of an element of a group relative to a fuzzy subgroup as introduced by Kim [26]. This notion generalizes the usual order of an element. We show that most of the basic properties of the order of an element in standard group theory are valid in the theory of fuzzy subgroups if the order of an element is replaced by fuzzy order.

We denote the identity element of a group G by e, the order of an element x of G by $o(x)$, and the greatest common divisor of integers m and n by (m, n).

Proposition 1.6.1. *Let G be a finite group and let μ be a fuzzy subgroup of G. If $o(y)|o(x)$ and $x, y \in \langle z \rangle$ for some $z \in G$, then $\mu(x) \leq \mu(y)$.*

Proof. Let $o(y) = k$. Then $o(x) = kq$ for some $q \in \mathbb{N}$. Now $y = z^i$ and $x = z^j$ for some $i, j \in \mathbb{Z}$. Hence $z^{ik} = e = z^{jkq}$. Thus $y = x^q$. Hence $\mu(y) = \mu(x^q) \geq \mu(x)$.   □

We now recall some basic results concerning the orders of elements of a group.

Theorem 1.6.2. *Let G be a group and let $x, y, z \in G$. Then the following assertions hold.*
   (1) *If $x^m = e$, then $o(x)|m$, where $m \in \mathbb{Z}$.*
   (2) *$o(x^m) = o(x)/(o(x), m)$, where $m \in \mathbb{Z}$.*
   (3) *If $(o(x), o(y)) = 1$ and $xy = yx$, then $o(xy) = o(x) \cdot o(y)$.*
   (4) *If $z = y^{-1}xy$, then $o(z) = o(x)$.*
   (5) *If $o(z) = mn$ with $(m, n) = 1$, then $z = xy = yx$ for some $x, y \in G$ with $o(x) = m$ and $o(y) = n$. Furthermore, such an expression for z is unique.*

We next define the fuzzy order of an element of a group and investigate some of its properties. Specifically, we show that Theorem 1.6.2 is valid if the orders of elements are replaced by their fuzzy orders.

Definition 1.6.3. *Let μ be a fuzzy subgroup of a group G and let $x \in G$. If there exists a positive integer n such that $\mu(x^n) = \mu(e)$, then the least such positive integer is called the **fuzzy order** of x with respect to μ and written as $FO_\mu(x)$. If no such n exists, x is said to be of **infinite fuzzy order** with respect to μ.*

Equality of $o(x)$ and $o(y)$ does not imply that of $FO_\mu(x)$ and $FO_\mu(y)$, as is shown in the following example.

Example 1.6.4. *Let $G = \{e, a, b, ab\}$ be the Klein four-group. Define the fuzzy subgroup μ of G by $\mu(e) = \mu(ab) = t_0$ and $\mu(a) = \mu(b) = t_1$, where $t_0 > t_1$. Clearly, $o(a) = o(ab) = 2$, but $FO_\mu(a) = 2$ and $FO_\mu(ab) = 1$. Also observe that in this example, $o(a)|o(ab)$, but $\mu(ab) > \mu(a)$. Thus Proposition 1.6.1 doesn't hold here since the elements a and ab do not lie in the same cyclic subgroup of G.*

Proposition 1.6.5. *Let μ be a fuzzy subgroup of a group G. For $x \in G$, if $\mu(x^m) = \mu(e)$ for some integer m, then $FO_\mu(x)|m$.*

Proof. Let $FO_\mu(x) = n$. By the Euclidean algorithm, there exist integers s and t such that $m = ns + t$, where $0 \leq t < n$. Then

$$\mu(x^t) = \mu(x^{m-ns}) = \mu(x^m(x^n)^{-s}) \geq \mu(x^m) \wedge \mu((x^n)^{-s})$$
$$\geq \mu(e) \wedge \mu(x^n) = \mu(e) \wedge \mu(e) = \mu(e).$$

Thus $\mu(x^t) = \mu(e)$. Hence $t = 0$ by the minimality of n.   □

If $o(x)$ is finite, then $FO_\mu(x)$ is clearly finite for all fuzzy subgroups μ of G. If $o(x)$ is infinite, then for each positive integer n, there exists a fuzzy subgroup μ_n of G such that $FO_{\mu_n}(x) = n$ as the following example shows.

Example 1.6.6. *Let x be an element of infinite order in the group G. For each positive integer n, define the fuzzy subgroup μ_n of G by*

$$\mu_n(y) = \begin{cases} t_0 & \text{if } y \in \langle x^n \rangle, \\ t_1 & \text{otherwise,} \end{cases}$$

where $t_0 > t_1$. Observe that $FO_{\mu_n}(x) = n$.

With this example in mind, the following corollary assumes added importance when one agrees that in the extended real number system ∞ is divisible by both ∞ and a positive real number.

Corollary 1.6.7. *Let μ be a fuzzy subgroup of a group G. Then $FO_\mu(x)|o(x)$ for all $x \in G$.*

Proposition 1.6.8. *Let μ be a fuzzy subgroup of a group G and let x and y be elements of G such that $(FO_\mu(x), FO_\mu(y)) = 1$ and $xy = yx$. If $\mu(xy) = \mu(e)$, then $\mu(x) = \mu(y) = \mu(e)$.*

Proof. Let $FO_\mu(x) = n$ and $FO_\mu(y) = m$. Then $\mu(e) = \mu(xy) \le \mu((xy)^m) = \mu(x^m y^m)$. Hence $\mu(x^m y^m) = \mu(e)$. Now $\mu(x^m) = \mu(x^m y^m y^{-m}) \ge \mu(x^m y^m) \wedge \mu(y^{-m}) = \mu(e) \wedge \mu(e) = \mu(e)$. Thus $\mu(x^m) = \mu(y^m) = \mu(e)$. Therefore, $n|m$ by Proposition 1.6.5. But $(n, m) = 1$. Thus $n = 1$, i.e., $\mu(x) = \mu(e)$. Similarly $\mu(y) = \mu(e)$. □

Proposition 1.6.8 is an improvement of the result that $\mu(xy^{-1}) = \mu(e)$ implies that $\mu(x) = \mu(y)$. By virtue of Corollary 1.6.7, the assumption of the proposition can be weakened as follows:

Corollary 1.6.9. *Let μ be a fuzzy subgroup of a group G, and let x and y be elements of G such that $(o(x), o(y)) = 1$ and $xy = yx$. If $\mu(xy) = \mu(e)$, then $\mu(x) = \mu(y) = \mu(e)$.*

Neither the assumption $(FO_\mu(x), FO_\mu(y)) = 1$ in Proposition 1.6.8 nor the assumption $(o(x), o(y)) = 1$ in Corollary 1.6.9 can be omitted. In fact, in Example 1.6.4, $\mu(a) = \mu(b) \ne \mu(e)$ and $\mu(ab) = \mu(e)$, but $FO_\mu(a) = FO_\mu(b) = o(a) = o(b) = 2$.

Theorem 1.6.10. *Let μ be a fuzzy subgroup of a group G. Let $FO_\mu(x) = n$, where $x \in G$. If m is an integer with $d = (m, n)$, then $FO_\mu(x^m) = n/d$.*

Proof. Let $FO_\mu(x^m) = t$. First, we have

$$\mu((x^m)^{n/d}) = \mu(x^{nk}) \text{ for some integer } k$$
$$\ge \mu(x^n) = \mu(e).$$

Thus $t|n/d$ by Proposition 1.6.5. Also, since $d = (m, n)$, there exist integers i and j such that $ni + mj = d$. We then have

$$\mu(x^{td}) = \mu(x^{t(ni+mj)}) = \mu(x^{nti} x^{mtj})$$
$$\ge \mu((x^n)^{ti}) \wedge \mu\left(((x^m)^t)^j\right)$$
$$\ge \mu(x^n) \wedge \mu((x^m)^t)$$
$$= \mu(e) \wedge \mu(e) = \mu(e).$$

This implies that $n|td$ by Proposition 1.6.5, i.e., $n/d|t$. Consequently, $t = n/d$. □

Proposition 1.6.11. *Let μ be a fuzzy subgroup of a group G. Let $FO_\mu(x) = n$ where $x \in G$. If m is an integer with $(n, m) = 1$, then $\mu(x^m) = \mu(x)$.*

Proof. Since $(n, m) = 1$, there exist integers s and t such that $ns + mt = 1$. We then have

$$
\begin{aligned}
\mu(x) = \mu(x^{ns+mt}) &= \mu((x^n)^s(x^m)^t) \\
&\geq \mu((x^n)^s) \wedge \mu((x^m)^t) \geq \mu(x^n) \wedge \mu(x^m) \\
&= \mu(e) \wedge \mu(x^m) = \mu(x^m) \geq \mu(x).
\end{aligned}
$$

$\square$

Proposition 1.6.11 is an improvement of Lemma 1.5.18.

Theorem 1.6.12. *Let μ be a fuzzy subgroup of a group G. Let $FO_\mu(x) = n$, where $x \in G$. If $i \equiv j \pmod{n}$, where $i, j \in \mathbb{Z}$, then $FO_\mu(x^i) = FO_\mu(x^j)$.*

Proof. Let $FO_\mu(x^i) = t$ and $FO_\mu(x^j) = s$. By the assumption, $i = j + nk$ for some integer k. We then have

$$
\begin{aligned}
\mu((x^i)^s) = \mu((x^{j+nk})^s) &= \mu((x^j)^s (x^n)^{ks}) \\
&\geq \mu((x^j)^s) \wedge \mu((x^n)^{ks}) \\
&\geq \mu(e) \wedge \mu(x^n) = \mu(e) \wedge \mu(e) = \mu(e),
\end{aligned}
$$

and so $t|s$. Similarly, $s|t$. Thus we have $t = s$. $\square$

Theorem 1.6.13. *Let μ be a fuzzy subgroup of a group G, and let x and y be elements of G such that $xy = yx$ and $(FO_\mu(x), FO_\mu(y)) = 1$. Then $FO_\mu(xy) = FO_\mu(x) \cdot FO_\mu(y)$.*

Proof. Let $FO_\mu(xy) = n$, $FO_\mu(x) = s$, and $FO_\mu(y) = t$. Then

$$
\begin{aligned}
\mu((xy)^{st}) = \mu(x^{st}y^{st}) &\geq \mu((x^s)^t) \wedge \mu((y^t)^s) \\
&\geq \mu(x^s) \wedge \mu(y^t) = \mu(e) \wedge \mu(e) = \mu(e).
\end{aligned}
$$

Thus $n|st$ by Proposition 1.6.5.

Now $\mu(e) = \mu((xy)^n) = \mu(x^n y^n)$. Since $n|st$, $n|s$ or $n|t$, say $n|t$. Then n and s are relatively prime. Hence $FO_\mu(x^n) = s$, by Proposition 1.6.11. By Theorem 1.6.10, $\mu(y^n) = \frac{t}{(n,t)}$. Since s and t are relatively prime, s and $\frac{t}{(n,t)}$ are relatively prime. Thus $(FO_\mu(x^n), FO_\mu(y^n)) = 1$. Therefore, $\mu(x^n) = \mu(y^n) = \mu(e)$ by Proposition 1.6.8, and so both s and t divide n by Proposition 1.6.5. Therefore, $st|n$ since $(s, t) = 1$. Hence we have $n = st$. $\square$

By virtue of Corollary 1.6.7, the assumption of Theorem 1.6.13 can be weakened as follows:

Corollary 1.6.14. *Let μ be a fuzzy subgroup of a group G, and let x and y be elements of G such that $xy = yx$ and $(o(x), o(y)) = 1$. Then $FO_\mu(xy) = FO_\mu(x) \cdot FO_\mu(y)$.*

In Theorem 1.6.13, even if μ were normal, the assumption $xy = yx$ could not be omitted as can be seen by the following example.

Example 1.6.15. *Define the fuzzy subset μ of the symmetric group S_4 of degree 4 by*

$$\mu(x) = \begin{cases} t_0 \text{ if } x = e, \\ t_1 \text{ otherwise}, \end{cases}$$

where $t_0 > t_1$. Then μ is a normal fuzzy subgroup of S_4. Now, let $x = (1\ 2)$ and $y = (2\ 3\ 4)$. Then $FO_\mu(x) = 2$, $FO_\mu(y) = 3$, $FO_\mu(xy) = FO_\mu(yx) = 4$, and $xy \neq yx$.

Theorem 1.6.16. *Let μ be a fuzzy subgroup of a group G. For $z \in G$, if $FO_\mu(z) = mn$ with $(m, n) = 1$, then there exist x and y in G such that $z = xy = yx$, $FO_\mu(x) = m$, and $FO_\mu(y) = n$. Furthermore, such an expression for z is unique in the sense of fuzzy grades, i.e., if (x, y) and (x_1, y_1) are such pairs, then $\mu(x) = \mu(x_1)$ and $\mu(y) = \mu(y_1)$.*

Proof. Since $(m, n) = 1$, there exist integers s and t such that $ms + nt = 1$. Here $(m, t) = (n, s) = 1$. Let $x = z^{nt}$ and $y = z^{ms}$. Then $z = xy = yx$ and by Theorem 1.6.10, $FO_\mu(x) = FO_\mu(z^{nt}) = m$ and $FO_\mu(y) = FO_\mu(z^{ms}) = n$. This proves the existence of x and y.

For the uniqueness, let (x, y) and (x_1, y_1) be pairs satisfying the assumption. Then, since $FO_\mu(x) = FO_\mu(x_1) = m$ and $FO_\mu(y) = FO_\mu(y_1) = n$, we obtain

$$\begin{aligned} \mu(x) &= \mu(x^{1-ms}) = \mu(x^{nt}) = \mu(x^{nt}y^{nt}) = \mu((xy)^{nt}) \\ &= \mu((x_1 y_1)^{nt}) = \mu(x_1^{nt} y_1^{nt}) = \mu(x_1^{nt}) \\ &= \mu(x_1^{1-ms}) = \mu(x_1). \end{aligned}$$

Similarly, $\mu(y) = \mu(y_1)$. This proves the uniqueness of (x, y). $\quad\square$

Theorem 1.6.17. *Let μ be a normal fuzzy subgroup of a group G. Then $FO_\mu(x) = FO_\mu(y^{-1}xy)$ for all $x, y \in G$.*

Proof. Let $x, y \in G$. Then we have $\mu(x^n) = \mu(y^{-1}x^n y) = \mu((y^{-1}xy)^n)$ for all $n \in \mathbb{Z}$. Thus $FO_\mu(x) = FO_\mu(y^{-1}xy)$. $\quad\square$

The next example shows that Theorem 1.6.17 is not valid if μ is not normal in G. The notation $\langle a, b | a^3 = b^2 = e, ba = a^2 b \rangle$ denotes the group generated by a and b, where a and b satisfy the indicated conditions. It is common to refer to this notation as a presentation of the group.

Example 1.6.18. *Let $D_3 = \langle a, b | a^3 = b^2 = e, ba = a^2 b \rangle$ be the dihedral group with six elements. Define a fuzzy subgroup μ of D_3 by*

$$\mu(x) = \begin{cases} t_0 \text{ if } x \in \langle b \rangle, \\ t_1 \text{ otherwise}, \end{cases}$$

where $t_0 > t_1$. Then $a^{-1}ba \notin \langle b \rangle$, and so $FO_\mu(b) = 1 \neq FO_\mu(a^{-1}ba)$.

1.7 Fuzzy Orders in Cyclic Groups

Lemma 1.7.1. *Let μ be a fuzzy subgroup of a cyclic group G and let a and b be any two generators of G. Then $FO_\mu(a) = FO_\mu(b)$.*

Proof. Suppose G is a finite cyclic group with $|G| = n$. Since G is generated by a and b, we get $o(a) = o(b) = n$. Further $b = a^m$ for some $m \in \mathbb{Z}$ and so we must have $(m, n) = 1$. Therefore, $FO_\mu(a) = FO_\mu(a^m) = FO_\mu(b) = n$ by Theorem 1.6.10. If G is an infinite group, then $b = a^{-1}$. □

Theorem 1.7.2. *Let μ be a fuzzy subgroup of a cyclic group G of finite order n. Then the following assertions hold for all $x, y \in G$.*
  (1) *If $o(x) = o(y)$, then $FO_\mu(x) = FO_\mu(y)$.*
  (2) *If $o(x)|o(y)$, then $FO_\mu(x)|FO_\mu(y)$.*
  (3) *If $o(x) > o(y)$, then $FO_\mu(x) \geq FO_\mu(y)$.*

Proof. Let $G = \langle a \rangle$. Let $x = a^s$, $y = a^t$, and $FO_\mu(a) = m$. By Lemma 1.7.1, m is independent of a particular choice of a generator a of G. Thus $o(x) = n/(s, n)$, $o(y) = n/(t, n)$, $FO_\mu(x) = m/(s, m)$, $FO_\mu(y) = m/(t, m)$, and $m|n$ by Corollary 1.6.7 and Theorem 1.6.10.
  (1) The result here follows from (2).
  (2) If $o(x)|o(y)$, then $(t, n)|(s, n)$, and so $(t, m)|(s, m)$ since $m|n$. Thus $FO_\mu(x)|FO_\mu(y)$.
  (3) If $o(x) > o(y)$, then $(s, n) < (t, n)$, and so $(s, m) \leq (t, m)$ since $m|n$. Hence $FO_\mu(x) \geq FO_\mu(y)$. □

Theorem 1.7.3. *Let μ be a fuzzy subgroup of a cyclic group G of finite order. Then the following assertions hold for all $x, y \in G$.*
  (1) *If $FO_\mu(x) = FO_\mu(y)$, then $\mu(x) = \mu(y)$.*
  (2) *If $FO_\mu(x)|FO_\mu(y)$, then $\mu(x) \geq \mu(y)$.*

Proof. Let $G = \langle a \rangle$. Let $x = a^s$, $y = a^t$, and $FO_\mu(a) = m$. By Lemma 1.7.1, m is independent of a particular choice of a generator a of G. Then $FO_\mu(x) = m/(s, m)$ and $FO_\mu(y) = m/(t, m)$, by Theorem 1.6.10. Let $s = h(s, m)$, $t = i(t, m)$, and $m = j(t, m) = k(s, m)$ for some $h, i, j, k \in \mathbb{Z}$. If $FO_\mu(x)|FO_\mu(y)$, then $(t, m)|(s, m)$. Thus t divides $si = h(s, m)i$ and m divides $sj = h(s, m)j$, and hence we have

$$
\begin{aligned}
\mu(x) &= \mu(a^s) \\
&= \mu(a^{s(iv+jw)}) \text{ for some } u, w \in \mathbb{Z} \text{ since } (i, j) = 1 \\
&= \mu(a^{siv}a^{sjw}) \geq \mu(a^{siv}) \wedge \mu(a^{sjw}) \\
&\geq \mu(a^t) \wedge \mu(a^m) = \mu(y) \wedge \mu(e) = \mu(y).
\end{aligned}
$$

This proves (2). Condition (1) follows from (2). □

Here is a counterexample to show that part (2) of Theorem 1.7.3 may fail to hold if G is not cyclic:

Example 1.7.4. *Let H be a cycle group of order 4 generated by x and let K be a cyclic group of order 2 generated by y. Let G be the direct product of H and K, $G = \{(u, v) | u \in H, v \in K\}$. Define the fuzzy subset μ of G as follows: $\mu(e, e) = 1$, $\mu(x^2, e) = 3/4$, $\mu(x, e) = \mu(x^3, e) = 1/2$ and $\mu(u, v) = 1/4$ for all other (u, v). Then the level subsets of μ are subgroups of G, and so μ is a fuzzy subgroup of G. Now $FO_\mu(x, e) = 4 > 2 = FO_\mu(e, y)$, but $\mu(x, e) = 1/2 > 1/4 = \mu(e, y)$. Thus $FO_\mu(e, y) | FO_\mu(e, x)$, but $\mu(y, e) = 1/4 < 1/2 = \mu(e, x)$.*

Theorem 1.7.3 is an improvement on Proposition 1.6.1. However, by combining Theorem 1.7.2 with Theorem 1.7.3, we have Proposition 1.6.1.

Corollary 1.7.5. *Let μ be a fuzzy subgroup of a cyclic group G of finite order. Then the following assertions hold for all $x, y \in G$.*
 (1) *If $o(x) = o(y)$, then $\mu(x) = \mu(y)$.*
 (2) *If $o(x) | o(y)$, then $\mu(x) \geq \mu(y)$.*

Remark 1.7.6. *Observe that Example 1.6.4 shows that the conclusions of Theorems 1.7.2 and Corollary 1.7.5 may fail to hold in finite non-cyclic groups.*

Remark 1.7.7. *A natural question that comes to mind after reading results from Section 1.7 is the following: Let μ be a fuzzy subgroup of a finite cyclic group G and let x, y be any two elements of G. If $FO_\mu(x) \geq FO_\mu(y)$, then does it follow that $\mu(y) \geq \mu(x)$? The answer is **"no"** as can be seen by the following example:*

Example 1.7.8. *Let $G = \{e, a, a^2, a^3, \dots, a^6\}$ be a cyclic group of order 6 generated by an element a. Define the fuzzy subgroup μ of G by $\mu(e) = 1, \mu(a^2) = \mu(a^4) = 3/4, \mu(z) = 1/4$ for all other $z \in G$. Then one can see that μ is a fuzzy subgroup of G. Now $FO_\mu(a^2) = 3$ and $FO_\mu(a^3) = 2$. But $\mu(a^2) = 3/4$ and $\mu(a^3) = 1/4$.*

Open Question

• Characterize groups G which satisfy conditions (1) and (2) of Theorem 1.7.3.

References

1. S. Abou-Zaid, On fuzzy subgroups, *Fuzzy Sets and Systems* 55 (1993) 237-240.
2. N. Ajmal and A. S. Prajapati, Fuzzy cosets and fuzzy normal subgroups, *Inform. Sci.* 64 (1992) 17-25.
3. M. Akgül, Some properties of fuzzy groups, *J. Math. Anal. Appl.* 133 (1988) 93-100.
4. J. M. Anthony and H. Sherwood, Fuzzy subgroups redefined, *J. Math. Anal. Appl.* 69 (1979) 124-130.

5. J. M. Anthony and H. Sherwood, A characterization of fuzzy subgroups, *J. Math. Anal. Appl.* 69 (1979) 297-305.
6. M. Asaad, Groups and fuzzy subgroups, *Fuzzy Sets and Systems* 39 (1991) 323-328.
7. S. K. Bhakat and P. Das, A note on fuzzy Archimedean ordering, *Fuzzy Sets and Systems* 91 (1997) 91-94.
8. K. R. Bhutani, Fuzzy Sets, Fuzzy Relations and Fuzzy Groups: Some Inter-relations, *Inform. Sci.* 73(1993), 107-115.
9. R. Biswas, Fuzzy subgroups and antifuzzy subgroups. *Fuzzy Sets and Systems* 35 (1990) 121-124.
10. D-G Chen and W-X Gu, Product structure of the fuzzy factor groups, *Fuzzy Sets and Systems* 60 (1993) 229-232.
11. D-G Chen and W-X Gu, Generating fuzzy factor groups and fundamental theorem of isomorphism, *Fuzzy Sets and Systems* 82 (1996) 357-360.
12. I. Chon, Fuzzy subgroups as products, *Fuzzy Sets and Systems*, to appear.
13. P. S. Das, Fuzzy groups and level subgroups, *J. Math. Anal. Appl.* 84 (1981) 264-269.
14. V. N. Dixit, S. K. Bhambri, and P. Kumar, Union of fuzzy subgroups, *Fuzzy Sets and Systems* 78 (1996) 121-123.
15. V. N. Dixit, R. Kumar, and N. Ajamal, Level subgroups and union of fuzzy subgroups, *Fuzzy Sets and Systems* 37 (1990)359-371.
16. M. S. Eroglu, The homomorphic image of a fuzzy subgroup is always a fuzzy subgroup, *Fuzzy Sets and Systems* 33 (1989) 255-256.
17. L. Filep, Structure and construction of fuzzy subgroups of a group, *Fuzzy Sets and Systems* 51 (1992) 105-109.
18. L. Filep and I. Gy. Maurer, Compatible fuzzy relations and groups, *Studia Scientiarum Mathematicarum Hungarica* 24 (1989) 345-348.
19. L. Fuchs, *Infinite Abelian Groups* I, *Pure and Applied Math.*, Vol. 36, Academic Press, New York 1970.
20. W-X Gu and D-G Chen, LP-fuzzy groups, *J. Fuzzy Math.* 1 (1993) 343-357.
21. W-X Gu and D-G Chen, A fuzzy subgroupoid which is not a fuzzy group, *Fuzzy Sets and Systems* 62 (1994) 115-116.
22. K. C. Gupta and S. Ray, Protogroups generated by fuzzy sets, *Inf. Sci.* 73 (1993) 1-16.
23. K. C. Gupta and S. Ray, The Schnitt axiom and unions of fuzzy subgroups, *Inf. Sci.* 70 (1993) 213-220.
24. A. K. Ibrahim and S. A. Khatab, *Pruc. Pakistan Acad. Sci.* 27 (1990) 46-53.
25. A. Jain and N. Ajmaal, A new approach to the theory of fuzzy groups, *J. Fuzzy Math.* 12 (2004) 341-355.
26. J. G. Kim, Fuzzy orders relative to fuzzy subgroups, *Inform. Sci.* 80 (1994) 341-348.
27. I. J. Kumar, P. K. Saxena, and P. Yadav, Fuzzy normal subgroups and fuzzy quotients, *Fuzzy Sets and Systems* 46 (1992) 121-132.
28. S, Kundu, The correct form of a recent result on level-subgroups of a fuzzy group, *Fuzzy Sets and Systems* 97 (1998) 261-263.
29. X. Li and G. Wang, The S_H-interval-valued fuzzy group, *Fuzzy Sets and Systems* 112 (2000) 319-325.
30. W. J. Liu, Fuzzy invariant subgroups and fuzzy ideals, *Fuzzy Sets and Systems* 8 (1982)133-139.

31. T. Lu and W. Gu, A note on fuzzy group theorems, *Fuzzy Sets and Systems* 61 (1994) 245-247.

32. B. B. Makamba, Direct product and isomorphism of fuzzy subgroups, *Inf. Sci.* 65 (1992) 33-43.

33. D. S. Malik and J. N. Mordeson, Fuzzy subgroups of abelian groups, *Chinese J. Math.* (Taipei) 19 (1991) 129-145.

34. D. S. Malik, J. N. Mordeson, and P. S. Nair, Fuzzy generators and fuzzy direct sums of abelian groups, *Fuzzy Sets and Systems* 50 (1992) 193-199.

35. N. N. Morsi and S. E. Yehia, Fuzz-quotient groups, *Inf. Sci.* 81 (1994) 177-191.

36. N. P. Mukherjee and P. Bhattacharya, Fuzzy normal subgroups and fuzzy cosets, *Inform. Sci.* 34 (1984) 225-239.

37. M. T. A. Osman, On the direct product of fuzzy subgroups, *Fuzzy Sets and Systems* 12 (1984) 87-91.

38. M. T. A. Osman, On some product of fuzzy groups, *Fuzzy Sets and Systems* 24 (1987) 79-86.

39. A. K. Ray, Quotient group of a group generated by a subgroup and a fuzzy subset, *J. Fuzzy Math.* 7 (1999) 459-463.

40. A. K. Ray, On product of fuzzy subgroups, *Fuzzy Sets and Systems* 105 (1999) 181-183.

41. S. Ray, Generated and cyclic fuzzy groups, *Inf. Sci.* 69 (1993) 185-200.

42. S. Ray, Modified TL- subgroups of a group, *Fuzzy Sets and Systems* 91 (1997) 375-387.

43. A. Rosenfeld, Fuzzy groups, *J. Math. Anal. Appl.* 35 (1971) 512-517.

44. B. K Sarma and T. Ali, Weak and strong homomorphisms of groups, *J. Fuzzy Math.* 12 (2004) 357-368.

45. P. K. Saxena, Fuzzy subgroups as union of two fuzzy subgroups, *Fuzzy Sets and Systems* 57 (1993) 209-218.

46. S. Sebastian and S. Babunder, Existence of fuzzy subgroups of every level-cardinality up to $\aleph_0$, *Fuzzy Sets and Systems* 67 (1994) 365-368.

47. S. Sebastian and S. Babunder, Fuzzy groups and group homomorphisms, *Fuzzy Sets and Systems* 81 (1996) 397-401.

48. H. Sherwood, Products of fuzzy subgroups, *Fuzzy Sets and Systems* 11 (1983) 79-89.

49. F.I. Sidky, Three-valued fuzzy subgroups, *Fuzzy Sets and systems* 87 (1997) 369-372.

50. J. Tang and X. Zhang, Product Operations in the category of L-fuzzy groups *J. Fuzzy Math.* 9 (2001) 1 - 10.

51. Y. Yu, A theory of isomorphisms of fuzzy groups, *Fuzzy Syst. and Math.* 2 (1988) 57-68.

52. Y. Yu, J. N. Mordeson, and C. S. Cheng, Elements of L-Algebra, Lecture Notes in Fuzzy Mathematics and Computer Science, Center for Research in Fuzzy Mathematics and Computer Science, Creighton University, 1994.

53. L. A. Zadeh, Fuzzy sets, *Inform. Control* 8 (1965) 338-353.

54. S. M. A. Zaidi and Q. A. Ansari, Some results of categories of L-fuzzy subgroups, *Fuzzy Sets and Systems* 64 (1994) 249-256.

55. J. Zhou, Su-Yun Li and Shu-You Li, LP-fuzzy normal subgroups and fuzzy quotient groups, *J. Fuzzy Math.* 5 (1997) 27-40.

2

Fuzzy Caley's Theorem and Fuzzy Lagrange's Theorem

We begin our discussion with properties of normal fuzzy subgroups. Fuzzy analogs of some group theoretic concepts such as cosets, characteristic subgroups, conjugate fuzzy subgroups, quotient groups are then presented. We show a number of results using these fuzzified notions. The order of a fuzzy subgroup is discussed as is the notion of a solvable fuzzy subgroup. The fuzzification of Caley's Theorem and Lagrange's Theorem are also presented. In this chapter, we consider primarily, although not exclusively, the work presented in [4, 10, 11].

2.1 Properties of Normal Fuzzy Subgroups

In Section 1.3, the notion of normal fuzzy subgroups and some of their properties were discussed. In this section, we continue the discussion of normal fuzzy subgroups and their fuzzy quotient groups.

Let S be a groupoid, i.e., a set which is closed under a binary relation denoted multiplicatively. A function $\mu : S \to [0,1]$ is called a **fuzzy subgroupoid** of S if $\mu(xy) \geqslant \mu(x) \wedge \mu(y) \forall x, y \in S$. Let N be a subgroup of a group G. We write $N \lhd G$ to denote that N is a normal subgroup of the group G.

Lemma 2.1.1. *If μ is a fuzzy subgroupoid of a finite group G, then μ is a fuzzy subgroup.*

Proof. Let $x \in G, x \neq e$. Since G is finite, x has finite order, say $n > 1$. Thus $x^n = e$ and so $x^{-1} = x^{n-1}$. Now using the definition of a fuzzy subgroupoid repeatedly, we have that $\mu(x^{-1}) = \mu(x^{n-1}) = \mu(x^{n-2}x) \geqslant \mu(x^{n-2}) \wedge \mu(x) \geq \mu(x) \wedge ... \wedge \mu(x) = \mu(x)$. Hence μ is a fuzzy subgroup of G. $\qquad \square$

Lemma 2.1.2. *Let μ be a fuzzy subgroup of G. Let $x \in G$. Then $\mu(xy) = \mu(y)$ $\forall y \in G$ if and only if $\mu(x) = \mu(e)$.*

Proof. Suppose that $\mu(xy) = \mu(y)$ $\forall y \in G$. Then by letting $y = e$, we get that $\mu(x) = \mu(e)$.

Conversely, suppose that $\mu(x) = \mu(e)$. Then $\mu(y) \leqslant \mu(x) \forall\, y \in G$ and so $\mu(xy) \geqslant \mu(x) \wedge \mu(y) = \mu(y)$. Also, $\mu(y) = \mu(x^{-1}xy) \geqslant \mu(x) \wedge \mu(xy) = \mu(xy)$. Hence $\mu(xy) = \mu(y)$. $\qquad\square$

Recall that a fuzzy subgroup μ of a group G is called normal if $\mu(xy) = \mu(yx)$ $\forall x, y \in G$. Recall also from group theory that elements x, y of G are called conjugates if there exists $z \in G$ such that $x = zyz^{-1}$. The notion of conjugacy is an equivalence relation and the equivalence classes are called **conjugacy classes.** .

Theorem 2.1.3. *Let μ be a fuzzy subgroup of a group G. Then μ is normal if and only if μ is constant on the conjugacy classes of G.*

Proof. Suppose that μ is normal. Then $\mu(y^{-1}xy) = \mu(xyy^{-1}) = \mu(x)$ $\forall x, y \in G$. Conversely, suppose that μ is constant on each conjugacy class of G. Then $\mu(xy) = \mu(xyxx^{-1}) = \mu(x(yx)x^{-1}) = \mu(yx)$ $\forall x, y \in G$. Hence μ is normal. $\qquad\square$

We now give an alternative formulation of the notion of normal fuzzy subgroup in terms of "commutators" of a group. First, we recall that if G is a group and $x, y \in G$, then the element $x^{-1}y^{-1}xy$ is usually denoted by $[x, y]$ and is called the **commutator** of x and y. If x and y commute with each other, then $[x, y] = e$. If H and K are subgroups of a group G, then $[H, K]$ is defined to be the subgroup generated by $\{[x, y] \mid x \in H, y \in K\}$.

The motivation behind the following theorem is that $N \triangleleft G$ if and only if $[N, G] \subseteq N$.

Theorem 2.1.4. *Let μ be a fuzzy subgroup of G. Then μ is normal if and only if $\mu([x, y]) \geq \mu(x) \forall x, y \in G$.*

Proof. Suppose μ is a normal fuzzy subgroup of G. Let $x, y \in G$. Then $\mu(x^{-1}y^{-1}xy) \geqslant \mu(x^{-1}) \wedge \mu(y^{-1}xy) = \mu(x) \wedge \mu(x) = \mu(x)$.

Conversely, assume that μ satisfies the inequality. Then for all $x, z \in G$, we have $\mu(x^{-1}zx) = \mu(zz^{-1}x^{-1}zx) \geq \mu(z) \wedge \mu([z, x]) = \mu(z)$ Thus $\mu(x^{-1}zx) \geqslant \mu(z) \forall z, x \in G$. Hence by Theorem 1.3.1 and Definition 1.3.2, μ is normal. $\quad\square$

We now give an example illustrating the above ideas.

Example 2.1.5. *Let G be the group of all symmetries of a square. Then G is a group of order 8 generated by a rotation of $\pi/2$ radians and a reflection along a diagonal of the square. Denote the elements of G by $\{e, \pi_{90}, \pi_{180}, \pi_{270}, h, v, d_1, d_2\}$, where e is the identity, $\pi_{90}, \pi_{180}, \pi_{270}$ are* *rotations through* 90, 180, 270 *degrees, respectively, h and v are reflections about the horizontal and vertical axes, respectively, and d_1, d_2 are reflections about the diagonals. It then follows that the conjugacy classes of G are* $\{e\}, \{\pi_{180}\}, \{\pi_{90}, \pi_{270}\}, \{h, v\}, \{d_1, d_2\}$. *Let* $H = \{e, \pi_{180}\}$ *and* $K = \{e, \pi_{90}, \pi_{180}, \pi_{270}\}$. *Then H and K are normal subgroups of G and in fact, H is the center of G. Thus we have a chain of normal subgoups given by* $\{e\} \subset H \subset K \subset G$.

Now we construct a fuzzy subgroup of G whose level subgoups are the members of the above chain. Let $t_i \in [0, 1]$, $i = 0, 1, 2, 3$, *be such that* $t_0 > t_1 > t_2 > t_3$. *Define* $\mu : G \to [0, 1]$ *as follows:* $\forall x \in G$,
$\mu(e) = t_0$, $\mu(x) = t_1$ *if* $x \in H \backslash \{e\}$, $\mu(x) = t_2$ *if* $x \in K \backslash H$, $\mu(x) = t_3$ *if* $x \in G \backslash K = t_3$.

Then by Theorem 2.1.3, μ is a normal fuzzy subgroup of G. Furthermore, it is clear that μ is constant on the conjugacy classes of G.

The following result follows from Theorem 1.3.12. However, we give another proof of it.

Theorem 2.1.6. *Let μ be a normal fuzzy subgroup of G. Then $\mu_* \lhd G$. Define the fuzzy subset $\mu^\#$ of G/μ_* as follows:* $\forall x\mu_* \in G/\mu_*$,
$$\mu^\#(x\mu_*) = \mu(x).$$
Then $\mu^\#$ is a normal fuzzy subgroup of G/μ_*. On the other hand, if $N \lhd G$ and $\mu_1^\#$ is a normal fuzzy subgroup of G/N such that $\mu_1^\#(xN) = \mu_1^\#(N)$ only when $x \in N$, then there is a normal fuzzy subgroup μ of G such that, in the above notation, $\mu_* = N$ and $\mu^\# = \mu_1^\#$.

Proof. Since μ is a normal fuzzy subgroup of G, it follows from Theorem 1.3.4 that $\mu_* \lhd G$. Further, if $x\mu_* = y\mu_*$ for some $x, y \in G$, then $y^{-1}x \in \mu_*$ and so
$$\mu(y^{-1}x) = \mu(e).$$
By Lemma 2.1.2, $\mu(x) = \mu(y)$ and so $\mu^\#(x\mu_*) = \mu^\#(y\mu_*)$. Therefore, $\mu^\#$ is well-defined. It follows easily that $\mu^\#$ is a fuzzy subgroup of G/μ_* We now show that $\mu^\#$ is normal. Let $x, y \in G$. Then
$$\mu^\#(x\mu_* y\mu_*) = \mu^\#(xy\mu_*) = \mu(xy)$$
$$= \mu(yx) \quad (\text{since } \mu \text{ is normal})$$
$$= \mu^\#(y\mu_* x\mu_*).$$
Hence $\mu^\#$ is normal.

Conversely, given the normal fuzzy subgroup $\mu_1^\#$ of G/N. Let μ be the fuzzy subset of G defined as follows: $\forall x \in G$, $\mu(x) = \mu_1^\#(xN)$. It follows easily that μ is a fuzzy subgroup of G. Let $x, y \in G$. Then
$$\mu(yxy^{-1}) = \mu_1^\#(yxy^{-1}N)$$
$$= \mu_1^\#(yNxNy^{-1}N)$$
$$= \mu_1^\#(xN) \quad (\text{since } \mu_1^\# \text{ is normal})$$
$$= \mu(x).$$
Hence μ is constant on the conjugacy classes of G and so by Theorem 2.1.3, μ is a normal fuzzy subgroup of G.

Furthermore, if $n \in N$, we have that
$$\mu(n) = \mu_1^{\#}(nN) = \mu_1^{\#}(N) = \mu(e).$$
Thus $N \subseteq \mu_*$. Let $x \in \mu_*$. Then $\mu(x) = \mu(e)$ and so $\mu_1^{\#}(xN) = \mu_1^{\#}(N)$. Consequently, $x \in N$. Thus $\mu_* \subseteq N$. Therefore, $N = \mu_*$. It now follows that $\mu_1^{\#} = \mu^{\#}$. $\square$

Proposition 2.1.7. *If μ is a normal fuzzy subgroup of G, then for all $x \in G$, $x\mu(xz) = x\mu(zx) = \mu(z) \; \forall z \in G$.*

Proof. By Theorem 2.1.3, since μ is normal, μ is constant on the conjugacy classes of G. Hence by the definition of $x\mu$ and the fact that $\mu(xz) = \mu(zx)$, we get $x\mu(zx) = x\mu(xz) = \mu(z)$. Similarly, $x\mu(zx) = \mu(z)$. $\square$

Proposition 2.1.7 is analogous to the result in group theory that if $N \lhd G$, then $Nx = xN \; \forall x \in G$.

If N is a normal subgroup of a group G, then the cosets of G with respect to N form a group (called the quotient group G/N). We now recall Theorem 1.3.12.

Let μ be a normal fuzzy subgroup of G. Let G/μ be the set of all the fuzzy cosets of μ. Then G/μ is a group under the composition $x\mu \circ y\mu = (xy)\mu \; \forall x, y \in G$.

Define the fuzzy subset $\mu^{(*)}$ of G/μ by
$$\mu^{(*)}(x\mu) = \mu(x) \quad \forall x \in G.$$
Then $\mu^{(*)}$ is a normal fuzzy subgroup of G/μ.

Definition 2.1.8. *Let μ be a normal fuzzy subgroup of G. Then the fuzzy subgroup $\mu^{(*)}$ of G/μ defined above is called the **fuzzy quotient group** determined by μ.*

Theorem 2.1.9. *Let μ be a normal fuzzy subgroup of G. Define the function $\theta : G \rightarrow G/\mu$ as follows: $\forall x \in G, \theta(x) = x\mu$. Then θ is a homomorphism with kernel $\mu_* = \{x \in G \mid \mu(x) = \mu(e)\}$.*

Proof. Let $x, y \in G$. Then $\theta(xy) = xy\mu = x\mu \circ y\mu = \theta(x) \circ \theta(y)$.

Hence θ is a homomorphism. Let $H = \{h \in G \mid h\mu = e\mu\}$. Then $h \in \mathrm{Ker}(\theta) \Leftrightarrow \theta(h) = e\mu \Leftrightarrow h\mu = e\mu \Leftrightarrow h \in H$. Thus $\mathrm{Ker}(\theta) = H$. Now $h \in H \Leftrightarrow h\mu(x) = e\mu(x) \forall x \in G \Leftrightarrow \mu(h^{-1}x) = \mu(x) \forall x \in G \Leftrightarrow \mu(h^{-1}) = \mu(e)$ (by Lemma 2.1.2) $\Leftrightarrow \mu(h) = \mu(e) \Leftrightarrow h \in \mu_*$. Thus $\mathrm{Ker}(\theta) = \mu_*$. $\square$

Corollary 2.1.10. *Let μ be a normal fuzzy subgroup of G. Then $G/H = G/\mu_* \simeq G/\mu$.*

The following result follows from Theorems 1.3.14, 1.3.15 and 1.4.8.

Theorem 2.1.11. *Let μ be a normal fuzzy subgroup of G. Then every (normal) fuzzy subgroup of G/μ corresponds in a natural way to a (normal) fuzzy subgroup of G.*

Let μ be a fuzzy subgroup of a finite group G. Then clearly G/μ is a finite set.

Definition 2.1.12. *Let μ be a fuzzy subgroup of a finite group G. Then the cardinality of G/μ is called the **index** of μ in G, written $[G : \mu]$.*

It is clear from Corollary 2.1.10 that the index of μ in G divides the order of G. However, in Corollary 2.1.10, μ is normal. In Section 2.3, we prove that the index of any fuzzy subgroup of a finite group divides the order of the group. Also, Corollary 2.1.10 provides the details to Theorem 1.3.12(3).

2.2 Characteristic Fuzzy Subgroups and Abelian Fuzzy Subgroups

We prove a number of results about fuzzy groups involving the concepts of fuzzy cosets and fuzzy normal subgroups. These results are analogs of standard results from group theory. Also, we introduce analogs of some group-theoretic concepts such as characteristic subgroups. We prove that if μ is a fuzzy subgroup of a group G such that the fuzzy index of μ is the smallest prime dividing the order of G, then μ is a normal fuzzy subgroup.

Definition 2.2.1. *Let μ be a fuzzy subgroup of a group G and θ a function from G into itself. Define the fuzzy subset μ^θ of G by,*

$$\forall x \in G, \qquad \mu^\theta(x) = \mu(x^\theta), \quad where \ \ x^\theta = \theta(x).$$

For a group G, a subgroup K is called a characteristic subgroup if $K^\theta = K$ for every automorphism θ of G, where K^θ denotes $\theta(K)$.

We now define an analog of this notion.

Definition 2.2.2. *A fuzzy subgroup μ on a group K is called a **fuzzy characteristic subgroup** of G if $\mu^\theta(x) = \mu(x)$ for every automorphism θ of G and all $x \in G$.*

Theorem 2.2.3. *Let μ be a fuzzy subgroup of a group G. Then the following assertions hold.*

(1) If θ is a homomorphism of G into itself, then μ^θ is a fuzzy subgroup of G.

(2) If μ is a fuzzy characteristic subgroup of G, then μ is a normal.

Proof. (1) Let $x, y \in G$. Then
$$\mu^\theta(xy) = \mu((xy)^\theta)$$
$$= \mu(x^\theta y^\theta)$$
since θ is a homomorphism. Since μ is a fuzzy subgroup of G,
$$\mu(x^\theta y^\theta) \geqslant \mu(x^\theta) \wedge \mu(y^\theta)$$
$$= \mu^\theta(x) \wedge \mu^\theta(y).$$
Thus
$$\mu^\theta(xy) \geqslant \mu^\theta(x) \wedge \mu^\theta(y).$$
Also,
$$\mu^\theta(x^{-1}) = \mu((x^{-1})^\theta) = \mu((x^\theta)^{-1}) = \mu(x^\theta) = \mu^\theta(x).$$
Hence μ^θ is a fuzzy subgroup of G.

(2) Let $x, y \in G$. To prove that μ is normal, we have to show that
$$\mu(xy) = \mu(yx).$$
Let θ be the function of G into itself defined by
$$\theta(z) = x^{-1} z x \quad \forall z \in G.$$
It is a standard result that θ is an automorphism of G (called the inner automorphism induced by x). Now since μ is a fuzzy characteristic subgroup of G, $\mu^\theta = \mu$. Thus
$$\mu(xy) = \mu^\theta(xy)$$
$$= \mu((xy)^\theta)$$
$$= \mu(x^{-1}(xy)x)$$
$$= \mu(yx).$$
Hence μ is normal.   □

Theorem 2.2.3 (2) is an analog of the result that a characteristic subgroup of a group is normal. We now obtain some other analogs.

Definition 2.2.4. *Let μ_1, μ_2 be two fuzzy subgroups of a group G. We say that μ_1 is **conjugate** to μ_2 if there exists $y \in G$ such that $\forall x \in G$, $\mu_1(x) = \mu_2(y^{-1}xy)$.*

Clearly, if μ_1 and μ_2 are fuzzy subgroups of a group G such that μ_1 is conjugate to μ_2, then μ_1 and μ_2 are **conjugate fuzzy subgroups** of G as defined previously.

It is easy to verify that the relation of conjugacy is an equivalence relation on the set of all fuzzy subgroups of a group. Consequently, the set of all fuzzy subgroups of a group is a union of pairwise disjoint classes of fuzzy subgroups each consisting of fuzzy subgroups which are equivalent to one another. We now obtain an expression giving the number of distinct conjugates of a fuzzy subgroup. First we give some preliminaries.

If μ is a fuzzy subgroup of a group G and $g \in G$, recall that we denote by μ^g the fuzzy subset of G defined by: $\mu^g(u) = \mu(g^{-1}ug) \forall u \in G$. Then from Theorem 2.2.3 (1), it follows that μ^g is a fuzzy subgroup of G. Let $g \in G$ and let θ be the automorphism of G defined by $\theta(x) = g^{-1}xg \; \forall x \in G$. Then $\mu^g(x) = \mu(g^{-1}xg) = \mu(\theta(x)) = \mu^\theta(x) \; \forall x \in G$. Hence $\mu^g = \mu^\theta$.

Recall that if μ is a fuzzy subgroup of a group G, the normalizer of μ is the set given by $N(\mu) = \{g \in G \mid \mu(gy) = \mu(yg) \; \forall y \in G\}$. We now note that $N(\mu) = \{g \in G \mid \mu^g = \mu\}$: It suffices to show that $\mu(gy) = \mu(yg)$ $\forall y \in G \Leftrightarrow \mu(g^{-1}yg) = \mu(y) \; \forall y \in G$. We have that $\mu(g^{-1}yg) = \mu(y) \; \forall y \in G \Leftrightarrow \mu(g^{-1}(gy)g) = \mu(gy) \; \forall y \in G \Leftrightarrow \mu(yg) = \mu(gy) \; \forall y \in G$.

Theorem 2.2.5 (2) below illustrates the motivation behind the term "normalizer", and it also shows the analogy with the fact that a subgroup H of a group G is normal in G if and only if the normalizer of H in G is equal to G itself. The reader may wish to compare the proof of Theorem 2.2.5 with that of Theorems 1.3.7 and 1.3.8.

Theorem 2.2.5. *Let μ be a fuzzy subgroup of a group G. Then the following assertions hold.*

(1) *$N(\mu)$ is a subgroup of G.*

(2) *μ is normal if and only if $N(\mu) = G$.*

(3) *The number of distinct conjugates of μ is equal to the index of $N(\mu)$ in G, provided that G is a finite group.*

Proof. (1) Let $g, h \in N(\mu)$. Then $\mu^{gh} = (\mu^g)^h = \mu^h = \mu$.

Hence $gh \in N(\mu)$. Let $g \in N(\mu)$. We show that $g^{-1} \in N(\mu)$. For all $y \in G, \mu(gy) = \mu(yg)$ and so $\mu((gy)^{-1}) = \mu((yg)^{-1})$. Thus $\forall y \in G, \mu(y^{-1}g^{-1}) = \mu(g^{-1}y^{-1})$ and so $\mu(yg^{-1}) = \mu(g^{-1}y)$, where the latter equality holds since y and y^{-1} are equally arbitrary. Thus $g^{-1} \in N(\mu)$. Hence $N(\mu)$ is a subgroup of G.

(2) Let μ be normal and $g \in G$. Then for all $u \in G$, we have
$$\begin{aligned}
\mu^g(u) &= \mu(g^{-1}ug) \\
&= \mu((g^{-1}u)g) \\
&= \mu(g(g^{-1}u)) \quad \text{(since } \mu \text{ is normal)} \\
&= \mu(u).
\end{aligned}$$
Thus $\mu^g = \mu$ and so $g \in N(\mu)$. Therefore, $N(\mu) = G$.

Conversely, suppose $N(\mu) = G$. Let $x, y \in G$. To prove that μ is normal, we show that
$$\mu(xy) = \mu(yx).$$
Now
$$\begin{aligned}
\mu(xy) &= \mu(xyxx^{-1}) \\
&= \mu(x(yx)x^{-1}) \\
&= \mu^{x^{-1}}(yx) \\
&= \mu(yx),
\end{aligned}$$
where the last equality follows since $N(\mu) = G$ and so $x^{-1} \in N(\mu)$. Hence $\mu^{x^{-1}} = \mu$. Thus μ is normal.

(3) The proof of this result is based on the same technique used to prove the corresponding result for groups. Consider the decomposition of G,

$$G = x_1 N(\mu) \cup x_2 N(\mu) \cup \ldots \cup x_k N(\mu)$$

as a union of pairwise disjoint cosets of $N(\mu)$, where k is the number of distinct cosets, that is, the index of $N(\mu)$ in G. Let $x \in N(\mu)$ and $i \in \mathbb{N}$ be such that $1 \leqslant i \leqslant k$. Then for all $g \in G$,

$$\begin{aligned}
\mu^{x_i x}(g) &= \mu((x_i x)^{-1} g (x_i x)) \\
&= \mu(x^{-1}(x_i^{-1} g x_i) x) \\
&= \mu^x(x_i^{-1} g x_i) \\
&= \mu(x_i^{-1} g x_i) \quad \text{(since } x \in N(\mu)) \\
&= \mu^{x_i}(g).
\end{aligned}$$

Thus we have,

$$\mu^{x_i x} = \mu^{x_i} \quad \forall x \in N(\mu),\, 1 \leqslant i \leqslant k.$$

Hence any two elements of G which lie in the same coset $x_i N(\mu)$ give rise to the same conjugate μ^{x_i} of μ. Now we show that two distinct cosets give two distinct conjugates of μ. For this, suppose that

$$\mu^{x_i} = \mu^{x_j},$$

where $j \neq i$ and $1 \leqslant i, j \leqslant k$. Thus

$$\mu^{x_i}(g) = \mu^{x_j}(g) \quad \forall g \in G,$$
$$\mu(x_i^{-1} g x_i) = \mu(x_j^{-1} g x_j) \quad \forall g \in G.$$

Choose $g = x_j h x_j^{-1}$. Then

$$\mu(x_i^{-1} x_j h x_j^{-1} x_i) = \mu(x_j^{-1} x_j h x_j^{-1} x_j) \quad \forall h \in G$$
$$\Rightarrow \mu((x_j^{-1} x_i)^{-1} h (x_j^{-1} x_i)) = \mu(h) \quad \forall h \in G$$
$$\Rightarrow \quad \mu^{x_j^{-1} x_i}(h) = \mu(h) \quad \forall g \in G$$
$$\Rightarrow x_j^{-1} x_i \in N(\mu).$$
$$\Rightarrow x_i N(\mu) = x_j N(\mu).$$

However, if $i \neq j$ this is not possible since $\{x_1 N(\mu), x_2 N(\mu), \ldots, x_k N(\mu)\}$ is a partition of G. Hence the number of distinct conjugates of μ is equal to the index of $N(\mu)$ in G.   □

We now state a result which we use in a number of places. Its proof appears in the proof of Theorem 2.1.9. In this proof, the normality of μ is not used.

Lemma 2.2.6. Let μ be a fuzzy subgroup of a finite group G. Let $K = \{x \in G \mid x\mu = e\mu\}$. Then K is a subgroup of G. In fact, $K = \mu_*$.

It is a standard result in group theory that if G is a group and H, K are subgroups of G, H normal in G, then $H \cap K$ is normal in K. We now derive an analog for fuzzy subgroups.

Proposition 2.2.7. Let μ be a fuzzy subgroup of G and ν be a normal fuzzy subgroup of G. Then $(\mu \cap \nu)|_{\mu_*}$ is a normal fuzzy subgroup of the subgroup μ_*.

Proof. By Corollary 1.2.7, μ_* is a subgroup of G. By Theorem 1.2.12, we have that $\mu \cap \nu$ is a fuzzy subgroup of G. Therefore, $(\mu \cap \nu)|_{\mu_*}$ is a fuzzy subgroup of μ_*. We now show that $(\mu \cap \nu)|_{\mu_*}$ is a normal fuzzy subgroup of μ_*. Let $x, y \in \mu_*$. Then xy and yx belong to μ_* since μ_* is a group. Hence

$$\mu(xy) = \mu(yx) = \mu(e)$$

by the definition of μ_*. Now

$$(\mu \cap \nu)(xy) = \mu(xy) \wedge \nu(xy) = \mu(yx) \wedge \nu(yx) = (\mu \cap \nu)(yx).$$

Hence $(\mu \cap \nu)|_{\mu_*}$ is a normal fuzzy subgroup of μ_*. $\qquad\square$

If H is a subgroup of a group G, then H is normal in G if and only if $Hx = xH \ \forall x \in G$. As an analog, we prove the following result.

Proposition 2.2.8. *Let μ be a fuzzy subgroup of a group G. Then μ is normal if and only if $\forall x \in G$, $\mu x = x\mu$.*

Proof. Suppose μ is normal. Then for all $x, y \in G$, $\mu x(y) = \mu(yx^{-1}) = \mu(x^{-1}y) = x\mu(y)$. Thus $\mu x = x\mu \ \forall x \in G$.

Conversely, suppose that $\mu x = x\mu$ for all x in G. Now $\mu(xy) = (x^{-1}\mu)(y) = (\mu x^{-1})(y) = \mu(yx) \ \forall x, y$ in G. Hence μ is normal. $\qquad\square$

We restrict ourselves in the subsequent discussion to fuzzy left cosets only. Corresponding results for fuzzy right cosets could be obtained easily. Consequently, from now on we call a fuzzy left coset a "fuzzy coset" and denote it as $x\mu$ for all $x \in G$.

It is a well-known result in group theory that subgroup of index 2 is a normal subgroup. We prove here an analog of a generalization of this result.

Theorem 2.2.9. *If μ is a fuzzy subgroup of a finite group G such that the fuzzy index of μ is p, the smallest prime dividing the order of G, then μ is a normal fuzzy subgroup of G.*

Proof. By Corollary 1.2.7, μ_* is a subgroup of G. Further, by Lemma 2.2.6 and [[9], Theorem 4.10] μ_* has index p in G, that is, μ_* has p distinct (left) cosets, say, $\{x_i\mu_* \mid 1 \leqslant i \leqslant p\}$. For all $x \in G$, define the function π_x of G/μ_* into itself by $\forall g \in G, \pi_x(g\mu_*) = xg\mu_*$. Then π_x is a one-to-one function of G/μ_* onto itself. Let $P(G) = \{\pi_x \mid x \in G\}$. Now consider the permutation representation of G on the cosets of μ_* given by the function π of G onto $P(G)$, where $\pi(x) = \pi_x \ \forall x \in G$.

As is well known, π is an isomorphism of G into the symmetric group $\mathrm{Sym}(p)$, since the index of μ_* in G is p. Furthermore, the kernel of the map π is the core of μ_*, written $\mathrm{Core}(\mu_*)$, that is, the intersection of all the conjugates $g^{-1}\mu_* g$, $g \in G$. By the fundamental theorem of homomorphism of groups and using Lagrange's theorem, we have that the order of $G/\mathrm{Core}(\mu_*)$ divides $p!$, which is the order of $\mathrm{Sym}(p)$. Since $G/\mu_* \cong (G/\mathrm{Core}(\mu_*))/(\mu_*/\mathrm{Core}(\mu_*))$,

$$|G/\mathrm{Core}(\mu_*)| = |G/\mu_*||\mu_*/\mathrm{Core}(\mu_*)|.$$

Now the order of G/μ_* is p. Thus it follows that the order of $\mu_*/\mathrm{Core}(\mu_*)$ divides $(p-1)!$. Now since the order of μ_* divides the order of G, we obtain that $\mu_* = \mathrm{Core}(\mu_*)$, otherwise we get a contradiction of the fact that p is the smallest prime dividing the order of G. Since $\mathrm{Core}(\mu_*)$ is always a normal subgroup of G, it follows that μ_* is a normal subgroup of G. Now consider the quotient group G/μ_*. Since the order of G/μ_* is p, G/μ_* is Abelian. Hence for all $x, y \in G$, we have

$$(x\mu_*)(y\mu_*) = (y\mu_*)(x\mu_*)$$
$$\Rightarrow \qquad xy\mu_* = yx\mu_*$$
$$\Rightarrow \qquad xy = yxh$$

for some $h \in \mu_*$. Hence

$$\mu(xy) = \mu(yxh)$$
$$\geqslant \mu(yx) \wedge \mu(h)$$
$$\geqslant \mu(yx) \wedge \mu(e)$$
$$= \mu(yx).$$

Thus $\mu(xy) \geqslant \mu(yx)$.

Similarly, we obtain that

$$\mu(yx) \geqslant \mu(xy)$$

Hence $\mu(xy) = \mu(yx) \ \forall x, y \in G$. Therefore, μ is normal. $\qquad\square$

Corollary 2.2.10. *If μ is a fuzzy subgroup of G such that the fuzzy index of μ is 2, then μ is normal.*

We now derive an analog of the following result from group theory which states that if θ is a homomorphism of a group G into itself whose kernel is N, then θ induces a homomorphism from G/N into itself.

Theorem 2.2.11. *Let μ be a normal fuzzy subgroup of G and let θ be a homomorphism of G into itself which leaves invariant the subgroup μ_*. Then θ induces a homomorphism $\overline{\theta}$ of G/μ into itself defined by*
$$\overline{\theta}(x\mu) = \theta(x)\mu \ \forall x \in G.$$

Proof. First we show that $\overline{\theta}$ is well defined. To show this, suppose $x, y \in G$ are such that

$$x\mu = y\mu.$$

Then it suffices to prove that

$$\theta(x)\mu = \theta(y)\mu.$$

Since $x\mu = y\mu$, we have that

$$x\mu(x) = y\mu(x), \ x\mu(y) = y\mu(y)$$
$$\Rightarrow \qquad \mu(e) = \mu(y^{-1}x), \ \mu(x^{-1}y) = \mu(e)$$
$$\Rightarrow \qquad \mu(y^{-1}x) = \mu(x^{-1}y)$$
$$\Rightarrow \qquad y^{-1}x, x^{-1}y \in \mu_*.$$

Since by hypothesis $\theta(\mu_*) = \mu_*$, we get that $\theta(y^{-1}x)$ and $\theta(x^{-1}y)$ also belong to μ_*. Thus we have

$$\mu(\theta(y^{-1}x)) = \mu(\theta(x^{-1}y)) = \mu(e).$$

Now let $g \in G$. Then

$$\theta(x)\mu(g) = \mu(\theta(x^{-1})g)$$
$$= \mu(\theta(x^{-1})\theta(y)\theta(y^{-1})g)$$
$$\geqslant \mu(\theta(x^{-1})\theta(y)) \wedge \mu(\theta(y^{-1})g)$$

$$= \mu(\theta(x^{-1}y)) \wedge \theta(y)\mu(g)$$
$$= \mu(e) \wedge \theta(y)\mu(g)$$
$$= \theta(y)\mu(g).$$

Thus

$$\theta(x)\mu(g) \geqslant \theta(y)\mu(g).$$

Similarly, we have that

$$\theta(x)\mu(g) \leq \theta(y)\mu(g).$$

Since $g \in G$ is arbitrary, we have

$$\theta(x)\mu = \theta(y)\mu.$$

Therefore, $\overline{\theta}$ is well defined. Next we show that $\overline{\theta}$ is a homomorphism. Let $x, y \in G$. Since θ is a homomorphism, $\theta(xy) = \theta(x)\theta(y)$. Thus $\theta(xy)\mu = \theta(x)\theta(y)\mu$. Hence $\overline{\theta}(xy\mu) = \theta(x)\mu \circ \theta(y)\mu$. Thus $\overline{\theta}(x\mu \circ y\mu) = \overline{\theta}(x\mu) \circ \overline{\theta}(y\mu.)$ Therefore, $\overline{\theta}$ is a homomorphism. □

In Theorem 2.2.11, we have assumed μ to be normal instead of assuming only that μ is a fuzzy subgroup. This has been done to ensure that the law of composition of fuzzy cosets is well defined. This fact is used in the proof of Theorem 2.2.11 to show that $\overline{\theta}$ is a homomorphism. However, it is clear from the proof that to show $\overline{\theta}$ is well defined it is not necessary to assume μ to be normal.

Corollary 2.2.12. *With the same hypothesis as in Theorem 2.2.11, the homomorphism $\overline{\theta}$ is an automorphism if θ is an automorphism and G is finite.*

Proof. Since G has finite order, it is easy to see that θ has finite order. Suppose that θ has order k. Then $\theta^k = \iota$, where ι denotes the identity map. Now since by Theorem 2.2.11, we have that $\overline{\theta}$ is a homomorphism, it remains to prove that θ is one-to-one. For this purpose, let $x, y \in G$ be such that $\overline{\theta}(x\mu) = \overline{\theta}(y\mu)$. Then $\theta(x)\mu = \theta(y)\mu$. Thus $\overline{\theta}(\theta(x)\mu) = \overline{\theta}(\theta(y)\mu)$. This implies by definition of $\overline{\theta}$ that $\theta^2(x)\mu = \theta^2(y)\mu$. Iterating, we obtain that $\iota(x)\mu = \theta^k(x)\mu = \theta^k(y)\mu = \iota(y)\mu$ and so $x\mu = y\mu$. Hence $\overline{\theta}$ is one-to-one. □

Corollary 2.2.13. *With the same hypothesis as in Theorem 2.2.11, the function θ is an automorphism of G if $\overline{\theta}$ is an automorphism and $\mu_* = \{e\}$.*

Proof. Let $x, y \in G$ be such that $\theta(x) = \theta(y)$. Then it follows that

$$\theta(x)\mu = \theta(y)\mu$$

That is,

$$\overline{\theta}(x\mu) = \overline{\theta}(y\mu).$$

Hence, $x\mu = y\mu$ since $\overline{\theta}$ is one-to-one by the hypothesis. Thus

$$x\mu(y) = y\mu(y),$$

which implies that

$$\mu(x^{-1}y) = \mu(e).$$

Therefore, $x^{-1}y \in \mu_*$ and so $x^{-1}y = e$ since $\mu_* = \{e\}$ by the hypothesis. Thus $x = y$. Hence θ is one-to-one. □

We now consider a variation of Corollary 2.2.13. The motivation of the following result arises from the standard theorem in group theory that says that if θ is an automorphism of G and N is a normal subgroup of G such that $N^\theta \subseteq N$, then θ induces an automorphism of the quotient group G/N. We prove the following result.

Theorem 2.2.14. *Let μ be a normal fuzzy subgroup of G and θ be an automorphism of G such that $\mu^\theta = \mu$. Then θ induces an automorphism $\overline{\theta}$ of G/μ defined by*

$$\overline{\theta}(x\mu) = \theta(x)\mu \ \forall x \in G.$$

Proof. Let $x, y \in G$. Then $x\mu = y\mu \Leftrightarrow x\mu^\theta = y\mu^\theta \Leftrightarrow x\mu^\theta(g) = y\mu^\theta(g) \forall g \in G \Leftrightarrow \mu^\theta(x^{-1}g) = \mu^\theta(y^{-1}g) \forall g \in G \Leftrightarrow \mu(\theta(x^{-1}g)) = \mu(\theta(y^{-1}g)) \forall g \in G \Leftrightarrow \mu(\theta(x^{-1})\theta(g)) = \mu(\theta(y^{-1})\theta(g)) \forall g \in G \Leftrightarrow \theta(x)\mu(\theta(g)) = \theta(y)\mu(\theta(g)) \forall g \in G \Leftrightarrow \theta(x)\mu = \theta(y)\mu$ (since θ is an automorphism) $\Leftrightarrow \overline{\theta}(x\mu) = \overline{\theta}(y\mu)$. Thus $\overline{\theta}$ is well defined and one-to-one. Clearly, $\overline{\theta}$ maps G/μ onto itself. The proof of the fact that $\overline{\theta}$ is a homomorphism is analogous to the corresponding part of the proof of Theorem 2.2.11 and we omit the details. $\qquad\square$

The next result gives a relationship between fuzzy cosets of a fuzzy subgroup and the cosets of a subgroup of the given group.

Recall Theorem 1.3.10. Let μ be a fuzzy subgroup of a finite group G.
Then for $x, y \in G$, we have that

$$x\mu_* = y\mu_* \ \Leftrightarrow \ x\mu = y\mu.$$

Thus $[G : \mu] = [G : \mu_*]$. This shows that there is a natural one-to-one correspondence between the (left) cosets of μ_* in G and the fuzzy cosets of μ in G. Thus we see that the subgroup μ_* plays a key role in the analysis of fuzzy cosets.

2.3 Fuzzy Caley's Theorem and Fuzzy Lagrange's Theorem

In this section, we present the fuzzy analogs of Cayley's Theorem and Lagrange's Theorem. We develop here the notions of the "order" of a fuzzy group and fuzzy Abelian group. We also introduce the concept of a fuzzy solvable group. We obtain some analogs of group theoretical results related to these ideas introduced here.

Recall that if μ is a fuzzy subgroup of G and $x \in G$, then

$$\mu(xy) = \mu(y) \ \forall y \in G \ \Leftrightarrow \ \mu(x) = \mu(e).$$

Recall also that if μ is a normal fuzzy subgroup of G, then G/μ is a group under the operation $: x\mu \circ y\mu = xy\mu \ \forall x, y \in G$.

Furthermore, if $\bar{\mu}$ is the fuzzy subset of G/μ defined by $\bar{\mu}(x\mu) = \mu(x)$ $\forall x \in G$, then $\bar{\mu}$ is a fuzzy subgroup of G/μ. Let μ be a normal fuzzy subgroup of a group G. Then μ_* is a normal subgroup of G. Define the fuzzy subset $\hat{\mu}$ of G/μ_* as follows: $\hat{\mu}(xH) = \mu(x)\forall x \in G$. Then $\hat{\mu}$ is a normal fuzzy subgroup of G/μ_*.

Let μ be a fuzzy subgroup of G. Then $\mu_* = \{x \in G \mid x\mu = e\mu\}$.
We now present a fuzzy Caley's Theorem. Much of its proof has been seen in Theorems 1.3.12 and 2.2.11. We note that we have the usual Caley's theorem if we let $\mu = 1_{\{e\}}$ in the next result.

Theorem 2.3.1. *Let μ be a normal fuzzy subgroup of a group G. Then there exists a homomorphism π of G onto G/μ that induces a natural permutation representation of G/μ. Furthermore, the kernel of π is μ_*.*

Proof. For all $x \in G$, define a function $\pi_x : G/\mu \to G/\mu$ by $\forall y \in G$, $\pi_x(y\mu) = xy\mu$. We now show that $\forall x \in G, \pi_x$ is a permutation of G/μ. Let $x \in G$. We have that $\pi_x(y\mu) = \pi_x(z\mu) \Leftrightarrow xy\mu = xz\mu \Leftrightarrow xy\mu(g) = xz\mu(g)$ $\forall g \in G$ $\Leftrightarrow \mu(y^{-1}x^{-1}g) = \mu(z^{-1}x^{-1}g)$ $\forall g \in G \Leftrightarrow \mu(y^{-1}u) = \mu(z^{-1}u)$ $\forall u \in G$ (since $x^{-1}g$ is also an arbitrary element of G) $\Leftrightarrow y\mu(u) = z\mu(u)\forall u \in G \Leftrightarrow y\mu = z\mu$. Thus π_x is a one-to-one function of G/μ into itself $\forall x \in G$. Thus $\forall x \in g, \pi_x$ is a permutation of G/μ since it follows easily that π_x maps G/μ onto itself.

Now let $\text{Sym}(G/\mu)$ denote the symmetric group on G/μ (that is, the group of all permutations of G/μ. Let $\pi : G \to \text{Sym}(G/\mu)$ be the function defined by $\pi(x) = \pi_x$ $\forall x \in G$. We show that π is a homomorphism. Let $x, y \in G$. Then for all $g \in G$, we have that

$$\pi_{xy}(g\mu) = (xy)g\mu = x(yg)\mu = \pi_x(yg\mu) = \pi_x(\pi_y(g\mu)) = \pi_x \circ \pi_y(g\mu)$$

Thus $\pi_{xy} = \pi_x \circ \pi_y$. Hence $\pi(xy) = \pi(x)\pi(y)$. Thus π is a homomorphism. We now prove that the kernel of π is μ_*. $x \in \text{Ker}(\pi) \Leftrightarrow \pi(x) = \pi_e \Leftrightarrow \pi_x = \pi_e \Leftrightarrow \pi_x(g\mu) = \pi_e(g\mu)\forall g \in G \Leftrightarrow xg\mu = g\mu\forall g \in G$
$\Leftrightarrow x\mu g\mu = g\mu\forall g \in G \Leftrightarrow x\mu = e\mu \Leftrightarrow x\mu(g) = e\mu(g)\forall g \in G \Leftrightarrow \mu(x^{-1}g) = \mu(g)\forall g \in G$
$\Leftrightarrow \mu(x^{-1}) = \mu(e) \Leftrightarrow \mu(x) = \mu(e) \Leftrightarrow x \in \mu_*$. Therefore, $\text{Ker}(\pi) = \mu_*$. $\square$

In the next result, we present another type of fuzzy Caley's Theorem. Let G be a group and $\forall x \in G$ define the function f_x of G into itself by $\forall y \in G, f_x(y) = xy$. Then it is known that f_x is a permutation of G. Let $P(G) = \{f_x \mid x \in G\}$. Then $P(G)$ is a group under composition of functions and $G \cong P(G)$ under the isomorphism f such that $f(x) = f_x$ $\forall x \in G$. This result is Caley's Theorem.

Theorem 2.3.2. *Let μ be a fuzzy subgroup of a group G. Let $S(\mu)$ denote the set of fuzzy singletons $\{x_t \mid x_t \subseteq \mu, x \in G\}$. Let f be the isomorphism of G onto $P(G)$ such that $f(x) = f_x\forall x \in G$. Then $S(\mu)$ is a completely regular semigroup with identity under max-min composition and f induces an isomorphism of $S(\mu)$ onto the semigroup $S(f(\mu)) = \{(f_x)_t \mid 0 \leq t \leq \mu(x)\}$.*

Proof. Clearly, $S(\mu)$ is a semigroup with identity $e_{\mu(e)}$. In fact, $S(\mu)$ is completely regular since it is the disjoint union of groups $\{x_t \mid x \in \mu_t\}$ for $0 \le t \le \mu(e)$. Let $x \in G$. Since $f(x) = f_x, f(x_t)(f_y) = \vee\{x_t(z) \mid f(z) = f_y\} = t$ if $f(x) = f_y$, i.e., $f_x = f_y$ and 0 otherwise. Thus $f(x_t) = (f_x)_t$. Now $f(xy) = f(x)f(y)$ for all $x, y \in G$. Hence $f(x_r y_s) = f((xy)_{r \wedge s}) = f(xy)_{r \wedge s} = (f_{xy})_{r \wedge s} = (f_x \circ f_y)_{r \wedge s} = (f_x)_r(f_y)_s$. Since f is one-to-one, $x_r = y_s \Leftrightarrow x = y$ and $r = s \Leftrightarrow f(x) = f(y)$ and $r = s \Leftrightarrow f_x = f_y$ and $r = s \Leftrightarrow (f_x)_r = (f_y)_s \Leftrightarrow f(x_r) = f(y_s)$.   □

We now develop the concept of fuzzy order. We note that Example 4.1.12 shows that the converse of the next result does not hold in general.

Proposition 2.3.3. *Let μ be a fuzzy subgroup of a group G. If $\mu([x, y]) = \mu(e) \forall x, y \in G$, then μ is normal in G.*

Proof. Recall $[x, y] = x^{-1}y^{-1}xy$. We have shown earlier that μ is normal if and only if μ is constant on the conjugacy classes of G. Thus we have that

$$\mu \text{ is normal}$$
$$\Leftrightarrow \mu(x^{-1}y^{-1}x) = \mu(y^{-1}) \; \forall x, y \in G$$
$$\Leftrightarrow \mu(x^{-1}y^{-1}xyy^{-1}) = \mu(y^{-1}) \; \forall x, y \in G$$
$$\Leftrightarrow \mu([x, y] \, y^{-1}) = \mu(y^{-1}) \; \forall x, y \in G$$
$$\Leftarrow \mu([x, y]) = \mu(e) \; \forall x, y \in G$$

by Lemma 2.1.2. (Note that $[x, y]$ does not remain fixed as y varies.)   □

Corollary 2.3.4. *Let μ be a fuzzy subgroup of a group G. If $\mu([x, y]) = \mu(e) \forall x, y \in G$, then μ_* is a normal subgroup of G. Furthermore, the quotient group G/μ_* is Abelian.*

Proof. The fact that μ_* is a normal subgroup of G follows from Proposition 2.3.3 and Theorem 1.3.4. Let G' denote the commutator subgroup of G, i.e., the subgroup generated by all the $[x, y]$ for $x, y \in G$. Since $\mu([x, y]) = \mu(e) \; \forall x, y \in G$, $G' \subseteq \mu_*$. Consequently, G/μ_* is Abelian.   □

Now we describe the motivation behind the two concepts of the order of a fuzzy subgroup and that of a fuzzy Abelian subgroup of a group G. Let K be a subgroup of a group G. Consider the set

$$H = \{x \in G \mid 1_K(x) = 1_K(e)\}.$$

Recall that for any subgroup M of G, the **index** of M in G, denoted by $[G : M]$, is equal to the number of distinct (left) cosets of M in G, which is also equal to $o(G)/o(M)$, where $o(G)$ denotes the order of G. Now in the above situation it is clear that $H = K$, and the index of K in G is indeed equal to the index of H in G. Furthermore, if $[G : H] = r$, then we have that

$o(H) = o(K) = \frac{o(G)}{r}$. Also, we note that H is Abelian if and only if K is Abelian. These considerations motivate us to consider the following analogs in the case of a fuzzy subgroup μ of G.

Definition 2.3.5. *Let μ be a fuzzy subgroup of a finite group G. Then the* **order** *of μ, written $o(\mu)$, is defined to be $o(\mu) = \frac{o(G)}{r}$, where r is the index of μ.*

Definition 2.3.6. *Let μ be a fuzzy subgroup of a finite group G. Then μ is called* **fuzzy Abelian** *if μ_* is an Abelian subgroup of G.*

Now μ_* is always a subgroup of G, but it is not necessarily Abelian. It follows that

$$o(\mu) = \frac{o(G)}{[G : \mu_*]}.$$

Hence $o(\mu) = o(\mu_*)$.

If H, K are subgroups of a finite group G such that $o(H) = o(K)$ and $H \subseteq K$, then trivially $H = K$. However, if μ, ν are fuzzy subgroups of a finite group G such that $o(\mu) = o(\nu)$ and $\mu \subseteq \nu$, then it is always not necessarily the case that $\mu = \nu$, as the following example shows.

Example 2.3.7. *Let $G = \{e, a, b, c\}$ be the Klein 4-group, where e is the identity of G. Let $t_0, t_1, t_2 \in [0, 1]$ be such that*

$$t_0 > t_1 > t_2.$$

Define $\nu : G \to [0, 1]$ as follows:

$$\nu(e) = t_0, \; \nu(a) = t_1, \; \nu(b) = t_2, \; \nu(ab) = t_2.$$

It follows easily that ν is a fuzzy subgroup of G. Let $s_0, s_1, s_2 \in [0, 1]$ be such that

$$s_0 > s_1 > s_2,$$

where $t_0 < s_0$, $t_1 < s_1$, $t_2 < s_2$. Define $\mu : G \to [0, 1]$ as follows:

$$\mu(e) = s_0, \; \mu(a) = s_1, \; \mu(b) = s_2, \; \mu(ab) = s_2.$$

It again follows easily that μ is a fuzzy subgroup of G. Furthermore, it follows that $\nu \subseteq \mu$. From the definitions of ν and μ, it is clear that $\nu_* = \mu_*$. Thus by using (13) we get that

$$o(\nu) = o(\nu_*) = o(\mu).$$

Hence we see that ν and μ have the same orders, $\nu \subseteq \mu$, and yet clearly $\nu \neq \mu$.

The above example illustrates the fact that not all properties of a fuzzy subgroup could be expected to be analogs of results from group theory.

The subgroup ν defined in Example 2.3.7 is also an example of a fuzzy Abelian subgroup, and the same is the case with the fuzzy subgroup μ defined therein. Since a fuzzy subgroup ν of a group G is called fuzzy Abelian if ν_* is Abelian, examples of fuzzy subgroups which are not fuzzy Abelian are easily constructed. For example, let G be any group containing a subgroup H that is not Abelian. Define the fuzzy subset μ of G by $\mu(x) = t_0 \forall x \in H$ and $\mu(x) = t_1 \forall x \in G\backslash H$, where $0 \leq t_1 < t_0 \leq 1$. Then μ is a fuzzy subgroup of G and $\mu_* = H$. Thus μ is not fuzzy Abelian.

We now prove an analog of a result from group theory that if U, V are two subgroups of a group G such that $U \subseteq V$ and V is Abelian, then U is Abelian. We have the following result.

Proposition 2.3.8. *Let μ, ν be fuzzy subgroups of a group G such that $\nu \subseteq \mu$, $\nu(e) = \mu(e)$, and μ is fuzzy Abelian. Then ν is fuzzy Abelian.*

Proof. By hypothesis μ_* is Abelian. Let $y \in \nu_*$. Then we have that $\nu(y) = \nu(e) = \mu(e)$. By hypothesis, $\nu(y) \leq \mu(y)$. Since $\mu(y) \leq \mu(e)$, $\mu(y) = \mu(e)$ and so $y \in \mu_*$. Thus $\nu_* \subseteq \mu_*$ and consequently ν_* is Abelian. Hence, ν is fuzzy Abelian.   □

Proposition 2.3.9. *A fuzzy subgroup of order p^2 is fuzzy Abelian if p is a prime.*

Proof. Let μ be a fuzzy subgroup of group G such that $o(\mu) = p^2$. By (13) we have $o(\mu) = o(\mu_*)$. Now it is a standard result in group theory that a group of order p^2 is Abelian. Thus μ_* is Abelian and consequently μ is fuzzy Abelian.   □

Let μ and ν be fuzzy subgroups of G such that $\nu \subseteq \mu$. Since $o(\mu) = o(\mu_*), o(\nu) = o(\nu_*), [G : \mu] = [G : \mu_*]$, and $[G : \nu] = [G : \nu_*]$, we are motivated to define the **index** of ν in μ, written $[\mu : \nu]$, to be $[\mu_* : \nu_*]$. The following theorem is one of several results one could call a Fuzzy Lagrange's Theorem.

Theorem 2.3.10. *Let μ and ν be fuzzy subgroups of a finite group G such that $\nu \subseteq \mu$. Then the order of ν divides the order of μ. In fact, $o(\mu) = o(\nu)[\mu : \nu]$.*

Proof. As in the proof of Proposition 2.3.8, we have that $\nu_* \subseteq \mu_*$. Thus by Lagrange's theorem, we have that $o(\mu_*) = o(\nu_*)[\mu_* : \nu_*]$. However, $o(\nu) = o(\nu_*)$, $o(\mu) = o(\mu_*)$, and $[\mu : \nu] = [\mu_* : \nu_*]$. Hence the desired result holds.   □

We now develop a fuzzy analog of solvability. First we need a preliminary result.

Lemma 2.3.11. *Let μ, ν be normal fuzzy subgroups of G such that $\nu \subseteq \mu$ and $\nu(e) = \mu(e)$. Then there exists a normal fuzzy subgroup η of the quotient group G/ν_* such that $\eta(x\nu_*) = \mu(x) \ \forall x \in G$.*

Proof. By Theorem 1.3.4, we have that ν_* and μ_* are normal subgroups of G. Since $\nu \subseteq \mu$, by arguing as in the proof of Proposition 2.3.8, we have that $\nu_* \subseteq \mu_*$. Define the fuzzy subset η of G/ν_* by $\eta(x\nu_*) = \mu(x) \ \forall x \in G$. We claim that η is well defined. For suppose that $x\nu_* = y\nu_*$. Then $y^{-1}x \in \nu_*$ and so $y^{-1}x \in \mu_*$ since $\nu_* \subseteq \mu_*$. Consequently, we have that $x\mu_* = y\mu_*$. Now by Theorem 2.1.6, μ induces a unique function $\mu^\#$ from G/μ_* into $[0, 1]$ such that $\mu^\#(x\mu_*) = \mu(x) \ \forall x \in G$. Therefore, we have that $\forall x, y \in G, \eta(x\nu_*) = \mu(x) = \mu^\#(x\mu_*) = \mu^\#(y\mu_*) = \mu(y) = \eta(y\nu_*)$ since $x\mu_* = y\mu_*$. Thus η is well defined. It follows easily that η is a fuzzy subgroup and that η is normal. $\square$

Let μ, ν be normal fuzzy subgroups of G such that $\nu \subseteq \mu$. Then ν_* is a normal subgroup of G. Let f be the natural homomorphism of G onto G/ν_*. Then $\forall x \in G, f(\mu)(x\nu_*) = \vee\{\mu(y) | f(y) = x\nu_*\} = \vee\{\mu(y) | y \in \nu_*\} = \mu(x) = (\mu/\nu)(x)$. Hence $f(\mu) = \mu/\nu$.

Let $x \in G$. Then $x\nu_* \in (\mu/\nu)_* \iff (\mu/\nu)(x\nu_*) = (\mu/\nu)(e\nu_*) \iff \mu(x) = \mu(e) \iff x \in \mu_*$. Thus $(\mu/\nu)_* = \mu_*/\nu_*$.

Definition 2.3.12. *With the hypothesis of Lemma 2.3.11, the fuzzy subgroup η defined above is called the **fuzzy quotient group** μ/ν.*

Now consider a chain of normal fuzzy subgroups of G given by $\nu_1 \supseteq \nu_2 \supseteq ... \supseteq \nu_k$, where $\nu_i(e) = \nu_1(e)$, $i = 1, ..., k$. Let $H_i = (\nu_i)_*, i = 1, ..., k$. Then we have that H_i is a normal subgroup of G since ν_i is normal. By Lemma 2.3.11, it follows that the chain $\nu_1 \supseteq \nu_2 \supseteq ... \supseteq \nu_k$ yields another chain of normal fuzzy subgroups: $\eta_1 \supseteq \eta_2 \supseteq ... \supseteq \eta_k$, where

$$\eta_i = \nu_i/\nu_{i+1}$$

and

$$H_i/H_{i+1} = \{xH_{i+1} \mid x \in G, \eta_i(x) = \eta_i(e)\}.$$

The above discussion motivates the following definition.

Definition 2.3.13. *A fuzzy subgroup ν of a group G is called **fuzzy solvable** if there exists a chain of normal fuzzy subgroups $\nu = \nu_1 \supseteq \nu_2 \supseteq ... \supseteq \nu_k$ with $\nu_k(x) = \nu_k(e)$ only when $x = e$, and $\nu_i(e) = \nu_1(e)$, $1 \le i \le k$, such that there is a corresponding chain of normal fuzzy subgroups $\eta_1 \supseteq \eta_2 \supseteq ... \supseteq \eta_k$, where $\eta_i = \nu_i/\nu_{i+1}$ and η_i is fuzzy Abelian, $1 \le i \le k$, $\nu_{k+1} = e_{\nu_1(e)}$.*

The chain of normal fuzzy subgroups given by $\nu = \nu_1 \supseteq \nu_2 \supseteq ... \supseteq \nu_k$ yields a chain of normal subgroups $H_1 \supseteq H_2 \supseteq ... \supseteq H_k$ of the group G, where H_i is defined by $H_i = (\nu_i)_*, i = 1, ..., k$. Clearly $H_k = \{e\}$.

If in the chain of fuzzy subgroups given by $\nu = \nu_1 \supseteq \nu_2 \supseteq ... \supseteq \nu_k$, we have that $\nu_1(x) = \nu_1(e)$ $\forall x \in G$, then $H_1 = G$. In this case, we have a chain of subgroups

$$H_1 = G \supseteq H_2 \supseteq ... \supseteq H_k = \{e\},$$

where H_i/H_{i+1} is Abelian. Consequently, G is solvable in this case. In the general situation when H_1 is not necessarily equal to G, it is clear that H_1 is always a solvable group.

For a solvable group, it is a standard result that any subgroup and any quotient group are both solvable. For a fuzzy solvable fuzzy subgroup, we present one analog since we consider these issues in greater detail in Chapter 3.

Theorem 2.3.14. *Let G be a group. Let μ, ν be normal fuzzy subgroups of G such that $\nu \subseteq \mu$ and $\nu(e) = \mu(e)$. If μ is fuzzy solvable, then ν is fuzzy solvable.*

Proof. Since μ is solvable, there is a chain of normal fuzzy subgroups $\mu = \mu_1 \supseteq \mu_2 \supseteq ... \supseteq \mu_k$ satisfying the conditions in Definition 2.3.13. Consider the chain of fuzzy subgroups $\nu = \nu_1 \supseteq \mu_2 \cap \nu \supseteq \mu_3 \cap \nu.... \supseteq \mu_k \cap \nu$. Clearly each $\mu_n \cap \nu$ is normal in G. Now $x, y \in (\mu_i \cap \nu)_* \Leftrightarrow x, y \in (\mu_i)_* \cap \nu_*$. This implies that $x, y \in (\mu_i)_*$ and $x, y \in \nu_*$ which further implies that $x^{-1}y^{-1}xy \in (\mu_{i+1})_*$ and $x^{-1}y^{-1}xy \in \nu_*$ since $(\mu_i)_*/(\mu_{i+1})_*$ is Abelian. Thus $xy^{-1}xy \in (\mu_{i+1})_* \cap \nu_* = (\mu_{i+1} \cap \nu)_*$ and so $(\mu_i \cap \nu)_*/(\mu_{i+1} \cap \nu)_*$ is Abelian. Thus $(\mu_i \cap \nu)/(\mu_{i+1} \cap \nu)$ is fuzzy Abelian and so ν is fuzzy solvable. $\qquad\square$

We next present another type of fuzzy Lagrange's theorem using the notion of fuzzy order as developed in Section 1.6 and references [8, 9].

It follows that if $\{x \in G \mid \mu(x) = \mu(e)\} = \{e\}$, then $FO_\mu(x) = o(x)$ for all $x \in G$. For suppose $\mu(x^n) = \mu(e)$ for some $x \in G$ and some positive integer n. Then $x^n = e$ by assumption.

Definition 2.3.15. *Let G be a group and μ a fuzzy subgroup of G. If there exists a positive integer n such that for all $x \in G$, $\mu(x^n) = \mu(e)$, then the smallest such positive integer is called the **fuzzy order** of μ, written $O(\mu)$. If no such positive integer exists, then μ is said to be of **infinite fuzzy order**.*

It follows in Definition 2.3.15, $O(\mu)$ is the least common multiple of the $FO_\mu(x)$ for $x \in G$ if this least common multiple exists. We also note that $O(\mu) = 1$ if and only if $\mu = \mu(e)_G$.

Example 2.3.16. *Consider the group $\mathbb{Z}$ under usual addition. Define the fuzzy subset μ of $\mathbb{Z}$ as follows: $\mu(x) = t_0$ if $x \in \langle 6 \rangle$ and $\mu(x) = t_1$ otherwise, where $t_0 > t_1$. Then μ is a fuzzy subgroup of $\mathbb{Z}$. If $x \in 2 + \langle 6 \rangle$ or $x \in 4 + \langle 6 \rangle$, then $FO_\mu(x) = 3$. If $x \in 3 + \langle 6 \rangle$, then $FO_\mu(x) = 2$. If $x \in \langle 6 \rangle$, then $FO_\mu(x) = 1$. Also, $O(\mu) = 6$.*

The following theorem provides a relation between fuzzy orders of fuzzy subgroups and orders of groups. It gives us another type of fuzzy Lagrange's Theorem.

Theorem 2.3.17. *Let G be a finite group and μ a fuzzy subgroup of G. Then $O(\mu)|o(G)$.*

Proof. For all $x \in G, FO_\mu(x)|o(G)$ by Proposition 1.6.5. Thus $o(G)$ is a common multiple of $FO_\mu(x)$ for all $x \in G$. Since $O(\mu)$ is the least common multiple of the $FO_\mu(x)$, it follows that $O(\mu)|o(G)$. $\qquad\square$

Example 2.3.18. *Let $G = \{e, a, b, c\}$ be the Klein-4 group. Define the fuzzy subset μ of G as follows: $\mu(e) = 1, \mu(a) = \mu(b) = \mu(c) = 1/2$. Then μ is a fuzzy subgroup of G. Now $O(\mu) = 2$ and $[G : \mu] = [G : \mu_*] = 4$. Hence $O(\mu)|o(G)$, but $o(G) \neq O(\mu)[G : \mu]$.*

Proposition 2.3.19. *Let G be a finite group and μ a fuzzy subgroup of G. If $O(\mu) = p^n m$, where p is prime and m and n are relatively prime positive integers, then there is an $x \in G$ such that $FO_\mu(x) = p^k$ for each nonnegative integer $k \leq n$.*

Proof. . Since $O(\mu)$ is the least common multiple of the $FO_\mu(x)$ for $x \in G$, there is an $x \in G$ such that $FO_\mu(x) = p^n$. Hence the desired result follows easily. $\qquad\square$

Corollary 2.3.20. *Let G be a finite Abelian group and μ a fuzzy subgroup of G. Then the following assertions hold.*
(1) If $O(\mu) = mn$ for some positive integers m and n, then there is an $x \in G$ such that $FO_\mu(x) = m$.
(2) $O(\mu) = \vee\{FO_\mu(x) \mid x \in G\}$.

Proof. The desired result follows from Proposition 2.3.19 and Theorem 1.6.13(2). $\qquad\square$

Lemma 2.3.21. *Let G be a finite group and μ and ν fuzzy subgroups of G. If $\nu \subseteq \mu$ and $\mu(e) = \nu(e)$, then $FO_\mu(x)|FO_\nu(x)$ for all $x \in G$ such that $FO_\nu(x)$ is finite.*

Proof. Let $FO_\nu(x) = n$. Then $\mu(e) = \nu(e) = \nu(x^n) \leq \mu(x^n)$. Since $\mu(e) \geq \mu(x^n)$, $\mu(x^n) = \mu(e)$. Thus $FO_\mu(x)|n$ by Proposition 1.6.5. $\qquad\square$

We now obtain another fuzzy analogue of Lagrange's theorem. The converse of the analogue does not hold in general although it holds for special cases. The converse of this analogue will be discussed in Section 7.2.

Theorem 2.3.22. *Let G be a finite group and μ and ν fuzzy subgroups of G. If $\nu \subseteq \mu$ and $\mu(e) = \nu(e)$, then $O(\mu)|O(\nu)$.*

Proof. We may assume that $O(\nu)$ is finite. Then $FO_\nu(x)$ is finite for all $x \in G$ and the number of elements of $H = \{FO_\nu(x) \mid x \in G\}$ is finite since G is finite. Thus $FO_\mu(x)$ is finite for all $x \in G$ by Lemma 2.3.21 and the number of elements of $K = \{FO_\mu(x) \mid x \in G\}$ is finite. Thus $O(\nu)$ and $O(\mu)$ are the least common multiples of elements of H and K, respectively. Hence $O(\mu)|O(\nu)$ by Lemma 2.3.21.   □

References

1. P. Bhattacharya, Fuzzy subgroups: Some characterization, *J. Math. Anal. Appl.*, 128(1987) 241-252.
2. P. Bhattacharya, Fuzzy subgroups: Some characterizations II, *Inform. Sci.*, 38(1986) 293-297.
3. P. Bhattacharya and N. P. Mukherjee, Fuzzy groups and fuzzy relations, *Inform. Sci.* 36 (1985) 267-282.
4. P. Bhattacharya and N. P. Mukherjee, Fuzzy groups: some group theoretic analogs. II, *Inform. Sci.* (1987) 77-91.
5. L. Biacino and G. Gerla, Closure systems and *L*-subalgebras, *Inform. Sci.* 33 (1984)181-195.
6. P. S. Das, Fuzzy groups and level subgroups, *J. Math. Anal. Appl.* 84 (1981) 264-269.
7. W.B.V. Kandasamy and D. Meiyappan, Fuzzy symmetric subgroups and conjugate fuzzy subgroups of a group, *Journal of Fuzzy Math.* 6 (1998) 905-913.
8. J. G. Kim, Fuzzy orders relative to fuzzy subgroups, *Inform. Sci* 80 (1994) 341-348.
9. J. G. Kim, Orders of fuzzy subgroups and fuzzy *p*-subgroups, *Fuzzy Sets and Systems* 61 (1994) 225-230.
10. N. P. Mukherjee and P. Bhattacharya, Fuzzy normal subgroups and fuzzy cosets, *Inform. Sci.* 34 (1984) 225-239.
11. N. P. Mukherjee and P. Bhattacharya, Fuzzy groups: Some group theoretic analogs, *Inform. Sci.* 39 (1986) 247-268.
12. A. Rosenfeld, Fuzzy groups, *J. Math. Anal. Appl.* 35 (1971) 512-517.
13. L. A. Zadeh, Fuzzy sets, *Inform. and Control* 8 (1965) 338-353.
14. Y. Zhang, Some properties on fuzzy subgroups, *Fuzzy Sets and Systems* 119 (2001) 427-438.
15. Y. Zhang and K. Zou, Normal fuzzy subgroups and conjugate fuzzy subgroups, *Journal of Fuzzy Math.* 1 (1993) 571-585.

3

Nilpotent, Commutator, and Solvable Fuzzy Subgroups

As the title suggests, this chapter is concerned with nilpotent, commutator, and solvable fuzzy subgroups. We give two approaches for the development of these notions, namely one via an ascending chain of fuzzy subgroups and one via descending chain of fuzzy subgroups. The results of this chapter are mainly from [5, 8, 18].

3.1 Commutative Fuzzy Subsets and Nilpotent Fuzzy Subgroups

In this section, we introduce the notion of a commutative fuzzy subset and the notion of a nilpotent fuzzy subgroup. We show that the nilpotence of a group can be completely characterized by the nilpotence of its fuzzy subgroups.

For the relevant classical group-theoretic results and definitions, the reader is referred to [6, 17].

Let μ be a fuzzy subset of a groupoid G. The **normalizer** $N(\mu)$ of μ in G is the set $\{x \in G \mid \mu(xy) = \mu(yx)$ for all $y \in G\}$ and μ is called **normal** in G if $N(\mu) = G$. If G is a group, then $\mu(e)$ is called the **tip** of μ.

Definition 3.1.1. *Let μ be a fuzzy subset of a semigroup G. Let $Z(\mu) = \{x \in G \mid \mu(xy) = \mu(yx)$ and $\mu(xyz) = \mu(yxz)$ for all $y, z \in G\}$. Then μ is called commutative in G if $Z(\mu) = G$.*

We adopt the terminology from [8] and call $Z(\mu)$ in Definition 3.1.1 the **centralizer** of μ in G .

If G has a right identity, then the equality $\mu(xy) = \mu(yx)$ in Definition 3.1.1 is redundant. If G is a semigroup, we let $Z(G)$ denote the center of G. It is clear that $Z(G) \subseteq Z(\mu) \subseteq N(\mu)$ and that $Z(G) \neq Z(\mu) \neq N(\mu)$ is possible. In the following example, recall that the notation $\langle a, b \mid a^3 = e = b^2, ba = a^2 b \rangle$ denotes the group generated by a and b, where a and b satisfy the properties $a^3 = e = b^2$ and $ba = a^2 b$.

Example 3.1.2. *Let* $G = D_3 \otimes D_3$, *where* D_3 *is the dihedral group*
$\langle a, b \mid a^3 = e = b^2, \, ba = a^2 b \rangle$.
   Let μ *be the fuzzy subset of* G *defined as follows:* $\forall x \in G$,

$$\mu(x) = \begin{cases} t_0 & \text{if } x \in \{e\} \otimes \langle a \rangle, \\ t_1 & \text{if } x \in (\langle a \rangle \otimes \{e\}) \backslash (\{e\} \otimes \langle a \rangle), \\ t_2 & \text{otherwise,} \end{cases}$$

where $t_0 > t_1 > t_2$. *Since* $\langle a \rangle$ *is a normal subgroup of* $D_3, xy \in \langle a \rangle$ *if and only if* $yx \in \langle a \rangle$ *for all* $x, y \in D_3$. *This allows us to show that* $N(\mu) = G$. *Let* $(x, y), (u, v) \in G$. *Then* $(xu, yv) \in \langle a \rangle \otimes \{e\} \Leftrightarrow (ux, vy) \in \langle a \rangle \otimes \{e\}$ *and* $(xu, yv) \in \{e\} \times \{a\} \Leftrightarrow (ux, vy) \in \{e\} \otimes \langle a \rangle$. *Hence* $(xu, yv) \in G \backslash ((\langle a \rangle \otimes \{e\}) \cup (\{e\} \otimes \langle a \rangle)) \Leftrightarrow (ux, vy) \in G \backslash ((\langle a \rangle \otimes \{e\}) \cup \{e\} \otimes \langle a \rangle))$. *Thus it follows that* $N(\mu) = G$. *Now* $\mu((e, a)(a, a^2 b)(a, b)) = \mu(a^2, e) = t_1$ *and* $\mu((a, a^2 b)(e, a)(a, b)) = \mu(a^2, a) = t_2$. *Hence* $Z(\mu) \neq N(\mu)$.

Lemma 3.1.3. *Let* μ *be a fuzzy subset of a semigroup* G. *Then* $x \in Z(\mu)$ *if and only if* $\mu(xy_1, ..., y_n) = \mu(y_1 x y_2 ... y_n) = ... = \mu(y_1 y_2 ... y_n x)$ *for all* $y_1, y_2, ..., y_n \in G$.

Proof. We prove the result by induction on n. Suppose $x \in Z(\mu)$. Then $\mu(xy_1 y_2) = \mu(y_1 x y_2)$ for all $y_1, y_2 \in G$. Assume $\mu(xy_1, ..., y_n) = \mu(y_1 x y_2 ... y_n) = ... = \mu(y_1 y_2 ... y_n x)$ for all $y_1, y_2, ..., y_n \in G$. Then
   $\mu(xy_1 ... (y_n y_{n+1})) = \mu(y_1 x y_2 ... (y_n y_{n+1})) = ... =$
   $\mu(y_1 y_2 ... x(y_n y_{n+1})) = \mu(y_1 y_2 ... (y_n y_{n+1})x)$
   and
   $\mu(x(y_1 y_2) ... y_n y_{n+1}) = \mu((y_1 y_2) x ... y_n y_{n+1}) = ... =$
   $\mu((y_1 y_2) ... y_n x y_{n+1}) = \mu((y_1 y_2) ... y_n y_{n+1} x)$
   for all $y_1, ..., y_n, y_{n+1} \in G$. This yields the desired result. $\square$

Theorem 3.1.4. *Let* μ *be a fuzzy subset of a semigroup* G. *Then* μ *is commutative in* G *if and only if* $\forall x_1, ..., x_n \in G, n \in \mathbb{N}, \mu(x_1 ... x_n) = \mu(x_{\pi(1)} ... x_{\pi(n)})$ *for all permutations* π *of* $\{1, ..., n\}$.

Proof. The proof follows from Lemma 3.1.3. $\square$

Lemma 3.1.5. *Let* μ *be a fuzzy subgroup of a group* G. *Then the following assertions hold.*
   (1) *For all* $x, y \in G, \mu(x) \neq \mu(y)$ *implies* $\mu(xy) = \mu(x) \wedge \mu(y)$.
   (2) *For all* $x, y \in G, \mu(xy^{-1}) = \mu(e)$ *implies* $\mu(x) = \mu(y)$.

Proof. (1) Suppose $\mu(x) > \mu(y)$. Then $\mu(y) = \mu(x^{-1} xy) \geq \mu(x^{-1}) \wedge \mu(xy) \geq \mu(x) \wedge \mu(x) \wedge \mu(y) = \mu(y)$. Thus $\mu(xy) = \mu(x) \wedge \mu(y)$. A similar argument holds if $\mu(y) > \mu(x)$.
   (2) Suppose $\mu(x) \neq \mu(y)$. Then $\mu(x) \neq \mu(y^{-1})$. Hence by (1), $\mu(e) = \mu(xy^{-1}) = \mu(x) \wedge \mu(y^{-1})$. Since $\mu(e) \geq \mu(z)$ for all $z \in G, \mu(x) = \mu(y^{-1}) = \mu(y)$, a contradiction. $\square$

Lemma 3.1.6. *Let μ be a fuzzy subgroup of a group G. Let $T = \{x \in G \mid \mu(xyx^{-1}y^{-1}) = \mu(e) \text{ for all } y \in G\}$. Then $T = Z(\mu)$.*

Proof. Let $x \in T$. Then $\forall y, z \in G$,

$$\mu((xyz)(yxz)^{-1}) = \mu(xyzz^{-1}x^{-1}y^{-1})$$
$$= \mu(xyx^{-1}y^{-1})$$
$$= \mu(e).$$

By Lemma 3.1.5 (2), $\mu(xyz) = \mu(yxz)$ for all $y, z \in G$ and so $x \in Z(\mu)$. Therefore, $T \subseteq Z(\mu)$. Conversely, if $x \in Z(\mu)$, then $\mu(xyx^{-1}y^{-1}) = \mu(e)$ for all $y \in G$ by Lemma 3.1.3. Thus $x \in T$. Hence $Z(\mu) \subseteq T$. □

Proposition 3.1.7. *Let μ be a fuzzy subgroup of a group G. Then $\mu(xyx^{-1}y^{-1}) = \mu(e)$ $\forall x, y \in G$ if and only if μ is commutative in G.*

Proof. The proof follows from Lemma 3.1.6. □

The next result is an analog of the standard result that the center $Z(G)$ of a group G is normal in G.

Proposition 3.1.8. *Let μ be a fuzzy subset of a semigroup G. If $Z(\mu)$ (resp. $N(\mu)$) is nonempty, then $Z(\mu)$ (resp. $N(\mu)$) is a subsemigroup of G. Moreover, if G is a group, then $Z(\mu)$ is a normal subgroup of G and $N(\mu)$ is a subgroup of G.*

Proof. Let $x_1, x_2 \in Z(\mu)$. Then for all $y, z \in G$, we have $\mu((x_1x_2)yz) = \mu(y(x_1x_2)z)$ by Lemma 3.1.3 and clearly $\mu((x_1x_2)y) = \mu(y(x_1x_2))$. Hence $x_1x_2 \in Z(\mu)$. Thus $Z(\mu)$ is a sub-semigroup of G if $Z(\mu)$ is nonempty.

Suppose G is a group. Then $Z(\mu)$ is nonempty since $e \in Z(\mu)$. If $x \in Z(\mu)$, then

$$\mu(x^{-1}yz) = \mu(x^{-1}yx^{-1}xz)$$
$$= \mu(xx^{-1}yx^{-1}z)$$
$$= \mu(yx^{-1}z)$$

for all $y, z \in G$ and so $x^{-1} \in Z(\mu)$. Hence $Z(\mu)$ is a subgroup of G by the first part of the proof. Next let $x \in Z(\mu)$ and $g \in G$. Then by Lemma 3.1.3,

$$\mu((g^{-1}xg)yz) = \mu(xg^{-1}gyz)$$
$$= \mu(xyz)$$
$$= \mu(xyg^{-1}gz)$$
$$= \mu(yg^{-1}xgz)$$
$$= \mu(y(g^{-1}xg)z)$$

for all $y, z \in G$ and so $g^{-1}xg \in Z(\mu)$. Thus $Z(\mu)$ is a normal subgroup of G if G is a group.

The argument that $N(\mu)$ is a sub-semigroup of G is similar to that for $Z(G)$. By Theorem 1.3.7, $N(\mu)$ is a subgroup of G when G is a group. □

We now consider homomorphic images and homomorphic preimages of the centralizer and the normalizer of a fuzzy subset.

Proposition 3.1.9. *Let f be a homomorphism of a semigroup G onto a semi-group H. Let μ and ν be fuzzy subsets of G and H, respectively.*

(1) If G and H are groups, then $f(Z(\mu)) \subseteq Z(f(\mu))$ and $f(N(\mu)) \subseteq N(f(\mu))$.

(2) $f^{-1}(Z(\nu)) = Z(f^{-1}(\nu))$ and $f^{-1}(N(\nu)) = N(f^{-1}(\nu))$.

Proof. (1) Let $x \in f(Z(\mu))$. Then there exists $u \in Z(\mu)$ such that $f(u) = x$. For all $y, z \in H$,

$$
\begin{aligned}
(f(\mu))(xyz) &= \vee\{\mu(a) \mid f(a) = xyz\} \\
&= \vee\{\mu(uvwk) \mid k \in \ker(f)\} \\
&= \vee\{\mu(vuwk) \mid k \in \ker(f)\} \\
&= \vee\{\mu(b) \mid f(b) = yxz\} \\
&= (f(\mu))(yxz),
\end{aligned}
$$

where $v, w \in G$ are such that $f(v) = y$ and $f(w) = z$. Thus $x \in Z(f(\mu))$. Hence $f(Z(\mu)) \subseteq Z(f(\mu))$. In a similar manner, we have that $f(N(\mu)) \subseteq N(f(\mu))$.

(2) Let $x \in f^{-1}(Z(\nu))$. Then for all $y, z \in G$,

$$
\begin{aligned}
(f^{-1}(\nu))(xyz) &= \nu(f(xyz)) \\
&= \nu(f(x)f(y)f(z)) \\
&= \nu(f(y)f(x)f(z)) \\
&= \nu(f(yxz)) \\
&= (f^{-1}(\nu))(yxz).
\end{aligned}
$$

Similarly, $(f^{-1}(\nu))(xy) = (f^{-1}(\nu))(yx)$. Thus $x \in Z(f^{-1}(\nu))$. Hence $f^{-1}(Z(\nu)) \subseteq Z(f^{-1}(\nu))$. Let $x \in Z(f^{-1}(\nu))$ and $f(x) = u$. Then for all $v, w \in H$,

$$
\begin{aligned}
\nu(uvw) &= \nu(f(x)f(y)f(z)) \\
&= \nu(f(xyz)) \\
&= (f^{-1}(\nu))(xyz) \\
&= (f^{-1}(\nu))(yxz) \\
&= \nu(f(yxz)) \\
&= \nu(f(y)f(x)f(z)) \\
&= \nu(vuw),
\end{aligned}
$$

where $y, z \in G$ are such that $f(y) = v$ and $f(z) = w$. Similarly, $\nu(uv) = \nu(vu)$. Thus $u \in Z(\nu)$, i.e., $x \in f^{-1}(Z(\nu))$. Hence $Z(f^{-1}(\nu)) \subseteq f^{-1}(Z(\nu))$. Thus $f^{-1}(Z(\nu)) = Z(f^{-1}(\nu))$. That $f^{-1}(N(\nu)) = N(f^{-1}(\nu))$ holds similarly. $\square$

The next result now follows:

Proposition 3.1.10. *Let f be a homomorphism of a semigroup G onto a semigroup H. Let μ and ν be fuzzy subsets of G and H, respectively. Then the following assertions hold.*

(1) If μ is commutative (normal) in G, then $f(\mu)$ is commutative (normal) in H, where G and H are groups.

(2) ν is commutative (normal) in H if and only if $f^{-1}(\nu)$ is commutative (normal) in G.

In Proposition 3.1.9(1), $f(Z(\mu)) \neq Z(f(\mu))$ and $f(N(\mu)) \neq N(f(\mu))$ in general even if μ is a fuzzy subgroup. Hence the converse of Proposition 3.1.10(1) does not hold in general, as the following example shows.

Example 3.1.11. *Let μ be the fuzzy subgroup of the dihedral group,*
$D_4 = \langle a, b \mid a^4 = e = b^2,\ ba = a^3 b \rangle$ *defined as follows:* $\forall x \in D_4$,

$$\mu(x) = \begin{cases} t_0 & \text{if } x \in \langle b \rangle \\ t_1 & \text{otherwise,} \end{cases}$$

where $t_0 > t_1$*. Let f be the natural homomorphism of D_4 onto D_4/D_4', where $D_4' = \{e, a^2\}$ is the commutator subgroup of D_4. We have $\mu(a(ab)) = \mu(a^2 b) = t_1 \neq t_0 = \mu(b) = \mu((ab)a)$ and $\mu(a^3(a^3 b)) = \mu(a^2 b) = t_1 \neq t_0 = \mu(b) = \mu((a^3 b)a^3)$. Hence $a, a^3 \notin N(\mu)$ and $a,\ a^3 \notin Z(\mu)$. Thus $f(a) \notin f(N(\mu))$ and $f(a) \notin f(Z(\mu))$ since $f^{-1}(f(a)) = \{a, a^3\}$. However, D_4/D_4' is Abelian. Hence $N(f(\mu)) = D_4/D_4' = Z(f(\mu))$. Thus $f(N(\mu)) \neq N(f(\mu))$ and $f(Z(\mu)) \neq Z(f(\mu))$.*

We now consider the centralizer and the normalizer of the intersection of fuzzy subsets.

Proposition 3.1.12. *(1) Let μ and ν be fuzzy subsets of a semigroup G. Then $Z(\mu) \cap Z(\nu) \subseteq Z(\mu \cap \nu)$ and $N(\mu) \cap N(\nu) \subseteq N(\mu \cap \nu)$. In particular, if μ and ν are commutative (invariant), then $(\mu \cap \nu)$ is commutative (invariant).*

(2) Let μ and ν be fuzzy subgroups of a group G such that $\mu(e) = \nu(e)$. Then $Z(\mu) \cap Z(\nu) = Z(\mu \cap \nu)$. In particular, μ and ν are commutative in G if and only if $\mu \cap \nu$ is commutative in G.

Proof. (1) The desired result follows easily.

(2) By Lemma 3.1.6,

$x \in Z(\mu \cap \nu)$

$\Leftrightarrow (\mu \cap \nu)(e) = (\mu \cap \nu)(xyx^{-1}y^{-1})$ for all $y \in G$

$\Leftrightarrow \mu(e) = \nu(e) = (\mu \cap \nu)(e) = \mu(xyx^{-1}y^{-1}) \wedge \nu(xyx^{-1}y^{-1})$ for all $y \in G$

$\Leftrightarrow \mu(xyx^{-1}y^{-1}) = \mu(e)$ and $\nu(xyx^{-1}y^{-1}) = \nu(e)$ for all $y \in G$

$\Leftrightarrow x \in Z(\mu)$ and $x \in Z(\nu)$

$\Leftrightarrow x \in Z(\mu) \cap Z(\nu)$.

Thus $Z(\mu) \cap Z(\nu) = Z(\mu \cap \nu)$. $\square$

We now study the centralizer and the normalizer of the sup-min product of two fuzzy subsets.

Let μ and ν be fuzzy subsets of a groupoid G. Recall that the product $\mu \circ \nu$ is the fuzzy subset of G defined by $\forall x \in G, (\mu \circ \nu)(x) = \vee\{\mu(a) \wedge \nu(b) \mid x = ab, a, b \in G\}$. The following result is a consequence of Theorem 1.3.1.

Lemma 3.1.13. *Let μ be a fuzzy subset of a group G. Let $S = \{x \in G \mid \mu(x^{-1}yx) = \mu(y)$ for all $y \in G\}$ and $T = \{x \in G \mid \mu(xyx^{-1}) = \mu(y)$ for all $y \in G\}$. Then $S = N(\mu) = T$.*

Proposition 3.1.14. *Let μ and ν be fuzzy subsets of a group G. Then $Z(\mu)Z(\nu) \subseteq Z(\mu \circ \nu)$ and $N(\mu) \cap N(\nu) \subseteq N(\mu \circ \nu)$.*

Proof. Let $x_1 \in Z(\mu)$ and $x_2 \in Z(\nu)$. Then for all $y, z \in G$,

$$
\begin{aligned}
(\mu \circ \nu)((x_1x_2)yz) &= \vee\{\mu(a) \wedge \nu(b) \mid x_1x_2yz = ab, a, b \in G\} \\
&= \vee\{\mu(x_1x_2yzb^{-1}) \wedge \nu(b) \mid b \in G\} \\
&= \vee\{\mu(x_2yx_1zb^{-1}) \wedge \nu(b) \mid b \in G\} \\
&= \vee\{\mu(a) \wedge \nu(b) \mid x_2yx_1z = ab, a, b \in G\} \\
&= \vee\{\mu(a) \wedge \nu(a^{-1}x_2yx_1z) \mid a \in G\} \\
&= \vee\{\mu(a) \wedge \nu(a^{-1}yx_1x_2z) \mid a \in G\} \\
&= \vee\{\mu(a) \wedge \nu(b) \mid yx_1x_2z = ab, a, b \in G\} \\
&= (\mu \circ \nu)(y(x_1x_2)z)
\end{aligned}
$$

by Lemma 3.1.3. Similarly, $(\mu \circ \nu)((x_1x_2)y) = (\mu \circ \nu)(y(x_1x_2))$. Hence $x_1x_2 \in Z(\mu \circ \nu)$. Thus $Z(\mu)Z(\nu) \subseteq Z(\mu \circ \nu)$.

Let $x \in N(\mu) \cap N(\nu)$. Then for all $y \in G$,

$$
\begin{aligned}
(\mu \circ \nu)(y) &= \vee\{\mu(a) \wedge \nu(b) \mid y = ab, a, b \in G\} \\
&= \vee\{\mu(x^{-1}ax) \wedge \nu(x^{-1}bx) \mid y = ab, a, b \in G\} \\
&\leq \vee\{\mu(c) \wedge \nu(d) \mid x^{-1}yx = cd, c, d \in G\} \\
&= (\mu \circ \nu)(x^{-1}yx)
\end{aligned}
$$

by Lemma 3.1.13, where the inequality holds since $y = ab \Rightarrow x^{-1}abx = cd \Leftrightarrow ab = xcdx^{-1} = (xcx^{-1})(xdx^{-1})$ and since $a = xcx^{-1}, b = xdx^{-1}$ implies $x^{-1}ax = c, x^{-1}bx = d$. Hence $(\mu \circ \nu)(x^{-1}yx) \leq (\mu \circ \nu)(x(x^{-1}yx)x^{-1}) = (\mu \circ \nu)(y)$ since $x^{-1} \in N(\mu) \cap N(\nu)$ by Proposition 3.1.8. Thus $(\mu \circ \nu)(x^{-1}yx) = (\mu \circ \nu)(y)$ for all $y \in G$. Therefore, $x \in N(\mu \circ \nu)$ by Lemma 3.1.13. Hence $N(\mu) \cap N(\nu) \subseteq N(\mu \circ \nu)$. $\square$

Corollary 3.1.15. *Let μ and ν be fuzzy subsets of a group G.*
(1) *If either μ or ν is commutative in G, then $\mu \circ \nu$ is commutative in G.*
(2) *If μ and ν are normal in G, then $\mu \circ \nu$ is normal in G.*

Proof. (1) Suppose μ is commutative in G. Then $G = GZ(\nu) = Z(\mu)Z(\nu) \subseteq Z(\mu) \circ Z(\nu) \subseteq G$.

(2) Suppose μ is normal in G. Then $G = GN(\nu) = N(\mu)N(\nu) \subseteq N(\mu) \circ N(\nu) \subseteq G$. $\qquad\square$

In Proposition 3.1.14, it is not the case that $N(\mu) \cap N(\nu) = N(\mu \circ \nu)$, in general, and the converse of Corollary 3.1.15(1) and (2) do not hold, even if μ and ν are fuzzy subgroups, as the next example demonstrates.

Example 3.1.16. *Let μ and ν be the fuzzy subgroups of the dihedral group $D_3 = \langle a, b \mid a^3 = e = b^2, ba = a^2 b \rangle$ defined as follows: $\forall x \in D_3$,*

$$\mu(x) = \begin{cases} t_0 & \text{if } x \in \langle a \rangle \\ t_1 & \text{otherwise} \end{cases} \quad \text{and} \quad \nu(x) = \begin{cases} t_0 & \text{if } x \in \langle ab \rangle \\ t_1 & \text{otherwise,} \end{cases}$$

where $t_0 > t_1$. Since $\langle a \rangle$ is a normal subgroup of D_3, the level sets of μ are normal subgroups of D_3. Thus $N(\mu) = D_3 = N(\mu \circ \nu)$. (Note $\mu \circ \nu$ is a constant on D_3.) We next show that $N(\nu) = \langle ab \rangle$. It follows that $\langle ab \rangle = \{e, ab\}$. Let $y \in D_3$. Then $aby = e \Rightarrow y = ab \Rightarrow yab = e \in \langle ab \rangle$. Also, $aby = ab \Rightarrow y = e \Rightarrow yab = ab \in \langle ab \rangle$. Now $aby \notin \langle ab \rangle \Leftrightarrow y \notin \langle ab \rangle \Leftrightarrow yab \notin \langle ab \rangle$. Hence $\nu(yab) = \nu(aby) \forall y \in D_3$. Thus it follows that $\langle ab \rangle \subseteq N(\nu)$. Now $\nu(ab) \neq \nu(a^2 b) = \nu(ba)$. Hence $a, b \notin N(\nu)$. Also $\nu((a^2 b)a^2) = \nu(baa^2) = \nu(b) \neq \nu(ab) = \nu(a^2(a^2 b))$. Thus $a^2, a^2 b \notin N(\nu)$. Therefore, $N(\nu) = \langle ab \rangle$. Hence μ and $\mu \circ \nu$ are normal in D_3. However, ν is not normal in D_3 and $N(\mu) \cap N(\nu) \neq N(\mu \circ \nu)$.

Example 3.1.17. *Let G and μ be as defined in Example 3.1.2. Let ν be the fuzzy subset of G defined as follows: $\forall x \in G$,*

$$\nu(x) = \begin{cases} t_0 & \text{if } x \in \langle a \rangle \otimes \{e\} \\ t_1 & \text{if } x \in (\{e\} \otimes \langle a \rangle) \backslash (\langle a \rangle \otimes \{e\}) \\ t_2 & \text{otherwise,} \end{cases}$$

where $t_0 > t_1 > t_2$. Since $\mu \circ \nu(a^i, a^j) = \mu \circ \nu((e, a^j)(a^i, e))$, it follows that $\mu \circ \nu(x, y) = t_0 \forall (x, y) \in \langle a \rangle \otimes \langle a \rangle$ and that $\mu \circ \nu(x, y) = t_2 \forall (x, y) \in D_3 \otimes D_3 \backslash (\langle a \rangle \otimes \langle a \rangle)$. Since $\langle a \rangle$ is a normal subgroup of D_3, it follows that $\langle a \rangle \otimes \langle a \rangle$ is a normal subgroup of $D_3 \otimes D_3 = G$. Thus the level sets of $\mu \circ \nu$ are normal subgroups of G. Hence $\mu \circ \nu$ is a normal fuzzy subgroup of G. Thus $N(\mu \circ \nu) = G$. Now for all $x, y, z \in D_3, xyz$ and yxz have the same number of b's (modulo 2) in their expressions. Therefore, it follows that $Z(\mu \circ \nu) = G$.

Proposition 3.1.18. *Let μ and ν be fuzzy subgroups of a group G such that $\mu \subseteq \nu$ and $\mu(e) = \nu(e)$. Then $Z(\mu) \subseteq Z(\nu)$. In particular, if ν is commutative in G, then μ is commutative in G.*

Proof. Application of Lemma 3.1.6 yields the desired result. $\qquad\square$

We next introduce the notions of the ascending central series of a fuzzy subgroup and a nilpotent fuzzy subgroup of a group. These are generalizations of the notions of the ascending central series of a group and a nilpotent group, respectively. Let μ be a fuzzy subgroup of a group G. Let $Z^0(\mu) = \{e\}$ and π_0 be the natural homomorphism of G onto $G/Z^0(\mu)$. Suppose that $Z^i(\mu)$ has been defined and that $Z^i(\mu)$ is a normal subgroup of G for $i \in \mathbb{N} \cup \{0\}$. Let π_i be the natural homomorphism of G onto $G/Z^i(\mu)$. Define $Z^{i+1}(\mu) = \pi_i^{-1}(Z(\pi_i(\mu)))$. Then $Z^{i+1}(\mu) \supseteq \mathrm{Ker}(\pi_i) = Z^i(\mu)$ for $i = 0, 1, ...$ The normality of $Z^{i+1}(\mu)$ in G follows by the correspondence theorem and Proposition 3.1.8.

Definition 3.1.19. *Let μ be a fuzzy subgroup of a group G. The **ascending central series** of μ is defined to be the ascending chain of normal subgroups of G,*

$$Z^0(\mu) \subseteq Z^1(\mu) \subseteq \text{....}$$

Definition 3.1.20. *A fuzzy subgroup μ of a group G is called **nilpotent** if there exists a nonnegative integer m such that $Z^m(\mu) = G$. The smallest such integer m is called the **class** of μ.*

Lemma 3.1.21. *Let μ be a fuzzy subgroup of a group G. Let $i \in \mathbb{N}$. If $xyx^{-1}y^{-1} \in Z^{i-1}(\mu)$ for all $y \in G$, then $x \in Z^i(\mu)$.*

Proof. Suppose that $xyx^{-1}y^{-1} \in Z^{i-1}(\mu)$ $(= \mathrm{Ker}(\pi_{i-1}))$ for all $y \in G$. Then clearly $e \in (\pi_{i-1})^{-1}(\pi_{i-1}(xyx^{-1}y^{-1}))$. Hence $(\pi_{i-1}(\mu))(\pi_{i-1}(xyx^{-1}y^{-1})) = \mu(e) = (\pi_{i-1}(\mu))(e)$ for all $y \in G$. Thus $\pi_{i-1}(x) \in Z(\pi_{i-1}(\mu))$ by Lemma 3.1.6 3.1.6. Hence $x \in Z^i(\mu)$. $\square$

We denote the ascending central series of a group G by

$$\{e\} = Z^0(G) \subseteq Z^1(G) \subseteq \text{....}$$

Proposition 3.1.22. *Let μ be a fuzzy subgroup of a group G. Then $Z^i(G) \subseteq Z^i(\mu)$ for all nonnegative integers i.*

Proof. We prove the result by induction on i. If $i = 0$ or 1, then the result is immediate. Assume $Z^{i-1}(G) \subseteq Z^{i-1}(\mu)$ for $i > 1$, the induction hypothesis. Let $x \in Z^i(G)$. Then $xyx^{-1}y^{-1} \in Z^{i-1}(G)$ for all $y \in G$. Hence $xyx^{-1}y^{-1} \in Z^{i-1}(\mu)$ for all $y \in G$ by the induction hypothesis. Thus $x \in Z^i(\mu)$ by Lemma 3.1.21. Hence the desired result holds. $\square$

Corollary 3.1.23. *Let μ be a fuzzy subgroup of a group G. If G is nilpotent of class m, then μ is nilpotent of class n for some nonnegative integer $n \leq m$.*

The converse of Corollary 3.1.23 is not true in general, as can be seen by the following example.

Example 3.1.24. *Let μ be the fuzzy subgroup of the dihedral group*
$D_3 = \langle a, b \mid a^3 = e = b^2, ba = a^2 b \rangle$ *defined as follows:* $\forall x \in D_3$,

$$\mu(x) = \begin{cases} t_0 & \text{if } x \in \langle a \rangle \\ t_1 & \text{otherwise,} \end{cases}$$

where $t_0 > t_1$. *Now* D_3 *is not nilpotent. We show that* μ *is nilpotent. It follows easily that*

$(D_3 \backslash \langle a \rangle)(D_3 \backslash \langle a \rangle) = \langle a \rangle$,
$\langle a \rangle (D_3 \backslash \langle a \rangle) = D_3 \backslash \langle a \rangle$,
$(D_3 \backslash \langle a \rangle) \langle a \rangle = D_3 \backslash \langle a \rangle$,
$\langle a \rangle \langle a \rangle = \langle a \rangle$.
Hence $xy \in \langle a \rangle \Leftrightarrow yx \in \langle a \rangle \ \forall x, y \in D_3$. *Thus* $Z(\mu) = D_3$.

The following proposition and theorem show that the nilpotence of a group can be completely characterized by the nilpotence of the fuzzy subgroups of the group.

Proposition 3.1.25. *Let G be a group. Then there exists a nontrivial fuzzy subgroup μ of G such that $Z^i(G) = Z^i(\mu)$ for every nonnegative integer i.*

Proof. Let μ be the fuzzy subgroup of G defined as follows: $\forall x \in G$,

$$\mu(x) = \begin{cases} 1 & \text{if } x \in Z^0(G), \\ \frac{1}{i+1} & \text{if } x \in Z^i(G) \backslash Z^{i-1}(G) \text{ for } i = 1, 2, ..., \\ 0 & \text{otherwise.} \end{cases}$$

We now show by induction on i that $Z^i(\mu) \subseteq Z^i(G)$ for every nonnegative integer i. If $i = 0$, then the result is immediate. Let $i > 0$ and assume $Z^{i-1}(\mu) \subseteq Z^{i-1}(G)$, the induction hypothesis. Then since μ has the sup property,

$x \in Z^i(\mu)$
$\Rightarrow \pi_{i-1}(x) \in Z(\pi_{i-1}(\mu))$
$\Rightarrow (\pi_{i-1}(\mu))(\pi_{i-1}(xyx^{-1}y^{-1})) = (\pi_{i-1}(\mu))(\pi_{i-1}(e)) = \mu(e)$ for all $y \in G$
$\Rightarrow \vee \{\mu(xyx^{-1}y^{-1}k) \mid k \in Z^{i-1}(\mu)\} = \mu(e)$ for all $y \in G$
$\Rightarrow \mu(xyx^{-1}y^{-1}k) = \mu(e)$ for all $y \in G$ and for some $k \in Z^{i-1}(\mu)$
$\Rightarrow \mu(xyx^{-1}y^{-1}k) = \mu(e)$ for all $y \in G$ and for some $k \in Z^{i-1}(G)$
$\Rightarrow \mu(xyx^{-1}y^{-1}) = \mu(k^{-1}) = \mu(k)$ for all $y \in G$ and for some $k \in Z^{i-1}(G)$
$\Rightarrow \mu(xyx^{-1}y^{-1}) \geq \frac{1}{i}$ for all $y \in G$
$\Rightarrow xyx^{-1}y^{-1} \in Z^{i-1}(G)$ for all $y \in G$
$\Rightarrow x \in Z^i(G)$

by Lemma 3.1.6 and the induction hypothesis. Thus $Z^i(\mu) \subseteq Z^i(G)$. By Proposition 3.1.22, $Z^i(G) \subseteq Z^i(\mu)$. Hence μ is a desired fuzzy subgroup of G. $\square$

Theorem 3.1.26. *Let G be a group. Then G is nilpotent if and only if all fuzzy subgroups of G are nilpotent.*

Proof. The proof follows from Corollary 3.1.23 and Proposition 3.1.25. $\square$

We now consider homomorphic images and homomorphic pre-images of nilpotent fuzzy subgroups.

Proposition 3.1.27. *Let f be a homomorphism of a group G onto a group H. Let μ and ν be fuzzy subgroups of G and H, respectively.*

(1) If μ is nilpotent of class m, then $f(\mu)$ is nilpotent of class n for some nonnegative integer $n \leq m$.

(2) ν is nilpotent of class m if and only if $f^{-1}(\nu)$ is nilpotent of class m.

Proof. (1) We show by induction on i that $f(Z^i(\mu)) \subseteq Z^i(f(\mu))$ for every nonnegative integer i. If $i = 0$ or 1, then the result is clear by Proposition 3.1.9(1). Let $i > 1$ and assume $f(Z^{i-1}(\mu)) \subseteq Z^{i-1}(f(\mu))$, the induction hypothesis. Let $h \in f(Z^i(\mu))$. Then there exists $x \in Z^i(\mu)$ such that $f(x) = h$ and for all $y \in G$,

$$\vee\{\mu(xyx^{-1}y^{-1}k) \mid k \in Z^{i-1}(\mu)\} = (\pi_{i-1}(\mu))(\pi_{i-1}(xyx^{-1}y^{-1})) = \mu(e)$$

by Lemma 3.1.21, where π_{i-1} is the natural homomorphism of G onto $G/Z^{i-1}(\mu)$. For all $y \in G$,

$$
\begin{aligned}
&(\pi'_{i-1}(f(\mu)))(\pi'_{i-1}(f(x)f(y)f(x)^{-1}f(y)^{-1})) \\
&= \vee\{f(\mu)(f(x)f(y)f(x)^{-1}f(y)^{-1}f(k)) \mid f(k) \in Z^{i-1}(f(\mu))\} \\
&\geq \vee\{(f(\mu))(f(x)f(y)f(x)^{-1}f(y^{-1})f(k)) \mid f(k) \in f(Z^{i-1}(\mu))\} \\
&\geq \vee\{(f(\mu))(f(xyx^{-1}y^{-1}k)) \mid k \in Z^{i-1}(\mu)\} \\
&= \vee\{\vee\{\mu(xyx^{-1}y^{-1}kg) \mid g \in \ker(f)\} \mid k \in Z^{i-1}(\mu)\} \\
&\geq \vee\{\mu(xyx^{-1}y^{-1}k) \mid k \in Z^{i-1}(\mu)\}
\end{aligned}
$$

by the induction hypothesis, where π'_{i-1} is the natural homomorphism of H onto $H/Z^{i-1}(f(\mu))$. Thus $(\pi'_{i-1}(f(\mu)))(\pi'_{i-1}(f(x)f(y)f(x)^{-1}f(y)^{-1})) = \mu(e) = (f(\mu))(e)$ for all $y \in G$. Therefore, $h = f(x) \in Z^i(f(\mu))$ by Lemma 3.1.6. Hence $f(Z^i(\mu)) \subseteq Z^i(f(\mu))$.

(2) We now show by induction on i that $Z^i(f^{-1}(\nu)) = f^{-1}(Z^i(\nu))$ for every positive integer i. If $i = 1$, the result follows by Proposition 3.1.9(2). Let $i > 1$ and assume $Z^{i-1}(f^{-1}(\nu)) = f^{-1}(Z^{i-1}(\nu))$, the induction hypothesis. Then for all $y \in G$,

$$(\pi'_{i-1}(f^{-1}(\nu)))(\pi'_{i-1}(xyx^{-1}y^{-1}))$$
$$= \vee\{(f^{-1}(\nu))(xyx^{-1}y^{-1}k) \mid k \in Z^{i-1}(f^{-1}(\nu))\}$$
$$= \vee\{\nu(f(xyx^{-1}y^{-1}k)) \mid k \in Z^{i-1}(f^{-1}(\nu))\}$$
$$= \vee\{\nu(f(xyx^{-1}y^{-1})f(k)) \mid k \in f^{-1}(Z^{i-1}(\nu))\}$$
$$= \vee\{\nu(f(xyx^{-1}y^{-1})h) \mid h \in Z^{i-1}(\nu)\}$$
$$= (\pi_{i-1}(\nu))(\pi_{i-1}(f(xyx^{-1}y^{-1})))$$

by the induction hypothesis, where π_{i-1} (resp. π'_{i-1}) is the natural homomorphism from H (resp. G) onto $H/Z^{i-1}(\nu)$ (resp. $G/Z^{i-1}(f^{-1}(\nu)))$. Thus

$$x \in Z^i(f^{-1}(\nu))$$
$$\Leftrightarrow \pi'_{i-1}(x) \in Z(\pi'_{i-1}(f^{-1}(\nu)))$$
$$\Leftrightarrow (\pi'_{i-1}(f^{-1}(\nu)))(\pi'_{i-1}(xyx^{-1}y^{-1})) = \nu(e) \quad \text{for all } y \in G$$
$$\Leftrightarrow (\pi_{i-1}(\nu))(\pi_{i-1}(f(xyx^{-1}y^{-1}))) = \nu(e) \quad \text{for all } y \in G$$
$$\Leftrightarrow f(x) \in \pi_{i-1}^{-1}(Z(\pi_{i-1}(\nu)))$$
$$\Leftrightarrow f(x) \in Z^i(\nu)$$
$$\Leftrightarrow x \in f^{-1}(Z^i(\nu))$$

by Lemma 3.1.6. Hence $Z^i(f^{-1}(\nu)) = f^{-1}(Z^i(\nu))$. $\square$

Corollary 3.1.28. *Let μ be a fuzzy subgroup of a group G such that μ_* is normal in G. If G/μ_* is nilpotent of class m, then μ is nilpotent of class n for some nonnegative integer $n \le m$.*

Proof. Suppose G/μ_* is nilpotent of class m. Let f be the natural homomorphism of G onto G/μ_*. Then $f(\mu)$ is nilpotent of class n for some nonnegative integer $n \le m$ by Corollary 3.1.23. Now $f^{-1}(f(\mu))(x) = f(\mu)(f(x)) = \vee\{\mu(z) \mid f(z) = f(x), z \in G\}$. Now $\forall x, z \in G$, $f(z) = f(x) \Leftrightarrow f(z^{-1}x) = e \Leftrightarrow z^{-1}x \in \text{Ker } f = \mu^* \Leftrightarrow \mu(z^{-1}x) = \mu(e) \Rightarrow \mu(z) = \mu(x)$. Hence it follows that $f^{-1}(f(\mu)) = \mu$. Thus μ is nilpotent of class n by Proposition 3.1.27(2). $\square$

The converse of Proposition 3.1.27(1) does not hold in general, as the following example shows.

Example 3.1.29. *Let μ be the fuzzy subset of the dihedral group $D_3 = \langle a, b \mid a^3 = e = b^2, ba = a^2b \rangle$ defined as follows: $\forall x \in D_3$,*

$$\mu(x) = \begin{cases} t_0 & \text{if } x = e, \\ t_1 & \text{otherwise,} \end{cases}$$

where $t_0 > t_1$. Then μ is a fuzzy subgroup of D_3. Now $a \notin Z(\mu)$ since $\mu(a(ba)b) = \mu((ab)^2) = \mu(e) = t_0 \ne t_1 = \mu(a) = \mu(ba^2b) = \mu((ba)ab)$. Thus $Z(\mu) = \{e\}$ since the only normal subgroups of D_3 are $\{e\}, \langle a \rangle$, and D_3

and $Z(\mu)$ is normal in D_3 by Proposition 3.1.8. Hence $Z^i(\mu) = \{e\}$ for all nonnegative integers i. Thus μ is not nilpotent. Let f be the natural homomorphism of D_3 onto $D_3/\langle a \rangle$. Then $Z(f(\mu)) = D_3/\langle a \rangle$ since $D_3/\langle a \rangle$ is Abelian. Hence $f(\mu)$ is nilpotent.

We now give an analog of Proposition 3.1.18.

Proposition 3.1.30. *Let μ and ν be fuzzy subgroups of a group G such that $\mu \subseteq \nu$ and $\mu(e) = \nu(e)$. If μ is nilpotent of class m, then ν is nilpotent of class n for some nonnegative integer $n \leq m$.*

Proof. We show by induction on i that $Z^i(\mu) \subseteq Z^i(\nu)$ for all nonnegative integers i. If $i = 0$ or 1, then the result follows by Proposition 3.1.18 3.1.18. Let $i > 1$ and assume $Z^{i-1}(\mu) \subseteq Z^{i-1}(\nu)$, the induction hypothesis. Let $x \in Z^i(\mu)$. Then for all $y \in G$,

$$
\begin{aligned}
(\pi'_{i-1}(\nu))(\pi'_{i-1}(xyx^{-1}y^{-1})) &= \vee\{\nu(xyx^{-1}y^{-1}k) \mid k \in Z^{i-1}(\nu)\} \\
&\geq \vee\{\nu(xyx^{-1}y^{-1}k) \mid k \in Z^{i-1}(\mu) \\
&\geq \vee\{\mu(xyx^{-1}y^{-1}k) \mid k \in Z^{i-1}(\mu)\} \\
&= (\pi_{i-1}(\mu))(\pi_{i-1}(xyx^{-1}y^{-1}) \\
&= \mu(e)
\end{aligned}
$$

by Lemma 3.1.6 and the induction hypothesis. Hence $x \in Z^i(\nu)$ by Lemma 3.1.6. Thus $Z^i(\mu) \subseteq Z^i(\nu)$.   □

3.2 Nilpotent Fuzzy Subgroups

The notion of a solvable fuzzy subgroup was introduced in [15]. In the previous section, the notion of a nilpotent fuzzy subgroup was introduced. In the previous section, an ascending series of subgroups of the underlying group was attached to a fuzzy subgroup to define nilpotency of the fuzzy subgroup. We define the commutator of a pair of fuzzy subsets of a group and use this technique to generate the descending central chain of fuzzy subgroups of a given fuzzy subgroup. We then propose a definition of a nilpotent fuzzy subgroup through its descending central chain. It is known that every Abelian group is nilpotent. If we accept the definition of an Abelian fuzzy group as proposed in [14], where a fuzzy group is called Abelian if its support is Abelian, then one can verify, as is done in Theorem 3.2.25, that Abelian fuzzy subgroups are nilpotent.

The power set $\mathcal{P}(S)$ of S can be embedded in $\mathcal{FP}(S)$ as a sublattice under the identification map $A \to 1_A$, where 1_A is the characteristic function of the subset A of S. For arbitrary λ in $\mathcal{FP}(G)$, recall that the least fuzzy subgroup of G containing λ is denoted by $\langle \lambda \rangle$. Then for all $x \in G$, $\langle \lambda \rangle(x) =$

$\vee\{\sigma(x_1) \wedge \sigma(x_2) \wedge \ldots \wedge \sigma(x_n) \mid x = x_1 x_2., ..x_n, x_i \in G, i = 1, 2, ...n; n \in \mathbb{N}\}$,
where $\sigma(y) = \lambda(y) \vee \lambda(y^{-1})$, $y \in G$.

The set $\mathcal{F}(G)$ of all fuzzy subgroups of G is a complete lattice in which the meet and the join of λ and μ are, respectively, $\lambda \cap \mu$ and $\langle \lambda \cup \mu \rangle$. The lattice $\mathcal{F}(G)$ contains the set of all subgroups of G as a sublattice.

For fuzzy subsets λ and μ of a group G, recall that the fuzzy subset $\lambda \circ \mu$ of G is defined by $\lambda \circ \mu(x) = \vee\{\lambda(a) \wedge \mu(b) \mid x = ab, a, b \in G\}$ for all $x \in G$.

A nonempty collection $\mathcal{D}$ of endomorphisms of a group G is called an **operator domain** on G. A fuzzy subset λ of G is called **admissible under** $\mathcal{D}$ or simply $\mathcal{D}$ **admissible** if for every $f \in \mathcal{D}, f(\lambda) \subseteq \lambda$. It follows that $f(\lambda) \subseteq \lambda$ if and only if $\lambda \subseteq f^{-1}(\lambda)$.

A fuzzy subgroup λ of G is **normal (characteristic, fully invariant)** if λ is admissible under all inner automorphisms (automorphisms, endomorphisms) of G.

The sup-min product of two $\mathcal{D}$-admissible fuzzy subsets is $\mathcal{D}$-admissible. Moreover, if λ is a $\mathcal{D}$-admissible fuzzy subset of G, then so is $\langle \lambda \rangle$; see [1]. Recall that if λ is a normal fuzzy subgroup and μ any fuzzy subgroup of G, then $\lambda \circ \mu$ is a fuzzy subgroup of G.

In a group G, the **commutator** of two elements a and b of G is the element $[a, b] = a^{-1}b^{-1}ab$ of G. If $A, B \subseteq G$, then the **commutator subgroup** of A and B is the subgroup $[A, B]$ of G generated by $\{[a, b] \mid a \in A, b \in B\}$. For all $x, y \in G$, let x^y denote $y^{-1}xy$. Then the following properties hold: $[a, b]^{-1} = [b, a], [ab, c] = [a, c]^b[b, c]$, and $f([a, b]) = [f(a), f(b)]$, where $a, b, c \in G$ and f is an endomorphism of G.

Let G be a group with identity e and let H be a subgroup of G. Let $Z_n(H) = [Z_{n-1}(H), H] \forall n \in N$, where $Z_0(H) = H$. Then $Z_n(H)$ is a subgroup of G $\forall n \in \mathbb{N}$. The chain $H = Z_0(H) \supseteq Z_1(H) \supseteq ... \supseteq Z_n(H) \supseteq ...$ is called the **descending central chain** of the subgroup H. The subgroup H is called **nilpotent** if $Z_n(H) = \{e\}$ for some $n \geq 0$. In fact, H is called **nilpotent of class** c if c is the smallest nonnegative integer such that $Z_c(H) = \{e\}$. The group G is said to be **nilpotent** if it is nilpotent as a subgroup of itself.

Definition 3.2.1. *Let λ and μ be fuzzy subsets of G. Let (λ, μ) be the fuzzy subset of G defined as follows: $\forall x \in G$,*
$$(\lambda, \mu)(x) = \begin{cases} \vee\{\lambda(a) \wedge \mu(b) \mid x = [a, b], a, b \in G\} & \text{if } x \text{ is a commutator of } G \\ 0 & \text{otherwise.} \end{cases}$$

*The **commutator** of λ and μ is the fuzzy subgroup $[\lambda, \mu]$ of G generated by (λ, μ).*

The following theorem shows that the notion of a commutator fuzzy subgroup contains the corresponding notion for crisp subgroups.

Theorem 3.2.2. *Let A and B be subsets of G. Then $[1_A, 1_B] = 1_{[A,B]}$.*

Proof. We first show that for any subset T of G, $\langle 1_T \rangle = 1_{\langle T \rangle}$, where $\langle T \rangle$ is the subgroup of G generated by T.

Let $\sigma(x) = 1_T(x) \vee 1_T(x^{-1}) \forall x \in G$. Then

$\langle 1_T \rangle(x) = \vee\{\sigma(a_1) \wedge \ldots \wedge \sigma(a_n) \mid x = a_1 \ldots a_n, a_i \in G, i = 1, \ldots, n; n \in \mathbb{N}\}, x \in G$.

Suppose $x \in \langle T \rangle$. Then there exists $a_1, \ldots, a_n \in G$ such that $x = a_1 \ldots a_n$ and for each i, either $a_i \in T$ or $a_i^{-1} \in T$. We then get $\sigma(a_i) = 1$ for each i and

$1 \geq \langle 1_T \rangle(x) \geq \{\sigma(a_1) \wedge \ldots \wedge \sigma(a_n) = 1.$

This yields $\langle 1_T \rangle(x) = 1 = 1_{\langle T \rangle}(x)$. Suppose $x \notin \langle T \rangle$. We show that $\langle 1_T \rangle(x) = 0$. If $\langle 1_T \rangle(x) > 0$, then there is a decomposition $x = a_1 \ldots a_n$ of x in G such that $0 < \sigma(a_1) \wedge \ldots \wedge \sigma(a_n)$. Thus for all i, $\sigma(a_i) = 1_T(a_i) \vee 1_T(a_i^{-1}) = 1$. Hence either $a_i \in T$ or $a_i^{-1} \in T$ for all i. Consequently, $x \in \langle T \rangle$,a contradiction. Hence $\langle 1_T \rangle(x) = 0 = 1_{\langle T \rangle}(x)$.

Now let A, B be any two subsets of G. Clearly, for all $x \in G$, $(1_A, 1_B)(x) = 1$ if $x \in \{[a, b] \mid a \in A, b \in B\}$ and 0 otherwise.

Thus $(1_A, 1_B) = 1_{\{[a,b] \mid a \in A, b \in B\}}$. By the above argument, we have that $[1_A, 1_B] = 1_{[A,B]}$.   □

The following theorem is concerned with the support and the tip of the commutator fuzzy subgroup of a pair of fuzzy subsets of G. Recall that μ^* denotes the support of a fuzzy subset μ of G.

Theorem 3.2.3. *Let λ and μ be fuzzy subsets of G. Then the following assertions hold.*

(1) $[\lambda, \mu]^* = [\lambda^*, \mu^*]$.

(2) *If r, s are the tips of λ, μ, respectively, then $t = r \wedge s$ is the tip of $[\lambda, \mu]$.*

Proof. (1) Let $x \in (\lambda, \mu)^*$. Then $(\lambda, \mu)(x) > 0$ and so there exist $a, b \in G$ such that $x = [a, b]$ and $\lambda(a) \wedge \mu(b) > 0$. Consequently, $a \in \lambda^*, b \in \mu^*$ and $x \in \{[l, m] \mid l \in \lambda^*, m \in \mu^*\}$. Thus $\text{supp}((\lambda, \mu)) \subseteq \{[l, m] \mid l \in \lambda^*, m \in \mu^*\}$. Suppose $x \in \{[l, m] \mid l \in \lambda^*, m \in \mu^*\}$. Then there exists $l \in \lambda^*, m \in \mu^*$ such that $x = [l, m]$ and $0 < \lambda(l) \wedge \mu(m) \leq (\lambda, \mu)([l, m])$. Hence $\{[l, m] \mid l \in \lambda^*, m \in \mu^*\} \subseteq (\lambda, \mu)^*$ and so $\{[l, m] \mid l \in \lambda^*, m \in \mu^*\} = (\lambda, \mu)^* \subseteq [\lambda, \mu]^*$. Since $[\lambda, \mu]^*$ is a subgroup of G and $[\lambda^*, \mu^*]$ is the smallest subgroup of G containing $\{[l, m] \mid l \in \lambda^*, m \in \mu^*\}$, we have $[\lambda^*, \mu^*] \subseteq [\lambda, \mu]^*$. On the other hand, $(\lambda, \mu)^* = \{[l, m] \mid l \in \lambda^*, m \in \mu^*\} \subseteq [\lambda^*, \mu^*]$ implies $(\lambda, \mu) \subseteq 1_{[\lambda^*, \mu^*]}$. Since $1_{[\lambda^*, \mu^*]}$ is a fuzzy subgroup of G and $[\lambda, \mu]$ is the smallest fuzzy subgroup of G containing (λ, μ), we have $[\lambda, \mu] \subseteq 1_{[\lambda^*, \mu^*]}$. Thus $[\lambda, \mu]^* \subseteq [\lambda^*, \mu^*]$. Hence $[\lambda, \mu]^* = [\lambda^*, \mu^*]$.

(2) Let u be the tip of $[\lambda, \mu]$. Let $x \in G$. If x is not a commutator in G, then $[\lambda, \mu](x) = 0 \leq t$. Suppose x is a commutator in G. Then $[\lambda, \mu](x) = \vee\{\lambda(a) \wedge \mu(b) \mid x = [a, b], a, b \in G\} \leq r \wedge s = t$. Thus $u \leq t$. Suppose $u < t$. Then $u < r$ and $u < s$ and so there exist $a, b \in G$ such that $u < \lambda(a), u < \mu(b)$. Hence $u < \lambda(a) \wedge \mu(b) \leq [\lambda, \mu]([a, b])$. However, this contradicts the fact that u is the tip of $[\lambda, \mu]$. Thus $u = t$.   □

Theorem 3.2.4. *Let* λ, μ *be fuzzy subsets of* G. *Then* $[\lambda, \mu] = [\mu, \lambda]$.

Proof. Let $x \in G$. We first show that
$$(\lambda, \mu)(x) = (\mu, \lambda)(x^{-1}).$$
If x is not a commutator in G, then x^{-1} is not a commutator and so
$(\lambda, \mu)(x) = 0 = (\mu, \lambda)(x^{-1})$. Suppose $x = [a, b]$ for some $a, b \in G$. Then
$$(\lambda, \mu)(x) = \vee\{\lambda(a) \wedge \mu(b) \mid x = [a, b], a, b \in G\}$$

$$= \vee\{\mu(b) \wedge \lambda(a) \mid x^{-1} = [a, b], a, b \in G\}$$
$$= (\mu, \lambda)(x^{-1}).$$
If $\sigma(x) = (\lambda, \mu)(x) \vee (\lambda, \mu)(x^{-1})$ and $\delta(x) = (\mu, \lambda)(x) \vee (\mu, \lambda)(x^{-1})$, then
$\sigma = \delta$ and hence $[\lambda, \mu] = \langle\sigma\rangle = \langle\delta\rangle = [\mu, \lambda]$.   □

The following two theorems are concerned with the admissibility of commutator fuzzy subgroups under operator domains on G.

Theorem 3.2.5. *If* λ *and* μ *are admissible fuzzy subsets of* G *under an operator domain* $\mathcal{D}$, *then* (λ, μ) *and* $[\lambda, \mu]$ *are* $\mathcal{D}$-*admissible.*

Proof. Since λ and μ are $\mathcal{D}$-admissible, $\lambda \subseteq f^{-1}(\lambda), \mu \subseteq f^{-1}(\mu)$ for all $f \in \mathcal{D}$.
Let $f \in \mathcal{D}$ and $x \in G$. If x is not a commutator in G, then
$$(\lambda, \mu)(x) = 0 \leq (\lambda, \mu)(f(x)).$$
Suppose $x = [a, b]$ for some $a, b \in G$. Then
$$(\lambda, \mu)(x) = \vee\{\lambda(a) \wedge \mu(b) \mid x = [a, b], a, b \in G\}$$
$$\leq \vee\{\lambda(f(a)) \wedge \mu(f(b)) \mid x = [a, b], a, b \in G\}$$
$$= \vee\{\lambda(f(a)) \wedge \mu(f(b)) \mid f(x) = [f(a), f(b)], a, b \in G\}$$
$$\leq \vee\{\lambda(c) \wedge \mu(d) \mid f(x) = [c, d], c, d \in G\}$$
$$= (\lambda, \mu)(f(x)).$$
Thus $(\lambda, \mu) \subseteq f^{-1}((\lambda, \mu))$. Hence (λ, μ) is $\mathcal{D}$-admissible. Thus $(\lambda, \mu) \subseteq f^{-1}((\lambda, \mu))$ and since $[\lambda, \mu]$ is generated by (λ, μ), we have that $(\lambda, \mu) \subseteq f^{-1}([\lambda, \mu])$. Hence $[\lambda, \mu] \subseteq f^{-1}([\lambda, \mu])$. Therefore, $[\lambda, \mu]$ is $\mathcal{D}$-admissible.   □

Theorem 3.2.6. *If* λ *and* μ *are normal (characteristic, fully invariant) fuzzy subgroups of* G, *then* $[\lambda, \mu]$ *is a normal (characteristic, fully invariant) fuzzy subgroup of* G *contained in* $\lambda \cap \mu$.

Proof. If λ and μ are normal (characteristic, fully invariant) fuzzy subgroups of G, it follows from Theorem 3.2.5 that $[\lambda, \mu]$ is a normal (characteristic, fully invariant) fuzzy subgroup of G. In each of the given cases, λ and μ are normal and thus they assume constant values on the conjugacy classes of G.

Since $\lambda \cap \mu$ is a fuzzy subgroup, it suffices to prove that $(\lambda, \mu) \subseteq \lambda \cap \mu$.
Let $x \in G$. If x is not a commutator in G, then $(\lambda, \mu)(x) = 0 \leq (\lambda \cap \mu)(x)$.
Suppose $x = [a, b]$ for some $a, b \in G$. Then $\lambda(x) = \lambda(a^{-1}(b^{-1}ab)) \geq \lambda(a^{-1}) \wedge \lambda(b^{-1}ab) = \lambda(a^{-1}) \wedge \lambda(a) = \lambda(a)$.

Similarly, $\mu(x) \geq \mu(b)$. Therefore, $\lambda(a) \wedge \mu(b) \leq \lambda(x) \wedge \mu(x) = (\lambda \cap \mu)(x)$.
Hence $(\lambda, \mu)(x) \leq (\lambda \cap \mu)(x)$.   □

Lemma 3.2.7. *If λ and μ are fuzzy subsets of G such that $\lambda \subseteq \mu$, then $[\lambda, \sigma] \subseteq [\mu, \sigma]$ for every fuzzy subset σ of G.*

Proof. Since $\lambda \subseteq \mu$, $\lambda(x) \leq \mu(x)$ for all $x \in G$. Let σ be any fuzzy subset of G and let $x \in G$. If x is not a commutator in G, then $(\lambda, \sigma)(x) = 0 = (\mu, \sigma)(x)$. Suppose x is a commutator in G. Then

$$(\lambda, \sigma)(x) = \vee\{\lambda(a) \wedge \sigma(b) \mid x = [a, b], a, b \in G\}$$
$$\leq \vee\{\mu(a) \wedge \sigma(b) \mid x = [a, b], a, b \in G\}$$
$$\subseteq (\mu, \sigma)(x).$$

Therefore, $(\lambda, \sigma) \subseteq (\mu, \sigma)$ and so $[\lambda, \sigma] \subseteq [\mu, \sigma]$. $\qquad\square$

Theorem 3.2.8. *Let λ and μ be normal fuzzy subgroups of G and let σ be any fuzzy subgroup of G. Then $[\lambda \circ \sigma, \mu] \subseteq [\lambda, \mu] \circ [\sigma, \mu]$ with equality holding if $\lambda(e) = \sigma(e)$.*

Proof. First we show that
$$(\lambda \circ \sigma, \mu) \subseteq [\lambda, \mu] \circ [\sigma, \mu].$$
Let $x \in G$. If x is not a commutator in G, then $(\lambda \circ \sigma, \mu)(x) = 0$ and the result is immediate. Suppose x is a commutator in G. Then

$(\lambda \circ \sigma, \mu)(x)$
$= \vee\{(\lambda \circ \sigma)(a) \wedge \mu(b) \mid x = [a, b] \in G\}$
$= \vee\{\vee\{\lambda(u) \wedge \sigma(v) \mid a = uv, u, v \in G\} \wedge \mu(b) \mid x = [a, b], a, b \in G\}$
$= \vee\{\vee\{(\lambda(u) \wedge \sigma(v)) \wedge \mu(b) \mid a = uv, u, v \in G\} \mid x = [a, b], a, b \in G\}$
$= \vee\{\vee\{(\lambda(u) \wedge \mu(b)) \wedge (\sigma(v) \wedge \mu(b)) \mid a = uv, u, v \in G\} \mid x = [a, b], a, b \in G\}$
$= \vee\{(\lambda(u) \wedge \mu(b)) \wedge (\sigma(v) \wedge \mu(b)) \mid x = [uv, b], u, v \in G\}$
$\leq \vee\{[\lambda, \mu]([u, b]) \wedge [\sigma, \mu]([v, b]) \mid x = [uv, b], u, v \in G\}$
$= \vee\{[\lambda, \mu]([u, b]^v) \wedge [\sigma, \mu]([v, b]) \mid x = [uv, b], u, v \in G\}$
(since $[\lambda, \mu]$ is normal)
$\leq \vee\{[\lambda, \mu](y) \wedge [\sigma, \mu](z) \mid x = yz, y, z \in G\}$,
(since $[uv, b] = [u, b]^v[v, b]$ and so $y = [u, b]^v, z = [v, b] \Rightarrow yz = [uv, b]$)
$= ([\lambda, \mu] \circ [\sigma, \mu])(x)$.
Since $[\lambda, \mu]$ is normal in G, $[\lambda, \mu] \circ [\sigma, \mu]$ is a fuzzy subgroup of G. Therefore, $[\lambda \circ \sigma, \mu] \subseteq [\lambda, \mu] \circ [\sigma, \mu]$.

Finally, suppose $\lambda(e) = \sigma(e)$. Then $\lambda \subseteq \lambda \circ \sigma$ and $\sigma \subseteq \lambda \circ \sigma$. Thus $[\lambda, \mu] \subseteq [\lambda \circ \sigma, \mu]$ and $[\sigma, \mu] \subseteq [\lambda \circ \sigma, \mu]$ by Lemma 3.2.7. Hence $[\lambda, \mu] \circ [\sigma, \mu] \subseteq [\lambda \circ \sigma, \mu]$ since $[\lambda \circ \sigma, \mu]$ is a fuzzy subgroup of G. Consequently, the desired equality holds. $\qquad\square$

We now consider homomorphic images and preimages of commutators of fuzzy subsets of groups.

Lemma 3.2.9. *Let f be a homomorphism of a group G into a group K. For all fuzzy subsets λ of G, $f(\langle\lambda\rangle) = \langle f(\lambda)\rangle$.*

Proof. Since $\lambda \subseteq \langle \lambda \rangle$, $f(\lambda) \subseteq f(\langle \lambda \rangle)$. Since $f(\langle \lambda \rangle)$ is a fuzzy subgroup of K, $\langle f(\lambda) \rangle \subseteq f(\langle \lambda \rangle)$.

Suppose $y \in K$ and $y = f(x)$, $x \in G$. We show that $\langle \lambda \rangle(x) \leq f(\langle \lambda \rangle)(y)$. Define

$$\sigma(t) = \lambda(t) \vee \lambda(t^{-1}), \forall\, t \in G$$

and

$$\delta(s) = f(\lambda)(s) \vee f(\lambda)(s^{-1}), \forall\, s \in K.$$

If $x = a_1 \ldots a_n$, then since $\lambda(t) \leq f(\lambda)(f(t))$ for all $t \in G$, we have that

$$\sigma(a_1) \wedge \ldots \wedge \sigma(a_n)$$
$$= \wedge \{\lambda(a_i) \vee \lambda(a_i^{-1}) \mid 1 \leq i \leq n\}$$
$$\leq \wedge \{f(\lambda)(f(a_i)) \vee f(\lambda)(f(a_i^{-1}))\} \mid 1 \leq i \leq n\}$$
$$= \wedge \{\delta(f(a_i)) \mid 1 \leq i \leq n\}$$
$$\leq \langle f(\lambda) \rangle(f(a_1) \ldots f(a_n))$$
$$= \langle f(\lambda) \rangle(y).$$

Hence $f(\langle \lambda \rangle)(y) = \vee \{\langle \lambda \rangle(x) \mid y = f(x)\} \leq \langle f(\lambda) \rangle(y)$. If $y \in K$ is such that there is no $x \in G$ such that $f(x) = y$, then $f(\langle \lambda \rangle)(y) = 0$. Therefore, $f(\langle \lambda \rangle) \subseteq \langle f(\lambda) \rangle$. $\qquad\square$

Corollary 3.2.10. *Let f be a homomorphism of a group G into a group K. If λ and μ are fuzzy subsets of G, then $f([\lambda, \mu])$ is the fuzzy subgroup of K generated by $f((\lambda, \mu))$.*

Proof. Since $[\lambda, \mu]$ is the fuzzy subgroup of G generated by the fuzzy subset (λ, μ), the result immediately follows from Lemma 3.2.9. In particular, $f([\lambda, \mu]) = f(\langle (\lambda, \mu) \rangle) = \langle f((\lambda, \mu)) \rangle$. $\qquad\square$

Theorem 3.2.11. *Let f be a homomorphism of a group G into a group K. Then for all fuzzy subsets λ and μ of G, $[f(\lambda), f(\mu)] = f([\lambda, \mu])$ in $\mathcal{F}(K)$.*

Proof. We first show that $f((\lambda, \mu) \subseteq [f(\lambda), f(\mu)]$. Let $y \in K$. If $f^{-1}(y) = \emptyset$, then $f((\lambda, \mu))(y) = 0 \leq [f(\lambda), f(\mu)](y)$. Suppose $y = f(x)$ for some $x \in G$. Then

$$(\lambda, \mu)(x) = \vee\{\lambda(a) \wedge \mu(b) \mid x = [a, b], a, b \in G\}$$
$$\leq \vee\{f(\lambda)(f(a)) \wedge f(\mu)(f(b)) \mid y = [f(a), f(b)], a, b \in G\}$$
$$\leq \vee\{f(\lambda)(c) \wedge f(\mu)(d) \mid y = [c, d], c, d \in G\}$$
$$= (f(\lambda), f(\mu))(y) \leq [f(\lambda), f(\mu)](y).$$

Hence

$$f((\lambda, \mu))(y) = \vee\{(\lambda, \mu)(x) \mid y = f(x), x \in G\} \leq [f(\lambda), f(\mu)](y).$$

We thus have that $f((\lambda, \mu)) \subseteq [f(\lambda), f(\mu)]$. Hence by Corollary 3.2.10, $f([\lambda, \mu]) \subseteq [f(\lambda), f(\mu)]$.

We next show that $(f(\lambda), f(\mu)) \subseteq f([\lambda, \mu])$. Let $y \in K$. If y is not a commutator in K, then

$$(f(\lambda), f(\mu))(y) = 0 \leq f([\lambda, \mu])(y).$$

Suppose y is a commutator in K. Then $y = [u, v]$ for some $u, v \in K$. If either $f^{-1}(u) = \emptyset$ or $f^{-1}(v) = \emptyset$, then $f(\lambda)(u) \wedge f(\mu)(v) = 0$. Otherwise, we have

$$f(\lambda)(u) \wedge f(\mu)(v)$$
$$= (\vee\{\lambda(s) \mid u = f(s)\}) \wedge (\vee\{\mu(t) \mid v = f(t)\})$$
$$= \vee\{\lambda(s) \wedge \mu(t) \mid u = f(s), v = f(t)\}$$
$$\leq \vee\{(\lambda, \mu)([s, t]) \mid y = f([s, t])\}$$
$$\leq \vee\{[\lambda, \mu]([s, t]) \mid y = f([s, t])\}$$
$$\leq \vee\{[\lambda, \mu](x) \mid y = f(x)\}$$
$$= f([\lambda, \mu])(y).$$

Thus $(f(\lambda), f(\mu)) \subseteq f([\lambda, \mu])$. Hence $[f(\lambda), f(\mu)] \subseteq f([\lambda, \mu])$. $\square$

Theorem 3.2.12. *Let f be a homomorphism of a group G into a group H. For all fuzzy subsets ν and ρ of K, $[f^{-1}(\nu), f^{-1}(\rho)] \subseteq f^{-1}([\nu, \rho])$ in $\mathcal{F}(G)$.*

Proof. Let $x \in G$. If x is not a commutator in G, then $(f^{-1}(\nu), f^{-1}(\rho))(x) = 0 \leq f^{-1}([\nu, \rho])(x)$. Suppose now that x is a commutator in G. Then
$$(f^{-1}(\nu), f^{-1}(\rho))(x)$$
$$= \vee\{\nu(f(a)) \wedge \rho(f(b)) \mid x = [a, b], a, b \in G\}$$
$$= \vee\{\nu(f(a)) \wedge \rho(f(b)) \mid f(x) = [f(a), f(b)], a, b \in G\}$$
$$\leq \vee\{\nu(c) \wedge \rho(d) \mid f(x) = [c, d], c, d \in G\}$$
$$= (\nu, \rho)(f(x)) \leq [\nu, \rho](f(x))$$
$$= f^{-1}([\nu, \rho])(x).$$

Therefore, $(f^{-1}(\nu), f^{-1}(\rho)) \subseteq f^{-1}([\nu, \rho])$. Since $f^{-1}([\nu, \rho])$ is a fuzzy subgroup of G, we obtain $[f^{-1}(\nu), f^{-1}(\rho)] \subseteq f^{-1}([\nu, \rho])$. $\square$

We next define the notion of the descending central chain of a fuzzy subgroup. Let λ be a fuzzy subgroup of a group G. Define $Z_0(\lambda) = \lambda$ and $Z_1(\lambda) = [\lambda, \lambda]$. Suppose $Z_{n-1}(\lambda)$ has been defined for $n \in \mathbb{N}$. Define $Z_n(\lambda) = [Z_{n-1}(\lambda), \lambda]$, $n \in \mathbb{N}$.

Theorem 3.2.13. *If H is a subgroup of G, then $Z_n(1_H) = 1_{Z_n(H)}$ for all $n \in \mathbb{N} \cup \{0\}$.*

Proof. The result follows from Theorem 3.2.2. $\square$

Suppose λ is a normal fuzzy subgroup of G. Let $n \in \mathbb{N}, n > 1$. In view of Theorem 3.2.6, we have that $Z_n(\lambda) = [Z_{n-1}(\lambda), \lambda] \subseteq Z_{n-1}(\lambda) \cap \lambda$. Hence $Z_n(\lambda) \subseteq Z_{n-1}(\lambda)$. The following theorem shows that this inclusion holds without the assumption that λ is normal.

Theorem 3.2.14. *Let λ be a fuzzy subgroup of G. Then $Z_n(\lambda) \subseteq Z_{n-1}(\lambda)$ for all $n \in \mathbb{N}$.*

Proof. We prove the result by induction on n. Let $x \in G$. If x is not a commutator in G, then $(\lambda, \lambda)(x) = 0 \leq \lambda(x)$. Suppose x is a commutator in G. Then
$$(\lambda, \lambda)(x) = \vee\{\lambda(a) \wedge \lambda(b) \mid x = [a, b], a, b \in G\}$$
$$= \vee\{\lambda(a^{-1}) \wedge \lambda(b^{-1}) \wedge \lambda(a) \wedge \lambda(b) \mid x = [a, b], a, b \in G\}$$

$$\leq \vee\{\lambda(a^{-1}b^{-1}ab) \mid x = [a,b], a, b \in G\}$$
$$= \lambda(x).$$

Thus $(\lambda, \lambda) \subseteq \lambda$. Since λ is a fuzzy subgroup, we have that $Z_1(\lambda) = [\lambda, \lambda] = [(\lambda, \lambda)] \subseteq \lambda = Z_0(\lambda)$. Therefore, the result holds for $n = 1$. Now suppose that $Z_k(\lambda) \subseteq Z_{k-1}(\lambda)$ for some $k \geq 1$, the induction hypothesis. Then by Lemma 3.2.7, $Z_{k+1}(\lambda) = [Z_k(\lambda), \lambda] \subseteq [Z_{k-1}(\lambda), \lambda] = Z_k(\lambda)$. Hence the desired result follows by induction. □

Definition 3.2.15. *Let λ be a fuzzy subgroup of G. Then the chain*
$$\lambda = Z_0(\lambda) \supseteq Z_1(\lambda) \supseteq \ldots \supseteq Z_n(\lambda) \supseteq \ldots$$
*of fuzzy subgroups of G is called the **descending central** chain of λ.*

Theorem 3.2.16. *If λ is a normal (characteristic, fully invariant) fuzzy subgroup of G, then each every subgroup in the descending central chain of λ is normal (characteristic, fully invariant) in G.*

Proof. The result follows from Theorem 3.2.6. □

Let $\mu_1, \mu_2, \ldots, \mu_n$ be fuzzy subgroups of a group G. Define the fuzzy subgroup $[\mu_1, \mu_2, \ldots, \mu_n]$ of G recursively as follows:
$$[\mu_1, \mu_2, \ldots, \mu_n] = [[\mu_1, \mu_2, \ldots, \mu_{n-1}], \mu_n], n \in \mathbb{N} \text{ and } n \geq 3.$$
If the μ_i are normal fuzzy subgroups of G, then $[\mu_1, \mu_2, \ldots, \mu_n]$ is a normal fuzzy subgroup of G contained $\cap_{i=1}^{n} \mu_i$ by Theorem 3.2.6.

Theorem 3.2.17. *Let λ be a normal fuzzy subgroup of G. If $\mu_1, \ldots, \mu_{n+1}$ are normal fuzzy subgroups of G such that $\mu_i = \lambda$ for $k + 1 (k \geq 0)$ distinct values of i, then $[\mu_1, \ldots, \mu_{n+1}] \subseteq Z_k(\lambda)$.*

Proof. We prove the result by induction on n. For $n = 0$, there is nothing to prove. If λ, μ are normal fuzzy subgroups of G, then $[\lambda, \mu] = [\mu, \lambda] \subseteq \lambda = Z_0(\lambda)$ and $[\lambda, \lambda] = Z_1(\lambda)$. Thus the result is true for $n = 1$ and for all possible values of k. Assume the result is true for $n - 1$ and for all possible values of k, the induction hypothesis. Let $\mu_1, \ldots, \mu_{n+1}$ be normal fuzzy subgroups of G such that $\mu_i = \lambda$ for $k + 1$ $(k \geq 0)$ distinct values of i. Suppose $k = 0$. If $\mu_{n+1} = \lambda$, then since $[\mu_1, \ldots, \mu_n]$ is a normal fuzzy subgroup of G, we have that $[\mu_1, \ldots, \mu_{n+1}] = [[\mu_1, \ldots, \mu_n], \lambda] \subseteq \lambda = Z_0(\lambda)$. If $\mu_{n+1} \neq \lambda$, then by the induction hypothesis, $[\mu_1, \ldots, \mu_n] \subseteq Z_0(\lambda)$ and thus $[\mu_1, \ldots, \mu_{n+1}] \subseteq [Z_0(\lambda), \mu_{n+1}] \subseteq Z_0(\lambda)$. Suppose $k \geq 1$. If $\mu_{n+1} \neq \lambda$, then by the induction hypothesis, $[\mu_1, \ldots, \mu_n] \subseteq Z_k(\lambda)$. Therefore, $[\mu_1, \ldots, \mu_{n+1}] \subseteq Z_k(\lambda)$. If $\mu_{n+1} = \lambda$, then again by the induction hypothesis, $[\mu_1, \ldots, \mu_n] \subseteq Z_{k-1}(\lambda)$. Thus $[\mu_1, \ldots, \mu_{n+1}] \subseteq [Z_{k-1}(\lambda), \lambda] = Z_k(\lambda)$. Hence the result holds by induction on n. □

Definition 3.2.18. *Let λ be a fuzzy subgroup of G. A chain*
$$\lambda = \lambda_0 \supseteq \lambda_1 \supseteq \ldots \supseteq \lambda_n \supseteq \ldots$$
*of fuzzy subgroups of G is called a **central chain** of λ if for all $n \in \mathbb{N}, \lambda_n(a) \wedge \lambda(b) \leq \lambda_{n+1}([a,b])$ for all $a, b \in G$.*

In Definition 3.2.18, the condition is equivalent to saying that $[\lambda_n, \lambda] \subseteq \lambda_{n+1}$ for all $n \geq 0$. Let t denote the tip of λ. If $\lambda_n = e_t$ for some $n \in \mathbb{N}$, then the series of fuzzy subgroups $\lambda = \lambda_0 \supseteq \lambda_1 \supseteq \ldots \supseteq \lambda_n = e_t$ is called a **central series** of λ.

Definition 3.2.19. *Let λ be a fuzzy subgroup of G with tip t. If the descending central chain $\lambda = Z_0(\lambda) \supseteq Z_1(\lambda) \supseteq \ldots \supseteq Z_n(\lambda) \supseteq \ldots$ of λ is such that $Z_n(\lambda) = e_t$ for some $n \in \mathbb{N}$, then λ is called **nilpotent**. If λ is nilpotent, then λ is called **nilpotent of class** c if c is the smallest nonnegative integer such that $Z_c(\lambda) = e_t$. In this case, the series $\lambda = Z_0(\lambda) \supseteq Z_1(\lambda) \supseteq \ldots \supseteq Z_c(\lambda) = e_t$ is called the **descending central series** of λ.*

We show in Example 3.2.31 that the definition of nilpotence here is not equivalent to the one given in Section 3.1. This can be seen from Example 3.1.24. In the remainder of this section, a fuzzy subgroup of a group is nilpotent if it satisfies Definition 3.2.19. The concept of nilpotency for a fuzzy subgroup of a group is not considered further in this book.

Theorem 3.2.20. *Let H be a subgroup of G. Then H is nilpotent if and only if 1_H is nilpotent.*

Proof. Consider the descending central chain
$$H = Z_0(H) \supseteq Z_1(H) \supseteq \ldots \supseteq Z_n(H) \supseteq \ldots$$
of H. By Theorem 3.2.13, $Z_n(1_H) = 1_{Z_n(H)}$ for all $n \in \mathbb{N}$. Therefore, $Z_n(1_H) = e_1$ if and only if $Z_n(H) = \{e\}$. The desired result now follows.   □

Example 3.2.21. *We now give a nontrivial example of a nilpotent fuzzy subgroup of the group S_4, the symmetric group on $\{1, 2, 3, 4\}$. Let*
$$D_4 = \langle (24), (1234) \rangle$$
$$= \{(1), (12)(34), (13)(24), (14)(23),$$
$$(13), (24), (1234), (1432)\}.$$
Then D_4 is a dihedral subgroup of S_4 with center $C = \{(1), (13)(24)\}$. Let λ be the fuzzy subset of S_4 defined by $\lambda(x) = 1$ if $x \in C$, $\lambda(x) = \frac{1}{2}$ if $x \in \langle 1234 \rangle \backslash C$, $\lambda(x) = \frac{1}{4}$ if $x \in D_4 \backslash \langle 1234 \rangle$, and $\lambda(x) = 0$ if $x \in S_4 \backslash D_4$. Clearly, λ is a fuzzy subgroup of S_4. It follows that the fuzzy subgroup $Z_1(\lambda)$ is given as follows:
$$Z_1(\lambda)((1)) = 1, \ Z_1(\lambda)((13)(24)) = \frac{1}{4} \text{ and } Z_1(\lambda)(x) = 0 \text{ if } x \in S_4 \backslash C.$$
It follows easily that $Z_2(\lambda) = e_1$. Thus λ is a nilpotent fuzzy subgroup of S_4 of class 2.

Example 3.2.22. *We now give an example of a normal fuzzy subgroup μ of S_4 that is not nilpotent. Let A_4 denote the alternating subgroup of S_4.*

Let $N = \{(1), (12)(34), (13)(24), (14)(23)\}$. Then N is a normal subgroup of S_4 and N is Abelian. Let μ be the fuzzy subset of S_4 defined by $\mu(x) = \frac{1}{2}$ if $x \in N$, $\mu(x) = \frac{1}{4}$ if $x \in A_4 \backslash N$, and $\mu(x) = 0$ if $x \in S_4 \backslash A_4$. Clearly, μ is a normal fuzzy subgroup of S_4. However, for $n \in \mathbb{N}$, $Z_n(\mu)$ is given by $Z_n(\mu)((1)) = \frac{1}{2}$, $Z_n(\mu)(x) = \frac{1}{4}$ if $x \in N \backslash \{(1)\})$, and $Z_n(\mu)(x) = 0$ if $x \in S_4 \backslash N$. Thus μ is not nilpotent.

Theorem 3.2.23. *Let λ be a fuzzy subgroup of G. Then λ is nilpotent if and only if λ has a central series.*

Proof. Let $t = \lambda(e)$. Suppose λ has a central series $\lambda = \lambda_0 \supseteq \lambda_1 \supseteq \ldots \supseteq \lambda_n = e_t$. We show by induction on i that $Z_i(\lambda) \subseteq \lambda_i$, $i = 0, 1, \ldots, n$. We have $Z_0(\lambda) = \lambda = \lambda_0$ and $Z_1(\lambda) = [\lambda_0, \lambda] \subseteq \lambda_1$. Assume $Z_i(\lambda) \subseteq \lambda_i$ for some $i = 0, 1, \ldots, n - 1$, the induction hypothesis. Then by Lemma 3.2.7, $Z_{i+1}(\lambda) = [Z_i(\lambda), \lambda] \subseteq [\lambda_i, \lambda] \subseteq \lambda_{i+1}$. Thus $Z_i(\lambda) \subseteq \lambda_i$ for $i = 0, 1, \ldots, n$ by induction. Therefore, $e_t \subseteq Z_n(\lambda) \subseteq \lambda_n = e_t$. Hence λ is nilpotent of class at most n.

Conversely, if λ is nilpotent of class c, then $\lambda = Z_0(\lambda) \supset Z_1(\lambda) \supset \cdots \supset Z_c(\lambda) = e_t$ is a central series of λ. □

Theorem 3.2.24. *Every fuzzy subgroup of a nilpotent group is nilpotent. In fact, if the group G is not nilpotent, then G has a nontrivial fuzzy subgroup that is not nilpotent.*

Proof. Suppose G is nilpotent of class c. Let λ be a fuzzy subgroup of G with tip t. For $i = 0, 1, \ldots, c$, define the fuzzy subset λ_i of G as follows: $\forall x \in G$,
$$\lambda_i(x) = \begin{cases} \lambda(x) & \text{if } x \in Z_i(G), \\ 0 & \text{otherwise.} \end{cases}$$
Then λ_i is a fuzzy subgroup of G and $\lambda = \lambda_0 \supseteq \lambda_1 \supseteq \cdots \supseteq \lambda_c$ is a finite chain of fuzzy subgroups of G such that $(\lambda_i)^* \subseteq Z_i(G)$ for all i. Clearly, $\lambda_c = e_t$. Let $a, b \in G$ and $0 \leq i \leq n - 1$. If $a \notin Z_i(G)$, then $\lambda_i(a) \wedge \lambda(b) = 0 \leq \lambda_{i+1}([a, b])$. Suppose $a \in Z_i(G)$. Then $[a, b] \in Z_{i+1}(G)$. Hence
$$\lambda_i(a) \wedge \lambda(b) = \lambda(a) \wedge \lambda(b) \leq \lambda([a, b]) = \lambda_{i+1}([a, b]).$$
Thus we have a central chain $\lambda = \lambda_0 \supseteq \lambda_1 \supseteq \ldots \supseteq \lambda_c = e_t$ of λ and λ is nilpotent of class at most c by Theorem 3.2.23.

Now suppose that G is not nilpotent. Consider the descending central chain
$$G = Z_0(G) \supseteq Z_1(G) \supseteq \ldots \supseteq Z_n(G) \supseteq \ldots$$
of G. If the chain is nonterminating, then $Z_n(G) \subset Z_{n-1}(G)$ for all $n \geq 1$. Define the fuzzy subset μ of G in this case as follows:

$\mu(x) = 1 - \frac{1}{2^n}$ if $x \in Z_{n-1}(G) \backslash Z_n(G)$, $n \geq 1$,

$\mu(x) = 1$ if $x \in \cap_{n=1}^{\infty} Z_n(G)$.

Then μ is a nontrivial fuzzy subgroup of G which is not nilpotent.

If the descending central chain terminates, then there is a nonnegative integer m such that
$$G = Z_0(G) \supset Z_1(G) \supset \ldots \supset Z_m(G) \neq \{e\},$$
and $Z_n(G) = Z_m(G)$ for all $n \geq m$. Define the fuzzy subset μ of G in this case by

$\mu(x) = 1 - \frac{1}{2^n}$ if $x \in Z_{n-1}(G) \backslash Z_n(G)$, $1 \leq n \leq m$,

$\mu(x) = 1 - \frac{1}{2^m}$ if $x \in G \backslash \{e\}$, and $\mu(e) = 1$.

Clearly, μ is a fuzzy subgroup of G. It follows that $Z_n(\mu)^* = Z_n(G) \neq \{e\}$ for all $n \geq 0$. Therefore, μ is not nilpotent. □

Theorem 3.2.25. *Every Abelian fuzzy subgroup of a group is nilpotent.*

Proof. Let λ be an Abelian fuzzy subgroup of G with tip t. Then $H = \lambda^*$ is an abelian subgroup of G. Consequently, $Z_1(\lambda)^* = [H, H] = \{e\}$. Hence $Z_1(\lambda) = e_t$. Thus λ is nilpotent of class 0 or 1. $\square$

Theorem 3.2.26. *Let λ be a nilpotent fuzzy subgroup of G. If μ is a fuzzy subgroup of G such that $\mu \subseteq \lambda$, then μ is nilpotent.*

Proof. If $\mu \subseteq \lambda$, we have that $Z_n(\mu) \subseteq Z_n(\lambda)$ for all $n \geq 0$ by Lemma 3.2.7. Therefore, if λ is nilpotent of class c, then μ is nilpotent of class at most c. $\square$

Theorem 3.2.27. *Let λ and μ be normal fuzzy subgroups of G. If λ and μ are nilpotent, then $\lambda \circ \mu$ is nilpotent.*

Proof. Since λ and μ are normal, we have by repeated use of Theorem 3.2.8 that for all $n \in \mathbb{N}$, $Z_n(\lambda \circ \mu)$ is contained in the sup-min product of the normal fuzzy subgroup $[\sigma_1, \ldots, \sigma_{n+1}]$ of G, where each σ_i is either λ or μ. Let r and s be the tips of λ and μ, respectively. Then $t = r \wedge s$ is the tip of $\lambda \circ \mu$. Suppose λ and μ are nilpotent of classes c and d, respectively. Let $n = c + d$. Then among the σ_i in $[\sigma_1, \ldots, \sigma_{n+1}]$, either at least $c + 1$ of them are equal to λ or at least $d + 1$ of them are equal to μ. Consequently, by Theorem 3.2.17,
$$[\sigma_1, \ldots, \sigma_{n+1}] \subseteq Z_c(\lambda) = e_r \text{ in the first case, and}$$
$$[\sigma_1, \ldots, \sigma_{n+1}] \subseteq Z_d(\mu) = e_s \text{ in the second.}$$
Therefore, $Z_n(\lambda \circ \mu) = e_t$. Hence $\lambda \circ \mu$ is nilpotent of class at most $c + d$. $\square$

The following example shows that the above result need not be true if λ and μ are not normal.

Example 3.2.28. *Define the fuzzy subsets λ and μ of S_3 as follows:*
$\lambda((1)) = 1, \lambda((12)) = \frac{1}{2}, \lambda(x) = 0$ *if* $x \in S_3 \backslash \{(1), (12)\}$,
$\mu((1)) = 1, \quad \mu(x) = \frac{1}{3}$ *if* $x \in \{(123), (132)\}$,
$\mu(x) = 0$ *if* $x \in \{(12), (13), (23)\}$.
Then λ and μ are Abelian fuzzy subgroups of G and therefore they are nilpotent. Even though λ, μ is a conormal pair, λ is not normal in S_3. The fuzzy subgroup $\lambda \circ \mu$ of S_3 is given by
$\lambda \circ \mu((1)) = 1, \lambda \circ \mu((12)) = \frac{1}{2}$,
$\lambda \circ \mu(x) = \frac{1}{3}$ *if* $x \in S_3 \backslash \{(1), (12)\}$.
It follows that for all $n \in \mathbb{N}$, $Z_n(\lambda \circ \mu) = \mu \neq e_1$. Hence $\lambda \circ \mu$ is not nilpotent.

In view of Theorem 3.2.27, we have the following result.

Theorem 3.2.29. *The set of all normal nilpotent fuzzy subgroups of G is a commutative semigroup of idempotent elements with identity e_1, where the operation is the sup-min product.*

We now examine homomorphic images and preimages of nilpotent fuzzy subgroups.

Theorem 3.2.30. *Let f be a homomorphism of the group G into the group K. If λ is a nilpotent fuzzy subgroup of G, then $f(\lambda)$ is a nilpotent fuzzy subgroup of K.*

Proof. By Theorem 3.2.11, we have that $Z_n(f(\lambda)) = f(Z_n(\lambda))$ for all $n \in \mathbb{N}$. Therefore, if λ is nilpotent of class c, then $f(\lambda)$ is nilpotent of class at most c. □

Let f be a homomorphism of the group G into the group K such that f has a nilpotent kernel. Let ν be a fuzzy subgroup of K. We show in the following example that if ν is nilpotent, then it does not necessarily follow that $f^{-1}(\nu)$ is nilpotent.

Example 3.2.31. *Let $G = S_3$, $K = \mathbb{Z}_2 = \{0, 1\}$ and let f be the homomorphism of G onto K with kernel $N = \{(1), (123), (132)\}$. Let ν be the fuzzy subgroup of K defined by $\nu(0) = t_0$, $\nu(1) = t_1$, where $t_0 > t_1$. Clearly, f has the nilpotent kernel N and ν is nilpotent of class 1. However, the fuzzy subgroup $f^{-1}(\nu)$ of G, which is given by*
$$f^{-1}(\nu)(x) = t_0 \text{ if } x \in N \text{ and } f^{-1}(\nu)(x) = t_1 \text{ if } x \in G \backslash N,$$
is not nilpotent. This can be seen by the following argument. Let $\mu = f^{-1}(\nu)$. Then $(\mu, \mu)(e) = t_0$. Now $\forall x, y \in G, yxa = xy$ implies either x or y is in $\langle a \rangle$ and $yxa^2 = xy$ implies either x or y is in $\langle a \rangle$. Also, such x and y exist since $baa = ab$ and $ba^2a^2 = a^2b$. Thus $(\mu, \mu)(a) = (\mu, \mu)(a^2) = t_1$. Now $\forall x, y \in G, yxb = xy, yxab = xy$, and $yxa^2b = xy$ are impossible. Hence $(\mu, \mu)(b) = (\mu, \mu)(ab) = (\mu, \mu)(a^2b) = 0$. Therefore, $(\mu, \mu) = [\mu, \mu] = Z_1(\mu) \neq e_{t_0}$. It now can easily be seen that $Z_2(\mu) = Z_1(\mu)$.

3.3 Solvable Fuzzy Subgroups

The commutator of a pair of fuzzy subsets of a group was defined in the previous section. The notion of a solvable fuzzy group was proposed in [15], where an ascending series of fuzzy subgroups of the groups was attached to the fuzzy subgroup to define solvability. We use the notion of commutators to generate the derived chain of an arbitrary fuzzy subgroup. This derived chain is a particular descending chain of fuzzy subgroups of the underlying group. An alternative definition of a solvable fuzzy subgroup is given and some properties of solvable fuzzy subgroups are derived [18].

For a fuzzy subset λ of G, recall that $\langle \lambda \rangle$ is the fuzzy subgroup of G such that $\forall x \in G$,

$$\langle \lambda \rangle(x) = \vee \{ \wedge \{\sigma(t_1), \sigma(t_2), ..., \sigma(t_n)\} \mid x = t_1 t_2 ... t_n, t_i \in G, i = 1, 2, ..., n;$$
$$n \in \mathbb{N}\},$$

where $\sigma(y) = \lambda(y) \vee \lambda(y^{-1})$, $y \in G$.

Recall that the commutator of two elements a and b of G is the element $[a, b] = a^{-1}b^{-1}ab$ of G. If A and B are subsets of G, then the commutator subgroup of A and B is the subgroup $[A, B]$ of G generated by $\{[a, b] \mid a \in A, b \in B\}$. For a subgroup H of G, the chain

$$H = H^{(0)} \supseteq H^{(1)} \supseteq ... \supseteq H^{(n)} \supseteq ...$$

of subgroups of G, where $H^{(n)} = [H^{(n-1)}, H^{(n-1)}]$, $n \in \mathbb{N}$, is called the **derived chain** of the subgroup H. The subgroup H is said to be **solvable** if $H^{(n)} = \{e\}$ for some $n \in \mathbb{N}$. The group G is said to be **solvable** if it is solvable as a subgroup of itself.

Let λ be a fuzzy subgroup of a group G. Define $\lambda^{(0)} = \lambda$. Let $n \in \mathbb{N}$ and suppose $\lambda^{(n-1)}$ has been defined for $n \geq 1$. Then define

$$\lambda^{(n)} = [\lambda^{(n-1)}, \lambda^{(n-1)}].$$

Theorem 3.3.1. *If H is a subgroup of G, then $(1_H)^{(n)} = 1_{H^{(n)}}$ for all $n \in \mathbb{N}$.*

Proof. The result follows from Theorem 3.2.2. □

Theorem 3.3.2. *Let λ be a fuzzy subgroup of G. Then for all $n \in \mathbb{N}$, $\lambda^{(n)} \subseteq \lambda^{(n-1)}$.*

Proof. We prove the result by induction on n. Let $x \in G$. If x is not a commutator in G, then $(\lambda, \lambda)(x) = 0 \leq \lambda(x)$. Suppose x is a commutator in G. Then

$$\begin{aligned}
(\lambda, \lambda)(x) &= \vee\{\lambda(a) \wedge \lambda(b) \mid x = [a, b], a, b \in G\} \\
&= \vee\{\lambda(a^{-1}) \wedge \lambda(b^{-1}) \wedge \lambda(a) \wedge \lambda(b) \mid x = [a, b], a, b \in G\} \\
&\leq \vee\{\lambda(a^{-1}b^{-1}ab) \mid x = [a, b], a, b \in G\} \\
&= \lambda(x).
\end{aligned}$$

Thus $(\lambda, \lambda) \subseteq \lambda$. Since λ is a fuzzy subgroup of G,

$$\lambda^{(1)} = [\lambda, \lambda] = [(\lambda, \lambda)] \subseteq \lambda = \lambda^{(0)},$$

and thus the result holds for $n = 1$. Suppose $\lambda^{(k)} \subseteq \lambda^{(k-1)}$ for $k \in \mathbb{N}$, the induction hypothesis. Then by Lemma 3.2.7, we have

$$\lambda^{(k+1)} = [\lambda^{(k)}, \lambda^{(k)}] \subseteq [\lambda^{(k-1)}, \lambda^{(k-1)}] = \lambda^{(k)}.$$

Hence the desired result holds by induction. □

Definition 3.3.3. *Let λ be a fuzzy subgroup of G. The chain $\lambda = \lambda^{(0)} \supseteq \lambda^{(1)} \supseteq ... \supseteq \lambda^{(n)} \supseteq ...$ of fuzzy subgroups of G is called the **derived chain** of λ.*

Theorem 3.3.4. *If λ is a normal (characteristic, fully invariant) fuzzy subgroup of G, then all of the fuzzy subgroups in the derived chain of λ are normal (characteristic, fully invariant) in G.*

Proof. The result follows from Theorem 3.2.6.   □

Definition 3.3.5. *Let λ be a fuzzy subgroup of G with tip t. If the derived chain*

$$\lambda = \lambda^{(0)} \supseteq \lambda^{(1)} \supseteq ... \supseteq \lambda^{(n)} \supseteq ...$$

*of λ is such that there is a nonnegative integer m such that $\lambda^{(m)} = e_t$, then λ is called **solvable**. If k is the smallest nonnegative integer such that $\lambda^{(k)} = e_t$, then the series*

$$\lambda = \lambda^{(0)} \supset \lambda^{(1)} \supset ... \supset \lambda^{(k)} = e_t$$

*is called the **derived series** of λ.*

Theorem 3.3.6. *Let H be a subgroup of G. Then H is solvable if and only if 1_H is solvable.*

Proof. Consider the derived chain

$$H = H^{(0)} \supseteq H^{(1)} \supseteq ... \supseteq H^{(n)} \supseteq ...$$

of H. By Theorem 3.3.1, $(1_H)^{(n)} = 1_{H^{(n)}}$ for all nonnegative integers n. Thus $(1_H)^{(n)} = 1_{\{e\}}$ if and only if $H^{(n)} = \{e\}$. The result now is clear.   □

Example 3.3.7. *The following is an example of a solvable fuzzy subgroup of the symmetric group S_4 on $\{1, 2, 3, 4\}$. Let*

$$D_4 = \langle (24), (1234) \rangle$$
$$= \{(1), (12)(34), (13)(24), (14)(23), (13), (24), (1234), (1432)\}.$$

Then D_4 is a dihedral subgroup of S_4 with center $C = \{(1), (13)(24)\}$. Let λ be the fuzzy subset of S_4 defined by

$$\lambda(x) = 1 \text{ if } x \in C$$
$$\lambda(x) = \frac{1}{2} \text{ if } x \in \langle (1234) \rangle \backslash C$$
$$\lambda(x) = \frac{1}{4} \text{ if } x \in D_4 \backslash \langle (1234) \rangle$$
$$\lambda(x) = 0 \text{ if } x \in S_4 \backslash D_4.$$

Clearly, λ is a fuzzy subgroup of S_4. It follows that the fuzzy subgroup $\lambda^{(1)}$ is given as follows:

$$\lambda^{(1)}((1)) = 1, \ \lambda^{(1)}((13)(24)) = \frac{1}{4}$$

and

$$\lambda^{(1)}(x) = 0 \text{ if } x \in S_4 \backslash C.$$

It follows easily that $\lambda^{(2)} = e_1$. Thus λ is a solvable fuzzy subgroup of S_4.

Definition 3.3.8. *Let λ be a fuzzy subgroup of G with tip t. A series*

$$\lambda = \lambda_0 \supseteq \lambda_1 \supseteq ... \supseteq \lambda_n = e_t$$

*of fuzzy subgroups of G is called a **solvable series** for λ if for all $i = 0, 1, ..., n-1$,*

$$\lambda_i(a) \wedge \lambda_i(b) \leq \lambda_{i+1}([a,b])$$

for all $a, b \in G$.

It follows that a fuzzy subgroup λ of G is solvable if and only if $[\lambda_i, \lambda_i] \subseteq \lambda_{i+1}$ for $i = 0, 1, ..., n-1$.

Theorem 3.3.9. *Let λ be a fuzzy subgroup of G. Then λ is solvable if and only if λ has a solvable series.*

Proof. Let λ be a fuzzy subgroup of G with tip t. Suppose λ has a solvable series

$$\lambda = \lambda_0 \supseteq \lambda_1 \supseteq ... \supseteq \lambda_n = e_t.$$

We show by induction that $\lambda^{(i)} \subseteq \lambda_i$, $i = 0, 1, ..., n$. We have $\lambda^{(0)} = \lambda = \lambda_0$ and $\lambda^{(1)} = [\lambda_0, \lambda_0] \subseteq \lambda_1$. Suppose $\lambda^{(i)} \subseteq \lambda_i$ for some $i = 0, 1, ..., n-1$, the induction hypothesis. Then by Lemma 3.2.7, we have that

$$\lambda^{(i+1)} = [\lambda^{(i)}, \lambda^{(i)}] \subseteq [\lambda_i, \lambda_i] \subseteq \lambda_{i+1}.$$

Thus $\lambda^{(i)} \subseteq \lambda_i$ for $i = 0, 1, ..., n$. Therefore,

$$e_t \subseteq \lambda^{(n)} \subseteq \lambda_n = e_t.$$

Hence λ is solvable. Conversely, if λ is solvable, then the derived series of λ is a solvable series for λ since for all $a, b \in G$, $\lambda^{(i+1)}([a,b]) = [\lambda^{(i)}, \lambda^{(i)}]([a,b]) = \vee\{\lambda^{(i)}(c) \wedge \lambda^{(i)}(d)|[a,b] = [c,d]\} \geq \lambda^{(i)}(a) \wedge \lambda^{(i)}(b), i = 0, 1, ... \quad \square$

Theorem 3.3.10. *Every fuzzy subgroup of a solvable group is solvable. In fact, if the group G is not solvable, then G has a nontrivial fuzzy subgroup that is not solvable.*

Proof. Suppose G is solvable. Let k be smallest nonnegative integer such that $G^{(k)} = \{e\}$. Let λ be a fuzzy subgroup of G with tip t. For $i = 0, 1, ..., k$, define λ_i by $\forall x \in G$,

$$\lambda_i(x) = \begin{cases} \lambda_i(x) \text{ if } x \in G^{(i)}, \\ \quad 0 \text{ otherwise.} \end{cases}$$

Then $\lambda = \lambda_0 \supseteq \lambda_1 \supseteq ... \supseteq \lambda_k$ is a finite chain of fuzzy subgroups of G such that $(\lambda_i)^* \subseteq G^{(i)}$ for all $i = 0, 1, ..., k$. Clearly $\lambda_k = e_t$. Let $a, b \in G$ and $0 \leq i \leq n-1$. If $a \notin G^{(i)}$ or $b \notin G^{(i)}$, then

$$\lambda_i(a) \wedge \lambda_i(b) = 0 \leq \lambda_{i+1}([a,b]).$$

If $a, b \in G^{(i)}$, then $[a, b] \in G^{(i+1)}$. Thus

$$\lambda_i(a) \wedge \lambda_i(b) = \lambda(a) \wedge \lambda(b)$$
$$\leq \lambda([a, b]) = \lambda_{i+1}([a, b]).$$

Hence $\lambda = \lambda_0 \supseteq \lambda_1 \supseteq ... \supseteq \lambda_k = e_t$ is a solvable series for λ and so λ is solvable.

Now suppose that G is not solvable. Consider the derived chain

$$G = G^{(0)} \supseteq G^{(1)} \supseteq ...$$

of G. If the chain is nonterminating, then $G^{(n)} \subset G^{(n-1)}$ for all $n \in \mathbb{N}$. In that case, let μ be the fuzzy subgroup of G defined by

$$\mu(x) = 1 - (1/2^n) \text{ if } x \in G^{(n-1)} \backslash G^{(n)}, \ n \geq 1,$$
$$\mu(x) = 1 \text{ if } x \in \cap_{i=1}^{\infty} G^{(i)}.$$

If the derived chain terminates (but not to $\{e\}$), then there is a nonnegative integer m such that

$$G = G^{(0)} \supset G^{(1)} \supset ... \supset G^{(m)} \neq (e),$$

and $G^{(n)} = G^{(m)}$ for all $n \geq m$. Define the fuzzy subset μ of G in this case by $\forall x \in G$,

$$\mu(x) = 1 - (1/2)^n \text{ if } x \in G^{(n-1)} \backslash G^{(n)}, \ n = 1, 2, ..., m,$$
$$\mu(x) = 1 - (1/2^{m+1}) \text{ if } x \in G^{(m)} \backslash \{e\},$$
$$\mu(e) = 1.$$

Clearly, μ is a nontrivial fuzzy subgroup of G. It follows that

$$(\mu^{(n)})^* = G^{(n)} \neq \{e\}$$

for each $n \geq 0$. Consequently, μ is not solvable.   □

Theorem 3.3.11. *Suppose λ is a solvable fuzzy subgroup of G. If μ is a fuzzy subgroup of G such that $\mu \subseteq \lambda$, then μ is solvable.*

Proof. Since $\mu \subseteq \lambda$, we have by Lemma 3.2.7 that $\mu^{(n)} \subseteq \lambda^{(n)}$ for all $n \geq 0$. Therefore, if λ is solvable, then μ is solvable.   □

We now examine the homomorphic images and preimages of solvable fuzzy subgroups.

Theorem 3.3.12. *Let f be a homomorphism of G into a group K. If λ is a solvable fuzzy subgroup of G, then $f(\lambda)$ is a solvable fuzzy subgroup of K.*

Proof. By Theorem 3.2.11(1), we have that $(f(\lambda))^{(n)} = f(\lambda^{(n)})$ for all non-negative integers n. Let e and e' be the identities of G and K, respectively. Since λ is solvable, $\lambda^{(n)} = e_t$ for some $n \geq 0$, where t is the tip of λ. Consequently, $(f(\lambda))^{(n)} = f(e_t) = e'_t$ and so $f(\lambda)$ is solvable.   □

Theorem 3.3.13. *Let f be a homomorphism of G into a group K. Suppose the kernel of f is solvable. If μ is solvable fuzzy subgroup of K, then $f^{-1}(\mu)$ is a solvable fuzzy subgroup of G.*

Proof. Let e and e' be the identities of G and K, respectively, and let N denote the kernel of f. Let

$$\mu = \mu^{(0)} \supseteq \mu^{(1)} \supseteq ... \supseteq \mu^{(n)} = e'_t$$

be the derived series of μ, where t is the tip of μ. For $0 \leq i \leq n$, set $\lambda_i = f^{-1}(\mu^{(i)})$. Then

$$f^{-1}(\mu) = \lambda_0 \supseteq \lambda_1 \supseteq ... \supseteq \lambda_n = t_N.$$

Let $N = N^{(0)} \supset N^{(1)} \supset ... \supset N^{(m)} = \{e\}$ be the derived series of N. For $1 \leq i \leq m$, let $\lambda_{n+i} = t_{N^{(i)}}$. Then $[\lambda_i, \lambda_i] = \lambda_{i+1}$ for $n \leq i < n + m$. By Theorem 3.2.12, we have that $[\lambda_i, \lambda_i] \subseteq \lambda_{i+1}$ for all $i = 0, 1, ..., n$. Thus

$$f^{-1}(\mu) = \lambda_0 \supseteq \lambda_1 \supseteq ... \supseteq \lambda_{n+m} = e_t$$

is a solvable series for $f^{-1}(\mu)$ and so $f^{-1}(\mu)$ is solvable.   □

References

1. W-X Gu, S-Y Li and D-G Chen, Fuzzy groups with operators, *Fuzzy Sets and Systems*, 66 (1994) 363-371.
2. K. C. Gupta and B. K. Sarma, Conormal fuzzy subgroups, *Fuzzy Sets and Systems*, 56 (1993) 317-322.
3. K. C. Gupta and B. K. Sarma, Operator domains of fuzzy groups, *Inform. Sci.* 84 (1995) 247-259.
4. K. C. Gupta and B. K. Sarma, Commutator fuzzy groups, *J. Fuzzy Math.* 4 (1996) 655-663.
5. K. C. Gupta and B. K. Sarma, Nilpotent fuzzy groups, *Fuzzy Sets and Systems* 101 (1999) 167-176.
6. M. Hall, Jr., *The Theory of Groups*, Macmillan, New York, 1959.
7. D. G. Kim, Characteristic fuzzy groups, *Comm. Korean Math Soc.* 18 (2003) 21- 29.
8. J. G. Kim, Commutative fuzzy sets and nilpotent fuzzy groups, *Inform. Sci.* 83 (1995) 161-174.
9. W. J. Liu, Fuzzy invariant subgroups and fuzzy ideals, *Fuzzy Sets and Systems* 8 (1982)133-139.
10. M.A.A. Mishref, Admissible and K-invariant fuzzy subgroups, *J. Fuzzy Math.* 6 (1998) 811- 819.

11. M.A.A Mishref, Restudy of fuzzy factor groups and fuzzy solvable groups, *J. Fuzzy Math.* 7 (1998) 311- 320.

12. N. N. Morsi, Note on normal fuzzy subgroups and fuzzy normal series of finite groups, *Fuzzy Sets and Systems*, 87 (1997) 255-256.

13. N. P. Mukherjee and P. Bhattacharya, Fuzzy groups: Some group-theoretic analogs, *Inform. Sci.* 39 (1986) 247-268.

14. S. Ray, Generated and cyclic fuzzy groups, *Inform. Sci.* 69 (1993) 185-200.

15. S. Ray, Solvable fuzzy groups, *Inform. Sci.* 75 (1993) 47-61.

16. A. Rosenfeld, Fuzzy groups, *J. Math. Anal.. Appl.* 35 (1971) 512-517.

17. J. J. Rotman, *An Introduction to the Theory of Groups,* 3rd ed., Allyn and Bacon, Boston, 1984.

18. B. K. Sarma, Solvable fuzzy groups, *Fuzzy Sets and Systems,* 106 (1999) 463-467.

19. S. I. Sidky and M.A.A. Mishref, S-fuzzy subgroups, *Bulletin Calcutta Math Soc.* 83 (1991) 19 - 24.

20. Wu Wang-ming, Normal fuzzy subgroups, *J. Fuzzy Math.* 1 (1981) 21-30 (in Chinese).

21. L. A. Zadeh, Fuzzy sets, *Inform. and Control* 8 (1965) 338-353.

4

Characterization of Certain Groups and Fuzzy Subgroups

Fuzzy subgroups of Hamiltonian, solvable, P-Hall, and nilpotent groups are examined. The notions of generalized characteristic fuzzy subgroups, fully invariant fuzzy subgroups, and characteristic fuzzy subgroups are introduced. It is shown that if G is a finite group all of whose Sylow subgroups are cyclic, then a fuzzy subgroup of a group G is normal if and only if it is a generalized fuzzy subgroup of G. Normal fuzzy subgroups, quasi-normal fuzzy subgroups, (p, q)-subgroups, fuzzy cosets, fuzzy conjugates and $SL(p, q)$-subgroups are also presented. The results of this chapter are mainly from the work of Asaad and Abou-Zaid [9, 11, 12, 13, 14].

4.1 Fuzzy Subgroups of Hamiltonian, Solvable, P-Hall, and Nilpotent Groups

Let G be a group. Recall that a fuzzy subgroup μ of G is called normal if $\mu(x) = \mu(y^{-1}xy)$ for all x, y in G.

We now consider groups whose fuzzy subgroups are normal.

Definition 4.1.1. *A group G is called **Hamiltonian** if G is not Abelian and every subgroup of G is normal.*

Note that the quaternion group $G = \langle x, y \mid x^2 = y^2, x^4 = e, y^{-1}xy = x^{-1} \rangle$ is an example of a Hamiltonian group.

Definition 4.1.2. *A group that is either Abelian or Hamiltonian is called **Dedekind**.*

Theorem 4.1.3. *Let G be a group. Then G is Dedekind if and only if every fuzzy subgroup of G is normal.*

Proof. Suppose that G is a Dedekind group. Let μ be a fuzzy subgroup of G. Then every level subgroup of G is normal since G is Dedekind. Thus by Theorem 1.3.3, μ is normal.

Conversely, suppose that every fuzzy subgroup of G is normal. Every subgroup H of a group G can be regarded as a level subgroup of some fuzzy subgroup μ of G. By assumption, μ is a normal fuzzy subgroup of G. By Theorem 1.3.3, H is a normal subgroup of G. Thus G is a Dedekind group.   □

Hamiltonian groups are characterized in the following theorem.

Theorem 4.1.4. *[[16], Theorem 12.5.4] A Hamiltonian group is the direct product of a quaternion group with an Abelian group in which every element is of odd order and an Abelian group of exponent two.*

The following corollary is an immediate consequence of Theorems 4.1.3 and 4.1.4.

Corollary 4.1.5. *Let G be a group. Then the following statements are equivalent.*

(1) G is a Hamiltonian group.

(2) G is the direct product of a quaternion group with an Abelian group in which every element is of odd order and an Abelian group of exponent two.

(3) G is not Abelian and every fuzzy subgroup of G is normal.

We next consider cyclic groups of prime power order.

Theorem 4.1.6. *Let G be a group of prime power order. Then G is cyclic if and only if there exists a fuzzy subgroup μ of G such that for all $x, y \in G$,*

(1) if $\mu(x) = \mu(y)$, then $\langle x \rangle = \langle y \rangle$,

(2) if $\mu(x) > \mu(y)$, then $\langle x \rangle \subset \langle y \rangle$.

Proof. Suppose that G is cyclic of order p^n, where p is a prime and $n \in \mathbb{N}$. Let μ be the fuzzy subset of G defined by $\forall x \in G$,

$$\mu(x) = t_i \qquad \text{if } o(x) = p^i, \ i = 1, 2, ..., n,$$
$$\mu(e) = t_0,$$

where $t_0 > t_1 > t_2 > ... > t_n$. We prove that μ is a fuzzy subgroup of G. Let $x, y \in G$. Since G is a cyclic group of prime power order, it follows that $\langle x \rangle \subseteq \langle y \rangle$ or $\langle y \rangle \subseteq \langle x \rangle$. Hence either $\langle xy \rangle \subseteq \langle y \rangle$ or $\langle xy \rangle \subseteq \langle x \rangle$. Therefore, $\mu(xy) \geqslant \mu(x) \wedge \mu(y)$. Clearly, $\mu(x^{-1}) = \mu(x)$. Thus μ is a fuzzy subgroup of G. Suppose $\mu(x) = \mu(y)$. Then by the definition of μ, $o(x) = o(y)$. Since G is a cyclic group of order p^n, it follows that $\langle x \rangle = \langle y \rangle$. Thus (1) holds. Suppose $\mu(x) > \mu(y)$. Then by the definition of μ, $o(y) > o(x)$. Since G is cyclic of order p^n, it follows that $\langle x \rangle \subset \langle y \rangle$. Thus (2) holds.

Conversely, suppose there exists a fuzzy subgroup μ of G such that (1) and (2) hold for all $x, y \in G$. If G has order p, then G is cyclic. Suppose the result is true for all groups of order p^m, where $m = 1, ..., n-1$ for $n \geq 2$, the induction hypothesis. Suppose $o(G) = p^n$ and let H be any proper subgroup of G. Then H is a subgroup of order p^m, where $1 \leqslant m < n$. Clearly, the restriction of μ to H is a fuzzy subgroup of H satisfying (1) and (2). Thus H is cyclic by the induction hypothesis. Hence all proper subgroups of G are cyclic. We prove that G has a unique maximal subgroup. If this is not the case, then there exist two distinct maximal subgroups $H_i = \langle x_i \rangle$, where $i = 1, 2$. If $\mu(x_1) > \mu(x_2)$, then $H_1 \subset H_2$. However, this is contradicts the fact that H_1 is maximal. Similarly, if $\mu(x_2) > \mu(x_1)$, we have a contradiction. Thus $\mu(x_1) = \mu(x_2)$. Hence by (1), $\langle x_1 \rangle = \langle x_2 \rangle$, a contradiction. Thus G has a unique maximal subgroup, say M. Hence every proper subgroup of G is contained in M. It now follows easily that $G = \langle y \rangle$ for any $y \in G \backslash M$. Thus G is cyclic. $\qquad\square$

We now consider fuzzy subgroups of groups of square free order. We recall from group theory that a subgroup H of a finite group G is called a **Hall** subgroup of G if $o(H)$ and $[G : H]$ are relatively prime.

Theorem 4.1.7. *Let G be a group of square free order. Let μ be a normal fuzzy subgroup of G. Then the following assertions hold for all $x, y \in G$.*

(1) $o(x) \mid o(y)$ *implies* $\mu(y) \leqslant \mu(x)$.

(2) $o(x) = o(y)$ *implies* $\mu(y) = \mu(x)$.

Proof. (1) Clearly, all Sylow subgroups of G are cyclic. Thus by [[27], Theorem 12.6.17, p. 356], $G = HK$, $H \cap K = \langle e \rangle$, where H is a cyclic normal Hall subgroup and K is a cyclic Hall subgroup. Since H and $G/H \cong K$ are solvable groups, we have that G is solvable [[27], Theorem 2.6.3, p. 39]. Clearly, $\langle x \rangle$ and $\langle y \rangle$ are Hall subgroups of G and $o(\langle x \rangle) \mid o(\langle y \rangle)$. Since $\langle y \rangle$ is cyclic, it follows that $\langle y \rangle$ has a subgroup H such that $o(H) = o(\langle x \rangle)$. Thus H and $\langle x \rangle$ are Hall subgroups of G of the same order. Since G is solvable, it follows from Hall's Theorem that $\langle x \rangle = g^{-1}Hg$ for some g in G [[24], Theorem 9.3.10, p. 228]. Hence $\langle x \rangle \subseteq g^{-1}\langle y \rangle g$ and so $x = g^{-1}y^i g$ for some $i \in \mathbb{N}$. Since μ is a normal fuzzy subgroup of G, $\mu(x) = \mu(g^{-1}y^i g) = \mu(y^i) \geqslant \mu(y)$. Thus $\mu(x) \geqslant \mu(y)$ and so (1) holds.

(2) Since $o(x) = o(y)$, we have that $o(x) \mid o(y)$ and $o(y) \mid o(x)$. By (1), we have that $\mu(y) \leqslant \mu(x)$ and $\mu(x) \leqslant \mu(y)$. Thus $\mu(x) = \mu(y)$ and so (2) holds. $\qquad\square$

Under the assumptions of Theorem 4.1.7, it is not the case that $o(x) < o(y) \Rightarrow \mu(x) \geqslant \mu(y)$. For example, let $S_3 = \{(1), (123), (132), (23), (13), (12)\}$ be the symmetric group on $\{1, 2, 3\}$. Define the fuzzy subset μ of S_3 as follows: $\mu((1)) = t_0, \mu((123)) = \mu((132)) = t_1, \mu((23)) = \mu((13)) = \mu((12)) = t_2$, where $t_0 > t_1 > t_2$. Then μ is a normal fuzzy subgroup of S_3. However, $o((23)) < o((123))$ and $\mu((23)) < \mu((123))$. Clearly, S_3 is not Abelian and is of order 6.

In [15], it was proved that if G is a finite Abelian group of order $p_1 p_2 ... p_r$, where the p_i are primes (not necessarily distinct) $(i = 1, 2, ..., r)$, then there exists a fuzzy subgroup μ of G such that the chain of level subgroups of μ,

$$\langle e \rangle = \mu_{t_0} \subset \mu_{t_1} \subset ... \subset \mu_{t_r} = G$$

is maximal, where $t_0 > t_1 > ... > t_r$.

A normal chain of G is a chain of subgroups, $\langle e \rangle = A_0 \subseteq ... \subseteq A_1 \subseteq A_r = G$, where A_{i-1} is a normal subgroup of A_i, $i = 1, 2, ..., r$. Two normal chains $\langle e \rangle = A_o \subseteq ... \subseteq A_r = G$ and $\langle e \rangle = B_0 \subseteq ... \subseteq B_s = G$ of G are equivalent if and only if $r = s$ and there exists $f \in S_r$, where S_r is the symmetric group of degree r such that

$A_i / A_{i-1} \cong B_{f(i)} / B_{f(i)-1}$ for $1 \leqslant i \leqslant r$.

A group G is called **simple** if it has contains no proper normal subgroups. A **composition chain** of G is a normal chain without repetition whose factors are all simple, [[27], p. 36].

We now give a generalization of the above result referred to in [15].

Theorem 4.1.8. *Let G be a group of order $p_1 p_2 ... p_r$, where p_i is a prime, $i = 1, 2, ..., r$, and the p_i's are not necessarily distinct. Then G is solvable if and only if there exists a fuzzy subgroup μ of G with exactly r level subgroups, $\mu_{t_0}, \mu_{t_1}, ..., \mu_{t_r}$, $\mathrm{Im}(\mu) = \{t_0, t_1, ..., t_r\}$, $t_0 > t_1 > ... > t_r$, such that the level subgroups form a composition chain.*

Proof. Suppose that G is solvable. The proof is by induction on r. If $r = 1$, then G is cyclic of prime order. Define the fuzzy subset μ of G by $\mu(e) = t_0$ and $\mu(x) = t_1$ for all $x \in G, x \neq e$, where $0 \leq t_1 < t_0 \leq 1$. Then μ is a fuzzy subgroup of G. Assume that $r \geqslant 2$. By [[16], Theorem 9.2.3, p. 139], $G = G_0 \supset G_1 \supset G_2 \supset ... \supset G_r = \langle e \rangle$, where G_i is a normal subgroup of G_{i-1} and $[G_{i-1} : G_i]$ is a prime, $i = 1, 2, ..., r$. By induction on r, it follows that there exists a fuzzy subgroup μ of G_1 such that $\mu_{t_0}, \mu_{t_1}, ..., \mu_{t_{r-1}}$ are the level subgroups of μ, where $\mathrm{Im}(\mu) = \{t_0, t_1, ..., t_{r-1}\}$, $t_0 > t_1 > ... > t_{r-1} > 0$, and the level subgroups form a composition chain. Set $\mu(x) = t_r$ for all $x \in G \backslash G_1$, where $t_r < t_{r-1}$. Then μ is a fuzzy subgroup of G with exactly r level subgroups of μ. Hence μ is the desired fuzzy subgroup of G.

Conversely, let μ be a fuzzy subgroup of G such that $\mu_{t_0}, \mu_{t_1}, ..., \mu_{t_r}$ are its level subgroups, $\mathrm{Im}(\mu) = \{t_0, t_1, ..., t_r\}$, $t_0 > t_1 > ... > t_r$, and the level subgroups form a composition chain. Then $\mu_{t_{i-2}} \lhd \mu_{t_i}$ and $[\mu_{t_i} : \mu_{t_{i-1}}]$ is a prime. By [[16], Theorem 9.2.3, p. 139], G is solvable. $\qquad \square$

We now make use of the Jordan-Holder Theorem, [[27], Theorem 2.5.8], which states that if G is a finite group, then any two composition chains of G are equivalent.

We have the following corollary.

Corollary 4.1.9. *Suppose that G is a finite group and that G has a composition chain $\langle e \rangle = A_0 \subset A_1 \subset ... \subset A_r = G_r$, where A_i / A_{i-1} is cyclic of prime order, $i = 1, 2, ..., r$. Then there exists a composition chain of level subgroups of some fuzzy subgroup μ of G and this composition chain is equivalent to $\langle e \rangle = A_0 \subset A_1 \subset ... \subset A_r = G$.*

Proof. Since G is a finite group and the factor groups in a composition chain from G to $\langle e \rangle$ are cyclic of prime order, we have by [[16], Theorem 9.2.3, p. 139] that G is solvable of order $p_1 p_2 ... p_r$, where the p_i's are (not necessarily distinct) primes. By Theorem 4.1.8, there exists a composition chain of level subgroups of some fuzzy subgroup μ of G. By the Jordan-Holder theorem, these two composition chains are equivalent.   □

We now consider Sylow subgroups and P-Hall subgroups. A group is called **periodic** if all its elements are of finite order. Let P be a set of primes and let $n \in \mathbb{N}$. Then n is called a P-number if every prime divisor of n is in P. A group G is called a P-**group** if G is periodic and $x \in G$ implies $o(x)$ is a P-number. Thus a finite group G is a P-group if and only if $o(G)$ is a P-number. A group is a p-group if it is a P-group with $P = \{p\}$.

Theorem 4.1.10. *(Dedekind Theorem) Let G be a finite group that is not Abelian and whose proper subgroups are all normal. Then $G = Q \otimes C \otimes E$, where Q is the quaternion group of order 8, C is an Abelian group in which every element is of odd order and E is an Abelian group of exponent 2 or 1.*

Definition 4.1.11. *A P-**Hall subgroup of** G is a P-subgroup of G whose index in G is not divisible by any prime in P. A **Sylow p-subgroup** of G is a maximal p-subgroup of G.*

A subgroup H of a finite group G is a Hall P-subgroup of G if H is a P-group and $[G : H]$ is a P'-number, where P' is the complement of P in the set $\mathcal{P}$ of all prime numbers, i. e., $P' = \mathcal{P} \backslash P$. We recall that if G is any group and $x, y \in G$, then $x^{-1}y^{-1}xy$ is usually denoted by $[x, y]$ and is called the commutator of x and y.

We recall from Lemma 3.1.5 that if μ is a fuzzy subgroup of G, then $\mu(xy^{-1}) = \mu(e)$ implies $\mu(x) = \mu(y)$ $\forall x, y \in G$. We also recall from Proposition 2.3.3 that if μ is a fuzzy subgroup of G, then $\mu([x, y]) = \mu(e)$ for all $x, y \in G$ implies μ is normal.

Example 4.1.12. *We now show that the converse of Proposition 2.3.3 is not true in general. Consider the symmetric group $G = S_3 = \langle a, b \mid a^3 = b^2 = e, ba^2 = ab \rangle$. Define the fuzzy subset μ of G as follows:* $\forall x \in G$,

$$\mu(x) = \begin{cases} 1 & \text{if } x = e, \\ \frac{1}{2} & \text{if } x \in \{a, a^2\}, \\ 0 & \text{otherwise.} \end{cases}$$

Since the level sets of μ are normal subgroups of G, μ is a normal fuzzy subgroup. However, μ does not satisfy the condition that $\mu([x, y]) = \mu(e) \forall x, y \in G$ since $\mu(b^{-1}a^{-1}ba) = \mu(a^2) = \frac{1}{2} \neq \mu(e)$. Now $\mu(e) = \mu(a^{-1}b^{-1}(ba))$. Thus $\mu(b^{-1}a^{-1}(ba)) \neq \mu(a^{-1}b^{-1}(ba))$. Hence $Z(\mu) \neq N(\mu)$.

Theorem 4.1.13. *Let G be a finite group. Then G is Abelian if and only if for every fuzzy subgroup μ of G, $\mu([x, y]) = \mu(e)$ for all $x, y \in G$.*

Proof. Suppose G is an Abelian group. Then for all x, y in G, $[x, y] = e$ and hence $\mu([x, y]) = \mu(e)$ for all fuzzy subgroups μ of G.

Conversely, suppose that for every fuzzy subgroup μ of G, $\mu([x, y]) = \mu(e)$ for all x, y in G. Let H be a proper subgroup of G and μ be a fuzzy subgroup of H. Clearly, by defining an extension μ_0 of μ to G by $\mu_0(x) = 0$ for all x in $G\backslash H$, it follows easily that μ_0 is a fuzzy subgroup of G. Hence by hypothesis, μ_0 satisfies the condition $\mu_0([x, y]) = \mu_0(e)$ for all x, y in G. Also, it is clear that the restriction of μ_0 to H (which is μ) satisfies the condition for all x, y in H. Thus it follows by induction on the order of G that H is Abelian. By Proposition 2.3.3, every fuzzy subgroup of G is normal. Thus it follows that every subgroup of G is normal. Now by Theorem 4.1.10, we have that if G is not Abelian, then $G = Q \otimes C \otimes E$, where Q is the quaternion group of order 8, C is an Abelian group in which every element is of odd order, and E is an Abelian group of exponent 2 or 1. Since every subgroup of G is Abelian and Q is not Abelian, we have that $E = C = \langle e \rangle$ and so $G = Q$. This is impossible since there exists a fuzzy subgroup μ of Q which does not satisfy the condition. For example, define the fuzzy subgroup μ of $Q = \langle a, b \mid a^4 = e, a^2 = b^2, ba = a^{-1}b \rangle$ as follows: $\forall x \in G$,

$$\mu(x) = \begin{cases} t_o \text{ if } x = e, \\ t_1 \text{ if } x = a^2, \\ t_2 \text{ otherwise,} \end{cases}$$

where $0 \leq t_2 < t_1 < t_0 \leq 1$. Then $\mu([a, b]) = \mu(a^2) = t_1 \neq \mu(e)$. Therefore, G is Abelian. $\qquad\square$

We next consider the existence of normal Hall subgroups. Let p be a prime. A subgroup P of G is called a Sylow p-subgroup (or a p-sylow subgroup) of G if P is a maximal p-subgroup of G. Let $Syl_p(G)$ denote the set of all Sylow p-subgroups of G.

Theorem 4.1.14. *Let G be a group of order divisible by at least two distinct primes. Let S be a Sylow p-subgroup of G. Then S is normal in G if and only if there exists a fuzzy subgroup μ of G such that $t_1 > t_2$, where $t_1 = \wedge\{\mu(x) \mid x \in H\}$ and $t_2 = \vee\{\mu(x) \mid x \in K\}$ and where $H = \{x \in G \mid x \text{ has order } p\}$ and $K = \{x \in G \mid x \text{ has order } p'\}, p'$ a prime $\neq p$.*

Proof. Suppose that S is a normal Sylow p-subgroup of G. Then $S = \{e\} \cup H$, where H is defined as above. Define the fuzzy subset of μ of G as follows: $\forall x \in G$,

$$\mu(x) = \begin{cases} t_1 \text{ if } x \in S, \\ t_2 \text{ if } x \notin S, \end{cases}$$

where $0 \leqslant t_2 < t_1 \leqslant 1$. Then it follows easily that μ is a fuzzy subgroup of G. We have that $\wedge\{\mu(x) \mid x \in H\} = \wedge\{\mu(x) \mid x \in S\} = t_1 > t_2 = \vee\{\mu(x) \mid x \notin S\} = \vee\{\mu(x) \mid x \in K\}$.

Conversely, suppose that μ is a fuzzy subgroup of G satisfying the condition in the theorem. Then $\mu_{t_1} = \{x \in G \mid \mu(x) \geqslant t_1\}$ is a level subgroup of μ. We prove that μ_{t_1} is a p-subgroup of G. Since every element of S is a p-element, we have that S is a subgroup of μ_{t_1}. Since the order of G is divisible by at least 2 distinct primes, there exists a prime $q(\neq p)$ such that $q \mid o(G)$. By Cauchy's Theorem [[17], Theorem II 5.2], G contains an element y of order q. We prove that $y \notin \mu_{t_1}$. If $y \in \mu_{t_1}$, then $\mu(y) \geqslant t_1 > t_2 = \vee\{\mu(x) \mid x \in K\} \geqslant \mu(y)$, a contradiction. Thus $q \nmid o(\mu_{t_1})$ for any prime $q(\neq p)$. Thus μ_{t_1} is a p-subgroup of G. Since $S \subseteq \mu_{t_1}$ and S is a Sylow p-subgroup of μ_{t_1}, we have $S = \mu_{t_1}$. It follows easily that μ_{t_1} is a normal subgroup of G. Since $o(x) = o(y^{-1}xy)$ for all x, y in G, we have that $x \in \mu_{t_1}$ implies $y^{-1}xy \in \mu_{t_1}$ for all y in G. $\square$

Corollary 4.1.15. *Let L be a proper P-Hall subgroup of G. Then L is a normal subgroup of G if and only if there exists a fuzzy subgroup μ of G such that $t_1 > t_2$, where $t_1 = \wedge\{\mu(x) \mid x$ is a non-trivial P-element of $G\}$ and $t_2 = \vee\{\mu(x) \mid x$ is a non-trivial P'-element of $G\}$.*

We now consider a class of nilpotent groups and (p, q)-subgroups. All groups in this section are assumed to be finite. We review briefly some definitions and results. We refer the reader to [10] for details. A group G is called a (p, q)-**group** if the following conditions hold:

(1) the factors of the order of G are of the form $p^i q^j, i, j \in \mathbb{N}$, where p and q are distinct primes,

(2) G is a minimal non-nilpotent group, i.e., a group which is not nilpotent and all its proper subgroups are nilpotent, and

(3) the derived group G' is the Sylow p-subgroup of G.

Let $\pi(G)$ denote the set of all primes dividing $|G|$. Recall that if G is any group and $x, y \in G$, then the $x^{-1}y^{-1}xy$ is usually denoted by $[x, y]$, and is called the commutator of x and y. $\mathrm{Exp}(G)$ denotes the exponent of G, i.e., the smallest $n \in \mathbb{N}$ such that $x^n = e \forall x \in G$. If H and K are two subgroups of a group G, then the subgroup $[H, K]$ is defined to be the subgroup generated by the set $\{[x, y] \mid x \in H, y \in K\}$. Let p be a prime. A subgroup P of G is a p-sylow subgroup (or a Sylow p-subgroup) of G if it is a maximal p-subgroup of G. Let $Syl_p(G)$ denote the set of all p-sylow subgroups of G.

Let A_4 denote the alternating group of degree 4. Then $A_4' = K_4$, the Klein 4-group, where A_4' is the commutator of A_4. In fact, $[K_4, A_4] = K_4$. Hence if we define the fuzzy subgroup μ of A_4 by $\forall x \in A_4$,

$$\mu(x) = \begin{cases} t_0 & \text{if } x = e, \\ t_1 & \text{if } x \in K_4 \backslash \langle e \rangle, \\ 0 & \text{otherwise,} \end{cases}$$

where $0 < t_1 < t_0 \leq 1$, then $\mu([x, y]) = t_1 \neq \mu(e)$ for some $x \in A_4$ and $y \in K_4$. This follows since A_4 is a minimal non-nilpotent group. In general, we have the following result.

Lemma 4.1.16. *If G is a minimal non-nilpotent group, then there exists a fuzzy subgroup μ of G such that $\mu([x, y]) \neq \mu(e)$ for some $x \in G$, $y \in G'$.*

Proof. Since G is a minimal non-nilpotent group, $G = PQ$, where $P \in Syl_p(G)$ and $Q \in Syl_q(G)$ for some distinct primes p and q, $P \lhd G$ and Q is cyclic and not normal in G, [27]. Thus it follows that $G' = P$. Define a fuzzy subset μ of G as follows: $\forall x \in G$,

$$\mu(x) = \begin{cases} t_0 & \text{if } x \neq e, \\ t_1 & \text{if } x \in G'\backslash\{e\}, \\ 0 & \text{otherwise}, \end{cases}$$

where $0 < t_1 < t_0 \leq 1$. Clearly, μ is a fuzzy subgroup of G. Also, we have $[G, G'] \neq \langle e \rangle$ else for all $y \in G' = P$ and $x \in Q$, we have $[x, y] = e$. Therefore, $G = P \otimes Q$ and so G is nilpotent, a contradiction. Therefore, there exist elements $y \in G'$ and $x \in G$ such that $[x, y] \neq e$. Thus $\mu([x, y]) = t_1 \neq \mu(e)$ since $[x, y] \in G'\backslash\langle e \rangle$. $\qquad\square$

Theorem 4.1.17. *Let G be a group. If every fuzzy subgroup μ of G satisfies the condition $\mu([x, y]) = \mu(e)$ for all $x \in G$ and $y \in G'$, then G is nilpotent.*

Proof. We prove the result by induction on the order of G. The result is clearly true if G is of order 1. Suppose G is of order $n > 1$ and the result is true for all groups of order less than n, the induction hypothesis. Let H be a proper subgroup of G and μ be a fuzzy subgroup H. By defining an extension map μ_0 of μ such that $\mu_0(x) = 0$ for all x in $G\backslash H$, it follows easily that μ_0 is fuzzy subgroup of G and hence by hypothesis μ_0 satisfies the condition $\mu_0([x, y]) = \mu_0(e)$ for all $x \in G$, $y \in G'$. Clearly, the restriction of μ_0 to H (which is μ) satisfies the condition $\mu_0([x, y]) = \mu_0(e)$ for all $x \in H$, $y \in H'$. Thus H is nilpotent by the induction hypothesis. Hence every proper subgroup of G is nilpotent. Therefore, if G is not nilpotent then, by Lemma 4.1.16, there exists a fuzzy subgroup ν of G such that $\nu([x, y]) \neq \nu(e)$ for some $x \in G$, $y \in G'$. However, this contradicts the hypothesis. Thus G is nilpotent. $\qquad\square$

The above theorem may be considered as a generalization of Theorem 4.1.14. The converse of Theorem 4.1.17 is not true in general. For that we have the following proposition.

Proposition 4.1.18. *Let G be a nilpotent group of class $\geqslant 3$. Then there exists a fuzzy subgroup μ of G such that $\mu([x, y]) \neq \mu(e)$ for some $x \in G$ and $y \in G'$.*

Proof. Since G is nilpotent of class $\geqslant 3$, $[G, G'] \neq \langle e \rangle$. Hence there exist elements $x_1 \in G$, $y_1 \in G'$ such that $[x_1, y_1] \neq e$. Define a fuzzy subset μ of G as follows: $\forall x \in G$,

$$\mu(x) = \begin{cases} t_0 & \text{if } x = e, \\ t_1 & \text{if } x \in G'\backslash\langle e \rangle, \\ 0 & \text{otherwise}, \end{cases}$$

where $0 < t_1 < t_0 \leq 1$. Clearly, μ is a fuzzy subgroup of G. Therefore, $\mu([x_1, y_1]) = t_1 \neq \mu(e)$. $\qquad\square$

Corollary 4.1.19. *Let G be a group. If G is nilpotent and every fuzzy subgroup μ of G is such that $\mu([x,y]) = \mu(e)\ \forall x \in G, y \in G'$, then G is nilpotent of class ≤ 2.*

Lemma 4.1.20. *[7, 10]. G is nilpotent if and only if G contains no (p,q)-subgroup for all $p, q \in \pi(G)$ with $p \neq q$.*

Theorem 4.1.21. *If for all fuzzy subgroups μ of G, $\mu([x,y]) = \mu(e)$ for all $x \in G, y \in G$ such that $|y| = prime$ or 4, then G is nilpotent.*

Proof. Suppose every fuzzy subgroup of G satisfies the condition. Assume G is not nilpotent. Then there exists a minimal nonnilpotent subgroup H of G such that $H = PQ$, $P \in Syl_p(H)$, $Q \in Syl_q(H)$, $p \neq q$ such that $P \lhd H$ and Q is cyclic by Lemma 4.1.20. Thus by Lemma 4.1.16, there exists a fuzzy subgroup μ of H such that μ is not normal and $\mu([x_1, y_1]) \neq \mu(e)$ for some $x_1 \in H$, $y_1 \in H' = P$. Now either p is odd or $p = 2$. If p is odd then $\text{Exp}(P) = p$. Thus $o(y_1) = p$. If $p = 2$, then $\text{Exp}(P)$ equals at most 4, [18]. Hence $o(y_1) = 2$ or 4. Let μ_0 be the extension of μ to G by defining $\mu_0(x) = 0$ for all $x \in G \backslash H$. Then $\mu_0([x_1, y_1]) \neq \mu_0(e)$ a contradiction to the hypothesis. $\square$

4.2 Characterization of Fuzzy Subgroups

We introduce the notions of generalized characteristic fuzzy subgroups, fully invariant fuzzy subgroups, and characteristic fuzzy subgroups. We prove that if G is a finite group all of whose Sylow subgroups are cyclic, then μ is a normal fuzzy subgroup of G if and only if μ is a generalized characteristic fuzzy subgroup of G.

A fuzzy subgroup μ of a group G is called a **generalized characteristic fuzzy subgroup** (GCFS) if for all $x, y \in G$, $o(x) = o(y)$ implies $\mu(x) = \mu(y)$, [2]. In Section 3.2 (see also [29]), a fuzzy subgroup μ of a group G is called a **fully invariant (characteristic)** fuzzy subgroup if $\mu \supseteq f(\mu)$ for every $f \in \text{End}(G)$ $(f \in \text{Aut}(G))$.

We study some relationships between GCFS, fully invariant fuzzy subgroups, and characteristic fuzzy subgroups.

Lemma 4.2.1. *Let f be a homomorphism of a group G into a group H and let μ and ν be fuzzy subgroups of G and H, respectively. Then $\nu \supseteq f(\mu)$ if and only if $\nu(f(x)) \geqslant \mu(x)$ for all x in G.*

Proof. $\nu \supseteq f(\mu) \Leftrightarrow \nu(f(x)) \geq f(\mu)(f(x))\forall x \in G \Leftrightarrow \nu(f(x)) \geq \vee\{\mu(z) \mid f(z) = f(x)\}\forall x \in G \Leftrightarrow \nu(f(x)) \geq \mu(x)\forall x \in G$. $\square$

Proposition 4.2.2. *Let μ be a fuzzy subgroup of a group G. Then the following statements are equivalent.*

(1) *μ is fully invariant (characteristic).*

(2) $\mu(f(x)) \geqslant \mu(x)$ for all x in G and for all f in $End(G)$ (f in $Aut(G)$).

(3) Every level subgroup of μ is a fully invariant (characteristic) subgroup of G.

Proof. We prove (2) $\Leftrightarrow$ (3). Suppose (2) holds. Let $f \in End(G)$ $(Aut(G))$ and $t \in [0, \mu(e)]$. Then $x \in \mu_t \Rightarrow \mu(x) \geq t \Rightarrow \mu(f(x)) \geq t \Rightarrow f(x) \in \mu_t$. Thus $f(\mu_t) \subseteq \mu_t$. Suppose (3) holds. Let $x \in G$ and let $\mu(x) = t$. Since $f(\mu_t) \subseteq \mu_t, f(x) \in \mu_t$ and so $\mu(f(x)) \geq t$. Thus $\mu(f(x)) \geq \mu(x)$.   □

Proposition 4.2.3. *Let G be a group and let $x, y \in G$. If μ is a GCFS of G and $o(x) \mid o(y)$, then $\mu(x) \geqslant \mu(y)$.*

Proof. Since $o(x) \mid o(y)$, $o(\langle x \rangle) \mid o(\langle y \rangle)$ and since $\langle y \rangle$ is a cyclic subgroup, it follows that $\langle y \rangle$ contains a subgroup H such that $o(H) = o(\langle x \rangle)$. Since H is cyclic, $H = \langle y_1 \rangle$ for some $y_1 \in \langle y \rangle$ and so $y_1 = y^i$ for some positive integer i. Now $o(x) = o(y_1)$ and so $\mu(x) = \mu(y_1) = \mu(y^i) \geqslant \mu(y)$ since μ is GCFS.   □

Proposition 4.2.4. *Let μ be a fuzzy subgroup of G. Then the following assertions hold.*

(1) *If μ is a GCFS, then μ is a fully invariant.*

(2) *If μ is a fully invariant, then μ is a characteristic.*

Proof. (1) Let μ be a GCFS and f be an endomorphism of G. It follows that $o(f(x)) \mid o(x) \ \forall x \in G$. Thus by Proposition 4.2.3, $\mu(f(x)) \geqslant \mu(x) \forall x \in G$. Hence by proposition 4.2.2, $\mu \supseteq f(\mu)$. That is, μ is fully invariant.

(2) The desired result follows since an automorphism is an endomorphism.
    □

The following examples show that the converse of Proposition 4.2.4 is not true.

Example 4.2.5. *Consider $G = C_2 \otimes S_3$, where $C_2 = \langle c \mid c^2 = e \rangle$ and $S_3 = \langle a, b \mid a^2 = b^3 = e, ab = b^2 a \rangle$. Then $Z(G) = C_2$ and hence C_2 is a characteristic subgroup of G. Define the fuzzy subset μ of G as follows: $\forall x \in G$,*

$$\mu(x) = \begin{cases} t_0 & \text{if } x = e, \\ t_1 & \text{if } x = c, \\ 0 & \text{otherwise,} \end{cases}$$

where $0 < t < t_0 \leq 1$. The chain of level subgroups of μ is $G \supset C_2 \supset \langle e \rangle$ and every subgroup in the chain is a characteristic subgroup of G. Hence by Proposition 4.2.2, μ is characteristic. However, μ is not a GCFS since $o(c) = o(a)$ and $\mu(c) \neq \mu(a)$. Furthermore, μ is not fully invariant since C_2 is a level subgroup of μ that is not a fully invariant subgroup of G.

Example 4.2.6. Let $G = S_4$, the symmetric group of degree 4. Let A_4 denote the alternating subgroup of G. It is known that
$N = \{e, (12)(34), (13)(24), (14)(23)\}$ is a normal subgroup of G. Define the fuzzy subset μ of G as follows: $\forall x \in G$,

$$\mu(x) = \begin{cases} t_0 & \text{if } x = e, \\ t_1 & \text{if } x \in N \setminus \langle e \rangle, \\ t_2 & \text{if } x \in A_4 \setminus N, \\ 0 & \text{otherwise}, \end{cases}$$

where $0 < t_2 < t_1 < t_0 \leq 1$. Clearly, the chain of level subsets of μ is $S_4 \supset A_4 \supset N \supset \langle e \rangle$. Hence μ is a fuzzy subgroup of G. Moreover $A_4 = G'$ (the commutator subgroup of G), $N = A_4'$ and $\langle e \rangle = N'$. Thus by [[27], Exercise 2.11.15 and Theorem 3.4.9], the level subgroups of μ are fully invariant subgroups of G. Therefore, by Proposition 4.2.2, μ is a fuzzy invariant fuzzy subgroup of G, but μ is not GCFS since $o((12)) = o((12)(34))$ and $\mu((12)) \neq \mu((12)(34))$.

Proposition 4.2.7. Let G be a group. If $\{\mu_i \mid i \in I\}$ is a family of GCFS (resp. characteristic or fully invariant) fuzzy subgroups of G, then $\mu = \cap_{i \in I} \mu_i$ is a GCFS (resp. characteristic or fully invariant) fuzzy subgroup of G.

Proof. By Theorem 1.2.13, μ is a fuzzy subgroup of G. First we prove that if each μ_i is a GCFS, then $\mu = \cap_{i \in I} \mu_i$ is a GCFS. Suppose $o(x) = o(y)$, where $x, y \in G$. Then $\mu_i(x) = \mu_i(y) \forall i \in I$. In fact, $\mu(x) = \cap_{i \in I} \mu_i(x) = \wedge \{\mu_i(x) \mid i \in I\} = \wedge \{\mu_i(y) \mid i \in I\} = \mu(y)$ and so μ is a GCFS. Now we show that if μ_i is a characteristic (fully invariant) fuzzy subgroup $\forall i \in I$, then $\mu = \cap_{i \in I} \mu_i$ is characteristic (fully invariant).

Let $f \in \text{Aut}(G)$ $(f \in \text{End}(G))$. Then
$f(\mu)(y) = f(\cap_{i \in I} \mu_i)(y)$
$$= \begin{cases} \vee\{(\cap_{i \in I} \mu_i)(x) \mid f(x) = y\} & \text{if } f^{-1}(y) \neq \emptyset, \\ 0 & \text{if } f^{-1}(y) = \emptyset. \end{cases}$$
$$= \begin{cases} \vee\{\wedge\{\mu_i(x) \mid i \in I\} \mid f(x) = y\} & \text{if } f^{-1}(y) \neq \emptyset, \\ 0 & \text{if } f^{-1}(y) = \emptyset. \end{cases}$$
$$\leqslant \begin{cases} \wedge\{\vee\{\mu_i(x) \mid f(x) = y\} \mid i \in I\} & \text{if } f^{-1}(y) \neq \emptyset, \\ 0 & \text{if } f^{-1}(y) = \emptyset. \end{cases}$$
$\leq \wedge\{f(\mu_i)(y) \mid i \in I\} = \cap_{i \in I} f(\mu_i)(y)$
$\leqslant \cap_{i \in I}(\mu_i)(y)$
(since μ_i is characteristic (fully invariant))
$= \mu(y)$. $\qquad\square$

We prove that if G is a group all of whose Sylow subgroups are cyclic, then μ is a normal fuzzy subgroup of G if and only if μ is a GCFS.

Remark 4.2.8. Let G be a group all of whose Sylow subgroups are cyclic. Then by [[27], Theorem 12.6.7], $G = HK$, $H \cap K = \langle e \rangle$ and H, K are cyclic Hall subgroups and H is normal in G. Furthermore, every element in G is of the form $h^i k^j$, where $H = \langle h \rangle$ and $K = \langle k \rangle$. Thus the following properties are immediate:

(1) G is a solvable group.

(2) If $o(H) = n$, $o(K) = m$, then $o(G) = nm$ and since G is solvable, it follows from Hall's theorem [[17], Theorem 7.14, p. 104] that any two subgroups of order m (or n) are conjugate. However, H is normal if G contains only one subgroup, namely H, of order n.

Theorem 4.2.9. *Suppose G is a group all of whose Sylow subgroups are cyclic. Let μ be a normal fuzzy subgroup of G. Then the following properties hold $\forall x, y \in G$.*

(1) If $o(x) \mid o(y)$, then $\mu(y) \leqslant \mu(x)$.

(2) If $o(x) = o(y)$, then $\mu(y) = \mu(x)$.

Proof. (1) Since all Sylow subgroups of G are cyclic, it follows from Remark 4.2.8 that $G = HK$ with $H \cap K = \langle e \rangle$, where K and H are cyclic Hall subgroups of G such that H is normal in G. Furthermore, every element of G is of the form $h^i k^j$, where $H = \langle h \rangle$ and $K = \langle k \rangle$, $k^m = e = h^n$, $(n, m) = 1$ and $o(G) = nm$. Suppose that $x, y \in G$ are such that $o(x) \mid o(y)$. Then we have the following three cases.

Case 1: Let $y \in H$. By Remark 4.2.8, H is the only normal cyclic subgroup of order n. Thus it follows that G has a unique subgroup of any order that divides n. Therefore, $H \supset \langle y \rangle \supset \langle x \rangle$. Hence $x = y^i$ for some integer i. Thus $\mu(x) = \mu(y^i) \geqslant \mu(y)$.

Case 2: Let $y \in K$. Consider the following two subcases.

Subcase 2.1: Suppose $y \in K$ and $x \in K$. Then $\langle y \rangle \supset \langle x \rangle$ and $x = y^j$ for some integer j. Therefore, $\mu(x) = \mu(y^j) \geqslant \mu(y)$.

Subcase 2.2: Suppose $y \in K$ and $x \notin K$. Then it follows that $x \in z^{-1} K z$ for some $z \in H$ by Remark 4.2.8. Therefore, there exists $x_1 \in K$ such that $x_1 = z x z^{-1}$. Hence $o(x) = o(x_1)$. Thus $\langle y \rangle \supset \langle x_1 \rangle$. Now $x_1 = y^i$ for some integer i. Since μ is normal, $\mu(x) = \mu(z x z^{-1}) = \mu(x_1) = \mu(y^i) \geqslant \mu(y)$.

Case 3: Suppose $y \notin H$ and $y \notin K$. If $o(y) = nm$, then G is cyclic and is generated by y. Thus $\mu(y) \leq \mu(x)$. Suppose $o(y) \neq nm$. We consider the following two subcases.

Subcase 3.1: Suppose $o(y) = m'$, $m' \mid m$. Then $z^{-1} K z \supset \langle y \rangle$ for some $z \in H$ and there exists $x_1 \in z^{-1} K z$ such that $x_1 = z_1^{-1} x z_1$ for some $z_1 \in H$. Thus $\langle y \rangle \supset \langle x_1 \rangle$. Hence $x_1 = y^i$ for some integer i. Therefore, $\mu(x) = \mu(z_1^{-1} x z_1) = \mu(x_1) = \mu(y^i) \geqslant \mu(y)$.

Subcase 3.2: Suppose $o(y) = m_1 n_1$, where $m_1 n_1 \mid mn$ and $(m_1, n_1) = 1$. Then by [[15], Lemma 3.2.1, p. 39], y has a unique representation $y = y_1 y_2 = y_2 y_1$, where $o(y_1) = n_1$, $o(y_2) = m_1$. Furthermore, both y_1 and y_2 are powers of y, i.e., $y_1 = y^{j_1}$, $y_2 = y^{j_2}$ for integers j_1, j_2. Now since $o(x) \mid o(y)$, we have that $o(x) = m_2 n_2$, where $m_2 \mid m_1$ and $n_2 \mid n_1$. Thus we have three different possibilities:

(3.2.1) $m_2 = 1$ and so $o(x) = n_2$: Since $\langle y_1 \rangle$ is a finite cyclic group of order n_1 and $n_2 \mid n_1$, it follows that $\langle y_1 \rangle$ has a unique subgroup of order n_2, namely $\langle x \rangle$. Therefore, $\langle y_1 \rangle \supset \langle x \rangle$ and so $x = y_1^k$ for some integer k. Thus $\mu(x) = \mu(y_1^k) = \mu((y^{j_1})^k) = \mu(y^{j_1 k}) \geqslant \mu(y)$.

(3.2.2) $n_2 = 1$ and so $o(x) = m_2$: Then $y_2 \in v^{-1} K v$ for some $v \in H$ and there exists $z_2 \in H$ such that $z_2^{-1} x z_2 \in v^{-1} K v$. Now since $o(z_2^{-1} x z_2) = o(x)$, it follows that $z_2^{-1} x z_2 \in \langle y_2 \rangle$. Hence $z_2^{-1} x z_2 = y_2^q$ for some integer q. Thus it follows that $\mu(x) = \mu(z_2^{-1} x z_2) = \mu(y_2^q) = \mu((y^{\alpha_2})^q) = \mu(y^{\alpha_2 q}) \geqslant \mu(y)$.

(3.2.3) $o(x) = m_2 n_2$, $m_2 \neq 1 \neq n_2$ and $(n_2, m_2) = 1$: Then by [[15], Lemma 3.2.1, p. 39], x has a unique representation $x = x_1 x_2 = x_2 x_1$, $o(x_1) = n_2$ and $o(x_2) = m_2$. Clearly, $H \supset \langle y_1 \rangle \supset \langle x_1 \rangle$. Hence $x_1 = y_1^{e_1}$ for some integer e_1. Also, $y_2 \in w_1^{-1} K w_1$ and $x_2 \in w_2^{-1} K w_2$ for some $w_1, w_2 \in H$. Thus there exists $u \in H$ such that $u^{-1} x_2 u \in w^{-1} K w_1$. Hence $u^{-1} x_2 u \in \langle y_2 \rangle$. Thus $u^{-1} x_2 u = y_2^{e_2}$ for some integer e_2. Therefore, we have that

$\mu(x) = \mu(x_1 x_2) \geqslant \mu(x_1) \wedge \mu(x_2)$

$\geqslant \mu(y_1^{e_1}) \wedge \mu(u^{-1} x_2 u)$ (since μ is normal)

$\geqslant \mu((y^{j_1})^{e_1}) \wedge \mu(y_2^{e_2})$

$\geqslant \mu(y^{j_1 e_1}) \wedge \mu(y^{j_2 e_2})$

$\geqslant \mu(y) \wedge \mu(y) \geqslant \mu(y)$.

Thus (1) holds.

(2) Since $o(x) = o(y)$, we have that $o(x) \mid o(y)$ and $o(y) \mid o(x)$. By (1), we have that $\mu(x) \geqslant \mu(y)$ and $\mu(x) \leqslant \mu(y)$. Hence $\mu(x) = \mu(y)$. Thus (2) holds. □

Corollary 4.2.10. *Let G be a group all of whose Sylow subgroups are cyclic. Let μ be a fuzzy subgroup of G. Then μ is normal if and only if μ is a GCFS.*

The following example shows that the normality assumption in Theorem 4.2.9 cannot be removed.

Example 4.2.11. *Let $G = S_3$, the symmetric group on $\{1, 2, 3\}$. Let $e = (1)$, $a = (12)$, $b = (123)$. Then $b^2 = (132)$, $ab = (13)$, $ab^2 = (23)$. Clearly, S_3 is a group all of whose Sylow subgroups are cyclic. Define the fuzzy subset μ of S_3 as follows: $\forall x \in G$,*

$$\mu(x) = \begin{cases} t_0 & \text{if } x = e, \\ t_1 & \text{if } x = a, \\ 0 & \text{otherwise,} \end{cases}$$

where $0 < t_1 < t_0 \leq 1$. Then μ is a fuzzy subgroup of G which is not normal. Also, μ is not GCFS since $\mu(a) \neq \mu(ab)$ and $o(a) = o(ab)$.

Corollary 4.2.12. *Let G be a dihedral group of order $2n$ with n odd and let μ be a fuzzy subgroup of G. Then μ is normal if and only if μ is GCFS.*

Proof. Now $G = \langle a, b \mid a^n = e = b^2, ba = a^{-1} b \rangle$. It follows that $\langle a \rangle$ is normal in G, $\langle a \rangle \cap \langle b \rangle = \langle e \rangle$ and $G = \langle a \rangle \langle b \rangle$. Hence it follows that all Sylow subgroups of G are cyclic. Thus by applying Corollary 4.2.10, we obtain the desired result. □

The following example shows that Corollary 4.2.12 is not true in general if the dihedral group G is of order $2n$, where n is even.

Example 4.2.13. *Let* $G = D_4$, *the dihedral group of order 8. For this group,* $n = 4$ *and* $G = \langle a, b \mid a^4 = e = b^2, ba = a^{-1}b \rangle$. *Define the fuzzy subset* μ *of* G *as follows:* $\forall x \in G$,

$$\mu(x) = \begin{cases} t_0 & \text{if } x = e, \\ t_1 & \text{if } x = a^2, \\ t_2 & \text{if } x = a, a^3, \\ 0 & \text{otherwise}, \end{cases}$$

where $0 < t_2 < t_1 < t_0 \leq 1$. *Clearly, the level subsets of* μ *form the chain,* $\langle e \rangle \lhd \langle a^2 \rangle \lhd \langle a \rangle \lhd G$, *which is a normal chain of subgroups. Hence* μ *is a normal fuzzy subgroup of* G, *but* μ *is not a GCFS since* $o(b) = o(a^2)$ *and* $\mu(b) \neq \mu(a^2)$.

4.3 Quasi-normal and Normal Fuzzy Subgroups

We now consider normal fuzzy subgroups, quasinormal fuzzy subgroups, (p, q)-subgroups, fuzzy cosets, fuzzy conjugates and $SL(p, q)$-subgroups.

We show that $(Q(G), \circ)$ is a commutative idempotent semisimple semigroup, where $Q(G)$ is the set of all quasinormal fuzzy subgroups of G and $\circ$ is the product operation of fuzzy subgroups. We prove some results about fuzzy conjugate subgroups. The concepts of the core and the closure of a fuzzy subgroup of a group G are introduced. We give some equivalent conditions for G/μ to be nilpotent of class at most n. We also give a characterization of groups which have normal p-complements in terms of certain fuzzy subgroups. Throughout the remainder of the chapter all groups are assumed to be of finite order.

Let G be a group and μ, ν be fuzzy subgroups of G. Recall that the product $\mu \circ \nu$ is defined as follows:

$\mu \circ \nu(x) = \vee\{\mu(y) \wedge \nu(z) \mid y, z \in G, x = yz\}$ for all $x \in G$.

The operation $\circ$ is associative. Also, it is commutative if and only if $\mu \circ \nu$ is a fuzzy subgroup of G for all fuzzy subgroups μ, ν of G. Moreover, we have the following result.

Theorem 4.3.1. *[4]* $(\mathcal{NF}(G), \circ)$ *is a commutative idempotent semisimple semigroup, where* $\mathcal{NF}(G)$ *is the set of all normal fuzzy subgroups of a group* G.

Definition 4.3.2. *A subgroup* H *of a group* G *is called* **quasinormal** *if* H *permutes with every subgroup of* G.

A **subnormal series** of a group G is a chain of subgroups $G = G_0 \supseteq G_1 \supseteq ... \supseteq G_n$ such that G_{i+1} is a normal subgroup of G_i, $i = 0, 1, ..., n-1$. A subgroup H is called **subnormal** in G if there is a subnormal series from G to H.

Theorem 4.3.3. *[19] If H is a quasinormal subgroup of G, then H is subnormal in G.*

Recall from Lemma 3.1.5 that $\forall x, y \in G, \mu(xy^{-1}) = \mu(e)$ implies $\mu(x) = \mu(y)$. Further, from Section 3.2 that $Z_0(G) = G$ and $Z_i(G) = [Z_{i-1}(G), G]$ for $i = 1, 2, ...$ and from Section 3.3 that $G^{(0)} = G$ and $G^{(i)} = [G^{(i-1)}, G^{(i-1)}]$ for $i = 1, 2, ...$ Recall also that for all $i \in \mathbb{N}$, $Z_i(G) \supseteq G^{(i)}$.

Theorem 4.3.4. *[24] The following statements are equivalent.*
(1) *G is nilpotent of class at most n,*
(2) *$Z_n(G) = \langle e \rangle$.*

Recall that a group G is called a (p, q)-group if the following conditions hold:
(1) The order of G involves only the prime factors p and q.
(2) G is a minimal non-nilpotent group, i.e., a group which is not nilpotent and all its proper subgroups are nilpotent.
(3) The derived group G' is the Sylow p-subgroup of G.

We also recall that if G is a group and $x, y \in G$, then the element $x^{-1}y^{-1}xy$ is usually denoted by $[x, y]$ and is called the commutator of x and y. $\text{Exp}(G)$ stands for exponent of G, the smallest positive integer n such that $x^n = e$ for all $x \in G$.

Definition 4.3.5. *The group of all nonsingular 2×2 matrices over the field Z_3 with determinate equal to 1 is called an $SL(2, 3)$ group.*

An $SL(2, 3)$ group contains only one Sylow 2-subgroup.

Theorem 4.3.6. *[8] Let $S \in Syl_2(G)$. Then $S \lhd G$ if and only if G contains no $(p, 2)$-subgroup for all $p \in \pi(G) \backslash \{2\}$, where $\pi(G)$ is the set of all distinct primes that divide $o(G)$).*

Definition 4.3.7. *A fuzzy subgroup μ of G is called **quasinormal** if its level subgroups are quasinormal subgroups of G.*

Proposition 4.3.8. *Every normal fuzzy subgroup of G is quasinormal.*

Proof. The result is immediate since any normal subgroup is quasinormal. $\square$

Example 4.3.9. *We show that the converse of Proposition 4.3.8 is not true. Consider the group* $G = \left\langle x, y \mid x^p = y^{p^2} = e, xyx^{-1} = y^{1+p} \right\rangle$, *where p is an odd prime and e is the identity element of G. It is known that $\langle x \rangle$ is a quasinormal subgroup, but not normal [26]. Define a fuzzy subset μ of G as follows:* $\forall x \in G$,

$$\mu(y) = \begin{cases} t_0 & \text{if } y = e, \\ t_1 & \text{if } y \in \langle x \rangle \backslash \langle e \rangle, \\ 0 & \text{otherwise}, \end{cases}$$

where $0 < t_1 < t_0 \leq 1$. Clearly, μ is a quasinormal fuzzy subgroup of G since its level subgroups $G, \langle x \rangle, \langle e \rangle$ are quasinormal subgroups of G. However, μ is not a normal fuzzy subgroup of G since $\langle x \rangle$ is not normal in G.

Let μ be a fuzzy subgroup of G and let $t \in [0, 1]$. Define the fuzzy subset $t\mu$ of G by $(t\mu)(x) = t\mu(x)$ for all $x \in G$. Let $\text{Im}(\mu) = \{t_0, t_1, ..., t_n\}$, where $t_0 > t_1 > ... > t_n$. We may write μ in the following form:

$$\mu = \sum_{i=0}^{n} t_i(1_{\mu_{t_i}} - 1_{\mu_{t_{i-1}}}), \text{ where } 1_{\mu_{t_0-1}}(x) = 0 \forall x \in G.$$

Lemma 4.3.10. *Let μ be a fuzzy subgroup of G and let H, K be subgroups of G. Let $t_i \in \text{Im}(\mu), i = 0, 1, ..., n$. Then the following assertions hold.*

(1) *Let $t \in [0, 1]$. Then $t\mu$ is a fuzzy subgroup of G. Furthermore, $t\mu$ and μ have the same level subgroups if $t \neq 0$.*

(2) $1_{HK} = 1_H \circ 1_K$.

(3) $(1_{\mu_{t_i}} - 1_{\mu_{t_{i-1}}}) \circ 1_K = 1_{\mu_{t_i}} \circ 1_K - 1_{\mu_{t_{i-1}}} \circ 1_K$.

(4) $(_{\mu_{t_i}} - 1_{\mu_{t_{i-1}}}) \circ \nu = 1_{\mu_{t_i}} \circ \nu - 1_{\mu_{t_{i-1}}} \circ \nu$ *for every fuzzy subgroup ν of G.*

(5) $(t_i 1_{\mu_{r_i}}) \circ 1_K = t_i(1_{\mu_{t_i}} \circ 1_K) = 1_{\mu_{t_i}} \circ (t_i 1_K)$.

(6) $[t_i(1_{\mu_{t_i}} - 1_{\mu_{t_{i-1}}}) + t_{i-1}(1_{\mu_{t_{i-1}}} - 1_{\mu_{t_{i-2}}})] \circ 1_K = [t_i(1_{\mu_{t_i}} - 1_{\mu_{t_{i-1}}}) \circ 1_K + t_{i-1}(1_{\mu_{t_{i-1}}} - 1_{\mu_{t_{i-2}}})] \circ 1_K$.

(7) $[\sum_{i=0}^{n} t_i(1_{\mu_{r_i}} - 1_{\mu_{t_{i-1}}})] \circ 1_K = \sum_{i=0}^{n}[t_i(1_{\mu_{t_i}} - 1_{\mu_{t_{i-1}}}) \circ 1_K]$.

(8) $[\sum_{i=0}^{n} t_i(1_{\mu_{r_i}} - 1_{\mu_{t_{i-1}}})] \circ \nu = \sum_{i=0}^{n}[t_i(1_{\mu_{t_i}} - _{\mu_{t_{i-1}}}) \circ \nu]$ *for every fuzzy subgroup ν of G.*

Proof. (1) This follows from definitions.

(2) Let $x \in G$. Then $1_{KH}(x) = 1 \Leftrightarrow x \in HK \Leftrightarrow x = x_1 x_2$, where $x_1 \in H, x_2 \in K \Leftrightarrow 1_H \circ 1_K(x_1 x_2) = 1 \Leftrightarrow 1_H \circ 1_k(x) = 1$. Also, $1_{HK}(x) = 0 \Leftrightarrow x \notin HK \Leftrightarrow x$ cannot be written as a product $x_1 x_2$, with $x_1 \in H, x_2 \in K \Leftrightarrow x = x_1 x_2(x_1 \in H \text{ and } x_2 \notin K)$ or $(x_1 \notin H \text{ and } x_2 \in K)$ or $(x_1 \notin H \text{ and } x_2 \notin K) \Leftrightarrow 1_H \circ 1_K(x) = 0$.

(3) Let $x \in G$. Then $(1_{\mu_{t_i}} - 1_{\mu_{t_{i-1}}}) \circ 1_K(x)$

$$= \begin{cases} & \text{if } x = yz, y \in \mu_{t_i} \backslash \mu_{t_{i-1}}, z \in K, \\ & 0 \text{ otherwise}, \end{cases}$$

$$= \begin{cases} 1 & \text{if } x = yz \ (y \in \mu_{t_i}, z \in K) \text{ and } (y \notin \mu_{t_{i-1}}, z \in K), \\ & 0 \text{ otherwise}, \end{cases}$$

$$= 1_{\mu_{t_i}} \circ 1_K(x) - 1_{\mu_{t_{i-1}}} \circ 1_K(x)$$
$$= (1_{\mu_{t_i}} \circ 1_K(x) - 1_{\mu_{t_{i-1}}} \circ 1_K)(x).$$

(4) - (8) : An argument similar to that in (3) establishes (4) - (8).

We note that in Lemma 4.3.10 parts $(3) - (8)$ are valid if we act from the right by $\circ$, e.g., (4) $\nu \circ (1_{\mu_{t_i}} - 1_{\mu_{t_{i-1}}}) = \nu \circ 1_{\mu_{t_i}} - \nu \circ 1_{\mu_{t_{i-1}}}$. $\qquad\square$

Proposition 4.3.11. *Let K be a subgroup of G. Then K is quasinormal if and only if 1_K is a quasinormal fuzzy subgroup of G. Moreover, K is quasi-normal if and only if $1_K \circ 1_H = 1_H \circ 1_K$ for all subgroups H of G.*

Proof. The result follows from Lemma 4.3.10 (2) and the definition of quasi-normality. $\qquad\square$

Proposition 4.3.12. *Let K be a subgroup of G. Then K is quasinormal if and only if $1_K \circ \mu = \mu \circ 1_K$ for all fuzzy subgroups μ of G.*

Proof. Let K be a subgroup of G such that $1_K \circ \mu = \mu \circ 1_K$ for every fuzzy subgroup μ of G. Then $1_K \circ 1_H = 1_H \circ 1_K$ for every subgroup H of G. Thus it follows by Lemma 4.3.10 (2) that $1_{KH} = 1_{HK}$. Hence $KH = HK$. Thus K is quasinormal. Conversely, let K be a quasinormal subgroup of G and μ be any fuzzy subgroup of G. Then for $t_i \in \mathrm{Im}(\mu), i = 1, ..., n$, where $1_{\mu_{t_0-1}}(x) = 0 \forall x \in G$,

$$\mu \circ 1_K = (\sum_{i=0}^{n} t_i(1_{\mu_{t_i}} - 1_{\mu_{t_{i-1}}})) \circ 1_K$$

$$= (\sum_{i=0}^{n} t_i 1_{\mu_{t_i}} - t_i 1_{\mu_{t_{i-1}}})) \circ 1_K$$

$$= \sum_{i=0}^{n}((t_i 1_{\mu_{t_i}}) \circ \lambda_K - (t_i 1_{\mu_{t_{i-1}}}) \circ 1_K) \qquad \text{(by Lemma 4.3.10)}$$

$$= \sum_{i=0}^{n}(t_i(1_K \circ 1_{\mu_{t_i}}) - t_i(1_K \circ \lambda_{\mu_{t_{i-1}}})) \qquad \text{(by Proposition 4.3.11)}$$

$$= 1_K \circ (\sum_{i=0}^{n} t_i(1_{\mu_{t_i}} - 1_{\mu_{t_{i-1}}})) \qquad \text{(by Lemma 4.3.10)}$$

$$= 1_K \circ \mu. \qquad\qquad\qquad\qquad\square$$

As in [6], define the relation $\sim$ on $\mathcal{F}(G)$ as follows: $\forall \mu, \mu' \in \mathcal{F}(G), \mu \sim \mu'$ if and only if for all x, y in G, $\mu(x) \geqslant \mu(y) \Leftrightarrow \mu'(x) \geqslant \mu'(y)$. Then $\sim$ is an equivalence relation on $\mathcal{F}(G)$. Moreover, if μ, μ' are fuzzy subgroups of G, then $\mu \sim \mu' \Leftrightarrow \mu$ and μ' have the same set of level subgroups.

Theorem 4.3.13. *Let μ be a fuzzy subgroup of G with $Im(\mu) = \{t_0, t_1, ..., t_n\}$. Then μ is quasinormal if and only if $\mu \circ \nu = \nu \circ \mu$ for all fuzzy subgroups ν of G.*

Proof. Suppose μ is a quasinormal fuzzy subgroup of G. Then the level subgroups μ_{t_i} of μ are quasinormal subgroups of G and so $1_{\mu_{t_i}}$ is quasinormal for $i = 0, 1, ..., n$. Therefore, $1_{\mu_{t_i}} \circ \nu = \nu \circ 1_{\mu_{t_i}}$ for every fuzzy subgroup ν of G and $\forall i = 0, 1, .., n$. Hence

$$\mu \circ \nu = (\sum_{i=0}^{n} t_i(1_{\mu_{t_i}} - 1_{\mu_{t_{i-1}}})) \circ \nu$$

$$= (\sum_{i=0}^{n} t_i(1_{\mu_{t_i}} - 1_{\mu_{t_{i-1}}}) \circ \nu) \qquad \text{(by Lemma 4.3.10)}$$

$$= \sum_{i=0}^{n} (t_i(1_{\mu_{t_i}} \circ \nu) - t_i(1_{\mu_{t_{i-1}}} \circ \nu)) \qquad \text{(by Lemma 4.3.10)}$$

$$= \sum_{i=0}^{n} (t_i(\nu \circ 1_{\mu_{t_i}}) - t_i(\nu \circ 1_{\mu_{t_{i-1}}})) \qquad \text{(by Proposition 4.3.12)}$$

$$= \sum_{i=0}^{n} (\nu \circ t_i(1_{\mu_{t_i}} - 1_{\mu_{t_{i-1}}})) \qquad \text{(by Lemma 4.3.10)}$$

$$= \nu \circ (\sum_{i=0}^{n} t_i(1_{\mu_{t_i}} - 1_{\mu_{t_{i-1}}})) \qquad \text{(by Lemma 4.3.10)}$$

$$= \nu \circ \mu.$$

Conversely, assume that μ is a fuzzy subgroup of G such that $\mu \circ \nu = \nu \circ \mu$ for every fuzzy subgroup ν of G. Then $\mu \circ 1_K = 1_K \circ \mu$ for every subgroup K of G. Define a fuzzy subgroup μ' of G as follows: for $0 < s < 1$, $\text{Im}(\mu') = \{s^0, s, s^2, ..., s^n\}$, where $\mu'_{s^0} = \mu_{t_0}, \mu'_s = \mu_{t_1}, \mu'_{s^2} = \mu_{t_2}, ..., \mu'_{s^n} = \mu_{t_n} = G$. Then μ and μ' have the same chain of level subgroups. Hence if μ' is quasinormal, then μ is by definition. Now $\mu' = \Sigma_{i=0}^{n} s^i(1_{\mu'_{s^i}} - 1_{\mu'_{s^{i-1}}})$, where $1_{\mu'_{s^{0-1}}}(x) = 0 \forall x \in G$. Thus for all $x \in G$,

$$\mu' \circ 1_K(x) = 1_K \circ \mu'(x)$$

$$\Rightarrow (\sum_{i=0}^{n} s^i(1_{\mu'_{ls^i}} - 1_{\mu'_{s^{i-1}}})) \circ 1_K(x) = 1_K \circ (\sum_{i=0}^{n} s^i(1_{\mu'_{s^i}} - 1_{\mu'_{s^{i-1}}}))(x)$$

$$\Rightarrow \sum_{i=0}^{n} s^i(((1_{\mu'_{s^i}} \circ 1_K - 1_K \circ 1_{\mu'_{s^i}}) - (1_{\mu'_{s^{i-1}}} \circ 1_K - 1_K \circ 1_{\mu'_{s^{i-1}}})))(x) = 0$$

$$\Rightarrow (((1_{\mu'_{s^i}} \circ 1_K - 1_K \circ 1_{\mu'_{s^i}}) - (1_{\mu'_{s^{i-1}}} \circ 1_K - 1_K \circ 1_{\mu'_{s^{i-1}}})))(x) = 0, i = 0, 1, ..., n$$

$$\Rightarrow (1_{\mu'_{s^i}} \circ 1_K - 1_K \circ 1_{\mu'_{s^i}})(x) = (1_{\mu'_{s^{i-1}}} \circ 1_K - 1_K \circ 1_{\mu'_{s^{i-1}}})(x), i = 0, 1, ..., n$$

$$\Rightarrow (1_{\mu'_{s^i}} \circ 1_K - 1_K \circ 1_{\mu'_{s^i}})(x) = 0, i = 0, 1, ..., n,$$

for otherwise there exists $x \in G$ such that $(1_{\mu'_{s^i}} \circ 1_K - 1_K \circ 1_{\mu'_{s^i}})(x) = 1$ or -1, say 1, and so $1_{\mu'_{s^i}} \circ 1_K(x) = 1$ and $1_K \circ 1_{\mu'_{s^i}}(x) = 0$ for all $0 \leqslant i \leqslant n$. Hence $1_G \circ 1_K(x) = 1$ and $1_K \circ 1_G(x) = 0$ and so $1_{GK}(x) = 1$, $1_{KG}(x) = 0$. Therefore, $x \in GK$ and $x \notin KG$, a contradiction. Thus it follows that

$$1_{\mu'_{s^i}} \circ 1_K = 1_K \circ 1_{\mu'_{s^i}}$$

and so

$$1_{\mu'_{s^i}K} = 1_{K\mu'_{s^i}} \text{ for all } 0 \leqslant i \leqslant n \qquad \text{(by Lemma 4.3.10(2)).}$$

Hence $\mu'_{s^i}K = K\mu'_{s^i}$ for every subgroup K of G and $0 \leqslant i \leqslant n$. Thus μ'_{s^i} is quasinormal subgroup for all $0 \leqslant i \leqslant n$. Hence μ_{t_i} is a quasinormal subgroup for all $0 \leqslant i \leqslant n$. Therefore, μ is quasinormal of G. $\qquad \square$

Recall that if μ and ν are two fuzzy subgroups of G, then $\mu \circ \nu$ is a fuzzy subgroup of G if and only if $\mu \circ \nu = \nu \circ \mu$. As a consequence of this result, we have the following two corollaries.

Corollary 4.3.14. *If μ is a quasinormal fuzzy subgroup and ν is any fuzzy subgroup of G such that $Im(\mu)$ is finite, then $\mu \circ \nu$ is fuzzy subgroup of ν.*

Corollary 4.3.15. *If μ and ν are quasinormal fuzzy subgroups of G with finite images, then $\mu \circ \nu$ is a quasinormal fuzzy subgroup of G.*

Proof. By Corollary 4.3.14, it follows that $\mu \circ \nu$ is a fuzzy subgroup of G. Let γ be a fuzzy subgroup of G. Then since $\circ$ is a associative and ν and μ are quasinormal,

$$(\mu \circ \nu) \circ \gamma = \mu \circ (\nu \circ \gamma)$$
$$= \mu \circ (\gamma \circ \nu)$$
$$= (\mu \circ \gamma) \circ \nu$$
$$= (\gamma \circ \mu) \circ \nu$$
$$= \gamma \circ (\mu \circ \nu).$$

Thus $\mu \circ \nu$ is quasinormal by Theorem 4.3.13. $\square$

Corollary 4.3.16. *Let $Q(G)$ denote the set of all quasinormal fuzzy subgroups of G. Then $(\mathbf{Q}(G), \circ)$ is a commutative idempotent semigroup.*

Proof. By Corollary 4.3.15, it follows that $\mathbf{Q}(G)$ is closed under operation $\circ$ and that $\circ$ is commutative. Since $\circ$ is an associative, $(\mathbf{Q}(G), \circ)$ is a semigroup. Also, for any fuzzy subgroup μ of G, $\mu \circ \mu = \mu$. $\square$

Definition 4.3.17. *Let S be a semigroup and M be a nonempty subset of S.*
*(1) M is called a **left (right) ideal** of S if $M \supseteq SM$ ($M \supseteq MS$). M is called an **ideal** if it is a left and right ideal.*
*(2) S is called **semisimple** if $M^2 = M$ for every ideal M of S.*

Definition 4.3.18. *Let $\mathcal{A}$ and $\mathcal{B}$ be two subsets of $\mathbf{Q}(G)$. Define $\mathcal{A} \circ \mathcal{B} = \{\mu \circ \nu \mid \mu \in \mathcal{A}, \nu \in \mathcal{B}\}$ and $\mathcal{A}^2 = \mathcal{A} \circ \mathcal{A}$.*

Theorem 4.3.19. *$(\mathbf{Q}(G), \circ)$ is a commutative idempotent semisimple semigroup.*

Proof. By Corollary 4.3.16, it suffices to show that $(\mathbf{Q}(G), \circ)$ is semisimple. Let $\mathcal{A}$ be an ideal in $\mathbf{Q}(G)$. Then $\mathcal{A} \supseteq \mathcal{A}^2$. Let $\mu \in \mathcal{A}$. Then $\mu = \mu \circ \mu \in \mathcal{A}^2$ and so $\mathcal{A}^2 \supseteq \mathcal{A}$. Thus $\mathcal{A}^2 = \mathcal{A}$ for every ideal $\mathcal{A}$ in $\mathbf{Q}(G)$. Hence $(\mathbf{Q}(G), \circ)$ is semisimple. $\square$

Note that $(\mathcal{NF}(G), \circ)$ is a subsemigroup of $(\mathbf{Q}(G), \circ)$, where $\mathcal{NF}(G)$ is the set of all normal fuzzy subgroups of G.

Definition 4.3.20. *A fuzzy subgroup is called* **subnormal** *if its level subgroups are subnormal.*

Theorem 4.3.21. *Any quasinormal fuzzy subgroup of G is subnormal.*

Proof. The result is immediate by Theorem 4.3.3 and the definitions of fuzzy quasi-normality and subnormality. $\quad\square$

We now show that the converse of Theorem 4.3.21 is not true: Let $G = A_4$, the alternating group on $\{1, 2, 3, 4\}$. Define the fuzzy subset μ of G as follows: $\forall x \in G$,

$$\mu(x) = \begin{cases} t_0 \text{ if } x = (1) = e, \\ t_1 \text{ if } x = (12)(34), \\ t_2 \text{ if } x \in N\backslash\langle(12)(34)\rangle, \\ 0 \text{ otherwise,} \end{cases}$$

where $0 \leq t_2 < t_1 < t_0 \leq 1$ and $N = \{(1), (12)(34), (13)(24), (23)(14)\}$. Then μ is a fuzzy subgroup since its level subsets are subgroups. Now the level subgroups of μ form a normal chain, $\langle(1)\rangle \lhd \langle(12)(34)\rangle \lhd N \lhd A_4 = G$. Thus μ is subnormal. However, μ is not quasinormal since $\langle(12)(34)\rangle$ is a level subgroup of μ that is not quasinormal.

Corollary 4.3.22. *Let G be a group of square free order and μ be a fuzzy subgroup of G. Then μ is quasinormal if and only if μ is normal.*

Proof. Suppose μ is quasinormal. Then the level subgroups of μ are quasinormal subgroups in G. Thus by Theorem 4.3.3, the level subgroups of μ are subnormal. However, since G in square free order, every level subgroup $\mu_t, t \in \text{Im}(\mu)$, is a Hall subgroup of G. Hence μ_t is normal for all $t \in \text{Im}(\mu)$ and thus μ is normal. The converse follows from Proposition 4.3.8. $\quad\square$

We now consider conjugate fuzzy subgroups.

Recall that if μ and ν are fuzzy subsets of a group G, then μ is said to be conjugate to ν if there exists $x \in G$ such that $\nu = x^{-1}\mu x$. The notation $\mu^x = x^{-1}\mu x$ is often used, where $\mu^x(g) = \mu(x^{-1}gx)\forall x \in G$. It follows that $(\mu^x)^y = \mu^{xy}$ for all x and y in G. This notion of conjugacy is an equivalence relation on $\mathcal{FP}(G)$. Suppose $\mu \in \mathcal{F}(G)$. Then $\mu^g \in \mathcal{F}(G)\forall g \in G$. Also, μ is normal if and only if μ^g is normal $\forall g \in G$. It follows easily that for all $x \in G$, μ and μ^x have, in order of inclusion, conjugate level subgroups of G. Moreover, the length of the chains of their level subgroups are equal.

Example 4.3.23. *Let $G = A_4$, the alternating group on $\{1, 2, 3, 4\}$. Then $G = \{(1), (12)(34), (13)(24), (14)(23), (123), (132), (142), (124), (234), (243), (134), (143)\}$. Define the fuzzy subset μ as follows: $\forall x \in G$,*

$$\mu(x) = \begin{cases} t_0 \text{ if } x = (1) = e, \\ t_1 \text{ if } x = (12)(34), \\ t_2 \text{ if } x \in \{(13)(24), (14)(23)\}, \\ 0 \text{ otherwise,} \end{cases}$$

where $0 < t_2 < t_1 < t_0 \leq 1$. Then μ is a fuzzy subgroup of G and the chain of level subgroups of μ is

$\langle(1)\rangle \lhd \langle(12)(34)\rangle \lhd \{(1), (12)(34), (13)(24), (23)(14)\} \lhd G$.

Let $x = (123)$. Then $\mu^x((12)(34)) = \mu((132)(12)(34)(123)) = \mu((13)(24)) = t_1$. We see it follows that

$\langle(1)\rangle \lhd \langle(13)(24)\rangle \lhd \{(1), (12)(34), (13)(24), (23)(14)\} \lhd G$.

is the chain of level subgroups of μ^x. The chains have the same length and the level subgroups of μ and μ^x, in order of inclusion, are conjugate.

Let μ be a fuzzy subgroup of G. Let $cl(\mu)$ denote the set of all fuzzy subgroups ν of G which are conjugate to μ.

Theorem 4.3.24. *If μ is a fuzzy subgroup of G, then $[G : N(\mu)] = |cl(\mu)|$.*

Proof. Let $x, y \in G$. Then $\mu^x = \mu^y$ if and only if $y^{-1}x \in N(\mu)$. Since $y^{-1}x \in N(\mu)$ if and only if $xN(\mu) = yN(\mu)$, the desired result now follows. □

We now obtain, for fuzzy subsets, an analog of a consequence of the following lemma.

Lemma 4.3.25. *[[27], Exercise 3.3.22(a)] Let f be a homomorphism of G onto a group H and let S is a subset of G. Then $f(cl(S)) = cl(f(S))$.*

Lemma 4.3.26. *Let f be a homomorphism of G onto a group H. If μ is a fuzzy subset of G, then $f(cl(\mu)) = cl(f(\mu))$.*

Proof. We have that

$\nu \in cl(f(\mu)) \Leftrightarrow \nu = (f(\mu))^{f(x)}$ for some $x \in G$

$\Leftrightarrow \nu(f(y)) = f(\mu)^{f(x)}(f(y)) \; \forall y \in G$

$\Leftrightarrow \nu(f(y)) = f(\mu)(f(x^{-1})f(y)f(x)) \forall y \in G \Leftrightarrow \nu(f(y)) = f(\mu)(f(x^{-1}yx)) \forall y \in G$

$\Leftrightarrow \nu(f(y)) = \vee\{\mu(z) \mid f(z) = f(x^{-1}yx), z \in G\} \forall y \in G \Leftrightarrow \nu(f(y)) = \vee\{\mu(z) \mid f(x^{-1}zx) = f(y), z \in G\} \forall y \in G$

$\Leftrightarrow \nu(f(y)) = \vee\{\mu(x^{-1}ux) \mid f(u) = f(y), u \in G\} \forall y \in G$

$\Leftrightarrow \nu(f(y)) = \vee\{\mu^x(u) \mid f(u) = f(y), u \in G\} \forall y \in G \Leftrightarrow \nu(f(y)) = f(\mu^x)(f(y)) \forall y \in G$

$\Leftrightarrow \nu = f(\mu^x) \Leftrightarrow \nu \in f(cl(\mu))$.

Let μ be a fuzzy subset of G. Define the subset $C(\mu)$ as follows: $C(\mu) = \{x \in G \mid \mu([x,y]) = \mu(e)$ for all $y \in G\}$. Clearly, $N(\mu) = \{x \in G \mid \mu(x^{-1}yx) = \mu(y) \; \forall \; y \in G\} = \{x \in G \mid \mu(xy) = \mu(yx) \; \forall \; y \in G\} \supseteq C(\mu)$. In general, $C(\mu) \neq N(\mu)$. For example, let $G = D_4 = \langle a, b \mid a^4 = 1 = b^2, ba = a^{-1}b \rangle$, the dihedral group of degree 4. Define the fuzzy subset μ as follows: $\forall x \in G$,

$$\mu(x) = \begin{cases} t_0 \text{ if } x = e, \\ t_1 \text{ if } x = a^2, \\ t_2 \text{ if } x \in \{a, a^3\}, \\ 0 \text{ otherwise,} \end{cases}$$

where $0 < t_2 < t_1 < t_0 \leq 1$. Then μ is a normal fuzzy subgroup since its level subsets are normal subgroups of G, and in fact, $\langle e \rangle \lhd \langle a^2 \rangle \lhd \langle a \rangle \lhd G$. Therefore, $N(\mu) = G$. Now $\mu([a, b]) = \mu(a^2) \neq t_0$. Thus $a, b \notin C(\mu)$. Now $C(\mu) = \langle a^2 \rangle$. Hence $N(\mu) \supset C(\mu)$. $\qquad\square$

Proposition 4.3.27. *If μ is a fuzzy subgroup of G, then $C(\mu)$ is a subgroup of G.*

Proof. Since $e \in C(\mu)$, $C(\mu)$ is not empty. Let $x, y \in C(\mu)$. Then $\mu([x, z]) = \mu(e)$ and $\mu([y, z]) = \mu(e)$ for all $z \in G$. We have that
$$\mu([xy, z]) = \mu([x, z]^y [y, z])$$
$$\geq ([x, z]^y) \wedge \mu([y, z])$$
$$\geq (\mu([x, z]^y) \quad \text{(since } \mu([y, z]) = \mu(e))$$
$$\geq \mu^y([x, z])$$
$$\geq \mu([x, z]) \quad \text{(since } y \in N(\mu))$$
$$= \mu(e).$$
Thus $xy \in C(\mu)$. Let $x \in C(\mu)$. Then $\mu([x, z]) = \mu(e)$ for all $z \in G$. Hence
$$\mu([x^{-1}, z]) = \mu(xz^{-1}x^{-1}z) = \mu(xz^{-1}x^{-1}zxx^{-1})$$
$$= \mu^{x^{-1}}(z^{-1}x^{-1}zx) \quad \text{(since } x^{-1} \in N_G(\mu))$$
$$= \mu([z, x])$$
$$= \mu([x, z]^{-1}) = \mu([x, z]) = \mu(e).$$
Therefore, $x^{-1} \in C(\mu)$. Thus $C(\mu)$ is a subgroup of G. $\qquad\square$

Definition 4.3.28. *The **core** of a fuzzy subgroup μ of G, written $core(\mu)$ or μ_G, is defined to be $core(\mu) = \cap_{x \in G} \mu^x$*

Proposition 4.3.29. *Let μ be a fuzzy subgroup of G. Then μ_G is a normal fuzzy subgroup of G.*

Proof. Since the intersection of any collection of fuzzy subgroups of G is a fuzzy subgroup of G, it suffices to show that μ_G is normal. Let $z \in G$. Then for all $y \in G$,
$$\mu_G^z(y) = \mu_G(z^{-1}yz) = \cap_{x \in G} \mu^x(z^{-1}yz) = \cap_{x \in G} \mu(x^{-1}z^{-1}yzx) =$$
$$\cap_{x \in G} \mu((zx)^{-1}y(zx)) = \cap_{x \in G} \mu^x(y) = \mu_G(y).$$
Thus μ_G is a normal fuzzy subgroup of G. $\qquad\square$

Proposition 4.3.30. *If μ, ν are fuzzy subgroups of G, ν is normal and $\mu \supseteq \nu$, then $\mu_G \supseteq \nu$.*

Proof. The result follows since $\mu^x \supseteq \nu^x = \nu$ for any $x \in G$ and so $\mu_G = \cap_{x \in G} \mu^x \supseteq \nu$. $\qquad\square$

Definition 4.3.31. *The **closure** of a fuzzy subgroup μ of G, written μ^G, is defined to be $\mu^G = \langle \{\mu^x \mid x \in G\} \rangle$, the fuzzy subgroup generated by all the conjugates of μ in G.*

It follows that $\mu^G = \langle \cup_{x \in G} \mu^x \rangle$.

The next result is immediate. Let $\text{Inn}(G)$ denote the set of all inner automorphisms of G.

Lemma 4.3.32. *If μ is a fuzzy subgroup of G, then $\mu^G = \cap_{\nu \supseteq f(\mu)} \nu$ for all $f \in \text{Inn}(G)$ and for all fuzzy subgroups ν of G.*

Proposition 4.3.33. *Let μ be a fuzzy subgroup of G. Then μ^G is a normal fuzzy subgroup of G.*

Proof. It suffices to show that $\mu^G \supseteq g(\mu^G)$ for all $g \in \text{Inn}(G)$. Let $g \in \text{Inn}(G)$. Then for all $f \in \text{Inn}(G)$ and for ν varying over $\mathcal{F}(G)$, we have that

$g(\mu^G)(x) = g(\cap_{\nu \supseteq f(\mu)} \nu)(x) \qquad \text{(Lemma 4.3.32)}$
$= \vee\{(\cap_{\nu \supseteq f(\mu)} \nu)(z) \mid g(z) = x, z \in G\}$
$= \vee\{\wedge\{\nu(z) \mid \nu \supseteq f(\mu), \nu \in \mathcal{F}(G)\} \mid g(z) = x, z \in G\}$
$\leq \wedge\{\vee\{\nu(z) \mid g(z) = x, z \in G\} \mid \nu \supseteq f(\mu), \nu \in \mathcal{F}(G)\}$
$\leq \cap_{\nu \supseteq f(\mu)} g(\nu)(x)$
$\leq (\cap_{g^{-1}(\gamma) \supseteq f(\mu)} \gamma)(x) \quad (g(\nu) = \gamma)$
$\leq \cap_{\gamma \supseteq g(f(\mu))} \gamma(x)$
$\leq \cap_{\gamma \supseteq h(\mu)} \gamma(x) \qquad \text{for all } h \in \text{Inn}G \text{ and } \gamma \in \mathcal{F}(G).$
$\leq \mu^G(x).$

Therefore, μ^G is normal. □

If μ is a normal fuzzy subgroup of G, recall that the set $G/\mu = \{x\mu \mid x \in G\}$ is a group with operation $(x\mu)(y\mu) = (xy)\mu$.

Lemma 4.3.34. *If μ is a fuzzy subgroup of G and $x, y \in G$, then $x\mu = y\mu$ if and only if $\mu(x^{-1}y) = \mu(y^{-1}x) = \mu(e)$.*

Proof. By Theorem 1.3.10, $x\mu = y\mu \Leftrightarrow x\mu_* = y\mu_* \Leftrightarrow y^{-1}x\mu_* = e\mu_* = x^{-1}y\mu_* \Leftrightarrow y^{-1}x\mu = e\mu = x^{-1}y\mu \Leftrightarrow \mu(x^{-1}yz) = \mu(z) = \mu(y^{-1}xz) \, \forall z \in G \Rightarrow \mu(x^{-1}y) = \mu(e) = \mu(y^{-1}x) \Leftrightarrow x^{-1}y, y^{-1}x \in \mu_* \Leftrightarrow x\mu_* = y\mu_* \Leftrightarrow x\mu = y\mu.$ □

Theorem 4.3.35. *Let μ be a fuzzy subgroup of G. Then $\mu(x^2) = \mu(e)$ for all $x \in G$ if and only if μ is normal and G/μ is an elementary Abelian 2 group.*

Proof. Suppose $\mu(x^2) = \mu(e)$ for all $x \in G$. Let $x, y \in G$. Then
$\mu(e) = \mu((xy)^2) = \mu(xyxy) = \mu((xyx^{-1})x^2y).$
Thus it follows by Lemma 3.1.5(2) that
$\mu(xyx^{-1}) = \mu((x^2y)^{-1}) = \mu(y^{-1}(x^2)^{-1})$
$\geqslant \mu(y^{-1}) \wedge \mu((x^2)^{-1}) = \mu(y^{-1}) \wedge \mu((x^{-1})^2)$
$\geqslant \mu(y) \wedge \mu(e)$
$\geqslant \mu(y).$

Thus μ is a normal fuzzy subgroup. Hence G/μ is a group. Now by Lemma 4.3.34, it follows that the statement $\mu(x^2) = \mu(e)$ is equivalent to the statement $x^2\mu = e\mu$. Thus for all $x\mu$ in G/μ, $(x\mu)^2 = (x\mu)(x\mu) = x^2\mu = e\mu$. Hence every element in G/μ is of order 1 or 2. Therefore, G/μ is an elementary Abelian 2-group. The converse follows by Lemma 4.3.34. □

Corollary 4.3.36. *Let μ be a fuzzy subgroup of G. Then $\mu(x^2) = \mu(e)$ for all $x \in G$ if and only if μ is normal and G/μ_* is an elementary abelian 2-group.*

Lemma 4.3.37. *Let μ be a normal fuzzy subgroup of G and H a subgroup of G. Then the restriction of μ to $H, \mu|_H$, is a normal fuzzy subgroup of H. Moreover, $H/\mu|_H$ is a subgroup of G/μ.*

Proof. It follows routinely that $\mu|_H$ is a normal fuzzy subgroup of H. By Theorem 1.3.12(3), $G/\mu \cong G/\mu_*$ and $H/\mu|_H \cong H/(\mu|_H)_*$. The natural homomorphism of G onto G/μ_* maps H onto the set of left cosets $\{h\mu_* | h \in H\}$. Now $(\mu|H)_* = H \cap \mu_*$ and $\{h\mu_* | h \in H\}$ is a group isomorphic to $H/(H \cap \mu_*)$. Thus making the appropriate identifications, we can consider $H/\mu|H$ as a subgroup of G/μ.

$\square$

Lemma 4.3.38. *If μ is a normal fuzzy subgroup of G, then any subgroup of G/μ can be written in the form $H/\mu|_H$, where H is a subgroup of G.*

Proof. Since any subset of G/μ is of the form $H/\mu|_H$, where H is a subset of G, it suffices to show that if $H/\mu|_H$ is a subgroup of G/μ, then H is a subgroup of G. Let $H/\mu|_H$ be a subgroup of G/μ and $a, b \in H$. Then $a\mu, b\mu \in H/\mu|_H$ and so $(a\mu)(b\mu)^{-1} \in H/\mu|_H$. Thus $ab^{-1}\mu \in H/\mu|_H$. Hence $ab^{-1} \in H$. Also, since $e\mu \in H/\mu|_H, e \in H$. Therefore, H is a subgroup of G. $\square$

In the remainder of the section, we write μ for $\mu|_H$ when there is no chance for confusion.

Theorem 4.3.39. *Let μ be a normal fuzzy subgroup of G and H, K be subgroups of G. If H and K are conjugates, then H/μ and K/μ are conjugate in G/μ.*

Proof. Consider the function $f : G \to G/\mu$ such that $f(x) = x\mu \; \forall x \in G$. Then f is a homomorphism from G onto G/μ. By Lemma 4.3.26, $H \in \text{cl}(K)$ implies $H/\mu \in f(\text{cl}(K)) = \text{cl}(f(K)) = \text{cl}(K/\mu)$ since H, K are conjugate subgroups in G. Thus H/μ and K/μ are conjugate in G/μ. $\square$

The converse of Theorem 4.3.39, is not true in general: Let $G = D_4$, the Dihedral group of degree 4, i.e., $G = \langle a, b \mid a^4 = e, b^2 = e, ba = a^{-1}b \rangle$. Define the fuzzy subset μ of G as follows: $\forall x \in G$,
$$\mu(x) = \begin{cases} t_0 & \text{if } x \in \langle a \rangle, \\ t_1 & \text{otherwise,} \end{cases}$$
where $0 \leq t_1 < t_0 \leq 1$. Clearly, μ is a normal fuzzy subgroup of G. Let $H = \{e, b, a^2, a^2b\}$ and $K = \langle b \rangle$. Then $K/\mu = \{\mu, b\mu\}$ and $H/\mu = \{\mu, b\mu, a^2\mu, a^2b\mu\}$. Now $t_1 = \mu(a^2e) = \mu(ea^2) = \mu(e)$. Also, $\mu((a^2b)^{-1}b) = \mu(ba^2b) = \mu(b^{-1}a^2b) = t_0$ since $b^{-1}a^2b \in \langle a \rangle$. By Lemma 4.3.34, $a^2\mu = \mu$ and $a^2b\mu = b\mu$. Hence $H/\mu = \{\mu, b\mu\} = K/\mu$. Thus H/μ and K/μ are trivially conjugate in G/μ. However, K and H are not conjugate in G since they have different orders.

The above example shows that $o(H/\mu) = o(K/\mu)$, but $o(H) \neq o(K)$. Also, if $L = \langle a \rangle$, then $o(L) = o(H)$. However, $o(L/\mu) \neq o(H/\mu)$ since $o(H/\mu) = 2$ and $o(L/\mu) = 1$.

In general, we have the following result.

Theorem 4.3.40. *If $G = D_{2^n}$, the dihedral group of degree 2^n, then there exists a normal fuzzy subgroup μ of G and subgroups H, K and L of G such that $o(H) = o(K)$, but $o(H/\mu) \neq o(K/\mu)$, $o(K) \neq o(L)$, and $o(K/\mu) = o(L/\mu)$.*

Proof. Now $G = \langle a, b \mid a^{2^n} = e, b^2 = e, ba = a^{-1}b \rangle$. Define the fuzzy subset μ of G as follows: $\forall x \in G$,

$$\mu(x) = \begin{cases} t_0 \text{ if } x \in \langle a \rangle, \\ t_1 \text{ otherwise,} \end{cases}$$

where $0 \leq t_1 < t_0 \leq 1$. Clearly, μ is a normal fuzzy subgroup of G. Let $H = \langle a \rangle$ and $L = \langle b \rangle$. Now $D'_{2^n} \cong Z_{2^{n-1}} \lhd G$, [17], and $D'_{2^n} \cap \langle b \rangle = \langle e \rangle$, $D'_{2^n} \langle b \rangle$ is a subgroup of G of order 2^n. If we let $K = D'_{2^n} \langle b \rangle$, then the desired result follows immediately. $\qquad\square$

Theorem 4.3.41. *Let μ be a normal fuzzy subgroup of G. Then the following conditions are equivalent:*
(1) *G/μ is nilpotent of class at most n.*
(2) *$\mu([x_1, x_2, ..., x_{n+1}]) = \mu(e)$ for all $x_i \in G, i = 1, 2, ..., n$.*
(3) *$\mu_* \supseteq G^{(n)}$.*

Proof. $(1) \Rightarrow (2)$: G/μ is a nilpotent of class at most $n \Rightarrow Z_n(G/\mu) = \{e\mu\}$ (Theorem 4.3.4) $\Rightarrow [x_1\mu, x_2\mu, ..., x_{n+1}\mu] = e\mu \Rightarrow [x_1, x_2, ..., x_{n+1}]\mu = e\mu$. Thus by Lemma 4.3.34, we have $\mu([x_1, x_2, ..., x_{n+1}] = \mu(e)$.

$(2) \Rightarrow (3)$: The result here is clear.

$(3) \Rightarrow (1)$: If $\mu_* \supseteq G^{(n)}$, then $\mu([x_1, x_2, ..., x_{n+1}]) = \mu(e)$ which implies, by Lemma 4.3.34, that $[x_1, x_2, ..., x_{n+1}]\mu = e\mu$. Thus $[x_1\mu, x_2\mu, ..., x_{n+1}\mu] = e\mu$ and hence $Z_n(G/\mu) = \{e\mu\}$. Thus G/μ is nilpotent of class at most n. $\qquad\square$

As an immediate consequence of this theorem, we have the following result.

Corollary 4.3.42. *Let μ be a normal fuzzy subgroup of G. Then the following statements are equivalent.*
(1) *G/μ is an Abelian group.*
(2) *$\mu_* \supseteq G^{(1)}$.*
(3) *$\mu([x_1, x_2]) = \mu(e)$ for all $x_i \in G$, $i = 1, 2$.*

We now characterize groups which have normal p-complements.

Theorem 4.3.43. *[14] Let p be a prime dividing $o(G)$. Then G has a normal p-complement if and only if G contains no (p, q)-subgroup for all $p, q \in \pi(G)$ with $p \neq q$.*

Theorem 4.3.44. *Let p be an odd prime number dividing $o(G)$. Then G has a normal p-complement if and only if there exists a fuzzy subgroup μ of G such that for all $x \in G$ with $o(x) = p$, $\mu(x)$ is fixed and $\mu(x) \neq \mu([x, y])$ for all $y \in G \backslash \{e\}$ such that p does not divide $o(y)$.*

Proof. Suppose G has a normal p-complement. Then there exist subgroups P, K of G such that $G = PK$, $K \lhd G$, $P \cap K = \langle e \rangle$ and $P \in Syl_p(G)$. Define a fuzzy subset μ of G as follows:$\forall x \in G$,

$$\mu(x) = \begin{cases} t_0 \text{ if } x = e, \\ t_1 \text{ if } x \in K \backslash \{e\}. \\ 0 \text{ otherwise,} \end{cases}$$

where $0 < t_1 < t_0 \leq 1$. Clearly, μ is a fuzzy subgroup and $[x, y] \in K$ for $x \in P$ and $y \in K$. Thus $\mu([x, y]) \geqslant t_1 > 0 = \mu(x)$. Hence $\mu([x, y]) \neq \mu(x)$. Conversely, suppose G has no normal p-complement. Then by Theorem 4.3.43, G contains a (p, q)-subgroup H of G, $p \neq q, p \neq 2$. Hence by definition of H, $H = PQ$, where $P \in Syl_p(H)$ and $Q \in Syl_q(H)$. Furthermore, $P \subset H$, Q is cyclic, $H' = P$ and $\text{Exp}(P) = p$. Thus for any $x \in P$ and $y \in Q, x \neq e \neq y$, $[x, y] \in P$. Since H is not nilpotent, it follows that there exist $x_0 \in P$ and $y_0 \in Q, x_0 \neq e \neq y_0$, such that $[x_0, y_0] \neq e$. Hence $o(x_0) = o([x_0, y_0]) = p$ since $\text{exp} P = p$. By the hypothesis $\mu(x_0) = \mu([x_0, y_0])$, a contradiction. □

The above theorem is not true in general if $p = 2$. For example, consider $G = SL(2, 3)$. Then G is a $(2, 3)$-group and hence has no normal 2-complement. Now G has only one element of order 2, say z. Moreover, $\langle z \rangle = Z(G)$, [23]. Thus for all $y \in G$, $[z, y] = e$. Define a fuzzy subset μ of G as follows: $\forall x \in G$,

$$\mu(x) = \begin{cases} t_0 \text{ if } x = e, \\ t_1 \text{ if } x = z, \\ 0 \text{ otherwise,} \end{cases}$$

where $0 < t_1 < t_0 \leq 1$. Then μ is a fuzzy subgroup of G and $\mu(z) \neq \mu([z, y])$. Thus we have a fuzzy subgroup μ of G which satisfies the hypothesis of Theorem 4.3.44, but G has no normal 2-complement.

We now consider the case $p = 2$.

Theorem 4.3.45. *Let G be a group of even order. Then G has a normal 2-complement if and only if there exists a fuzzy subgroup μ of G such that for all $x, y \in G$ with $o(x)|4$, $(o(x), o(y)) = 1$ and $\mu(x)$ is fixed implies $\mu(x) \neq \mu([x, y])$.*

Proof. The proof follows similarly as in the proof of Theorem 4.3.44 and by the fact that $\text{Exp}(S) = 4$, where $S \in Syl_2(G)$. □

Theorem 4.3.46. *Let G be a group of even order and $S \in Syl_2(G)$. Then S is normal in G if and only if there exists a fuzzy subgroup μ of G such that $\mu(x) \neq \mu([x, y])$ and $\mu(x) = $ fixed number for all $x, y \in G$, where $o(x)$ is an odd prime and y is a 2-element.*

Proof. Suppose $S \in Syl_2(G)$ and S is normal. Then by Theorem 4.3.6, G contains no $(p, 2)$-subgroup for all $p \in \pi(G) \backslash \{2\}$. Thus $G = SK$ and $(o(S), o(K)) = 1$. Define a fuzzy subset μ of G as follows:

$$\mu(x) = \begin{cases} t_0 \text{ if } x = e, \\ t_1 \text{ if } x \in S \backslash \langle e \rangle, \\ 0 \text{ otherwise} \end{cases}$$

where $0 < t_1 < t_0 \leq 1$. Clearly, μ is a fuzzy subgroup and $[x, y] \in S$ for all $x \in K$ and $y \in S$. Therefore, $\mu([x, y]) \geqslant t_1 > 0 = \mu(x)$. Hence $\mu([x, y]) \neq \mu(x)$. Conversely, suppose G has no normal subgroup S, where S in $Syl_2(G)$. Then by Theorem 4.3.6, G has a $(p, 2)$-subgroup H in G. Hence by definition of H, $H = PQ$, where $P \in Syl_p(L)$, P is normal in H, $Q \in Syl_2(H)$ and $p \equiv 1(2)$. Hence $o(P) = p$ else $o(P/p') = p^n, n \geqslant 2$. However, from the structure of a $(p, 2)$-subgroup, n must be the smallest positive integer such that $p^n \equiv 1(2)$. Thus $n = 1$, a contradiction. Now for all $x \in P$ and $y \in Q, x \neq e \neq y$, it follows that $[x, y] \in P$. Since H is not nilpotent, it follows that there exist $x_0 \in P$ and $y_0 \in Q, x_0 \neq e \neq y_0$, such that $[x_0, y_0] \neq e$. Hence $o([x_0, y_0]) = o(x_0) = p$. Thus $\mu([x_0, y_0]) = \mu(x_0)$, a contradiction. $\square$

References

1. S. Abou-Zaid, On normal fuzzy subgroups, *J. Fac. Ed. Ain Shams Univ. Cairo* 13(1988) 115-125.
2. S. Abou-Zaid, On generalized characteristic fuzzy subgroups of a finite group, *Fuzzy Sets and Systems* 43(1991) 235-241.
3. S. Abou-Zaid, On fuzzy subnormal and pronormal subgroup of a finite group, *Fuzzy Sets and Systems* 47(1992) 347-349.
4. S. Abou-Zaid, On fuzzy subgroups (short communication), *Fuzzy Sets and Systems* 55(1993) 237-240.
5. N. Ajmaal and K.V. Thomas, Quainormality and fuzzy subgroups, *Fuzzy Sets and Systems* 58 (1993) 217 - 225.
6. M. Akgul, Some properties of fuzzy groups, *J. Math. Anal. Appl.* 133(1988) 93-100.
7. M. Asaad, On the solvability, supersolvability, and nilpotency of finite groups, Annales Univ. Sci. Budapest XVI (1973) 115-124.
8. M. Asaad, Normal 2-subgroups of finite groups, *Afrika Mat.* 8(1986) 19-22.
9. M. Asaad, Groups and fuzzy subgroups, *Fuzzy Sets and Systems* 39(1991) 323-328.
10. M. Asaad, Groups without certain minimal non-nilpotent subgroups, *PU. M.A. Ser. A* 2 (3-4) (1991) 157-160.
11. M. Asaad and S. Abou-Zaid, Groups and fuzzy subgroups II (short communication), *Fuzzy Sets and Systems* 56 (1993) 375-377.
12. M. Asaad and S. Abou-Zaid, Fuzzy subgroups of nilpotent groups, *Fuzzy Sets and Systems* 60 (1993) 321-323.
13. M. Asaad and S. Abou-Zaid, Characterization of fuzzy subgroups, *Fuzzy Sets and Systems* 77 (1996) 247-251.
14. M. Asaad and S. Abou-Zaid, A contribution to the theory of fuzzy subgroups, *Fuzzy Sets and Systems* 77 (1996) 355-369.

15. P. S. Das, Fuzzy groups and level subgroups, *J. Math. Anal. Appl.* 84 (1981) 264-269.
16. M. Hall, *The Theory of Groups,* Macmillan, New York, 1959.
17. T. W. Hungerford, *Algebra,* Springer, New York, 1974.
18. B. Huppert, Endiche Gruppen I, Springer-Verlag, Berlin-Heidelberg-New York, 1967.
19. O. H. Kegel, Sylow-gruppen und subnormalteiler endlicher gruppen, *Math. Z.* 78 (1962) 205-221.
20. J-G Kim, On groups and fuzzy subgroups, *Fuzzy Sets and Systems* 67 (1994) 347 - 348.
21. R. Kumar, Fuzzy Sylow subgroups, *Fuzzy Sets and Systems* 46 (1992) 267 - 271.
22. W. J. Liu, Fuzzy invariant subgroups and fuzzy ideals, *Fuzzy Sets and Systems* 8 (1982) 133-139.
23. N. P. Mukherjee and P. Bhattacharya, Fuzzy normal subgroups and fuzzy cosets, *Inform. Sci.* 34 (1984) 225-239.
24. J. S. Rose, *A Course on Group Theory* (Cambridge Univ. Press, Cambridge, 1978).
25. A. Rosenfeld, Fuzzy groups, *J. Math. Anal. Appl.* 35 (1971) 512-517.
26. N. S. Narasimha Sastry and W. E. Deskins, Influence of normality conditions on almost minimal subgroups of a finite group, *J. Algebra* 52 (1978) 364-377.
27. W.R. Scott, *Group Theory* (Prentice-Hall, Englewood Cliffs, NJ, 1964).
28. F. I. Sidky and M.A. Mishref, Fuzzy cosets and cyclic abelian fuzzy subgroups, *Fuzzy Sets and Systems* 43 (1991) 243-250.
29. F. I. Sidky and M.A. Mishref, Fully invariant, characteristic and *S*-fuzzy subgroups, *Inform. Sci.* 55 (1991) 27-33.
30. L. A. Zadeh, Fuzzy sets, *Inform. and Control* 8 (1965) 338-353.

5

Free Fuzzy Subgroups and Fuzzy Subgroup Presentations

In this chapter, we define a notion of a free fuzzy subgroup and study its basic properties. We examine two approaches. The first approach is based on the work appearing in [18]. It uses the notion of levels and leads to the development of the notion of a presentation of a fuzzy subgroup. The definition of a presentation for a fuzzy subgroup provides a convenient method of defining fuzzy subgroups and opens the door for a development of a combinatorial group theory for fuzzy subgroups. The second approach follows along the lines of [28, 36]. The development here is presented in Section 5.3.

5.1 Free Fuzzy Subgroups

An important concept in the study of many mathematical theories is that of a free object. For instance, free objects appear in abstract algebra [25, 26, 27], mathematical logic, lattice theory [19], category theory, and in universal algebra [17]. Free objects are also of importance in computer science. They appear in logic programming and automated theorem proving [14, 24], as the so called Herbrand Universe. They figure prominently in algebraic approaches to the semantics of programming languages [20]. Consequently, a suitable concept of freeness for fuzzy objects may prove useful in the study of fuzzy counterparts of these theories. It is worth noting that there is a strong connection between fuzzy set theory, fuzzy logic, and computer science. For example, work in this area can be found in [29] on fuzzy logic programming, [15] on fuzzy programming languages, and [23, 35] on fuzzy logic and automated theorem proving.

In group theory, a notion related to that of a free object is the notion of a presentation. The notion of a presentation has proved useful in group theory. Presentations provide a convenient method of specifying defining properties of a group. They have also proved to be interesting in the study of combinatorial group theory. Problems in combinatorial group theory often use ideas from group theory, logic, and theoretical computer science. These fields have benefitted from the exchange of ideas.

Let X be a set and let x_t be a fuzzy singleton of X. We call x and t the **foot** and **level** of x_t, respectively.

Let Y be a set and $f : X \to Y$. Let μ be a fuzzy subset of X. Recall that $f(\mu)$ is the fuzzy subset of Y defined as follows: for all $y \in Y$,

$$f(\mu)(y) = \begin{cases} \vee\{\mu(x) \mid x \in f^{-1}(y)\} \text{ if } f^{-1}(y) \neq \emptyset, \\ 0 \text{ if } f^{-1}(y) = \emptyset. \end{cases}$$

If μ is a fuzzy subset of X and ν is a fuzzy subset of Y, then $\mu \times \nu$ is the fuzzy subset of $X \times Y$ defined by for all $x \in X, y \in Y$,

$$(\mu \times \nu)(x, y) = \mu(x) \wedge \nu(y).$$

We now recall some of the basic ideas concerning the construction of a free group. Let I be an index set. Let $X = \{x_i \mid i \in I\}$ be an alphabet of symbols and $X^{-1} = \{x_i^{-1} \mid i \in I\}$ be a set disjoint from X. Let $\Sigma = X \cup X^{-1}$ and define Σ^* to be the set of all finite sequences of symbols from Σ, including the empty sequence. Σ is sometimes called a **symmetrical alphabet**. Elements from Σ^* are called **words** over Σ. Define the relation $\sim$ on Σ^* as follows: $\forall w_1, w_2 \in \Sigma^*$, $w_1 \sim w_2$ if and only if w_1 can be transformed into w_2 by a finite sequence of insertions and/or deletions of subwords of the form $x_i x_i^{-1}$ or $x_i^{-1} x_i$, where $i \in I$. Then $\sim$ is an equivalence relation on Σ^*. Let $[u]$ denote the equivalence class of u with respect to $\sim$, where $u \in \Sigma^*$. Let $\Sigma^* / \sim = \{[u] \mid u \in \Sigma^*\}$. Define the binary operation on $\Sigma^* / \sim$ as follows: $\forall [u], [v] \in \Sigma^* / \sim$, $[u][v] = [uv]$, where uv represents the concatenation of the words of u, v. Then $\Sigma^* / \sim$ forms a group called the **free group** over X, and is denoted by $F(X)$. The identity of $F(X)$ is the equivalence class determined by the empty word. Since $F(X) = \langle X \rangle$, X is said to be the set of **generators** $F(X)$. Each equivalence class $[u]$ has a unique word in **reduced form**, i.e., one that contains no subwords of the form $x_i x_i^{-1}$ or $x_i^{-1} x_i$. The following theorems are well known. Their proofs can be found in [25, 27].

Theorem 5.1.1. *Every group is the homomorphic image of a free group.*

Theorem 5.1.2. *Let a group G be generated by a set $B = \{g_i \mid i \in I\}$ and let X be an alphabet $\{x_i \mid i \in I\}$. Then the function $m : X \to B$ defined by $m(x_i) = g_i \forall x_i \in X$, extends to a unique epimorphism $\hat{m} : F(X) \to G$, in the sense that $\hat{m}([x]) = m(x)$ for all $x \in X$.*

The elements of the set X are called the **defining generators** of the group G. By the first isomorphism theorem for groups, G is isomorphic to $F(X)/\text{Ker}(\hat{m})$. Hence G is completely determined (up to isomorphism) by specifying the alphabet X and the kernel of $\hat{m}$. $\text{Ker}(\hat{m})$ can itself be specified by giving a subset S of $\text{Ker}(\hat{m})$ whose normal closure is $\text{Ker}(\hat{m})$ and then by selecting a reduced word from each element of S. The set R of reduced words obtained in this manner is called a set of **defining relators** of G with respect to X. The pair $\langle X, R \rangle$ is called a **presentation** of G.

We now consider the notion of a free fuzzy subgroup. First recall that $x_t y_s = (xy)_{t \wedge s}$ and $(x_t)^{-1} = (x^{-1})_t$ for all fuzzy singletons x_t, y_s of G.

Definition 5.1.3. *Let G, H be groups and let μ, ν be fuzzy subgroups of G, H, respectively. We say ν is a **homomorphic image** of μ if there exists an epimorphism $h : G \to H$ such that $h(\mu) = \nu$. If h is an isomorphism, then we say μ and ν are **isomorphic**. (See Definition 1.4.9.)*

Every $w \in \Sigma^*$ can be written uniquely as $w = x_{i_1}^{\epsilon(1)} x_{i_2}^{\epsilon(2)} ... x_{i_k}^{\epsilon(k)}$ with $\epsilon(j) = \pm 1$, where $x^1 = x$. We call the set $\{i_1, i_2, ..., i_k\}$, the *I*-**index** set of w, and denote it by $I(w)$. The **inverse** of the word w is the word

$$w^{-1} = x_{i_k}^{-\epsilon(k)} x_{i_{k-1}}^{-\epsilon(k-1)} ... x_{i_1}^{-\epsilon(1)}.$$

We now define the concept of a free fuzzy subgroup.

Definition 5.1.4. *Let $T = \{t_i \in [0, 1] \mid i \in I\}$ and let $t \in [0, 1]$ be such that $t \geqslant \vee\{s \mid s \in T\}$. Define the fuzzy subset $f(X; T, t)$ by, for all $y \in F(X)$, $f(X; T, t)(y) = \vee\{\wedge\{t \wedge t_i \mid i \in I(w)\} \mid w \in y\}$. X is called the set of **generators**, T is called the set of **generating levels**, and t is called the **height** of $f(X; T, t)$. For all $i \in I, t_i$ is the **level** of x_i and x_i^{-1}.*

Theorem 5.1.5. *The fuzzy subset $f(X; T, t)$ of $F(X)$ is a fuzzy subgroup of $F(X)$.*

Proof. Set $\mu = f(X; T, t)$. Let $y_1, y_2 \in F(X)$ and let $w_1 \in y_1$, $w_2 \in y_2$. Then $w_1 w_2 \in y_1 y_2$ and

$$\begin{aligned} \mu(y_1 y_2) &= \vee\{\wedge\{t \wedge t_i \mid i \in I(w)\} \mid w \in y_1 y_2\} \\ &\geqslant \wedge\{t \wedge t_i \mid i \in I(w_1 w_2)\} \\ &= \wedge\{\wedge\{t \wedge t_{i_j} \mid i_j \in I(w_j)\} \mid j = 1, 2\}. \end{aligned}$$

Since w_1, w_2 are arbitrary, we have

$$\begin{aligned} \mu(y_1 y_2) &\geqslant \vee\{\wedge\{\wedge\{t \wedge t_i \mid i \in I(u_j)\} \mid j = 1, 2\} \mid u_1 \in y_1, u_2 \in y_2\} \\ &= (\vee\{\wedge\{t \wedge t_i \mid i \in I(u_1)\} \mid u_1 \in y_1\}) \wedge (\vee\{\wedge\{t \wedge t_i \mid i \in I(u_2)\} \mid u_2 \in y_2\}) \\ &= \mu(y_1) \wedge \mu(y_2). \end{aligned}$$

The proof that $\mu(y^{-1}) \geq \mu(y)$ for all $y \in G$ follows easily since for all $y \in F(X)$, $w \in y^{-1}$ if and only if $w^{-1} \in y$ and $I(w) = I(w^{-1})$. $\square$

We call $f(X; T, t)$ the **free fuzzy subgroup** of $F(X)$ over X, T, and t.

Theorem 5.1.6. *Every fuzzy subgroup is the homomorphic image of a free fuzzy subgroup.*

Proof. Let G be a group with identity element e and μ a fuzzy subgroup of G. For all $s \in \mu(G)$, let G_s and X_s be sets and $m_s : X_s \to G_s$ be a function such that the following conditions hold:

(1) G_s is a set of generators for the level subgroup μ_s of G,

(2) X_s is in one to one correspondence with G_s such that for all $r \in \mu(G)$ such that $r \neq s$, $X_s \cap X_r = \emptyset$,

(3) $m_s : X_s \to G_s$ of (2) is a bijection.

Set $X = \cup_{s \in \mu(G)} X_s$ and define $m : X \to G$ as follows: $\forall x \in X$, $m(x) = m_s(x)$ if $x \in X_s$.

Finally, define a function $h : F(X) \to G$

$$h([x_{i_1}^{\epsilon(1)}...x_{i_k}^{\epsilon(k)}]) = \prod_{j=1}^{k} m(x_{i_j})^{\epsilon(j)}, \tag{5.1.1}$$

where the product on the right is the group operation of G. Then h is a homomorphism from $F(X)$ onto G. Let $T = \{\mu(m(x_i)) \mid i \in I\}, t = \mu(e)$, and $\nu = f(X; T, t)$. We now show that $h(\nu) = \mu$.

Let $z \in G$. By the definition of a free fuzzy subgroup,

$$h(\nu)(z) = \vee\{\vee\{\wedge\{\mu(e) \wedge \mu(m(x_i)) \mid i \in I(w)\} \mid w \in y\} \mid h(y) = z\}. \tag{5.1.2}$$

However, for all $z \in G$, $y \in F(X)$ such that $h(y) = z$, and $w = x_{i_1}^{\epsilon(1)}...x_{i_k}^{\epsilon(k)} \in y$, $z = \prod_{j=1}^{k} x_{i_j}^{\epsilon(j)}$ by Equation (5.1.1). Also, by the definition of a fuzzy subgroup, $\mu(z) \geqslant \wedge\{\mu(m(x_{i_j})) \mid j = 1, ..., k\} = \wedge\{\mu(m(x_i)) \mid i \in I(w)\} = \wedge\{\mu(e) \wedge \mu(m(x_i)) \mid i \in I(w)\}$. By Equation (5.1.2), $h(\nu)(z) \leqslant \mu(z)$.

We now consider the reverse inequality. There exist symbols $x_{l_1}, ..., x_{l_n}$ in $X_{\mu(z)}$ such that $z = \prod_{j=1}^{n} m(x_{l_j})^{\epsilon(j)}$ in G. By the definition of $X_{\mu(z)}$,

$$\wedge\{\mu(e) \wedge \mu(m(x_{l_i})) \mid i = 1, ..., n\} \geqslant \mu(z). \tag{5.1.3}$$

Set $w = [x_{l_1}^{\epsilon(1)}...x_{l_n}^{\epsilon(n)}]$. Then $h(w) = z$. It now follows from Equations (5.1.2) and (5.1.3) that $h(\nu)(z) \geqslant \wedge\{\mu(e) \wedge \mu(m(x_q)) \mid q \in I(w)\} \geqslant \mu(z)$. Thus $h(\nu) = \mu$. $\square$

The remainder of this section considers an extension of Theorem 5.1.2. The extension needs the concept of a generating set for the fuzzy case. For any set S, let $S_P = \{x_t \mid x \in S, t \in [0, 1]\}$ be the set of all fuzzy singletons in S. If $Q \subseteq S_P$, then the foot of Q is the set $\text{foot}(Q) = \{x \in S \mid x_t \in Q\}$.

Definition 5.1.7. *Let ν be a fuzzy subgroup of G and let $J \subseteq [0, 1]$. A collection $\{B_i \mid i \in J\}$ of nonempty subsets of G_P is said to be **generating** for ν if the following conditions hold:*

(1) $\nu(e) \in J$ and $e_{s(e)} \in B_{s(e)}$, where e is the identity element of G;

(2) for every fuzzy singleton $x_t \in G_P$ and each $i \in J$, $x_t \in B_i$ implies $t = \nu(x) = i$;

(3) for all $x \in G$, there exist finitely many $x_1, ..., x_k \in foot(\cup_{i \geq \nu(x)} B_i)$ such that $x = \prod_{j=1}^{k} x_j^{\epsilon(j)}$, where every $\epsilon(j) \in \{-1, 1\}$.

The last condition in the above definition is much stronger than the requirement that $foot(\cup_{i \in J} B_i)$ should generate G. Also, it is clear that for every fuzzy subgroup ν, the collection $\{B_i \mid i \in \nu(G)\}$ defined by $B_{\nu(x)} = \{x_{\nu(x)} \mid x \in G\}$ is generating for ν. Thus every fuzzy subgroup has at least one such generating set. The next result follows from Definition 5.1.7.

Lemma 5.1.8. Let ν be a fuzzy subgroup of G and let $x \in G$. If $\{B_i \mid i \in J\}$ is generating for ν, then

(1) there exist finitely many $x_1, ..., x_k \in foot(\cup_{i \in J} B_i)$ such that

$$x = \prod_{j=1}^{k} x_j^{\epsilon(j)} \text{ and } \nu(x) = \nu(x_1) \wedge ... \wedge \nu(x_k),$$

(2) there exist finitely many fuzzy singletons

$(x_1)_{\nu(x_1)}, ..., (x_k)_{\nu(x_k)} \in \cup_{i \in J} B_i$

such that

$$x_{\nu(x)} = \prod_{j=1}^{k} ((x_j)_{\nu(x_j)})^{\epsilon(j)}.$$

Theorem 5.1.9. Let $\{B_i \mid i \in J\}$ be a generating set for the fuzzy subgroup ν of G. Suppose the following conditions hold.

(1) $\{X_i \mid i \in J\}$ is a pairwise disjoint collection of sets of symbols such that for all $i \in J$, there exists a bijection $m_i : X_i \to B_i$.

(2) $X = \cup_{i \in J} X_i$, and $m : X \to G$ is the function defined by $m(x) = foot(m_i(x))$, where $x \in X_i$.

Then there exists a unique epimorphism $h : F(X) \to G$ such that the following assertions hold.

(3) $h([x]) = m(x)$ for all $x \in X$,

(4) $h(f(X; T, t)) = \nu$, where $T = \{\nu(m(x)) \mid x \in X\}$ and $t = \nu(e)$.

Proof. We may assume without loss of generality that X and T are indexed by the same set I. Define $h : F(X) \to G$ by

$$h(w) = \prod_{j=1}^{k} m(x_{i_j})^{\epsilon(j)},$$

where $w = [x_{i_1}^{\epsilon(1)} ... x_{i_k}^{\epsilon(k)}] \in F(X)$. Then h is an epimorphism and clearly (3) holds. The uniqueness of h follows from Theorem 5.1.2.

To prove (4), let $z \in G$. Then $z = \prod_{j=1}^{k} z_j^{\epsilon(j)}$ for some $z_j \in \text{foot}(\cup_{i \in J} B)$, $\epsilon(j) \in \{-1, 1\}$, $j = 1, ..., k$. Using Lemma 5.1.8, we may assume that $\nu(z) = \wedge\{\nu(z_j) \mid j = 1, ..., k\}$.

By the definition of m, there exist $x_{i_1}, ..., x_{i_k} \in X$ such that $m(x_{i_j}) = z_j$ for all $j = 1, ..., k$ and $h([x_{i_j}^{\epsilon(j)} ... x_{i_k}^{\epsilon(k)}]) = z$. Then,

$$
\begin{aligned}
(f(X; T, t))(z) &= \vee\{\vee\{t \wedge t_i \mid i \in I(w) | w \in u | h(u) = z\}\} \\
&\geqslant \wedge\{t \wedge t_i \mid i \in I([x_{i_1}^{\epsilon(1)} ... x_{i_k}^{\epsilon(k)}])\} \\
&= \wedge\{\nu(e) \wedge \nu(m(x_i)) \mid i \in I([x_{i_1}^{\epsilon(1)} ... x_{i_k}^{\epsilon(k)}])\} \\
&= \wedge\{\nu(m(x_{ij})) \mid j = 1, ..., k\} \\
&= \wedge\{\nu(z_j) \mid j = 1, ... k\} \\
&= \nu(g).
\end{aligned}
$$

We now consider reverse inequality. Let $u \in F(X)$ and $w \in u$ be such that $h(u) = z$. Then $w = x_{i_1}^{\epsilon(1)} ... x_{i_n}^{\epsilon(n)}$. Thus $z = h([w]) = \prod_{j=1}^{n} m(x_{i_j})^{\epsilon(j)}$ and so

$$
\begin{aligned}
\nu(z) &\geqslant \wedge\{\nu(m(x_{i_j})) \mid j = 1, ..., n\} \\
&= \wedge\{\nu(e) \wedge \nu(m(x_{i_j})) \mid j = 1, ..., n\} \\
&= \wedge\{t \wedge t_i \mid i \in I(w)\}.
\end{aligned}
$$

Since u, w are arbitrary, we get $\nu(z) \geqslant \vee\{\vee\{\wedge\{t \wedge t_i \mid i \in I(w)\} \mid w \in u\} \mid h(u) = z\} = h(f(X; T, t))(z)$ and so the desired inequality holds. $\square$

We note that in the free fuzzy subgroup $f(X; T, t)$ of Theorem 5.1.9, the level of each x in X_i equals i. Thus knowing $\{X_i \mid i \in J\}$ renders T superfluous. Further, since $e_{\nu(e)} \in B(\nu(e))$, specification of $t = \nu(e)$ is also unnecessary. Since $X = \cup_{i \in J} X_i$, X, T, and t can all be recovered from the knowledge of $\{X_i \mid i \in J\}$. We therefore simplify the notation $f(X; T, t)$ to $f(\{X_i \mid i \in J\})$.

5.2 Presentations of Fuzzy Subgroups

The notion of a quotient is needed in order to define a presentation. The notion of a quotient together with the results of the previous section are used to extend the concept of a presentation to fuzzy subgroups. Let μ be a fuzzy subgroup of G and let N be a normal subgroup of G. Define the fuzzy subset μ_N of G/N by for all $x \in G/N$, $\mu_N(x) = \vee\{\mu(y) \mid y \in xN\}$. Then μ_N is a fuzzy subgroup of G/N by Theorem 1.3.13. We call μ_N the **quotient** of μ by N.

Let ν be a fuzzy subgroup G with $\{B_i \mid i \in J\}$ its generating set. There exists a collection $\{X_i \mid i \in J\}$ of sets of symbols and a collection of functions

$\{m_i{:}X_i \to B_i \mid i \in J$ and (1) and (2) of Theorem 5.1.9 hold$\}$. Let $X = \cup_{i \in J} X_i$ and let $h : F(X) \to G$ be an epimorphism satisfying (3) and (4) of Theorem 5.1.9. Set $\mu = \mu((X_i)_{i \in J})$, $N = \ker(h)$, and consider the quotient μ_N of μ by N. Define $\psi : F(X)/N \to G$ by for all $y \in F(X), \psi([y]) = h(y)$.

Clearly ψ is an isomorphism. Let $z \in G$. Then

$$\begin{aligned}
\psi(\mu_N)(z) &= \vee\{f_N(u) \mid \psi(u) = z\}\\
&= \vee\{\mu(w) \mid w \in u, \psi(u) = z\}\\
&= \vee\{\mu(w) \mid \psi([w]) = z\}\\
&= h(\mu)(z).
\end{aligned}$$

Since $h(\mu) = \nu$ by Theorem 5.1.9, it follows that μ_N and ν are isomorphic.

From the above discussion, we see that to specify the fuzzy subgroup ν of G (up to isomorphism), it suffices to specify a collection $\{X_i \mid i \in J\}$ of sets of symbols with each X_i being in one to one correspondence with B_i for some collection $\{B_i \mid i \in J\}$ generating ν, as well as the subgroup $N = \ker(h)$ of $F(X)$, where $X = \cup_{i \in J} X_i$. N may be specified using a subset S of N whose normal closure is N. Every element e_S of S may be specified by selecting the reduced word in e_S as its representative. Let R be the set of all such reduced words. Then ν is specified by the pair $\langle \{X_i \mid i \in J\}, R \rangle$. This pair is called a **presentation** of ν. The collection $\{X_i \mid i \in J\}$ is called the **generating collection**, and R is called the set of **relators** of ν with respect to $\{X_i \mid i \in J\}$.

Since the domain of $\mu_N, F(X)/N$, is isomorphic to the domain of ν, G, the presentation of the fuzzy subgroup ν of G yields a presentation for the domain of ν. Moreover, the two presentations have the same relators.

We now give two examples to illustrate the definitions and main theorems.

Example 5.2.1. *Define the fuzzy subgroup ρ of $\mathbb{Z}_{12}$ as follows:* $\forall x \in \mathbb{Z}_{12}$,

$$\rho(x) = \begin{cases} 1 & \text{if } x = 0 \ \text{ or } x = 6, \\ \frac{1}{2} & \text{if } x = 3 \ \text{ or } x = 9, \\ \frac{1}{3} & \text{otherwise.} \end{cases}$$

To obtain a presentation for ρ, we form the generating collection $\{B_i \mid i \in J\}$, where $J = \{1, \frac{1}{2}, \frac{1}{3}\}$, and a collection $\{X_i \mid i \in J\}$ as in Theorem 5.1.9:

$$\begin{aligned}
B_1 &= \{0_1, 6_1\}, \ X_1 = \{z, s\},\\
B_{\frac{1}{2}} &= \{3_{\frac{1}{2}}\}, \ X_{\frac{1}{2}} = \{t\},\\
B_{\frac{1}{3}} &= \{1_{\frac{1}{3}}\}, \ X_{\frac{1}{3}} = \{u\}.
\end{aligned}$$

Here z, s, t and u stand for zero, six, three, and one, respectively. For every $i \in J$, $m_i : X_i \to B_i$ is the obvious bijection. In particular, $m_1(z) = 0_1, m_1(s) = 6_1, m_{\frac{1}{2}}(t) = 3_{\frac{1}{2}}$, and $m_{\frac{1}{3}}(u) = 1_{\frac{1}{3}}$. Let $F(X)$ be the free group on $X = \cup_{i \in J} X_i$, where X is assumed to be indexed by itself. The free fuzzy subgroup μ of $F(X)$

is given by $\mu(y) = \wedge\{t \wedge t_i \mid i \in X(w)\}$, where w is the reduced word in the equivalence class $y \in F(X)$ and $t = 1 = \vee J$. We identify every such y with its reduced word for the sake of convenience. We now give some examples of computing the values of the fuzzy free subgroup μ at some points. For example,

$$\mu(u) = 1 \wedge \frac{1}{3} = \frac{1}{3},$$

$$\mu(utst^{-1}) = 1 \wedge \frac{1}{3} \wedge \frac{1}{2} \wedge 1 \wedge \frac{1}{2} = \frac{1}{3},$$

$$\mu(tt) = 1 \wedge \frac{1}{2} \wedge \frac{1}{2} = \frac{1}{2}.$$

The value of μ at the empty word is $1 = \vee J$.

The collection $\{X_i \mid i \in J\}$ constitutes the generators of the presentation. By Theorem 5.1.9, there exists a unique homomorphism h such that $F(X)/\ker(h) \simeq Z_{12}$. The kernel of h gives the relators of the (fuzzy) presentation. Condition (3) of Theorem 5.1.9 yields the relations

$$h(z) = 0, \qquad h(s) = 6, \qquad h(t) = 3, \qquad h(u) = 1.$$

From these it follows that the defining relations are

$$z = e, \qquad s^2 = e, \qquad s = t^2, \qquad t = u^3.$$

Thus the fuzzy subgroup ρ has the presentation

$$\left\langle z_1, s_1, t_{\frac{1}{2}}, u_{\frac{1}{3}} \mid z = e, s^2 = e, s = t^2, t = u^3 \right\rangle.$$

Example 5.2.2. Consider the dihedral group

$$D_4 = \left\langle a, b \mid a^4 = e, b^2 = e, ba = a^3b \right\rangle.$$

Let ρ be the fuzzy group of D_4 defined as follows:

$$\rho(x) = \begin{cases} 1 & \text{if } x = e, \\ \frac{1}{2} & \text{if } x = a^2b, \\ \frac{1}{3} & \text{if } x \in \{a^2, b\}, \\ \frac{1}{4} & \text{if } x \in \{a, a^3, ab, ba\}. \end{cases}$$

Then

$$
\begin{aligned}
B_1 &= \{e_1\}, & X_1 &= \{\bar{e}\} \\
B_{\frac{1}{2}} &= \{w_{\frac{1}{2}}\}, & X_{\frac{1}{2}} &= \{\bar{w}\}, \\
B_{\frac{1}{3}} &= \{x_{\frac{1}{3}}, y_{\frac{1}{3}}\}, & X_{\frac{1}{3}} &= \{\bar{x}, \bar{y}\} \\
B_{\frac{1}{4}} &= \{z_{\frac{1}{4}}\}, & X_{\frac{1}{4}} &= \{\bar{z}\},
\end{aligned}
$$

where $w = a^2b, x = a^2, y = b,$ and $z = a$. It follows that there is a unique homomorphism h satisfying the relations

$$h(\bar{e}) = e, \qquad h(\bar{w}) = a^2b, \qquad h(\bar{x}) = a^2,$$
$$h(\bar{y}) = b, \qquad h(\bar{z}) = a.$$

This yields the presentation

$$\left\langle e_1, w_{\frac{1}{2}}, x_{\frac{1}{3}}, y_{\frac{1}{3}}, z_{\frac{1}{4}} \mid w = xy, z^4 = e, y^2 = e, x = z^2, yz = z^3y \right\rangle,$$

where the overlines on the generators of the presentation are suppressed.

5.3 Constructing Free Fuzzy Subgroups

In this section, we give an outline of another approach for the construction of free fuzzy subgroups. The approach differs from that in Section 5.1. It follows along the lines of [28] and [36]. If X is a set and χ a fuzzy subset of X, we call (X, χ) a **fuzzy set** in this section. Let G be a group and μ a fuzzy subgroup of G. We call (G, μ) a **fuzzy group** in this section. Let X be a subset of G and let χ be a fuzzy subset of X such that $\chi \subseteq \mu$. We call (X, χ) a **fuzzy subset** of (G, μ). We say that (X, χ) **generates** (G, μ) if $G = \langle X \rangle$ and $\mu = \langle \chi \rangle$, where we consider χ to be the fuzzy subset of G obtained by extending χ from X to G such that $\chi(x) = 0$ for all $x \in G \backslash X$. If (X, χ) generates (G, μ), we write $(G, \mu) = \langle X, \chi \rangle$.

Let (X, μ) and (Y, ν) be fuzzy sets and f a function of X into Y. If $f(\mu) \subseteq \nu$, then f is called a **fuzzy function** of (X, μ) into (Y, ν). If $f(X) = Y$ and $f(\mu) = \nu$, then f is called **surjective**.

Let (G, μ) and (H, ν) be fuzzy groups and let f be a fuzzy function of (G, μ) into (H, ν) such that f is a homomorphism of G into H. If f is an epimorphism (isomorphism) of G onto H and $f(\mu) = \nu$, then f is called an **epimorphism (isomorphism)** of (G, μ) onto (H, ν).

Definition 5.3.1. *Let (F, μ) be a fuzzy group and (X, χ) a fuzzy set, where $X \subseteq F$. Then (F, μ) is called a **free fuzzy group** on (X, χ) provided the following conditions hold.*

(1) (X, χ) generates (F, μ).

(2) If (G, ν) is any fuzzy group having a generating fuzzy set (Y, η), then for any surjection $f : (X, \chi) \to (Y, \eta)$, there exists an epimorphism $f^* : (F, \mu) \to (G, \nu)$ such that $f^*(x) = f(x)$ for all $x \in X$.

We call the fuzzy set (X, χ) in Definition 5.3.1 a **free fuzzy basis** for the group (F, μ).

The proof of the next result follows from the known situation for free groups and from Definition 5.3.1. We note from Definition 5.3.1 that if $f^* : (F, \mu) \to (G, \nu)$ is an epimorphism, then $f^*(\mu) = \nu$.

Theorem 5.3.2. *Let (F_i, μ_i) be free fuzzy groups on fuzzy sets (X_i, χ_i) with $i = 1, 2$, respectively. Suppose f_1 is a one-to-one function of X_1 onto X_2 such that $f_1(\chi_1) = \chi_2$. Then there is an isomorphism of fuzzy groups from (F_1, μ_1) onto (F_2, μ_2).*

We now show that for any fuzzy set (X, χ), there exists a free fuzzy group on (X, χ). Results on free groups can be found in [16, 25, 27].

Let $\Sigma = X \cup X^{-1}$ and Σ^* be as in Section 5.1. Let $F(X) = \Sigma^* / \sim$ as in Section 5.1. We write $X^{\pm} = X \cup X^{-1}$. Let χ be a fuzzy subset of X and χ^{-1} the fuzzy subset of X^{-1} such that $\chi^{-1}(x^{-1}) = \chi(x)$ for all $x \in X$. A word is a finite sequence of symbols, $w = x_1^{\varepsilon_1} ... x_n^{\varepsilon_n}$, $n \geq 0$, $\varepsilon_i = \pm 1$, $x_i \in X$ for $i = 1, ..., n$, where $n \in \mathbb{N}$. If $n = 0$, then $w = e$, the empty word. Recall that the set Σ^* of all words is a monoid under concatenation.

Let $w = x_1^{\varepsilon_1}...x_n^{\varepsilon_n} \in \Sigma^*$ with $x_i \in X, \varepsilon_i = \pm 1, i = 1,...,n$. We call n the **length** of w and denote it by $|w|$. If e is the empty string, then e has length 0 and we write $|e| = 0$. Define the fuzzy subset ξ of Σ^* as follows: $\forall$ $w = x_1^{\varepsilon_1}...x_n^{\varepsilon_n} \in \Sigma^*$,

$$\xi(w) = \wedge\{\chi(x_i) \mid 0 \leq i \leq |w|\}.$$

Then the following proposition holds.

Proposition 5.3.3. *If the fuzzy set (Σ^*, ξ) is as above, then ξ is a fuzzy subgroupoid of Σ^*.*

Proposition 5.3.4. *Let μ be a fuzzy subset of a group G. Then $\langle \mu \rangle (x) = \vee\{r | x \in \langle \mu_r \rangle, r < \vee\mu\}$.*

Proof. Let μ^* be the fuzzy subset of G defined by $\forall x \in G, \mu^*(x) = \vee\{r | x \in \langle \mu_r \rangle, r < \vee\mu\}$. Suppose there exist $x, y \in G$ such that $\mu^*(xy) < \mu^*(x) \wedge \mu^*(y)$. Then $\mu^*(x) > \mu^*(xy)$ and $\mu^*(y) > \mu^*(xy)$. Thus there exist $t_1, t_2 \in [0,1]$ such that $t_1, t_2 > \mu^*(xy)$ and $x \in \langle \mu_{t_1} \rangle, y \in \langle \mu_{t_2} \rangle$. Suppose $t_2 \geq t_1$. Then $x, y \in \langle \mu_{t_2} \rangle \subseteq \langle \mu_{t_1} \rangle$. Since $t_1 > \mu^*(xy) = \vee\{t | xy \in \langle \mu_t \rangle, t \leq \vee\mu\}$, we have that $xy \notin \langle \mu_{t_1} \rangle$. However, this contradicts the fact that $\langle \mu_{t_1} \rangle$ is a subgroup of G. Since $x \in \langle \mu_t \rangle \Leftrightarrow x^{-1} \in \langle \mu_t \rangle$ for all $x \in G$, it follows that $\mu^*(x) = \mu^*(x^{-1})$ for all $x \in G$. Thus μ^* is a fuzzy subgroup of G. Let $x \in G$ and let $\mu(x) = t$. Then $x \in \langle \mu_t \rangle$. Thus $\mu^*(x) \geq t$ and so $\mu \subseteq \mu^*$. Therefore, $\langle \mu \rangle \subseteq \mu^*$. we have equality once we show that μ^* is the smallest fuzzy subgroup of G containing μ. Let θ be any fuzzy subgroup of G containing μ. Suppose that $\mu^*(x) > \theta(x)$ for some $x \in G$. Then there exists $t \in [0,1]$ such that $t > (x)$ and $x \in \langle \mu_t \rangle$. Thus $x = x_1 x_2 ... x_n$ for some $x_i \in \langle \mu_t \rangle$ such that $\mu(x_i) \geq t$ or $\mu(x_i^{-1}) \geq t, i = 1, 2, ..., n$. Since θ is a fuzzy subgroup of $G, \theta(x_i) = \theta(x^{-1})$ for $i = 1, 2, ..., n$. Hence $\theta(x_i) \geq \mu(x_i) \vee \mu(x_i^{-1}) \geq t$ for $i = 1, 2, ..., n$. Thus $\theta(x) = \theta(x_1 x_2 ... x_n) \geq \theta(x_1) \wedge \theta(x_2) \wedge ... \wedge \theta(x_n) \geq t$. However, this contradicts the fact that $\theta(x) < t$. Thus $\mu^* \subseteq \theta$. $\square$

We define the inverse w^{-1} of $w = x_1^{\varepsilon_1}...x_n^{\varepsilon_n}$ to be $w^{-1} = x_n^{-\varepsilon_n}...x_1^{-\varepsilon_1}$, $(wv)^{-1} = v^{-1}w^{-1}$ and $e^{-1} = e$.

An elementary transformation of a word w consists of inserting or deleting a part of the form xx^{-1} for $x \in X^{\pm 1}$.

Two words w and v are equivalent, $w \sim v$, if there is a chain of elementary transformations leading from w to v. Then the relation $\sim$ is an equivalence relation on W. We let $[w]$ denote the equivalence class of $w \in X^{\pm}$. The product of equivalence classes $[w]$ and $[v]$ is defined by $[w][v] = [wv]$. This product is well defined. This $F(X) = \Sigma^*/\sim$ is a group.

Define $g : \Sigma^* \rightarrow F(X)$ by $\forall w \in \Sigma^*, g(w) = [w]$. Then g is a homomorphism of Σ^* onto $F(X)$. The image of ξ under g is as follows: $\forall [w] \in F(X)$,

$$g(\xi)([w]) = \vee\{\xi(u) \mid g(u) = [w], u \in \Sigma^*\}$$
$$= \vee\{\wedge\{\chi(x_i) \mid 0 \leq i \leq n\} \mid u = x_1^{\varepsilon_1}...x_n^{\varepsilon_n}, g(u) = [w], u \in \Sigma^*\}.$$

Let $\mu = g(\xi)$. Then it follows that μ is a fuzzy subgroup of $F(X)$.

We now show that the fuzzy group $(F(X), \mu)$ above is a free fuzzy group on (X, χ).

We first show that $(F(X), \mu) = \langle X, \chi \rangle$. Since the function sending x to $[x]$ for $x \in X$ is clearly one-one, X can be regarded as a subset of $F(X)$ when x is identified with $[x]$. Thus $F(X) = \langle X \rangle$.

Next we show that $\mu = \langle \chi \rangle$, where for all $[w] \in F(X)$,

$$\mu([w]) = \vee\{\wedge\{\chi(x_i) \mid 0 \leq i \leq n\}\mid u = x_1^{\epsilon_1}...x_n^{\epsilon_n}, g(u) = [w], u \in \Sigma^*\}$$

and

$$\langle \chi \rangle([w]) = \vee\{k \mid [w] \in \langle \chi_k \rangle, k < \vee\chi\}.$$

by Proposition 5.3.4. For all $[w] \in F(X)$, $g(w) = [w]$ with $w \in \Sigma^*$. Let $t = \langle \chi \rangle([w])$. Then it follows that $[w] \in \langle \chi \rangle_t$. Since $\langle \chi \rangle_t = \langle \chi_{t-\epsilon} \rangle$ for any sufficiently small positive number ϵ, it follows that $[w] \in \langle \chi \rangle_t = \langle \chi_{t-\epsilon} \rangle$. Since there exist $x_1, ..., x_n \in \chi_{t-\epsilon}$ such that $w = x_1^{\epsilon_1}...x_n^{\epsilon_n}$, where $n \geq 0$ and $\varepsilon_i = \pm 1$, it follows that $\chi(x_i) \geq t - \epsilon$. Hence $\wedge\{\chi(x_i) \mid i = 1, ..., n\} \geq t - \epsilon$, where $w = x_1^{\epsilon_1}...x_n^{\epsilon_n}$. Thus it follows that $\mu([w]) = \vee\{\wedge\{\chi(x_i) \mid 0 \leq i \leq n\}\mid g(u) = [w], u = x_1^{\epsilon_1}...x_n^{\epsilon_n}, u \in \Sigma^*\} \geq t = \langle \chi \rangle([w])$. We now show that $\mu([w]) \leq \langle \chi \rangle([w])$.

For all $[w] \in F(X)$, it follows that

$$\mu([w]) = \vee\{\wedge\{\chi(x_i) \mid 0 \leq i \leq n\}\mid u = x_1^{\epsilon_1}...x_n^{\epsilon_n}, g(u) = [w], u \in \Sigma^*\}$$
$$= \vee\{k \mid k = \wedge\{\chi(x_i) \mid 0 \leq i \leq n\}\mid g(u) = [w], u = x_1^{\epsilon_1}...x_n^{\epsilon_n}, u \in \Sigma^*\}.$$

Now $w \in \langle \chi_k \rangle$ implies that there exist $x_i \in \chi_k$ such that $w = x_1^{\epsilon_1}...x_n^{\epsilon_n}$, where $n \in \mathbb{N}$ and $\varepsilon_i = \pm 1$, and $\chi(x_i) \geq k$ for all x_i with $i = 1, ..., n$.

Since $\wedge\{\chi(x_i) \mid 0 \leq i \leq n\} \leq \chi(x_i)$, it follows that

$$\mu([w]) = \vee\{k \mid k = \wedge\{\chi(x_i) \mid 0 \leq i \leq n\}\mid g(u) = [w], u = x_1^{\epsilon_1}...x_n^{\epsilon_n}, u \in \Sigma^*\}$$
$$\leq \vee\{k \mid [w] \in \langle \chi_k \rangle, k < \vee\chi\} = \langle \chi \rangle([w]).$$

Hence $\mu([w]) \leq \langle \chi \rangle([w])$. Therefore, $\mu = \langle \chi \rangle$ and so $(F, \mu) = \langle X, \chi \rangle$.

We now consider condition (2) of Definition 5.3.1.

Let (G, ν) be a fuzzy group having a generating fuzzy set (Y, η) such that $f : (X, \chi) \to (Y, \eta)$ is surjective. Then by setting $f(x^{-1}) = f(x)^{-1}$ with $x \in X$, φ induces a function of $X^{\pm 1}$ onto $Y^{\pm 1}$. Define an extension $\overline{f}$ of f from Σ^* onto G by $\forall w = x_1^{\epsilon_1}...x_n^{\epsilon_n}$, $\overline{f}(w) = f(x_1)^{\epsilon_1}...f(x_n)^{\epsilon_n}$, $x_i \in X$, $\varepsilon_i = \pm 1$ for $i = 1, ..., n$.

If w_1 and w_2 are equivalent, then $\overline{f}(w_1) = \overline{f}(w_2)$ and so $\overline{f}$ maps equivalent words onto the same element of G. Thus $\overline{f}$ induces a function $f^* : F(X) \to G$ defined by $f^*([w]) = \overline{f}(w)$. Clearly, f^* is an epimorphism of groups.

It remains to show that f^* is an epimorphism in a fuzzy sense, i.e., $f^*(\mu) = \nu$.

Suppose that $(G, \nu) = \langle Y, \eta \rangle$ and that $f : (X, \chi) \to (Y, \eta)$ is surjective, where $Y \subseteq G$. Then ν is the fuzzy subgroup of G generated by η and $f(\chi) = \eta$. Thus it suffices to show that $f^*(\mu) = \nu$, that is, to show that

$$\vee\{\vee\{k \mid [w] \in \langle\chi_k\rangle, k < \vee\chi\} \mid f^*([w]) = \overline{w}\}$$
$$= \vee\{k \mid \overline{w} \in \langle\eta_k\rangle, k < \vee\eta\} \text{ for all } \overline{w} \in G.$$

Let $\overline{w} \in G$ and $t = \nu(\overline{w})$. Then for all sufficiently small positive number ϵ, it follows that $\overline{w} \in \nu_t = \langle\eta_{t-\epsilon}\rangle$. Hence there exist $y_i \in \eta_{t-\epsilon}$ such that $\overline{w} = y_1^{\varepsilon_1}...y_n^{\varepsilon_n}$, $n \geq 0$, $\varepsilon_i = \pm 1$. Thus $\eta(y_i) \geq t - \epsilon$. For all $y_i \in Y$ with $\overline{w} = y_1^{\varepsilon_1}...y_n^{\varepsilon_n}$, there exist $x_i \in X$ such that $\varphi(x_i) = y_i$ and $w = x_1^{\varepsilon_1}...x_n^{\varepsilon_n}$ since f is surjective. Furthermore, it follows that

$$\vee\{\chi(x_i) \mid f(x_i) = y_i\} = \eta(y_i) \text{ for all } y_i = Y^{\pm 1}.$$

Thus $\vee\{k \mid w \in \langle\chi_k\rangle\} \geq t - \epsilon$. Therefore, it follows that

$$\vee\{\vee\{k \mid [w] \in \langle\chi_k\rangle, k < \vee\chi\} \mid f^*([w]) = \overline{w}\} \geq t \text{ for all } \overline{w} \in G.$$

Hence $f^*(\mu) \supseteq \nu$. We now show that $f^*(\mu) \subseteq \nu$.

Since f^* is an epimorphism of groups, for $\overline{w} \in G$ there exist $[w] \in F(X)$ such that $f^*([w]) = \overline{w}$. Set $t = \mu([w]) = \vee\{k \mid [w] \in \langle\chi_k\rangle, k < \vee\chi\}$. Then it follows that $[w] \in \mu_t = \langle\chi_{t-\epsilon}\rangle$ for sufficiently small positive number ϵ. Hence there exist $x_i \in \chi_{t-\epsilon}$ such that $w = x_1^{\varepsilon_1}...x_n^{\varepsilon_n}$, $n \in \mathbb{N}$, $\varepsilon_i = \pm 1$ and $\chi(x_i) \geq t - \epsilon, i = 1, ..., n$.

Since f is a fuzzy function, it follows that $\chi(x) \leq \eta(f(x))$ for all $x \in X$. Hence $t - \epsilon \leq \chi(x_i) \leq \eta(f(x_i))$ with $w = x_1^{\varepsilon_1}...x_n^{\varepsilon_n}$. Since $f(x_1)^{\varepsilon_1}...f(x_n)^{\varepsilon_n} = \overline{w}$ and ϵ is arbitrary, it follows that $t \leq f(\varphi(x_i))$. Thus $t \leq \vee\{k \mid \overline{w} \in \langle\eta_k\rangle\}$. Therefore, since $[w] \in F(X)$ is arbitrary and $f^*([w]) = \overline{w}$, it follows that

$$(f^*(\mu))(\overline{w}) = \vee\{\vee\{k \mid [u] \in \langle\chi_k\rangle, k < \vee\chi\} \mid f^*([u]) = \overline{w}\}$$
$$\leq \vee\{k \mid \overline{w} \in \langle\eta_k\rangle, k < \vee\eta\} = \nu(\overline{w}).$$

By the preceding inequalities, it follows that $f^*(\mu) = \nu$.

We thus have the following result.

Theorem 5.3.5. *Every fuzzy group is a homomorphic image of a free fuzzy group.*

The interested reader can find results on free products of fuzzy subgroups in [1] and [33].

5.4 Free (s,t]-Fuzzy Subgroups

In [8, 10, 30] the idea of a fuzzy point and its membership to and quasi-coincidence with a fuzzy set were used to define and study certain kinds of fuzzy topological spaces and $(\in, \in \vee q)$-fuzzy subgroups, respectively. Let G be a group. The purpose of this section is two-fold. First, we wish to extend the notion of an $(\epsilon, \epsilon \vee q)$-fuzzy subgroup of G introduced in [8, 10] to a more general situation. It is shown in [10] that μ is an $(\epsilon, \epsilon \vee q)$-fuzzy subgroup of G if and only if $\forall x, y \in G, \mu(xy^{-1}) \geq \mu(x) \wedge \mu(y) \wedge \frac{1}{2}$. Let $c : [0,1] \rightarrow [0,1]$. Then c is called a **fuzzy complement** if (1) $c(0) = 1$ and $c(1) = 0$ and (2) $\forall s, t \in [0,1], s \leq t \Rightarrow c(s) \geq c(t)$, [[22], p.52]. Let c be a fuzzy complement and let $\alpha \in [0,1]$. If $c(\alpha) = \alpha$, then α is called an **equilibrium** of c and c is said to have an equilibrium. It is known that if c has an equilibrium, then the equilibrium is unique [[22],Theorem 3.2, p. 57]. Also, $\forall s, t \in [0,1]$, $t > s$ and $t > c(s) \Rightarrow t > \alpha$. We assume throughout that c has an equilibrium α, i.e., there is an element $\alpha \in [0,1]$ such that $c(\alpha) = \alpha$. We define a c-quasi-fuzzy subgroup and show that μ is a c-quasi-fuzzy subgroup of G if and only if $\forall x, y \in G, \mu(xy^{-1}) \geq \mu(x) \wedge \mu(y) \wedge \alpha$. If c is is the standard fuzzy complement defined by $c(x) = 1 - x \; \forall x \in [0,1]$, then $\alpha = \frac{1}{2}$. Hence we have the extension referred to above.

Second, we study the notion of an $(s,t]$-fuzzy subgroup as introduced in [40]. We feel that this notion has possible applications for suitable choices of s and t in certain situations. The notion of an $(s,t]$-fuzzy subgroup is more general than that of a c-quasi-fuzzy subgroup. The latter can be obtained by letting $s = 0$. We present some basic results of $(s,t]$-fuzzy subgroups before considering the notion of a free $(s,t]$-fuzzy subgroups [13]. Many basic results concerning the above types of fuzzy subgroups have been presented in [2] - [8], [11, 21], [37] - [40].

Throughout this section, we assume that c is a fuzzy complement and that c has an equilibrium α. We also assume that G is a group.

Definition 5.4.1. *Let μ be a fuzzy subset of G. Then μ is called a c-quasi-fuzzy subgroup of G if $\forall x, y \in G$ and $\forall t, r \in [0,1]$, $x_t, y_r \subseteq \mu$ implies $(xy^{-1})_{t \wedge r} \subseteq \mu$ or $t \wedge r > (c\mu)(xy^{-1})$.*

Theorem 5.4.2. *Let μ be a fuzzy subset of G. Then μ is a c-quasi-fuzzy subgroup of G if and only if $\forall x, y \in G$, $\mu(xy^{-1}) \geq \mu(x) \wedge \mu(y) \wedge \alpha$.*

Proof. Suppose μ is a c-quasi fuzzy subgroup of G. Suppose $\exists x, y \in G$ such that $\mu(xy^{-1}) < \mu(x) \wedge \mu(y) \wedge \alpha$. Then $\mu(xy^{-1}) < \alpha$. Let $x_{t_0}, y_{r_0} \subseteq \mu$. Suppose $t_0 \wedge r_0 > \mu(xy^{-1})$. Then by hypothesis, $t_0 \wedge r_0 > (c\mu)(xy^{-1})$. Hence $t_0 \wedge r_0 > \alpha$. That is, $\mu(xy^{-1}) < \alpha < t_0 \wedge r_0$. Let $t \leq t_0$ and $r \leq r_0$ be such that $\mu(xy^{-1}) < t \wedge r < \alpha$. Now $x_t, y_r \subseteq \mu$ and by an argument as above, $t \wedge r > \alpha$, a contradiction. Hence either $\mu(xy^{-1}) \geq \mu(x) \wedge \mu(y) \wedge \alpha$, the desired result, or $t_0 \wedge r_0 \leq \mu(xy^{-1})$. In the latter case, we can take $t_0 = \mu(x)$ and $r_0 = \mu(y)$ and so $\mu(xy^{-1}) \geq \mu(x) \wedge \mu(y)$.

Conversely, suppose $\forall x, y \in G, \mu(xy^{-1}) \geq \mu(x) \wedge \mu(y) \wedge \alpha$. Suppose $x_t, y_r \subseteq \mu$. Suppose $\mu(xy^{-1}) < t \wedge r$. Then it suffices to show $t \wedge r > (c\mu)(xy^{-1})$. Suppose $t \wedge r \leq (c\mu)(xy^{-1})$. Then $\mu(xy^{-1}) < t \wedge r \leq (c\mu)(xy^{-1})$. Hence $\mu(xy^{-1}) < (c\mu)(xy^{-1})$ and so $\mu(xy^{-1}) < \alpha$. Thus $\mu(xy^{-1}) \geq \mu(x) \wedge \mu(y) \; (\geq t \wedge r)$ since $\mu(xy^{-1}) \geq \mu(x) \wedge \mu(y) \wedge \alpha$. Hence $\mu(xy^{-1}) \geq t \wedge r$. Thus $\mu(xy^{-1}) < t \wedge r$ and $\mu(xy^{-1}) \geq t \wedge r$, a contradiction. Therefore, $t \wedge r > (c\mu)(xy^{-1})$. $\square$

Proposition 5.4.3. *Suppose μ is a c-quasi-fuzzy subgroup of G. Let $x, y \in G$. If $\mu(x) > \mu(y)$, then $\mu(xy) \wedge \alpha = \mu(y) \wedge \alpha$.*

Proof. $\mu(y) = \mu(x^{-1}xy) \geq \mu(x^{-1}) \wedge \mu(xy) \wedge \alpha \geq \mu(x) \wedge \alpha \wedge \mu(xy) \wedge \alpha = \mu(x) \wedge \mu(xy) \wedge \alpha \geq \mu(x) \wedge \mu(x) \wedge \mu(y) \wedge \alpha \wedge \alpha = \mu(y) \wedge \alpha$.

Thus $\mu(y) \wedge \alpha \geq \mu(x) \wedge \mu(xy) \wedge \alpha \geq \mu(y) \wedge \alpha$. Thus $\mu(y) \wedge \alpha = \mu(x) \wedge \mu(xy) \wedge \alpha$. Since $\mu(x) > \mu(y) \wedge \alpha$, $\mu(y) \wedge \alpha = \mu(xy) \wedge \alpha$. $\qquad\square$

Proposition 5.4.4. *Suppose μ is an c-quasi-fuzzy subgroup of G. Let $x, y \in G$. If $\mu(x) \wedge \alpha > \mu(y) \wedge \alpha$, then $\mu(xy) \wedge \alpha = \mu(y) \wedge \alpha = \mu(y) = \mu(xy)$.*

Proof. As in the proof of the previous proposition, $\mu(y) \wedge \alpha = \mu(x) \wedge \mu(xy) \wedge \alpha$. Since $\mu(x) \wedge \alpha > \mu(y) \wedge \alpha$, $\mu(y) \wedge \alpha = \mu(xy)$. Now $\mu(x) \wedge \alpha > \mu(y) \wedge \alpha$ implies $\alpha \geq \mu(y)$. Thus $\mu(y) = \mu(xy)$. $\qquad\square$

Definition 5.4.5. *[40] Let G be a group and μ a fuzzy subset of G. Then for $s, t \in [0, 1]$, $s < t$, μ is called an $(s, t]$-fuzzy subgroup of G if the following conditions hold:*

(1) $\forall x, y \in G$, $\mu(xy) \vee s \geq \mu(x) \wedge \mu(y) \wedge t$,
(2) $\forall x \in G$, $\mu(x^{-1}) \vee s \geq \mu(x) \wedge t$.

Theorem 5.4.6. *[40] Let μ be a fuzzy subset of G. Then μ is an $(s, t]$-fuzzy subgroup of G if and only if μ_r is a subgroup of G for all $r \in (s, t]$.*

Theorem 5.4.7. *Let f be a homomorphism of a group G into a group K.*

(1) If μ is an $(s, t]$-fuzzy subgroup of G, then $f(\mu)$ is an $(s, t]$-fuzzy subgroup of K.

(2) If ν is an $(s, t]$-fuzzy subgroup of K, then $f^{-1}(\nu)$ is an $(s, t]$-fuzzy subgroup of G.

Proof. (1) Let $y_1, y_2 \in f(G)$. Then $f(\mu)(y_1 y_2) \vee s = \vee\{\mu(x) \mid f(x) = y_1 y_2, x \in G\} \vee s \geq \vee\{\mu(x_1 x_2) \mid f(x_1) = y_1, f(x_2) = y_2, x_1, x_2 \in G\} \vee s = \vee\{\mu(x_1 x_2) \vee s \mid f(x_1) = y_1, f(x_2) = y_2, x_1, x_2 \in G\}$

$\geq \vee\{\mu(x_1) \wedge \mu(x_2) \wedge t \mid f(x_1) = y_1, f(x_2) = y_2, x_1, x_2 \in G\} = \vee\{\mu(x_1) \wedge t \mid f(x_1) = y_1, x_1 \in G\} \wedge \vee\{\mu(x_2) \wedge t \mid f(x_2) = y_2, x_2 \in G\}$

$= \vee\{\mu(x_1) \mid f(x_1) = y_1, x_1 \in G\} \wedge \vee\{\mu(x_2) \mid f(x_2) = y_2, x_2 \in G\} \wedge t = f(\mu)(y_1) \wedge f(\mu)(y_2) \wedge t$.

Suppose either $y_1 \notin f(G)$ or $y_2 \notin f(G)$. Then either $f(\mu)(y_1) = 0$ or $f(\mu)(y_2) = 0$. Hence $f(\mu)(y_1 y_2) \vee s = 0 = f(\mu)(y_1) \wedge f(\mu)(y_2) \wedge t$.

Let $y \in f(G)$. Then $f(\mu)(y^{-1}) \vee s = \vee\{\mu(x^{-1}) \mid f(x^{-1}) = y^{-1}, x^{-1} \in G\} \vee s = \vee\{\mu(x^{-1}) \vee s \mid f(x^{-1}) = y^{-1}, x^{-1} \in G\}$

$\geq \vee\{\mu(x) \wedge t \mid f(x^{-1}) = y^{-1}, x^{-1} \in G\} = \vee\{\mu(x) \mid f(x^{-1}) = y^{-1}, x^{-1} \in G\} \wedge t = \vee\{\mu(x) \mid f(x) = y, x \in G\} \wedge t = f(\mu)(x) \wedge t$.

Suppose $y \notin f(G)$. Thus $y^{-1} \notin f(G)$. Hence $f(\mu)(y^{-1}) = 0 = f(\mu)(y)$. Thus $f(\mu)(y^{-1}) \vee s \geq f(\mu)(y) \wedge t$.

(2) Let $x_1, x_2 \in G$. Then $f^{-1}(\nu)(x_1 x_2) \vee s = \nu(f(x_1 x_2)) \vee s = \nu(f(x_1)f(x_2)) \vee s \geq \nu(f(x_1) \wedge \nu(f(x_2)) \wedge t = f^{-1}(\nu)(x_1) \wedge f^{-1}(\nu)(x_2) \wedge t$. Let

$x \in G$. Then $f^{-1}(\nu)(x^{-1}) \vee s = \nu(f(x^{-1})) \vee s = \nu(f(x)^{-1}) \vee s \geq \nu(f(x)) \wedge t = f^{-1}(\nu)(x) \wedge t$. $\square$

Proposition 5.4.8. *Suppose G is a group and $s, t \in [0,1]$ with $s < t$. Let H be a subgroup of G. Suppose μ is a fuzzy subset of G such that $\forall x \in G$, $\mu(x) \geq t$ if $x \in H$ and $\mu(x) \leq s$ if $x \notin H$. Then μ is an $(s,t]$-fuzzy subgroup of G.*

Proof. Let $x, y \in G$. Suppose $x, y \in H$. Then $\mu(xy) \vee s = \mu(xy) \geq t = \mu(x) \wedge \mu(y) \wedge t$. Also $\mu(x^{-1}) \vee s = \mu(x^{-1}) \geq t = \mu(x) \wedge t$. Suppose $x \in H$ and $y \notin H$. Then $xy \notin H$. Hence $\mu(xy) \vee s = s \geq \mu(x) \wedge \mu(y) \wedge t$ since $\mu(y) \leq s$. Suppose $x \notin H$ and $y \notin H$. We consider the cases $xy \notin H$ and $xy \in H$. Suppose $xy \notin H$. Then $\mu(xy) \vee s = s \geq \mu(x) \wedge \mu(y) \wedge t$. Suppose $xy \in H$. Then $\mu(xy) \vee s \geq t \geq \mu(x) \wedge \mu(y) \wedge t$. Since $x \notin H$, $x^{-1} \notin H$. Thus $\mu(x^{-1}) \vee s = s \geq \mu(x) \geq \mu(x) \wedge t$. $\square$

Proposition 5.4.9. *Let G be a group and μ an $(s,t]$-fuzzy subgroup of G. Then $\forall x \in G$, $(\mu(x) \vee s) \wedge t = (\mu(x^{-1}) \vee s) \wedge t$.*

Proof. Let $x \in G$. Then $\mu(x^{-1}) \vee s \geq \mu(x) \wedge t$ and $\mu(x) \vee s \geq \mu(x^{-1}) \wedge t$. Thus $(\mu(x^{-1}) \vee s) \wedge t \geq \mu(x) \wedge t$. Now $(\mu(x^{-1}) \vee s) \wedge t = (\mu(x^{-1}) \wedge t) \vee (s \wedge t) = (\mu(x^{-1}) \wedge t) \vee s$. Hence $(\mu(x^{-1}) \wedge t) \vee s \geq \mu(x) \wedge t$. Thus $(\mu(x^{-1}) \wedge t) \vee s \geq (\mu(x) \wedge t) \vee s$. Similarly, $(\mu(x) \wedge t) \vee s \geq (\mu(x^{-1}) \wedge t) \vee s$. Thus $(\mu(x) \wedge t) \vee s = (\mu(x^{-1}) \vee s) \wedge t$. But $(\mu(x) \wedge t) \vee s = (\mu(x) \vee s) \wedge (t \vee s) = (\mu(x) \vee s) \wedge t$. Hence $(\mu(x) \vee s) \wedge t = (\mu(x^{-1}) \vee s) \wedge t$. $\square$

Proposition 5.4.10. *Let μ be an $(s,t]$-fuzzy subgroup of a group G. Then $\forall x_1, ..., x_n \in G, \mu(x_1...x_n) \vee s \geq \mu(x_1) \wedge ... \wedge \mu(x_n) \wedge t$.*

Proof. The proof is by induction on n. By Definition 5.4.1, the result is true for $n = 2$. Suppose the result is true for $k \in \mathbb{N}, k \geq 2$, the induction hypothesis. Let $x_1, ..., x_{k+1} \in G$. Then $\mu(x_1...x_{k+1}) \vee s \geq \mu(x_1...x_k) \wedge \mu(x_{k+1}) \wedge t$. Thus $\mu(x_1...x_{k+1}) \vee s \geq (\mu(x_1...x_k) \vee s) \wedge \mu(x_{k+1}) \wedge t \geq (\mu(x_1) \wedge ... \wedge \mu(x_k) \wedge t) \wedge \mu(x_{k+1}) \wedge t = \mu(x_1) \wedge ... \wedge \mu(x_k) \wedge \mu(x_{k+1}) \wedge t$. $\square$

Proposition 5.4.11. *Let $\{\mu_i | i \in I\}$ be a collection of $(s,t]$-fuzzy subgroups of a group G. Then $\cap_{i \in I} \mu_i$ is an $(s,t]$-fuzzy subgroup of G.*

Proof. Let $x, y \in G$. Then $(\cap_{i \in I} \mu_i)(xy^{-1}) \vee s = \wedge\{\mu_i(x) \mid i \in I\} \vee s = \wedge\{\mu_i(xy^{-1}) \vee s \mid i \in I\} \geq$
$\wedge\{\mu_i(x) \wedge \mu_i(y) \wedge t \mid i \in I\} = (\wedge\{\mu_i(x) \mid i \in I\}) \wedge (\wedge\{\mu_i(y) \vee s \mid i \in I\} \wedge t) = (\cap_{i \in I} \mu_i)(x) \wedge (\cap_{i \in I} \mu_i)(y) \wedge t$. $\square$

Definition 5.4.12. *Let η be a fuzzy subset of a group G. Let $\prec \eta \succ$ denote the intersection of all $(s,t]$-fuzzy subgroups μ of G such that $\forall x \in G$, $\mu(x) \vee s \geq \eta(x) \wedge t$.*

$\forall x \in G$, let $\lambda(x) = \eta(x) \vee \eta(x^{-1})$.
$\forall z \in G$, let

$$r_z = \vee\{\lambda(z_1) \wedge ... \wedge \lambda(z_n) \mid z = z_1...z_n, z_i \in G, i = 1, ..., n; n \in \mathbb{N}\}.$$

Lemma 5.4.13. *Let η be a fuzzy subset of a group G. Let μ be an $(s, t]$-fuzzy subgroup of G such that $\forall x \in G, \mu(x) \vee s \geq \eta(x) \wedge t$. Then the following assertions hold $\forall x \in G$:*

(1) $\mu(x) \vee s \geq \lambda(x) \wedge t$;

(2) *If $x = x_1...x_n$ for $x_1, ..., x_n \in G$, then $\mu(x) \vee s \geq \lambda(x_1) \wedge ... \wedge \lambda(x_n) \wedge t$.*

Proof. (1) Since $\mu(x) \vee s \geq \eta(x) \wedge t$ and $\mu(x^{-1}) \vee s \geq \eta(x^{-1}) \wedge t$, $(\mu(x) \vee s) \vee (\mu(x^{-1}) \vee s) \geq (\eta(x) \wedge t) \vee (\eta(x^{-1}) \wedge t) = \lambda(x) \wedge t$. Hence $((\mu(x) \vee s) \wedge t) \vee ((\mu(x^{-1}) \vee s) \wedge t) \geq \lambda(x) \wedge t$. By Proposition 5.4.9, $(\mu(x) \vee s) \wedge t \geq \lambda(x) \wedge t$. Thus $\mu(x) \vee s \geq \lambda(x) \wedge t$.

(2) By Proposition 5.4.10, $\mu(x) \vee s \geq \mu(x_1) \wedge ... \wedge \mu(x_n) \wedge t$. Thus $\mu(x) \vee s \geq (\mu(x_1) \vee s) \wedge ... \wedge (\mu(x_n) \vee s) \wedge t \geq \lambda(x_1) \wedge ... \wedge \lambda(x_n) \wedge t$, where the latter inequality holds by (1). ∎

Theorem 5.4.14. *Let η be a fuzzy subset of a group G. Let δ be the fuzzy subset of G defined by $\forall x \in G$,*

$$\delta(x) = \begin{cases} t & if & r_x \geq t, \\ r_x & if & s < r_x \leq t, \\ 0 & if & r_x \leq s, \end{cases}$$

where r_x is defined immediately above. Then δ is a fuzzy subgroup of G and $\delta = \prec \eta \succ$.

Proof. We first show that δ is an fuzzy subgroup of G. Let $x, y \in G$. Then

$$r_{xy} = \vee\{\lambda(z_1) \wedge ... \wedge \lambda(z_n) | xy = z_1...z_n; z_i \in G, i = 1, ..., n; n \in \mathbb{N}\}$$
$$\geq \vee\{\lambda(x_1) \wedge ... \wedge \lambda(x_m) \wedge \lambda(y_1) \wedge ... \wedge \lambda(y_k) | x = x_1...x_m,$$
$$y = y_1...y_k, x_i, y_j \in G, i = 1, ..., m; j = 1, ..., k; m, n \in \mathbb{N}\}$$
$$= r_x \wedge r_y$$
$$\geq \delta(x) \wedge \delta(y).$$

Thus $\delta(xy) \geq \delta(x) \wedge \delta(y)$ if $s < r_{xy} \leq t$. Suppose $r_{xy} \geq t$. Since $\delta(z) \leq t \forall z \in G, \delta(xy) = t \geq \delta(x) \wedge \delta(y)$. Suppose $r_{xy} \leq s$. Then $r_x \wedge r_y \leq s$. Hence $\delta(xy) = 0$. Now either $r_x \leq s$ or $r_y \leq s$. Thus either $\delta(x) = 0$ or $\delta(y) = 0$. Hence $\delta(xy) \geq \delta(x) \wedge \delta(y)$.

For $z \in G$, $z = z_1...z_n$ if and only if $z^{-1} = z_n^{-1}...z_1^{-1}$. Thus it follows that $r_z = r_{z^{-1}}$. Let $x \in G$. Suppose $s < r_{x^{-1}} \leq t$. Then $\delta(x^{-1}) = r_{x^{-1}} = r_x = \delta(x)$. Suppose $r_{x^{-1}} \geq t$. Then $r_x \geq t$ and so $\delta(x^{-1}) = t = \delta(x)$. Suppose $r_{x^{-1}} \leq s$. Then $r_x \leq s$ and so $\delta(x^{-1}) = 0 = \delta(x)$. Hence δ is a fuzzy subgroup of G (and thus an $(s, t]$-fuzzy subgroup of G).

Suppose $\eta(x) \geq t$. Then $r_x \geq t$. Hence $\delta(x) = t = \eta(x) \wedge t$. Thus $\delta(x) \vee s \geq \eta(x) \wedge t$. Suppose $s < \eta(x) \leq t$. Then either $\delta(x) = t$ or $\delta(x) = r_x$. Hence $\delta(x) \geq r_x \wedge t \geq \eta(x) \wedge t$. Thus $\delta(x) \vee s \geq \eta(x) \wedge t$. Suppose $\eta(x) \leq s$. Then $\delta(x) \vee s \geq s > \eta(x) \wedge t$. Hence $\forall x \in G, \delta(x) \vee s \geq \eta(x) \wedge t$.

Now $\prec \eta \succ$ is the smallest $(s,t]$-fuzzy subgroup of G such that $\forall x \in G$, $\prec \eta \succ (x) \vee s \geq \eta(x) \wedge t$. Hence $\prec \eta \succ \subseteq \delta$.

Let μ be any $(s,t]$-fuzzy subgroup of G such that $\forall x \in G$, $\mu(x) \vee s \geq \eta(x) \wedge t$. In order to show, $\delta \subseteq \prec \eta \succ$ it suffices to show $\delta \subseteq \mu$. Let $x \in G$. By (2) of Lemma 5.4.13, it follows that $\mu(x) \vee s \geq r_x \wedge t$. Suppose $r_x \geq t$. Then $\mu(x) \vee s \geq r_x \wedge t \geq \delta(x) \wedge t$. Since $s < t \leq r_x$, $\mu(x) \geq \delta(x) \wedge t$. But $\delta(x) \leq t$. Thus $\mu(x) \geq \delta(x)$. Suppose $s < r_x \leq t$. Then $\mu(x) \vee s \geq r_x \wedge t = \delta(x) \wedge t$. Since $s < r_x$, $\mu(x) \geq \delta(x) \wedge t$. But $\delta(x) \leq t$. Thus $\mu(x) \geq \delta(x)$. Suppose $r_x \leq s$. Then $\delta(x) = 0 \leq \mu(x)$. Thus $\delta \subseteq \mu$. Hence $\delta \subseteq \prec \eta \succ$. Thus $\delta = \prec \eta \succ$.   □

Let G be a group and Y a subset of G. Recall that the notation $\langle Y \rangle$ is used to denote the subgroup of G generated by Y.

Definition 5.4.15. *Let η be a fuzzy subset of a group G. Define $\langle\langle \eta \rangle\rangle$ to be the intersection of all $(s,t]$-fuzzy subgroups of G which contain η.*

If η is a fuzzy subset of a group G, then $\langle\langle \eta \rangle\rangle$ is an $(s,t]$-fuzzy subgroup of G by Proposition 5.4.11. $\langle\langle \eta \rangle\rangle$ is called the $(s,t]$-fuzzy subgroup of G **generated** by η. Let η be a fuzzy subset of a group G. Recall that the notation $\langle \eta \rangle$ is used to denote the intersection of all fuzzy subgroups of G which contain η.

Proposition 5.4.16. *Let η be a fuzzy subset of a group G. Then $\prec \eta \succ \subseteq \langle\langle \eta \rangle\rangle \subseteq \langle \eta \rangle$.*

Proof. The proof follows from the definitions.   □

Definition 5.4.17. *Let G and H be groups and let μ and ν be $(s,t]$-fuzzy subgroups of G and H, respectively. Let f be a homomorphism of G onto H. Then ν (or (H,ν)) is called an $(s,t]$-**homomorphic image** of μ (or (G,μ)) if $\forall x \in G$, $f(\mu)(f(x)) = \nu(f(x))$ if $s < \nu(f(x)) \leq t$, $f(\mu)(f(x)) \wedge t = \nu(f(x)) \wedge t$ if $\nu(f(x)) > t$, and $f(\mu)(f(x)) \vee s = \nu(f(x)) \vee s$ if $\nu(f(x)) \leq s$.*

Let F be a group and μ a fuzzy subgroup of F. Let X be a subset of F and χ a fuzzy subset of X. Recall Definition 5.3.1; (F,μ) is called a free fuzzy group on (X,χ) provided the following conditions hold:

(1) X generates F and χ generates μ;

(2) Let G be a group and ν a fuzzy subgroup of G such that $G = \langle Y \rangle$ and $\nu = \langle \eta \rangle$. Then for any surjection $f : X \to Y$ such that $f(\chi) = \eta$, there exists an epimorphism $f^* : F \to G$ such that $f^*(x) = f(x)$ for all $x \in X$ and $f^*(\mu) = \nu$. We point out that the fuzzy subset χ of X in Definition 5.3.1 can be considered to be a fuzzy subset of F by letting $\chi(x) = 0$ for all $x \in G, x \notin X$. We make this type of assumption elsewhere without further comment. Let X be a set and χ a fuzzy subset of X. Then by Theorem 5.3.5, there exists a free fuzzy group on (X,χ) and if G is a group and ν a fuzzy subgroup of G, then (G,ν) is a homomorphic image of a free fuzzy group.

Definition 5.4.18. *Let F be a group and μ an $(s,t]$-fuzzy subgroup of F. Let X be a subset of F and χ a fuzzy subset of X. Then (F,μ) is called a free $(s,t]$-fuzzy group on (X,χ) provided the following conditions hold:*

(1) X generates F and χ generates μ, i. e., $F = \langle X \rangle$ and $\mu = \prec \chi \succ$.

(2) Let G be a group and ν an $(s,t]$-fuzzy group such that $G = \langle Y \rangle$ and $\nu = \prec \eta \succ$. Then for any surjection $f : X \to Y$ such that $f(\chi) = \eta$, there exists an epimorphism $f^* : F \to G$ such that $f^*(x) = f(x)$ for all $x \in X$ and ν is an $(s,t]$-homomorphic image of μ under f^*.

Theorem 5.4.19. *Let X be a set and χ a fuzzy subset of X. Then there exists a free $(s,t]$-fuzzy group on (X,χ).*

Proof. Let (F,μ) be a free fuzzy group on (X,χ). Let (G,ν) be an $(s,t]$-fuzzy group having a generating set (Y,η) and let $f : (X,\chi) \to (Y,\eta)$ be a surjection. Let $\nu' = \langle \eta \rangle$. Then there exists an epimorphism $f^* : F \to G$ such that $f^*(x) = f(x)$ for all $x \in X$ and $f^*(\mu) = \nu'$. Let $x \in G$. Suppose that $s < r_{f^*(x)} \le t$. Then $f^*(\mu)(f^*(x)) = \nu'(f^*(x)) = \langle \eta \rangle (f^*(x)) = r_{f^*(x)} = \prec \eta \succ (f^*(x)) = \nu(f^*(x))$. Suppose $r_{f^*(x)} > t$. Then $\langle \eta \rangle (f^*(x)) = r_{f^*(x)}$ and $\prec \eta \succ (f^*(x)) = t$. Thus $f^*(\mu)(f^*(x)) = \nu'(f^*(x)) = \langle \eta \rangle (f^*(x)) = r_{f^*(x)} > t$ and $\nu(f^*(x)) = \prec \eta \succ (f^*(x)) = t$. Hence $f^*(\mu)(f^*(x)) \wedge t = \nu(f^*(x)) \wedge t$. Suppose that $r_{f^*(x)} \le s$. Then $f^*(\mu)(f^*(x)) = \nu'(f^*(x)) = \langle \eta \rangle (f^*(x)) = r_{f^*(x)} \le s$ and $\nu(f^*(x)) = \prec \eta \succ (f^*(x)) = 0$. Thus $f^*(\mu)(f^*(x)) \vee s = s = \nu(f^*(x)) \vee s$. Hence (F,μ) is a free $(s,t]$-fuzzy group on (X,χ). $\square$

Let μ_0 be the fuzzy subgroup of G defined by $\forall x \in G, \mu_0(x) = \nu(x)$ if $s < r_x \le t$, $\mu_0(x) = t$ if $r_x > t$, and $\mu_0(x) = 0$ if $r_x \le s$. Let μ_1 be an $(s,t]$-fuzzy subgroup of F such that $\forall x \in G, \mu_1(x) = \mu_0(x)$ if $s < r_x \le t$, $\mu_0(x) \le \mu_1(x) \le \nu(x)$ if $r_x > t$ or if $r_x \le s$. Then (G,ν) is an $(s,t]$-fuzzy homomorphic image of (F,μ_1). Thus we have the following result.

Theorem 5.4.20. *Let G be a group and ν a fuzzy subgroup of G. Then (G,ν) is an $(s,t]$-homomorphic image of a free $(s,t]$-fuzzy group.*

Remark 5.4.21. We feel that the notion of an $(s,t]$-fuzzy subgroup has possible applications for suitable s and t. For example, in medical diagnosis the presence or absence of a disease may be determined by certain high values, t, or certain low values, s, [22, 31]. The use of free algebraic structures in medical diagnosis can be found in [31]. An interesting problem would be to determine applications of $(s,t]$-fuzzy substructures of algebraic structures.

References

1. N. Ajmal, The free product is not associative, *Fuzzy Sets and Systems* 60 (1993) 241 - 244.
2. S. K. Bhakat, ($\in \vee q$)-level subset, *Fuzzy Sets and Systems* 103 (1999) 529-533.

3. S. K. Bhakat, $(\in, \in \vee q)$-fuzzy quasinormal, normal and maximal subgroups, *Fuzzy Sets and Systems* 112 (2000) 299-312.

4. S. K. Bhakat, On fuzzy commutativity, *J. Fuzzy Math.* 6 (1998) 915-921.

5. S. K. Bhakat, q-fuzzy cyclic subgroup, *J. Fuzzy Math.* 7 (1999) 521 - 529.

6. S. K. Bhakat, Fuzzy order and $(\epsilon, \epsilon \vee q)$-fuzzy subgroup, *J. Fuzzy Math.* (2000) 13 - 26

7. S. K. Bhakat, $(\in, \in \vee q)$-fuzzy cyclic subgroup, *J. Fuzzy Math.* 8 (2000) 597-606.

8. S. K. Bhakat and P. Das, On the definition of a fuzzy subgroup, *Fuzzy Sets and Systems* 51 (1992) 235-241.

9. S. K. Bhakat and P. Das, q-similtudes and q-fuzzy partitions, *Fuzzy Sets and Systems* 51 (1992) 195-202.

10. S. K. Bhakat and P. Das, (α, β)-fuzzy mappings, *Fuzzy Sets and Systems* 56 (1993) 89 - 95.

11. S. K. Bhakat and P. Das, $(\in, \in \vee q)$-fuzzy subgroup, *Fuzzy Sets and Systems* 80 (1996) 359-368.

12. S. K. Bhakat and P. Das, Fuzzy subrings and ideals redefined, *Fuzzy Sets and Systems*, 81 (1996) 383-393.

13. K. R. Bhutani, R. Crist and J. N. Mordeson, Free $(s, t]$-Fuzzy Subgroups, Submitted.

14. C. C. Chang and R.C.T. Lee, *Symbolic Logic and Mechanical Theorem Proving*, Academic Press, New York, 1973.

15. C. L. Chang, Interpretation and execution of fuzzy programs, in: L.A. Zadeh et al., Eds., *Fuzzy Sets and Their Applications to Cognitive and Decision Processes* (Academic Press, New York, 1975) 191-218.

16. D. E. Cohen, *Combinatorial Group Theory: a topological approach*, Cambridge University Press, Cambridge (1989).

17. P. M. Cohn, *Universal Algebra*, Harper & Row, New York, 1965.

18. M. Garzon and G. C. Muganda, Free fuzzy groups and fuzzy group presentations, *Fuzzy Sets and Systems* 48 (1992) 249-255.

19. G. Gratzer, *Lattice Theory*, Freeman, San Francisco, CA, 1971.

20. I. Guessarian, A survey on classes of interpretations and some of their applications, in: M. Nivat, J.C. Reynolds, Eds., *Algebraic Methods in Semantics*, Cambridge University Press, London, 1985, 383-409.

21. D. S. Kim, $\mathfrak{D}$-admissible $(\epsilon, \epsilon \vee q)$-fuzzy subgroups, preprint.

22. G. J. Klir and B. Yuan, *Fuzzy Sets and Fuzzy Logic: Theory and Applications*, Prentice Hall P T R, Upper Saddle River, New Jersey 1995.

23. R. C. T. Lee, Fuzzy logic and the resolution principle, *J. Assoc. Comput. Mach.* 19 (1972) 109-119.

24. J. W. Lloyd, *Foundations of Logic Programming*, Springer-Verlag, New York, 1984.

25. R. Lyndon and P.E. Schupp, *Combinatorial Group Theory*, Springer-Verlag, Berlin, 1978.

26. R. C. Lyndon, *Notes on Logic*, Van Nostrand, New York, 1966.

27. W. Magnus, A. Karrass, D. Solitar, *Combinatorial Group Theory*, Dover Publications, New York, 1976, 85-95.

28. D. S. Malik, J. N. Mordeson, and M. K. Sen, Free fuzzy submonoids and fuzzy autamata, *Bull. Calcutta. Math. Soc.* 88 (1996) 145-150.

29. T. P. Martin, J.F. Baldwin and B.W. Pilsworth, The implementation of Fprolog-A fuzzy Prolog interpreter, *Fuzzy Sets and Systems* 23 (1987) 119-129.

30. P. P. Ming and L. Y. Ming, Fuzzy topology I: Neighbourhood structure of a fuzzy point and Moore-Smith convergence, *J. Math. Anal. Appl.* 76 (1980) 571-599.
31. J. N. Mordeson, D. S. Malik and S-C Cheng, Fuzzy Mathematics in Medicine, *Studies in Fuzziness and Soft Computing*, Physica-Verlag, A Springer-Verlag Company, 55, 2000.
32. G. C. Muganda and M. Garzon, On the structure of fuzzy groups, *First International Conference on Fuzzy Theory and Technology*, Ed.: Paul P. Wang, Proceedings (1992) 250-255.
33. S. Ray, The free product of fuzzy subgroups, *Fuzzy Sets and Systems* 50 (1992) 225 - 235.
34. A. Rosenfeld, Fuzzy groups, *J. Math. Anal. Appl.* 35 (1971) 512-517.
35. D. Schwartz, Axioms for a theory of semantic equivalence, *Fuzzy Sets and Systems* 21 (1987) 319-349.
36. Y. Suzuki, On the construction of free fuzzy groups, *J. Fuzzy Math.* 2 (1994) 1-15.
37. Bingxue Yao, (λ, μ)-fuzzy normal subgroups and (λ, μ)-fuzzy quotient subgroups, preprint.
38. Bingxue Yao, $(\epsilon, \epsilon \vee q)$-fuzzy quotient subgroups, preprint.
39. Xuechai Yuan, Cheng Zheng, Yonghong Ren, Generalized fuzzy groups and many-valued implications, *Fuzzy Sets and Systems* 138 (2003) 205 - 211.
40. Xuechai Yuan, Cheng Zheng, Yonghong Ren, $(\overline{\beta}, \overline{\alpha})$-fuzzy subgroup, R-fuzzy subgroup and $(\lambda, \mu]$-fuzzy subgroup, *Fuzzy Sets and Systems*, to appear.
41. L. A. Zadeh, Fuzzy sets, *Inform. Control* 8 (1965) 338-353.

6

Fuzzy Subgroups of Abelian Groups

Some of the best examples of algebraic structure theory come from commutative group theory. Commutative group theory is also a principal reason for the study of module theory. In this chapter, we present structure results for fuzzy subgroups of Abelian groups. Many of the results in this chapter have been applied to the study of fuzzy group subalgebras and fuzzy field extensions [3, 11, 14, 15, 16, 21].

Throughout this chapter G is an (additive) Abelian group. Let H be a subgroup of G. Recall that H is called torsion if every element of H has finite order and H is called torsion-free if no element of H has finite order other than 0. The results in this chapter come mainly from [8, 9, 13, 16, 17]. For other approaches the reader is referred to [1, 5, 20].

6.1 Minimal Generating Sets and Direct Sums

Let $\mu \in \mathcal{F}(G)$. We introduce the concept of a minimal generating set for μ when the support of μ is a direct sum of cyclic groups. We show that if the support of μ is cyclic, then μ has a minimal generating set. However, we show that this result does not hold in general when the support of μ is a direct sum of two or more cyclic groups. Consequently, although a subgroup of a group G that is a direct sum of cyclic groups is a again a direct sum of cyclic groups, the corresponding result for fuzzy subgroups of G need not hold.

Let I denote a nonempty index set. By the summation $x = \sum_{i \in I} x_i$, we mean all but a finite number of the x_i are zero. If $\{\mu_i \mid i \in I\}$ is a collection of fuzzy subgroups of G, then define the fuzzy subset $\sum_{i \in I} \mu_i$ of G as follows: $\forall x \in G$,

$$(\sum_{i \in I} \mu_i)(x) = \vee\{\wedge\{\mu_i(x_i) | i \in I\} | x = \sum_{i \in I} x_i\}.$$

This definition corresponds to the definition of weak product given in Definition 1.5.2. Thus the results in Chapter 1 concerning the weak product hold here with the appropriate changes in terminology and notation. We speak of the weak direct sum rather than the weak direct product. We replace $\prod$ and $\otimes$ with $\oplus$.

Definition 6.1.1. *Let $\mu \in \mathcal{F}(G)$ and let*

$$\mathcal{F}(\mu) = \{\nu \in \mathcal{F}(G) \mid \nu \subseteq \mu\}.$$

*Let χ be a fuzzy subset of G such that $\chi \subseteq \mu$. Define $\langle \chi \rangle$ to be the intersection of all $\gamma \in \mathcal{F}(\mu)$ such that $\chi \subseteq \gamma$. Then χ is called a **generating set** for $\langle \chi \rangle$ in μ.*

Clearly, $\langle \chi \rangle$ is a fuzzy subgroup of G. In fact, $\langle \chi \rangle$ is the smallest fuzzy subgroup of G in $\mathcal{F}(\mu)$ containing χ.

Definition 6.1.2. *Let $\mu \in \mathcal{F}(G)$. Let $\mathcal{S}$ denote a set of fuzzy singletons such that if $x_a, x_b \in \mathcal{S}$, then $a = b > 0$ and $x \neq 0$. Define the fuzzy subset $\chi(\mathcal{S})$ of G as follows: $\forall x \in G$*

$$\chi(\mathcal{S})(x) = \begin{cases} a \text{ if } x_a \in \mathcal{S} \\ 0 \text{ otherwise.} \end{cases}$$

*Set $\langle \mathcal{S} \rangle = \langle \chi(\mathcal{S}) \rangle$. Then $\mathcal{S}$ is called a **minimal generating set** for μ if $\mu = \langle \mathcal{S} \rangle \cup 0_{\mu(0)}$ and $\mu \supset \langle \mathcal{S} \backslash \{x_a\} \rangle \cup 0_{\mu(0)}$ for all $x_a \in \mathcal{S}$.*

Theorem 6.1.3. *Let $\mu \in \mathcal{F}(G)$. If $x_1, x_2, \ldots, x_n \in G$ are such that $\wedge_{i=1}^{n-1} \mu(x_i) > \mu(x_n)$, then $\mu(x_1 + x_2 + \ldots + x_n) = \mu(x_n)$.*

Proof. Let $x = x_1 + x_2 + \ldots + x_{n-1}$. Then $\mu(x) \geq \wedge_{i=1}^{n} \mu(x) = \wedge_{i=1}^{n-1} \mu(x_i) > \mu(x_n)$. Hence $\mu(x_n) = \mu(-x + x + x_n) \geq \mu(x) \wedge \mu(x + x_n) \geq \mu(x) \wedge \mu(x_n) = \mu(x_n)$. Thus $\mu(x) \wedge \mu(x + x_n) = \mu(x_n)$. Since $\mu(x) > \mu(x_n)$, $\mu(x + x_n) = \mu(x_n)$. $\blacksquare$

Theorem 6.1.4. *Let $\mu \in \mathcal{F}(G)$ and $\mu(G) \backslash \{0\} = \{a_0, a_1, \ldots, a_n\}$. Suppose that $a_{i-1} > a_i$ for all $i \in I \backslash \{0\}$, where $I = \{0, 1, \ldots, n\}$. If there exists a subgroup H_i of μ_{a_i} such that $\mu_{a_i} = \mu_{a_{i-1}} \oplus H_i$ for all $i \in I \backslash \{0\}$, then there exists $\mu_i \in L(G)$, $i \in I$, such that $\mu = \oplus_{i \in I} \mu_i$, $\mu_0^* = \mu_{a_0}$, and $\mu_i^* = H_i$ for all $i \in I \backslash \{0\}$.*

Proof. Since $\mu_{a_i} = \mu_{a_{i-1}} \oplus H_i$ for all $i \in I \backslash \{0\}$, it follows that $\mu_{a_n} = \mu_{a_0} \oplus H_1 \oplus \ldots \oplus H_n$. Define the fuzzy subset μ_i of G as follows: $\forall x \in G$, $\mu_i(0) = a_0$, $\mu_i(x) = a_i$ if $x \in H_i \backslash \{0\}$, and $\mu_i(x) = 0$ otherwise, $i \in I$, where $H_0 = \mu_{a_0}$. Then $\mu_i \in \mathcal{F}(G)$ and $\mu_i^* = H_i$, $i \in I$. Clearly, $\mu_0 = \mu$ on H_0 and $\mu_i = \mu$ on H_i since $H_i \subseteq \mu_{a_i}$ and $H_i \cap \mu_{a_{i-1}} = \{0\}$, $i \in I \backslash \{0\}$. Let $x \in \mu^*$. Then $x = x_0 + x_1 + \ldots + x_n$, where $x_i \in H_i$, $i \in I$. Hence $\mu(x) = \mu_0(x) \wedge \mu_1(x) \wedge \ldots \wedge \mu_n(x)$ by Theorem 6.1.3. Now

$$(\textstyle\sum_{i\in I}\mu_i)(x) = \vee\{\wedge_{i=0}^n \mu(y_i)|x = y_0 + y_1 + \ldots + y_n, y_i \in \mu_i^*, i \in I\}$$
$$= \mu(x_0) \wedge \mu(x_1) \wedge \ldots \wedge \mu(x_n)$$

since the expression $x = x_0 + x_1 + \ldots + x_n$ is unique. Hence $\mu = \sum_{i\in I}\mu_i$. Thus $\mu = \oplus_{i\in I}\mu_i$ by Theorem 1.5.9. □

Theorem 6.1.5. *Let $\mu \in \mathcal{F}(G)$ and $\mu(G)\backslash\{0\} = \{a_0, a_1, \ldots, a_n\}$. Suppose that $a_{i-1} > a_i$ for all $i \in I\backslash\{0\}$, where $I = \{0, 1, \ldots, n\}$. If $p\mu^* = \{0\}$ for p a prime, then μ is a weak direct sum of fuzzy subgroups whose supports are cyclic.*

Proof. Since $p\mu^* = \{0\}$, $\mu_{a_{i-1}}$ is pure in μ_{a_i} and $\mu_{a_{i-1}}$ is a direct summand of μ_{a_i}, say $\mu_{a_{i-1}} \oplus H_i$ for some subgroup H_i of G, $i \in I\backslash\{0\}$. Thus by Theorem 6.1.4, $\mu^* = \mu_{a_0} \oplus H_1 \oplus \ldots \oplus H_n$ and $\mu = \oplus_{i\in I}\mu_i$ for fuzzy subgroups μ_i of G such that $\mu_i^* = H_i$, $\mu_0^* = \mu_{a_0} = H_0$. Since $p\mu^* = \{0\}$, each H_i is a weak direct sum of cyclic groups, say $H_i = \oplus_{j_i\in J_i}K_{j_i}$, $j_i \in J_i$ (an index set), $i = 0, 1, \ldots, n$. Define the fuzzy subsets ξ_{j_i} of G as follows: $\forall x \in G$,

$$\xi_{j_i}(x) = \begin{cases} \mu_i(x) \text{ if } x \in K_{j_i}, \\ 0 \qquad \text{otherwise,} \end{cases}$$

$j_i \in J_i$, $i = 0, 1, \ldots, n$. Then ξ_{j_i} is an fuzzy subgroup of G and $\xi_{j_i}^* = K_{j_i}$ is cyclic. Since every ξ_{j_i} is the same constant a_i on $K_{j_i}\backslash\{0\}$ and every μ_i equals a_i on $H_i\backslash\{0\}$, $\mu_i = \sum_{j_i\in J_i}\xi_{j_i} = \oplus_{j_i\in J_i}\xi_{j_i}$. Thus $\mu = \oplus_{i\in I}(\oplus_{j_i\in J_i}\xi_{j_i})$. □

Proposition 6.1.6. *Let $\mu \in \mathcal{F}(G)$ and let χ and η be fuzzy subsets of G. If $\chi, \eta \subseteq \mu$, then $\chi + \eta \subseteq \mu$.*

Proof. Let $z \in G$. Then $(\chi+\eta)(z) = \vee\{\chi(x)\wedge\eta(y) \mid z = x+y\} \leq \vee\{\mu(x)\wedge\mu(y) \mid z = x + y\} \leq \vee\{\mu(x + y) \mid z = x + y\} = \mu(z)$. □

Proposition 6.1.7. *Let $\mu \in \mathcal{F}(G)$. Let χ be an fuzzy subset of G. If $\chi \subseteq \mu$, then $-\chi \subseteq \mu$.*

Proof. By Definition 1.2.1, $(-\chi)(z) = \chi(-z) \leq \mu(-z) = \mu(z)$. □

For any fuzzy singleton x_a of G, it follows easily that $-(x_a) = (-x)_a$.

Theorem 6.1.8. *Let $\mu \in \mathcal{F}(G)$. Let χ be a fuzzy subset of G such that $\chi \subseteq \mu$. Let σ be the fuzzy subset of G defined by $\forall y \in G$, $\sigma(y) = \vee\{(e_1(x_1)_{a_1} + \ldots + e_n(x_n)_{a_n})(y) \mid x_i \in G, a_i \in L, \chi(x_i) = a_i, e_i = \pm 1, i = 1, \ldots, n; n \in \mathbb{N}\}$. Then $\sigma = \langle\chi\rangle$.*

Proof. By Proposition 6.1.7, $e_i(x_i)_{a_i} \subseteq \langle\chi\rangle$ for $i = 1, \ldots, n$. Thus $\sigma \subseteq \langle\chi\rangle$ since $e_1(x_1)_{a_1} + \ldots + e_n(x_n)_{a_n} \subseteq \langle\chi\rangle$ by Proposition 6.1.6. To show that $\langle\chi\rangle \subseteq \sigma$, it suffices to show that $\sigma \in \mathcal{F}(\mu)$ and $\chi \subseteq \sigma$. Let $x \in G$ and $\chi(x) = a$. Then clearly $x_a(x) \leq \sigma(x)$ since $x_a = e(x_a)$ with $e = 1$ and $n = 1$. Hence $\chi \subseteq \sigma$. Let $x, y \in G$. Then $\sigma(x)$ and $\sigma(y)$ are supremums of numbers of the form $(e_1(x_1)_{a_1} + \ldots + e_n(x_n)_{a_n})(x)$ and $(f_1(y_1)_{b_1} + \ldots + f_m(y_m)_{b_m})(y)$, respectively. Suppose that $\sigma(x) > 0$ and $\sigma(y) > 0$. Then $\exists$ sequences $a_j = a_{1_j} \wedge \ldots \wedge a_{n_j}$ and $b_k = b_{1_k} \wedge \ldots \wedge b_{m_k}$ such that $a_j \to \sigma(x)$ and $b_k \to \sigma(y)$. If $x \in \langle x_1, \ldots, x_n \rangle$ and $y \in \langle y_1, \ldots, y_m \rangle$, then $x + y \in \langle x_1, \ldots, x_n, y_1, \ldots, y_m \rangle$. Thus $\sigma(x + y) \geq \vee \{a_j \wedge b_k \mid j, k = 1, 2, \ldots\} = (\vee\{a_j \mid j = 1, 2, \ldots\}) \wedge (\vee\{b_k \mid k = 1, 2, \ldots\}) = \sigma(x) \wedge \sigma(y)$. Clearly, $\sigma(x) = \sigma(-x)$. Hence $\sigma \in \mathcal{F}(\mu)$. $\square$

Corollary 6.1.9. *Let $x \in G$ and $a \in [0, 1]$. Then $\forall \, y \in G$, $\langle x_a \rangle (y) = a$ if $y \in \langle x \rangle$ and $\langle x_a \rangle (y) = 0$ if $y \notin \langle x \rangle$.*

Proof. $\langle x_a \rangle (y) = \vee\{(e_1(x)_a + \ldots + e_n(x)_a)(y) \mid e_i = \pm 1, i = 1, \ldots, n; n \in \mathbb{N}\} = \vee\{(m(x)_a)(y) \mid m \in \mathbb{Z}\}$. $\square$

Corollary 6.1.10. *Let $\mu \in \mathcal{F}(G)$. Let $\mathcal{S}$ be a set of fuzzy singletons such that $\forall x_a \in \mathcal{S}$, $0 < a \leq \mu(x)$. Then $\langle \mathcal{S} \rangle^* = \langle \mathcal{S}^* \rangle$.*

Proof. $x \in \langle \mathcal{S} \rangle^* \Leftrightarrow \langle \mathcal{S} \rangle (x) > 0 \Leftrightarrow \vee\{(e_1(x_1)_{a_1} + \ldots + e_k(x_k)_{a_k})(x) \mid (x_i)_{a_i} \in \mathcal{S}, e_i = \pm 1, i = 1, \ldots, k, k \in \mathbb{N}\} > 0 \Leftrightarrow x = e_1 x_1 + \ldots + e_k x_k$ for some $(x_i)_{a_i} \in \mathcal{S}$, $i = 1, \ldots, k \Leftrightarrow x \in \langle \mathcal{S}^* \rangle$. $\square$

Recall that by a sum of the form $\sum_{u \in \text{Im}(\mu)} m_u (n_u x)_{a_u}$, where $m_u, n_u \in \mathbb{N} \cup \{0\}$, we mean only a finite number of m_u are different from 0.

Lemma 6.1.11. *Let $\mu \in \mathcal{F}(G)$. Suppose that $\mu^* = \langle x \rangle$ for some $x \in \mu^*$. For all $u \in \mu(G)\backslash\{0\}$, let n_u denote the smallest positive integer such that $\mu_u = \langle n_u x \rangle$ if $\mu_u \neq \{0\}$ and let $a_u = \mu(n_u x)$. Then the following assertions hold:*
  (1) $u = a_u$;
  (2) $\mu = \langle \{(n_u x)_{a_u} | u \in \mu(G)\} \rangle$.

Proof. (1) $a_u = \mu(n_u x) \geq u$. Let $z \in \mu^* \backslash \{0\}$ be such that $\mu(z) = u$. Then $z \in \mu_u$ and so $\exists m \in \mathbb{Z}$ such that $z = m(n_u x)$. Thus $u = \mu(z) \geq \mu(n_u x) = a_u$. Hence $u = a_u$.

(2) Clearly, $\langle \{(n_u x)_{a_u} | u \in \mu(G)\} \rangle \subseteq \mu$. Let $z \in \mu^*$. Now $\langle \{(n_u x)_{a_u} | u \in \mu(G)\} \rangle (z) = \vee\{(\sum_{u \in \mu(G)} m_u (n_u x)_{a_u})(z) \mid m_u \in \mathbb{Z}\}$. Hence if $\mu(z) = u$, then $z \in \mu_u$. Thus $z = m(n_u x)$ for some $m \in \mathbb{Z}$. Clearly, then

$$\mu(z) \leq \langle \{(n_u x)_{a_u} | u \in \mu(G)\} \rangle (z).$$

Let χ be a fuzzy subset of G such that $\chi \subseteq \mu$, where $\mu \in \mathcal{F}(G)$. Let $\langle \mathcal{S}(\chi) \rangle$ denote the intersection of all fuzzy subgroups of G in $\mathcal{F}(\mu)$ such that $x_a \subseteq \mu$ $\forall \, x_a \in \mathcal{S}(\chi)$. Then clearly $\langle \mathcal{S}(\chi) \rangle = \langle \chi \rangle$. $\square$

Theorem 6.1.12. *Let $\mu \in \mathcal{F}(G)$. Suppose that $\mu^* = \langle x \rangle$ for some $x \in \mu^*$. For all $u \in \mu(G)$, let n_u denote the smallest positive integer such that $\mu_u = \langle n_u x \rangle$. Then $\{(n_u x)_u \mid u \in \mu(G)\backslash\{0\}\}$ is a minimal generating set for μ.*

Proof. Let $I = \mathrm{Im}(\mu)\backslash\{0\}$. By Lemma 6.1.11, $\{(n_u x)_u \mid u \in I\}$ is a set of generators of μ. Let $v \in I$ and $\nu = \langle\{(n_u x)_u | u \in I\}\backslash\{(n_v x)_v\}\rangle$. Then $\nu \in L(\mu)$. Now $v = \mu(n_v x)$ and $\nu(n_v x) = \vee\{u_1 \wedge \ldots \wedge u_n\} \mid n_v x = \sum_{u \in I\backslash\{v\}} m_u(n_u x)$, $m_u \in \mathbb{Z}$, $m_u \neq 0$ for $u \in \{u_1, \ldots, u_n\}$, $n \in \mathbb{N}\} = \vee\{u| n_v x = m_u(n_u x), u < v, u \in I\}$ (if $\exists u \in I$ such that $u < v$) $< v$, where $=$ and $<$ occur since I is well-ordered under $<$. Thus $\nu \subset \mu$. Now suppose there does not exist $u \in I$ such that $u < v$. Suppose that $\nu(n_v x) > 0$. Then $n_v x = \sum_{i=1}^{k} m_{u_i}(n_{u_i} x)$ for some n_{u_i}, m_{u_i}, $(u_i > v)$, $i = 1, \ldots, k$. Hence $n_v x \in \langle n_u x\rangle = \mu_u$ for some $u > v$, a contradiction. Thus $\nu(n_v x) = 0$. Hence $\nu \subset \mu$. $\qquad\square$

Proposition 6.1.13. *Let $\mu \in \mathcal{F}(G)$ and let $\{\mu_j \mid j \in J\} \subseteq \mathcal{F}(\mu)$, where J is a nonempty index set. Suppose that $\mu^* = \oplus_{j \in J}(\mu_j)^*$, where $\forall j \in J$, $(\mu_j)^* = \langle x_j\rangle$ for some $x_j \in G$. Let $\mathcal{S}_j = \{((n_j)_{u_j} x_j)_{u_j} \mid u_j \in \mu_j(G), (n_j)_{u_j} \in \mathbb{N}\}$ be a minimal generating set for μ_j, $j \in J$. Then the following assertions hold:*

(1) *$\cup_{j \in J}\mathcal{S}_j$ is a minimal generating set for $\oplus_{j \in J}\mu_j$;*

(2) *$\cup_{j \in J}\mathcal{S}_j$ is a minimal generating set for μ if and only if $\mu = \oplus_{j \in J}\mu_j$.*

Proof. Let $\nu = \oplus_{j \in J}\mu_j$. Then $\nu(0) = \mu_j(0) \,\forall j \in J$. Let $z \in (\mu_m)^*$ for some $m \in J$. Then $\nu(z) = \vee\{\wedge\{\mu_j(z_j)|j \in J\} \mid z = \sum_{j \in J} z_j, z_j \in (\mu_j)^* \,\forall j \in J\} = \wedge\{\mu_m(z) \wedge \mu_j(0) \mid j \in J\backslash\{m\}\} = \mu_m(z)$ since the representation $z = \sum_{j \in J} z_j$ is unique and $\nu(0) = \mu_j(0) \,\forall j \in J$. That is, $\nu = \mu_m$ on $(\mu_m)^*$. Let $y_j \in \mathcal{S}_m$. Then $\langle\cup_{j \in J}\mathcal{S}_j\rangle > = \oplus_{j \in J}\mu_j$ and so $\langle\cup_{j \in J}\mathcal{S}_j\backslash\{y_j\}\rangle (y) = (\oplus_{j \in J, m \neq j} \langle\mathcal{S}_j\rangle \oplus \langle\mathcal{S}_m\backslash\{y_j\}\rangle)(y) = \langle\mathcal{S}_m\backslash\{y_j\}\rangle (y) < \langle\mathcal{S}_m\rangle (y) = \nu(y)$. Thus $\cup_{j \in J}\mathcal{S}_j$ is a minimal generating set for ν. $\qquad\square$

Proposition 6.1.13 shows that if μ^* is a direct sum of cyclic groups, then the question of when μ has a minimal set of generators is equivalent to that of when μ is a direct sum of fuzzy subgroups whose supports are cyclic.

Example 6.1.14. *Let $G = \mathbb{Z}(p) \oplus \mathbb{Z}(p^2)$ and let $H = \langle(1, p)\rangle$. Define the fuzzy subsets μ, ν, and γ of G by $\mu(x) = 1$ if $x \in H$, $\mu(x) = \frac{1}{2}$ otherwise; $\nu((0,0)) = 1$, $\nu(x) = \frac{1}{2}$ if $x \in \mathbb{Z}(p) \oplus \{0\}\backslash\{(0,0)\}$, $\nu(x) = 0$ otherwise; $\gamma(0,0) = 1$, $\gamma(x) = \frac{1}{2}$ if $x \in \{0\} \oplus Z(p^2)\backslash\{(0,0)\}$, $\gamma(x) = 0$ otherwise. Then μ, ν, γ are fuzzy subgroups G such that $\nu, \gamma \subseteq \mu$ and $\mu^* = \nu^* \oplus \gamma^*$. However, $\mu \neq \nu + \gamma$ since $\mu((1, p)) = 1$, while $(\nu + \gamma)((1, p)) = \frac{1}{2}$. Now $\mathrm{Im}(\nu) = \{1, \frac{1}{2}, 0\} = \mathrm{Im}(\gamma)$, $\nu_1 = \{(0,0)\}$, $\nu_{\frac{1}{2}} = \mathbb{Z}(p) \oplus \{0\}$, $\gamma_1 = \{(0,0)\}$, $\gamma_{\frac{1}{2}} = \{0\} \oplus \mathbb{Z}(p^2)$. Thus $\{n_{\frac{1}{2}}(1,0)_{\frac{1}{2}}\}$ and $\{m_{\frac{1}{2}}(0,1)_{\frac{1}{2}}\}$ are minimal generating sets for ν and γ respectively, where $n_{\frac{1}{2}} = 1 = m_{\frac{1}{2}}$. Now $\{(1,0)_{\frac{1}{2}}\} \cup \{(0,1)_{\frac{1}{2}}\}$ is a minimal generating set for $\nu + \gamma$, but not a generating set for μ since $\mu \neq \nu + \gamma$. However, $\mu = 1_H \oplus \gamma$ and so μ has a minimal generating set.*

Theorem 6.1.15. *Suppose that $G = \oplus_{i=1}^{m} G_i$, where G_i is a cyclic subgroup of G for $i = 1, \ldots, m$. Suppose that $\exists$ a cyclic subgroup H of G such that H is not contained in any cyclic direct summand of G. Then $\exists$ an fuzzy subgroup μ of G such that μ is not a direct sum of fuzzy subgroups of G whose supports are cyclic.*

Proof. Define the fuzzy subset μ of G by $\mu(x) = 1$ if $x \in H$ and $\mu(x) = \frac{1}{2}$ otherwise. Then μ is a fuzzy subgroup of G and $\mu^* = G$. By hypothesis $\exists \mu_i \in \mathcal{F}(\mu)$ for $i = 1, \ldots, m$ such that $\mu^* = \oplus_{i=1}^m \mu_i^*$ and $\mu_i^* = \langle c_i \rangle$ for some $c_i \in G$, $i = 1, \ldots, m$. We show that $\mu \neq \sum_{i=1}^m \mu_i$ so that $\mu \neq \oplus_{i=1}^m \mu_i$. Let $z \in G$ be such that $H = \langle z \rangle$. Then $(\sum_{i=1}^m \mu_i)(z) = \vee\{\wedge\{\mu_i(z_i)|i = 1, \ldots, m\} \mid z = z_1 + \ldots + z_m\}$. Thus $(\sum_{i=1}^m \mu_i)(z) = 0$ if $z_i \notin \mu_i^*$ for some i, while $\mu(z) = 1$. Suppose that $(\sum_{i=1}^m \mu_i)(z) = 1$. Then $\exists z_1, \ldots, z_m \in G$ such that $z = z_1 + \ldots + z_m$ and $\mu_i(z_i) = 1$ for $i = 1, \ldots, m$. Since $\mu_i \subseteq \mu$, $z_i \in H \cap \mu_i^*$, $i = 1, \ldots, m$. Suppose that $|H| = r$. Then z has order r and so z_i has order r for some i. Since $z_i \in H \cap \mu_i^*$, $H = \langle z_i \rangle \subseteq \mu_i^*$. Thus H is contained in a direct summand of G, a contradiction. Suppose that $|H| = \infty$. Then $z_i = q_i z$ for some integers q_i, $i = 1, \ldots, m$. If $q_i \neq 0 \neq q_j$ for $i \neq j$, then $\mu_i^* \cap \mu_j^* \supseteq \langle q_i z \rangle \cap \langle q_j z \rangle \supset \{0\}$, a contradiction. Hence $\exists i$ such that $\forall j, j \neq i, q_j = 0, q_i = \pm 1$. Thus $H \subseteq \mu_i^*$, a contradiction. Therefore, $(\sum_{i=1}^m \mu_i)(z) \neq 1$ and so $\sum_{i=1}^m \mu_i \subset \mu$. $\square$

Example 6.1.16. *Let $G = G_1 \oplus G_2$, where G_i is a cyclic subgroup of G for $i = 1, 2$. Suppose that $|G_1| = \infty$ and $|G_2| = p^r$ for some prime p. Then we show that $\exists$ a fuzzy subgroup μ of G such that μ is not a direct sum of fuzzy subgroups of G whose supports are cyclic. Let $G_i = \langle x_i \rangle$, $i = 1, 2$. Let $H = \langle (px_1, x_2) \rangle$. Let K be a subgroup of G such that $H \cap K = \{(0, 0)\}$. If $(s_1 x_1, s_2 x_2) \in K$ for $s_1, s_2 \in \mathbb{Z}$, then $(p^{r+1} s_1 x_1, 0) \in H \cap K$ and so $s_1 = 0$. Thus if H is contained in a direct summand of G, say $G = H' \oplus K$ with $H \subseteq H'$, then $K \subseteq G_2$. Hence $K = G_2$ since G_2 is a torsion subgroup of G. Also H' is cyclic and of infinite order. Let $H' = \langle (t_1 x_1, t_2 x_2) \rangle$. Then $\exists q \in \mathbb{Z}$ such that $(px_1, x_2) = q(t_1 x_1, t_2 x_2)$. Hence $qt_1 = p$ and $p \sqrt{qt_2}$. Thus $q = \pm 1$ and so $t_1 = \mp p$. Hence $H = H'$. But then $(x_1, 0) \notin H' \oplus K$, a contradiction. That is, H is not contained in a direct summand of G. Hence the desired result follow from Theorem 6.1.15.*

Example 6.1.17. *Let $G = G_1 \oplus G_2$, where G_i is a cyclic subgroup of G, $i = 1, 2$ and $|G_1| = p^r$, $|G_2| = p^s$ for $s - r \geq 2$. Then we show that $\exists$ a fuzzy subgroup μ of G such that μ is not a direct sum of fuzzy subgroups of G whose supports are cyclic. Let $G_i = \langle x_i \rangle$, $i = 1, 2$. Let $H = \langle (x_1, p^k x_2) \rangle$ for $0 < k < s - r$. Let K and J be any cyclic subgroups of G such that $|K| = p^r$, $|J| = p^s$, and $G = K \oplus J$. Now $J = \langle (c, d) \rangle$ for some $c \in G_1$ and $d \in G_2$. Now $|H| = p^{s-k}$ and $s - k > r$. Thus $H \not\subseteq K$. Suppose that $H \subseteq J$. Then $\exists q \in \mathbb{Z}$ such that $q(c, d) = (x_1, p^k x_2)$. Now $o(qc) = p^r$. Thus $p \nmid q$. Hence $o(qd) = o(d) = p^s$, a contradiction. Thus H is not contained in a direct summand of G. The result now follows from Theorem 6.1.15.*

6.2 Independent Generators

It is convenient to have available the concept of linear independence of a set of fuzzy singletons for the purpose of selecting a basis for a direct sum of cyclic fuzzy subgroups.

Let $n \in \mathbb{N}$ and x_a be a fuzzy singleton in G. Then nx_a denotes,

$$\underbrace{x_a + \ldots + x_a}_{n \text{ times}} = (nx)_a.$$

Definition 6.2.1. *Let $\mu \in \mathcal{F}(G)$. A system of fuzzy singletons*
$\{(x_1)_{a_1}, \ldots, (x_k)_{a_k}\}$, *where $0 < a_i \le \mu(x_i)$ for $i = 1, \ldots, k$ is said to be*
linearly independent *in μ if $n_1(x_1)_{a_1} + \ldots + n_k(x_k)_{a_k} = 0_a$ implies $n_1 x_1 =$*
$\ldots = n_k x_k = 0$ where $n_i \in \mathbb{Z}$ for $i = 1, \ldots, k$ and $a \in (0, 1]$. A system of fuzzy
singletons is called **linearly dependent** *if it is not independent. An arbitrary*
system $\mathcal{S}$ of fuzzy singletons is independent in μ if every finite subsystem of
$\mathcal{S}$ is independent.

We let $\mathcal{S}$ denote a system of fuzzy singletons such that $\forall\, x_a \in \mathcal{S},\, 0 < a \le$
$\mu(x)$. Let $\mathcal{S}^* = \{x|\ x_a \in \mathcal{S}\}$ and $\mathcal{S}_a = \mu_a \cap \mathcal{S}^* \,\forall\, a \in \{a \in (0, 1] \mid a \le \mu(0)\}$.

Proposition 6.2.2. *Let $\mu \in \mathcal{F}(G)$. The following conditions are equivalent*
on $\mathcal{S}$.

(1) *$\mathcal{S}$ is independent in μ;*
(2) *$\mathcal{S}^*$ is independent in μ^*;*
(3) *$\mathcal{S}_a$ is independent in μ_a for all a such that $0 < a \le \mu(0)$.*

Proof. (1) $\Rightarrow$ (2). Let $n_1 x_1 + \ldots + n_k x_k = 0$, where $x_1, \ldots, x_k \in \mathcal{S}^*$. Then
$n_1(x_1)_{a_1} + \ldots + n_k(x_k)_{a_k} = 0_a$, where $(x_i)_{a_i} \in \mathcal{S}$ for $i = 1, \ldots, k$ and $a = a_1 \wedge \ldots \wedge a_k$. Thus $n_1 x_1 = \ldots = n_k x_k = 0$.

(2) $\Rightarrow$ (3). The result is immediate since $\mathcal{S}_a \subseteq \mathcal{S}^* \,\forall\, a \in \{a \in (0, 1] \mid a \le \mu(0)\}$.

(3) $\Rightarrow$ (1). Suppose that $n_1(x_1)_{a_1} + \ldots + n_k(x_k)_{a_k} = 0_a$, where $(x_i)_{a_i} \in \mathcal{S}$ for $i = 1, \ldots, k$. Then $a = a_1 \wedge \ldots \wedge a_k$, $n_1 x_1 + \ldots + n_k x_k = 0$ and $x_i \in \mu_a$ for $i = 1, \ldots, k$. Thus $n_1 x_1 = \ldots = n_k x_k = 0$. $\qquad\square$

Definition 6.2.3. *Let $\mu \in \mathcal{F}(G)$. An independent system $\mathcal{M}$ of fuzzy sin-*
gletons in μ is said to be **maximal** *if $\nexists$ an independent system $\mathcal{S}$ of fuzzy*
singletons in μ such that $\mathcal{M} \subset \mathcal{S}$.

Proposition 6.2.4. *Let $\mu \in \mathcal{F}(G)$. Every independent system $\mathcal{S}$ in μ can be*
extended to a maximal independent system.

Proof. By Proposition 6.2.2, $\mathcal{S}^*$ is independent in μ^*. By Zorn's Lemma, $\mathcal{S}^*$
can be extended to a maximal independent system M in μ^*. Let $\mathcal{M} = \{x_a|\ x \in M, \mu(x) = a\}$. Since $x \in \mu^*$, $a > 0$. Thus $\mathcal{M}^* = M$ and so $\mathcal{M}$ is independent.
Clearly, $\mathcal{M}$ is maximal since $\mathcal{M}^*$ is maximal. $\qquad\square$

Theorem 6.2.5. *Let $\mu \in \mathcal{F}(G)$. $\mathcal{S}$ is independent in μ if and only if the fuzzy*
subgroup of G generated by $\mathcal{S}$ in μ is a direct sum of fuzzy subgroups of G
whose supports are cyclic, i.e., for $\mathcal{S} = \{(x_i)_{a_i} | 0 < a_i \le \mu(x_i), i \in I\}$, $\langle \mathcal{S} \rangle =$
$\oplus_{i \in I} \langle (x_i)_{a_i} \rangle$.

Proof. $\langle \mathcal{S} \rangle = \sum_{i \in I} \langle (x_i)_{a_i} \rangle$. Now $\mathcal{S}$ is independent in $\mu \Leftrightarrow \mathcal{S}^*$ is independent
in $\mu^* \Leftrightarrow \langle \mathcal{S}^* \rangle = \oplus_{i \in I} \langle x_i \rangle \Leftrightarrow \langle \mathcal{S} \rangle = \oplus_{i \in I} \langle (x_i)_{a_i} \rangle$ since $\langle (x_i)_{a_i} \rangle^* = \langle x_i \rangle \,\forall\, i \in I$. $\qquad\square$

6.3 Primary Fuzzy Subgroups

Definition 6.3.1. *Let $\mu \in \mathcal{F}(G)$. Then μ is called a p-**primary** fuzzy subgroup of G if $\exists$ a prime p such that $\forall$ fuzzy singletons $x_a \subseteq \mu$ with $a > 0$, $\exists$ $n \in \mathbb{N}$ such that $p^n(x_a) = 0_a$.*

Let $\mu \in \mathcal{F}(G)$. Then μ is p-primary if and only if μ^* is p-primary. Also μ is p-primary if and only if μ_a is p-primary $\forall a \in \{a \in (0,1] \mid a \leq \mu(0)\}$.

For H a subgroup of G and p prime, let H_p denote the p-primary component of H.

Proposition 6.3.2. *Let $\mu \in \mathcal{F}(G)$. Let p be a prime. Define the fuzzy subset $\mu^{(p)}$ of G by $\forall x \in G$, $\mu^{(p)}(x) = \mu(x)$ if $x \in (\mu^*)_p$ and $\mu^{(p)}(x) = 0$ otherwise. Then $\mu^{(p)}$ is a p-primary fuzzy subgroup of G and $(\mu^{(p)})^* = (\mu^*)_p$. Furthermore, $(\mu^{(p)})_a = (\mu_a)_p \ \forall \ a \in [0,1]$, $a \leq \mu(0)$.* □

Proposition 6.3.3. *Let $\mu \in \mathcal{F}(G)$. Let p be a prime. Then $\mu^{(p)}$ is the unique maximal p-primary fuzzy subgroup of G in μ.* □

For p a prime, we call $\mu^{(p)}$ the **p-primary component** of μ.

Lemma 6.3.4. *Let $\mu \in \mathcal{F}(G)$. Let $\mu_i \in \mathcal{F}(\mu)$, $i \in I$. Suppose that $\mu^* = \oplus_{i \in I} \mu_i^*$ and that μ^* is torsion. Suppose that $\forall \ x \in \mu_i^*$, $\mu(x) = \mu_i(x) \ \forall \ i \in I$. If the order of the elements of μ_i^* are relatively prime to those of μ_j^* ($i \neq j$, i, $j \in I$), then $\mu = \oplus_{i \in I} \mu_i$.*

Proof. Let $x \in \mu^*$. Then $x = \sum_{i \in I} x_i$, where $x_i \in \mu_i^*$ and all but a finite number of $x_i = 0$. By the assumption concerning the orders of the elements of μ_i^*, $\forall \ i \in I$, $\exists \ k_i \in \mathbb{N}$ such that $k_i x = k_i x_i$ and k_i, $o(x_i)$ are relatively prime. Hence $\langle x_i \rangle = \langle k_i x_i \rangle$ and so $\mu(x_i) = \mu(k_i x_i) = \mu(k_i x) \geq \mu(x) \ \forall \ i \in I$. Thus $(\sum_{i \in I} \mu_i)(x) = \vee\{\wedge\{\mu_i(x_i') \mid x = \sum_{i \in I} x_i', \ x_i' \in \mu_i^*\} \mid i \in I\} = \wedge\{\mu_i(x_i) \mid x = \sum_{i \in I} x_i\}$ (since $\mu^* = \oplus_{i \in I} \mu_i^*$) $= \wedge\{\mu(x_i) \mid x = \sum_{i \in I} x_i\} \geq \mu(x)$. Clearly, $(\sum_{i \in I} \mu_i)(x) \leq \mu(x)$. Thus $\mu = \sum_{i \in I} \mu_i$. Hence $\mu = \oplus_{i \in I} \mu_i$ by Theorem 1.5.9. □

Definition 6.3.5. *Let $\mu \in \mathcal{F}(G)$. μ is called a **torsion fuzzy subgroup** of G if $\forall$ fuzzy singletons $x_a \subseteq \mu$ with $a > 0$, $\exists n \in \mathbb{N}$ such that $n(x_a) = 0_a$.*

Proposition 6.3.6. *Let $\mu \in \mathcal{F}(G)$. The following assertions hold.*

(1) μ is torsion if and only if μ^* is a subgroup of the torsion subgroup of G.

(2) μ is torsion if and only if μ_a is torsion for all a such that $0 < a \leq \mu(0)$.

(3) There exists a unique maximal fuzzy subgroup τ of G such that $\tau \subseteq \mu$ and τ is torsion. □

Theorem 6.3.7. *Let μ be a torsion fuzzy subgroup of G. Then μ is a direct sum of primary fuzzy subgroups of G.*

Proof. Since μ^* is torsion, $\mu^* = \oplus_p (\mu^*)_p$, where $(\mu^*)_p$ is the p-primary component of μ^*. By Propositions 6.3.2 and 6.3.3, $\mu^{(p)}$ is the p-primary component of μ and $(\mu^{(p)})^* = (\mu^*)_p$. Thus $\mu^* = \oplus_p (\mu^{(p)})^*$. By Lemma 6.3.4, $\mu = \oplus_p \mu^{(p)}$. □

Theorem 6.3.8. *Let $\mu, \nu \in \mathcal{F}(G)$ such that $\mu \subseteq \nu$. Then there is a unique maximal fuzzy subgroup γ of G such that $\mu \subseteq \gamma \subseteq \nu$ and $\forall x_a \subseteq \gamma$ with $a > 0$, there exists $n \in \mathbb{N}$ and $b \in (0,1]$ such that $nx_b \subseteq \mu$.*

Proof. There exists a unique maximal subgroup T of ν^* such that $T \supseteq \mu^*$ and T/μ^* is torsion. Define the fuzzy subset γ of G by $\forall x \in G$, $\gamma(x) = \nu(x)$ if $x \in T$ and $\gamma(x) = 0$ if $x \notin T$. Then γ is a fuzzy subgroup of G and $\mu \subseteq \gamma \subseteq \nu$. Let $x_a \subseteq \gamma$ with $a > 0$. Then $x \in T$ and so there exists $n \in \mathbb{N}$ such that $nx \in \mu^*$. Thus $nx_b \subseteq \mu$ for $b = \mu(nx) > 0$. Let δ be any fuzzy subgroup of G such that $\mu \subseteq \delta \subseteq \nu$. If $x_a \subseteq \delta$ with $a > 0$ and $nx_b \subseteq \mu$ for some n and $b > 0$, then $\delta(x) \leq \nu(x) = \gamma(x)$. Hence γ is the desired unique maximal fuzzy subgroup of G. □

The fuzzy subgroup γ of G in Theorem 6.3.8 is called the **torsion closure** of μ in ν.

Theorem 6.3.9. *Let $\mu, \nu \in \mathcal{F}(G)$ be such that $\mu \subseteq \nu$. Let γ be an fuzzy subgroup of G such that $\mu \subseteq \gamma \subseteq \nu$. Then γ is the torsion closure of μ in ν if and only if γ^*/μ^* is the torsion subgroup of ν^*/μ^* and $\gamma = \nu$ on γ^*.*

Proof. Suppose that γ is the torsion closure on μ in ν. Let $x \in \gamma^*$. Then $x_a \subseteq \gamma$ for some $a > 0$. Hence $\exists n$ and $b \in (0,1]$ such that $nx_b \subseteq \mu$. Then $nx \in \mu^*$. Thus γ^*/μ^* is torsion. Let $x \in \nu^*$ be such that $\exists n$ such that $nx \in \mu^*$. Then $x_a \subseteq \nu$ with $a > 0$ and $nx_b \subseteq \mu$ for some $b \in (0,1]$. Thus $x_a \subseteq \gamma$ and so $x \in \gamma^*$. Hence γ^*/μ^* is the torsion subgroup of ν^*/μ^*. Clearly, $\gamma = \nu$ on γ^*. Conversely, suppose that γ^*/μ^* is the torsion subgroup of ν^*/μ^* and $\gamma = \nu$ on γ^*. Let $x_a \subseteq \gamma$ with $a > 0$. Then $x \in \gamma^*$. Thus $\exists n \in \mathbb{N}$ and $b \in (0,1]$ such that $nx_b \subseteq \mu$. Now suppose that $x_a \in \nu$ with $a > 0$ and $n_b x \subseteq \mu$ for some n and b. Then $x \in \gamma^*$. Hence $\gamma(x) = \nu(x) \geq a > 0$. Thus $x_a \subseteq \gamma$. Hence γ is the torsion closure of μ in ν. □

6.4 Divisible and Pure Fuzzy Subgroups

One of the most important types of subgroup of an Abelian group is a divisible group. The structure of a divisible group is completely known. A divisible subgroup of an Abelian group is a direct summand of that group. The notion of a pure subgroup is very useful in Abelian group theory. In this section, we study the notion of divisible and pure subgroups in a fuzzy setting.

Definition 6.4.1. *Let $\mu \in \mathcal{F}(G)$. Then μ is called **divisible** if $\forall x_a \subseteq \mu$ with $a > 0$, and $\forall n \in \mathbb{N}$, $\exists y_a$ such that $y_a \subseteq \mu$ such that $n(y_a) = x_a$.*

Proposition 6.4.2. *Let $\mu \in \mathcal{F}(G)$. Then μ is divisible if and only if μ_a is divisible $\forall a \in \{a | 0 < a \le \mu(0)\}$.*

Proof. Suppose μ is divisible. Let $a \in (0, \mu(0)]$ and $n \in \mathbb{N}$. Let $x \in \mu_a$. Then $x_a \subseteq \mu$. Thus $\exists y_a \in \mu$ such that $n(y_a) = x_a$. Hence $ny = x$. Since $\mu(x) \ge a$, $x \in \mu_a$. Thus μ_a is divisible. Conversely, suppose μ_a is divisible $\forall a \in (0, \mu(0)]$. Let $x_a \subseteq \mu$ with $a > 0$ and $n \in \mathbb{N}$. Since $x \in \mu_a, \exists\, y \in \mu_a$ such that $ny = x$. Hence $(ny)_a = x_a$ or $n(y_a) = x_a$. Clearly $y_a \subseteq \mu$ and so μ is divisible.   □

Proposition 6.4.3. *Let $\mu \in \mathcal{F}(G)$. If μ is divisible, then μ^* is divisible.*

Proof. Let $x \in \mu^*$ and $n \in \mathbb{N}$. Then $\mu(x) = a > 0$ for some $a \in (0, 1]$. Thus $x_a \subseteq \mu$. Since μ is divisible, $\exists y_a \subseteq \mu$ such that $n(y_a) = x_a$. Hence $ny = x$. Since $\mu(y) \ge a > 0$, $y \in \mu^*$. Thus μ^* is divisible.

    □

Proposition 6.4.4. *Let $\mu \in \mathcal{F}(G)$. If μ^* is divisible and μ is constant on $\mu^* \backslash \{0\}$, then μ is divisible.*

Proof. Let $x_a \subseteq \mu$ with $a > 0$ and $n \in \mathbb{N}$. Then $x \in \mu^*$ and so $\exists\, y \in \mu^*$ such that $x = ny$. If $y = 0$, then $x = 0$ and the result follows. Let $y \ne 0$. Since μ is constant on $\mu^* \backslash \{0\}$, $\mu(y) = \mu(x) \ge a$. Thus $y_a \subseteq \mu$ and $x_a = n(y_a)$. Hence μ is divisible.   □

Let $n \in \mathbb{N}$ and $x_a \subseteq 1_{\{0\}}$. Then $x = 0$. Hence $\exists\, y_a \subseteq 1_{\{0\}}$ such that $n(y_a) = x_a$. Thus $1_{\{0\}}$ is divisible.

Let T denote the torsion subgroup of G.

Proposition 6.4.5. *Let $\mu \in \mathcal{F}(G)$. $\forall x, y \in G$ and $n \in \mathbb{N}$, $ny = x$ implies that $\mu(x) = \mu(y)$ for all divisible fuzzy subgroups μ of G if and only if G is torsion-free.*

Proof. Suppose that the condition concerning μ holds. Let $x \in T$. Then $\exists n \in \mathbb{N}$ such that $nx = 0$. Since $1_{\{0\}}$ is divisible, $1_{\{0\}}(x) = 1_{\{0\}}(0) = 1$. Hence $x = 0$. Thus $T = \{0\}$. Conversely, suppose that G is torsion-free. Let μ be any divisible fuzzy subgroup of G. Suppose that $nx = y$. Let $a = \mu(x)$. Then $n(y_a) = x_a$. Now $\exists y' \in G$ such that $ny'_a = x_a$ and $y' \subseteq \mu$. Thus $ny' = x$. Since G is torsion-free $y = y'$ and so $y_a \subseteq \mu$. Thus $a = \mu(x) = \mu(ny) \ge \mu(y) \ge a$. Hence $\mu(x) = \mu(y)$.   □

Theorem 6.4.6. *Let $\mu \in \mathcal{F}(G)$. Let $G = \mathbb{Q}$, where $\mathbb{Q}$ denotes the additive group of rational numbers. Then μ is divisible if and only if μ is constant on G.*

Proof. Suppose that μ is divisible. Let $x \in G$. Then $\exists n \in \mathbb{N}$ such that $nx = m \in \mathbb{Z}$. By Proposition 6.4.5, $\mu(x) = \mu(m)$. Now $m \cdot 1 = m$ and so $\mu(m) = \mu(1)$ by Proposition 6.4.5. Thus $\mu(x) = \mu(1)$. Conversely, suppose that μ is constant on G. Let $x_a \subseteq \mu$ and $n \in \mathbb{N}$. Now $\exists y \in G$ such that $ny = x$ and so $n(y_a) = x_a$. Since μ is constant on G, $\mu(y) = \mu(x)$. Hence $y_a \subseteq \mu$. Thus μ is divisible.   □

Proposition 6.4.7. *Let $G = \mathbb{Z}(p^\infty)$. Then for all divisible fuzzy subgroups μ of G, $\forall x, y \in G\backslash\{0\}$, $\forall n \in \mathbb{N}$, $ny = x$ implies $\mu(x) = \mu(y)$.*

Proof. Let $x, y \in G\backslash\{0\}$, $n \in \mathbb{N}$ and $ny = x$. If $\mu(x) = 0$, then $\mu(x) = \mu(y)$. Suppose that $\mu(x) = a > 0$. Then $x_a \subseteq \mu$. Now $\exists y_a \subseteq \mu$ such that $n(y_a) = x_a$. Thus $ny = x$. Hence $a = \mu(x) = \mu(ny) \geq \mu(y) \geq a$. Thus $\mu(y) = \mu(x)$. Hence it suffices to show that if $nw = x$ and $nz = x$, then $\mu(w) = \mu(z)$ since we would have $\mu(y') = \mu(y) = \mu(x)$ whenever $ny' = x$. Let $w = u/p^r$ and $z = v/p^s$, where u, p are relatively prime and v, p are relatively prime. Assume that $r \leq s$. Suppose that $n(u/p^r) = n(v/p^s)$ for some $n \in \mathbb{N}$. If $p^r | n$ in $\mathbb{Z}$, then $n(u/p^r) = n(v/p^s) = 0$ in $\mathbb{Z}(p^\infty)$ and so $x = 0$, a contradiction. Thus $n = p^t n'$, where $0 \leq t < r$ and n' and p are relatively prime. Hence $n'(u/p^{r-t}) = n'(v/p^{s-t})$. Thus $0 = n'u = n'(v/p^{s-r})$ in $\mathbb{Z}(p^\infty)$. If $r < s$, then $p | n'v$ in $\mathbb{Z}$, which is impossible. Hence $r = s$. Now $\exists a, b \in \mathbb{Z}$ such that $1 = au + bp^r$ and so $1/p^r = a(u/p^r) + b = au/p^r$ in $\mathbb{Z}(p^\infty)$. Thus $v/p^s = v/p^r = va(u/p^r)$. Hence $\mu(v/p^s) = \mu(va(u/p^r)) \geq \mu(u/p^r)$. By symmetry, $\mu(u/p^r) \geq \mu(v/p^s)$. $\square$

We note that if $G = \mathbb{Z}(p^\infty)$, then $\mu(x) = \mu(y)$ for all fuzzy subgroups μ of G and $\forall x, y \in G\backslash\{0\}$.

Theorem 6.4.8. *Let $\mu \in \mathcal{F}(G)$, where $G = \mathbb{Z}(p^\infty)$. Then μ is divisible if and only if μ is constant on $G\backslash\{0\}$.*

Proof. Suppose that μ is divisible. The subgroups of G are chained, say $\{0\} \subset H_1 \subset \ldots \subset H_i \subset \ldots$, where H_i is a cyclic group of order p^i. Let $x \in G\backslash\{0\}$. Then x has order p^i for some $i \in \mathbb{N}$. Let $y = p^{i-1}x$. By Proposition 6.4.7, $\mu(x) = \mu(y)$. Now y has order p and $H_1 = \langle y \rangle$. By comments preceding the theorem, μ is constant, say a, on $H_1\backslash\{0\}$. Hence $\mu(x) = a$. Conversely, suppose that μ is constant on $G\backslash\{0\}$. Let $x_a \subseteq \mu$ with $a > 0$ and let $n \in \mathbb{N}$. There exists $y \in G$ such that $ny = x$. Hence $ny_a = x_a$. However, $\mu(y) = \mu(x) \geq a$. Thus $y_a \subseteq \mu$. Hence μ is divisible. $\square$

Let

$$\mathcal{F}'(\mu) = \mathcal{F}(\mu)\backslash\{1_{\{0\}}\}.$$

Let $n \in \mathbb{N}$ and let μ be a fuzzy subgroup of G. Define the fuzzy subset $n\mu$ of G by $\forall x \in G, n\mu(x) = \vee\{\mu(y) | x = ny, y \in G\}$. Then $n\mu$ is a fuzzy subgroup of G. In [20], μ is called divisible if $\mu = n\mu \forall n \in \mathbb{N}$. Suppose $0 < a \leq \mu(0)$. Then $x \in (n\mu)_a \Leftrightarrow (n\mu)(x) \geq a \Leftrightarrow \vee\{\mu(y) | x = ny, y \in G\} \geq a \Leftrightarrow \exists y \in G$ such that $x = ny$ and $\mu(y) \geq a$ (assuming μ has the sup property) $\Leftrightarrow x \in n\mu_a$. Thus if μ has the sup property, $(n\mu)_a = n\mu_a$. Hence if μ has the sup property the two notions of divisibility are equivalent as can be seen as follows. If μ has the sup property, μ is divisible with respect to Definition 6.4.1$\Leftrightarrow \mu_a$ is divisible (Proposition 6.4.2) $\forall a \in [0, \mu(0)] \Leftrightarrow n\mu_a = \mu_a \forall a \in [0, \mu(0)] \Leftrightarrow (n\mu)_a = \mu_a \forall a \in [0, \mu(0)] \Leftrightarrow \mu = n\mu$.

Definition 6.4.9. *Let $\mu \in \mathcal{F}(G)$. Then μ is called **reduced** if $\nexists \nu \in \mathcal{F}'(\mu)$ such that ν is divisible.*

Proposition 6.4.10. *Let $\mu \in \mathcal{F}(G)$. Then μ is reduced if and only if there exists a such that $0 < a \leq \mu(0)$ and μ_a is reduced.*

Proof. Suppose that μ is not reduced. Then $\exists \nu \in \mathcal{F}'(\mu)$ such that ν is divisible. If $\nu(0) \neq \mu(0)$, then define the fuzzy subset ν' of G by $\nu'(0) = \mu(0)$ and $\nu'(x) = \nu(x) \ \forall x \in G \backslash \{0\}$. Then $\nu' \in \mathcal{F}'(\mu)$ and ν' is divisible. Hence we may assume that $\nu(0) = \mu(0)$. By Proposition 6.4.2, ν_a is divisible for all a such that $0 < a \leq \mu(0)$. Since $\nu_a \subseteq \mu_a$, μ_a is not reduced for all a such that $0 < a < \mu(0)$. Conversely, suppose that for all a such that $0 < a \leq \mu(0)$, μ_a is not reduced. Then $\exists$ a divisible subgroup $D^{(a)}$ of μ_a for all a such that $0 < a \leq \mu(0)$. Let $D = D^{(a)}$ for $a = \mu(0)$. Then $D \subseteq \mu_a$ for all a such that $0 < a \leq \mu(0)$. Since $D \subseteq \mu_*$, $\mu(x) = \mu(0) \ \forall x \in D$. Define the fuzzy subset ν of G by $\forall x \in D$, $\nu(x) = \mu(0)$ and $\nu(x) = 0$ otherwise. Then ν is an fuzzy subgroup of G such that $\nu \subseteq \mu$. Now $\forall x_a \subseteq \nu$ with $a > 0$ and $\forall n \in \mathbb{N}$, $\exists y \in D$ such that $ny = x$. Since $\nu(y) = \mu(0) \geq a$, $y_a \subseteq \nu$. Thus ν is divisible. Hence μ is not reduced. $\qquad \square$

Corollary 6.4.11. *Let $\mu \in \mathcal{F}(G)$. μ is reduced if and only if μ_* is reduced.*

Proof. Suppose μ is reduced. Then μ_a is reduced for some a such that $0 < a \leq \mu(0)$. Now $\mu_* \subseteq \mu_a$ and so μ_* is reduced. Conversely, suppose that μ_* is reduced. Let $a = \mu(0)$. $\qquad \square$

Definition 6.4.12. *Let $\mu \in \mathcal{F}(G)$. Let $\nu \in \mathcal{F}(\mu)$. Then ν is said to be **pure** in μ if $\forall x_a \subseteq \nu$ with $a > 0$, $\forall n \in \mathbb{N}$, $\forall y_a \subseteq \mu$, $n(y_a) = x_a$ implies $\exists z_a \subseteq \nu$ such that $n(z_a) = x_a$.*

Proposition 6.4.13. *Let $\mu \in \mathcal{F}(G)$. Let $\nu \in \mathcal{F}(\mu)$. Then ν is pure in μ if and only if ν_a is pure in μ_a for all a such that $0 < a \leq \nu(0)$.*

Definition 6.4.14. *Let χ be an fuzzy subset of G and let $n \in \mathbb{N}$. Define the fuzzy subset $n\chi$ of G by $\forall x \in G, (n\chi)(x) = 0$ if $x \notin nG$ and $(n\chi)(x) = \vee\{\chi(y)|y \in G, x = ny\}$.*

Proposition 6.4.15. *Let $\mu \in \mathcal{F}(G)$. Let $n \in \mathbb{N}$. Then*
   (1) $n\mu(0) = \mu(0)$;
   (2) $n\mu \subseteq \mu$;
   (3) $n\mu$ *is a fuzzy subgroup of G;*
   (4) *If μ has the sup property, then $n\mu(G) \subseteq \mu(G)$.*

Proof. (1) $(n\mu)(0) = \vee\{\mu(y)|0 = ny\} = \mu(0)$.
   (2) If $x \notin nG$, then $(n\mu)(x) = 0 \leq \mu(x)$. Suppose that $x \in nG$ and $x = ny$ for some $y \in G$. Then $\mu(x) = \mu(ny) \geq \mu(y)$. Hence $(n\mu)(x) = \vee\{\mu(y)|y \in G, x = ny\} \leq \mu(x)$.
   (3) Let $x, y \in G$. If either $x \notin nG$ or $y \notin nG$, then $(n\mu)(x) \wedge (n\mu)(y) = 0 \leq (n\mu)(x - y)$. Suppose that $x \in nG$ and $y \in nG$. Then $x - y \in nG$ and $(n\mu)(x-y) = \vee\{\mu(w)|w \in G, x-y = nw\} \geq \vee\{\mu(u-v)|u, v \in G, x = nu, y = \,$

$nv\} \geq \vee\{\mu(u) \wedge \mu(v)|x = nu, y = nv\} = (\vee\{\mu(u)|x = nu\}) \wedge (\vee\{\mu(v)|y = nv\})$
$= (n\mu)(x) \wedge (n\mu)(y)$. Thus $n\mu$ is an fuzzy subgroup of G.

(4) Let $a \in (n\mu)(G)$. Then $(n\mu)(x) = a$ for some $x \in G$. Now $(n\mu)(x) = \vee\{\mu(y)|x = ny, y \in G\} = a$. Since μ has the sup property, $\exists y \in G$ such that $\mu(y) = a$. Hence $a \in \mu(G)$. Thus $(n\mu)(G) \subseteq \mu(G)$. $\qquad\square$

Suppose that μ^* is torsion-free. If $x \in n\mu^*, n \in \mathbb{N}$, then $(n\mu)(x) = \mu(w)$ for some unique $w \in \mu^*$ such that $x = nw$.

Proposition 6.4.16. *Let $\mu \in \mathcal{F}(G)$ and let $n \in \mathbb{N}$. Then the following assertions hold.*

(1) $(n\mu)^* = n\mu^*$;

(2) $n\mu_a \subseteq (n\mu)_a$ $\forall a$ such that $0 < a \leq \mu(0)$. If either μ has the sup property or μ^* is torsion free, then $n\mu_a = (n\mu)_a$ $\forall a$ such that $0 < a \leq \mu(0)$.

(3) Let $\nu \in \mathcal{F}(\mu)$ and $\nu(0) = \mu(0)$. If μ has the sup property and ν is pure in μ, then $n\nu_a = (n\nu)_a$ for all a such that $0 < a \leq \mu(0)$.

Proof. (1) $x \in (n\mu)^* \Leftrightarrow (n\mu)(x) > 0 \Leftrightarrow x = nw$ for some $w \in \mu^* \Leftrightarrow x \in n\mu^*$.

(2) Let $x \in n\mu_a$. Then $x = nw$ for some $w \in \mu_a$. Hence $(n\mu)(x) \geq a$. Thus $x \in (n\mu)_a$. Conversely, let $x \in (n\mu)_a$. Now $(n\mu)(x) = \vee\{\mu(y)|x = ny, y \in G\}$. Thus $\exists y \in G$ such that $x = ny$ and $\mu(y) \geq a$. Hence $x \in n\mu_a$.

(3) Let $a \in \{a|0 < a \leq \mu(0)\}$. Since ν is pure in μ, ν_a is pure in μ_a. Thus $n\nu_a = \nu_a \cap n\mu_a$. Let $x \in (n\nu)_a$. Then $(\nu \cap n\mu)(x) \geq (n\nu)(x) \geq a$. Thus $x \in (\nu \cap n\mu)_a = \nu_a \cap (n\mu)_a = \nu_a \cap n\mu_a = n\nu_a$. Hence $(n\nu)_a \subseteq n\nu_a$. $\qquad\square$

Proposition 6.4.17. *Let $\mu \in \mathcal{F}(G)$ and let $\nu \in \mathcal{F}(\mu)$ be such that $\nu(0) = \mu(0)$. Then the following assertions.*

(1) Suppose that μ has the sup property. If ν is pure in μ, then $\forall n \in \mathbb{N}$, $n\nu = \nu \cap n\mu$.

(2) Suppose that μ has the sup property. If $\forall n \in \mathbb{N}, n\nu = \nu \cap n\mu$, then ν is pure in μ.

(3) Suppose that μ^* is torsion-free. Then ν is pure in μ if and only if $\forall n \in \mathbb{N}, n\nu = \nu \cap n\mu$.

Proof. ν is pure in $\mu \Leftrightarrow \forall a$ such that $0 < a \leq \mu(0)$, ν_a is pure in $\mu_a \Leftrightarrow \forall a$ such that $0 < a \leq \mu(0), \forall n \in \mathbb{N}, n\nu_a = \nu_a \cap n\mu_a \Leftrightarrow \forall a$ such that $0 < a \leq \mu(0)$, $\forall n \in \mathbb{N}, (n\nu)_a = \nu_a \cap (n\mu)_a \Leftrightarrow \forall a$ such that $0 < a \leq \mu(0), \forall n \in \mathbb{N}, n\nu_a = (\nu \cap n\mu)_a \Leftrightarrow \forall n \in \mathbb{N}, n\nu = \nu \cap n\mu$. $\qquad\square$

Let ν be a fuzzy subgroup of G. In [20], ν is said to to be pure in G if $n\nu = \nu \cap n(1_G)\forall n \in \mathbb{N}$. It follows easily that $n(1_G) = 1_{nG}$. Hence it follows from Propositions 6.4.13, 6.4.16 and 6.4.17 that the notions of purity in G are equivalent if μ has the sup property. The notion of p-pure fuzzy subgroups is also studied in [20].

Proposition 6.4.18. *Let $\mu \in \mathcal{F}(G)$ and let $\nu, \gamma \in \mathcal{F}(\mu)$ be such that $\gamma \subseteq \nu$ and $\gamma(0) = \nu(0) = \mu(0)$. Then the following assertions hold.*

(1) *If γ is pure in ν and ν is pure in μ, then γ is pure in μ.*

(2) *If ν is divisible, then ν is pure in μ.*

(3) *Suppose that μ is divisible. Then ν is pure in μ if and only if ν is divisible.*

Proof. (1) Suppose that $ny_a = x_a$, where $y_a \subseteq \mu, x_a \subseteq \gamma$, and $n \in \mathbb{N}$. Since $x_a \subseteq \nu, \exists z_a \subseteq \nu$ such that $nz_a = x_a$. Hence $\exists\ w_a \subseteq \gamma$ such that $nw_a = x_a$.

(2) ν_a is divisible $\forall a$ such that $0 < a \leq \nu(0)$. Thus ν_a is pure in μ_a $\forall a$ such that $0 < a \leq \mu(0)$. Hence ν is pure in μ.

(3) μ_a is divisible $\forall a$ such that $0 < a \leq \mu(0)$. Thus ν_a is pure in μ_a $\forall a$ such that $0 < a \leq \mu(0) \Leftrightarrow \nu_a$ is divisible $\forall a$ such that $0 < a \leq \mu(0)$. That is, ν is pure in $\mu \Leftrightarrow \nu$ is divisible. $\square$

Proposition 6.4.19. *Let $\mu \in \mathcal{F}(G)$. Let $\nu \in \mathcal{F}(\mu)$ be such that $\nu(0) = \mu(0)$. If $\forall x_a \subseteq \mu$ such that $x_a \not\subseteq \nu$, $\exists\ n \in \mathbb{N}$ such that $n(x_a) \subseteq \nu$, then ν is pure in μ.*

Proof. If $\exists a$ such that $0 < a \leq \mu(0)$, $\exists\ x \in \mu_a$ such that $x \notin \nu_a$ and $nx = z \in \nu_a$ for some $z \in \nu_a$ and for some $n \in \mathbb{N}$, then we have a contradiction of the hypothesis. Thus μ_a/ν_a is torsion-free and so ν_a is pure in μ_a $\forall\ a$ such that $0 < a \leq \mu(0)$, ([4], p. 114). Thus ν is pure in μ. $\square$

Proposition 6.4.20. *Let $\mu \in \mathcal{F}(G)$. Suppose that μ^* is torsion-free. Let $\{\nu_j \mid j \in J\}$ be a collection of fuzzy subgroups of G such that $\nu_j \subseteq \mu$, ν_j is pure in μ, and $\nu_j(0) = \mu(0)\ \forall j \in J$. Then $\cap_{j \in J}\nu_j$ is pure in μ.*

Proof. Let $x \in \mu^*$. Then $(\cap_{j \in J}n\nu_j)(x) = \wedge\{n\nu_j(x)|j \in J\} = 0$ if $x \notin n\mu^*$ and $(\cap_{j \in J}n\nu_j)(x) = \wedge\{n\nu_j(x)|j \in J\} = \wedge\{\nu_j(y)|j \in J\}$ if $x = ny \in n\mu^*$. Now $(n \cap_{j \in J} \nu_j)(x) = 0$ if $x \notin n\mu^*$ and $(n \cap_{j \in J} \nu_j)(x) = (\cap_{j \in J}\nu_j)(y) = \wedge\{\nu_j(y)|j \in J\}$ if $x = ny \in n\mu^*$. Thus $n \cap_{j \in J} \nu_j = \cap_{j \in J}n\nu_j$. Now $(\cap_{j \in J}(\nu_j \cap n\mu))(x) = \wedge\{(\nu_j \cap n\mu)(x) \mid j \in J\} = \wedge\{\nu_j(x) \wedge (n\mu)(x) \mid j \in J\} = \wedge\{\nu_j(x) \mid j \in J\} \wedge (n\mu)(x) = (\cap_{j \in J}\nu_j)(x) \wedge (n\mu)(x) = ((\cap_{j \in J}\nu_j) \cap (n\mu))(x)$. Hence $\cap_{j \in J}(\nu_j \cap n\mu) = (\cap_{j \in J}\nu_j) \cap (n\mu)$. $\forall j \in J, \forall n \in \mathbb{N}, n\nu_j = \nu_j \cap n\mu$. Thus $\forall n \in \mathbb{N}$, $\cap_{j \in J}n\nu_j = \cap_{j \in J}(\nu_j \cap n\mu)$ and so $n \cap_{j \in J} \nu_j = (\cap_{j \in J}\nu_j) \cap n\mu$. $\square$

Proposition 6.4.21. *Let $\mu \in \mathcal{F}(G)$. Suppose that μ^* is torsion-free. Let $\nu \in \mathcal{F}(\mu)$ be such that $\nu(0) = \mu(0)$. Then ν is contained in a unique smallest pure fuzzy subgroup in μ.*

Proof. Let $\mathcal{B} = \{\gamma|\ \nu \subseteq \gamma \subseteq \mu, \gamma$ is a pure fuzzy subgroup of G in $\mu\}$. Then $\mu \in \mathcal{B}$ and so $\mathcal{B} \neq \emptyset$. The desired result now follows since $\cap_{\gamma \in \mathcal{B}}\gamma$ is an fuzzy pure subgroup of G in μ by Proposition 6.4.20. $\square$

Lemma 6.4.22. *Let $\{\nu_j|j \in J\}$ and $\{\gamma_j|j \in J\}$ be chains of fuzzy subgroups of G. Let $n \in \mathbb{N}$. Then the following assertions hold.*

(1) $n(\cup_{j \in J}\nu_j) = \cup_{j \in J}n\nu_j$.

(2) $\cup_{j \in J}(\nu_j \cap \gamma_j) = (\cup_{j \in J}\nu_j) \cap (\cup_{j \in J}\gamma_j)$.

Proof. (1) Let $x \in G$. Then $n(\cup_{j\in J}\nu_j)(x) = 0$ if $x \notin nG$ and $n(\cup_{j\in J}\nu_j)(x)$ $= \vee\{\cup_{j\in J}\nu_j(y) \mid x = ny\} = \vee\{\vee\{\nu_j(y) \mid j \in J\} \mid x = ny\}$ if $x \in nG$. Now $(\cup_{j\in J}n\nu_j)(x) = \vee\{(n\nu_j)(x) \mid j \in J\} = 0$ if $x \notin nG$ and $(\cup_{j\in J}n\nu_j)(x) =$ $\vee\{(n\nu_j)(x) \mid j \in J\} = \vee\{\vee\{\nu_j(y) \mid x = ny\} \mid j \in J\}$ if $x \in nG$. Thus we have the desired result.

(2) Let $x \in G$. Then $(\cup_{j\in J}(\nu_j \cap \gamma_j))(x) = \vee\{(\nu_j \cap \gamma_j)(x) \mid j \in J\} =$ $\vee\{\nu_j(x) \wedge \gamma_j(x) \mid j \in J\} = (\vee\{\nu_j(x) \mid j \in J\}) \wedge (\vee\{\gamma_j(x) \mid j \in J\}) =$ $(\cup_{j\in J}\nu_j)(x) \wedge (\cup_{j\in J}\gamma_j)(x) = ((\cup_{j\in J}\nu_j) \cap (\cup_{j\in J}\gamma_j))(x)$. $\square$

Proposition 6.4.23. *Let $\mu \in \mathcal{F}(G)$. Let $\{\mu_j | j \in J\}$ be a chain of fuzzy subgroups of G such that $\mu_j \subseteq \mu, \mu_j$ is pure in μ, and $\mu_j(0) = \mu(0) \ \forall j \in J$. Suppose that either μ and $\cup_{j\in J}\mu_j$ have the sup property or μ^* is torsion-free. Then $\cup_{j\in J}\mu_j$ is a pure fuzzy subgroup of G in μ.*

Proof. Let $n \in \mathbb{N}$. Then $n\cup_{j\in J}\mu_j = \cup_{j\in J}n\mu_j = \cup_{j\in J}(\mu_j \cap n\mu) = (\cup_{j\in J}\mu_j) \cap n\mu$ by Lemma 6.4.22 and Proposition 6.4.17. $\square$

Proposition 6.4.24. *Let $\mu \in \mathcal{F}(G)$. Let $\nu, \gamma, \delta \in \mathcal{F}(\mu)$ be such that $\delta \subseteq \nu$ and $\delta(0) = \mu(0)$. Suppose that $\mu = \nu \oplus \gamma$. Then the following assertions hold.*
(1) $\forall n \in \mathbb{N}, n\mu = n\nu \oplus n\gamma$.
(2) $\delta + \gamma = \delta \oplus \gamma$ *and* $\nu \cap (\delta \oplus \gamma) = \delta$.

Proof. (1) Let $x \in \mu^*$. Then $(n\mu)(x) = \vee\{\mu(y) | y \in G, x = ny\} = \vee\{(\nu \oplus \gamma)(x)$ $\mid x = ny\} = \vee\{\nu(y_1) \wedge \gamma(y_2) \mid x = ny, y = y_1 + y_2$ for $y_1 \in \nu^*, y_2 \in \gamma^*\}$ since $\mu^* = \nu^* \oplus \gamma^*$ by Theorem 1.5.9. Now $(n\nu \oplus n\gamma)(x) = (n\nu(x_1) \wedge (n\gamma)(x_2)$ such that $x_1 \in \nu^*, x_2 \in \gamma^*$. Thus $(n\nu \oplus n\gamma)(x) = (\vee\{\nu(y_1) | x_1 = ny_1\}) \wedge$ $(\vee\{\gamma(y_2) | x_2 = ny_2\})$. Hence $n\mu \subseteq n\nu \oplus n\gamma$. Clearly, $n\nu \oplus n\gamma \subseteq n\mu$.

(2) For all $x \in G$, $0 \le (\delta \cap \gamma)(x) \le (\nu \cap \gamma)(x) = 0$. Thus $\delta + \gamma = \delta \oplus \gamma$. Now $\forall x \in G, (\nu \cap (\delta \oplus \gamma))(x) = \nu(x) \wedge (\delta \oplus \gamma)(x) = \nu(x) \wedge \delta(x_1) \wedge \gamma(x_2)$ if $x = x_1 + x_2$ for $x_1 \in \delta^*, x_2 \in \gamma^*$ and $(\nu \cap (\delta \oplus \gamma))(x) = 0$ otherwise. Consider the former case. Suppose that $x \notin \nu^*$. Then $(\nu \cap (\delta \oplus \gamma))(x) = 0 = \delta(x)$. Suppose that $x \in \nu^*$. Then $x_2 = 0$ and so $(\nu \cap (\delta \oplus \gamma))(x) = \nu(x) \wedge \delta(x) \wedge \gamma(0)$ $= \nu(x) \wedge \delta(x) = \delta(x)$. $\square$

Proposition 6.4.25. *Let $\mu \in \mathcal{F}(G)$. Let ν and $\gamma \in \mathcal{F}(\mu)$. If $\mu = \nu \oplus \gamma$, then ν and γ are pure in μ.*

Proof. Suppose that $ny_a = x_a$, where $n \in \mathbb{N}, y_a \subseteq \mu$, and $x_a \subseteq \nu$. Since $\mu = \nu \oplus \gamma$, $\mu_a = \nu_a \oplus \gamma_a$. Thus $\nu_a \cap n\mu_a = n\nu_a$. Now $ny_a = (ny)_a$. Hence $ny = x, y \in \mu_a$, and $x \in \nu_a$. Thus $\exists z \in \nu_a$ such that $nz = x$. Hence $nz_a = x_a$. Thus ν is pure in μ. $\square$

Example 6.4.26. *Let $G = \mathbb{Z}(p^\infty) \oplus \mathbb{Z}(p^\infty)$. Define the fuzzy subsets μ, ν, and γ of G by $\forall x \in G, \mu((0,0)) = 1, \mu(x) = \frac{1}{2} + \frac{1}{n}$ if $x \in (\langle \frac{1}{p^{n-1}} \rangle \oplus \{0\}) \setminus (\langle \frac{1}{p^{n-2}} \rangle \oplus \{0\})$ for $n = 2, 3, \ldots, \mu(x) = \frac{1}{2}$ if $x \notin \mathbb{Z}(p^\infty) \oplus \{0\}; \nu((0,0)) = 1, \nu(x) = \frac{1}{2}$ if $x \in \{0\} \oplus \mathbb{Z}(p^\infty) \setminus \{(0,0)\}, \nu(x) = 0$ otherwise; $\gamma((0,0)) = 1, \gamma(x) = \frac{1}{2}$ if*

$x \in \left\langle (\frac{1}{p}, \frac{1}{p}), \ldots, (\frac{1}{p^i}, \frac{1}{p^i}), \ldots \right\rangle \backslash \{(0,0)\}$, $\gamma(x) = 0$ *otherwise. Then* $\mu, \nu,$ *and* γ
are fuzzy subgroups of G *such that* $\nu, \gamma \subseteq \mu$ *and* $\mu^* = G$, $\nu^* = \{0\} \oplus \mathbb{Z}(p^\infty)$,
$\gamma^* = \left\langle (\frac{1}{p}, \frac{1}{p}), \ldots, (\frac{1}{p^i}, \frac{1}{p^i}), \ldots \right\rangle$. *Since* $(\nu+\gamma)^* = \nu^*+\gamma^* = \nu^* \oplus \gamma^*$, $\nu+\gamma = \nu \oplus \gamma$
by Theorem 1.5.9. Since ν^* *and* γ^* *are divisible and* ν, γ *are constants on*
$\nu^* \backslash \{(0,0)\}$, $\gamma^* \backslash \{(0,0)\}$, *respectively,* ν *and* γ *are divisible by Proposition 6.4.4.*
Now $(\nu \oplus \gamma)^* = G$ *which is divisible. Let* $(u, v) \in G \backslash \{(0,0)\}$. *Then* $(\nu \oplus \gamma)(u, v)$
$= \nu((0, v-u)) \wedge \gamma(u, u)) = \frac{1}{2}$. *That is,* $(\nu \oplus \gamma)$ *is constant on* $(\nu \oplus \gamma)^* \backslash \{(0,0)\}$.
Hence $\nu \oplus \gamma$ *is divisible.* $\nexists$ *a nontrivial fuzzy subgroup* δ *of* G *such that* $\mu =$
$(\nu \oplus \gamma) \oplus \delta$ *else* $(\nu \oplus \gamma)^* \oplus \delta^* = \mu^* = (\nu \oplus \gamma)^*$ *and so* $\delta^* = \{(0,0)\}$. *Clearly,*
$\mu \neq \nu + \gamma$. *That is, we have shown that a divisible fuzzy subgroup need not be*
a direct summand (of μ*) even though the corresponding result holds for crisp*
Abelian groups.

6.5 Invariants of Fuzzy Subgroups

The description of a class of groups by means of invariants is one of the major
goals of group theory. If G is a finitely generated commutative group, then
G is a direct sum of cyclic groups of order infinity and powers of primes
and thus has a complete system of invariants, [[4], p. 81]. In this section,
we determine a complete system of invariants for those fuzzy subgroups μ of
G which are direct sums of fuzzy subgroups whose supports are cyclic. Our
invariants satisfy the criteria that they are easily described quantities and are
uniquely determined by the fuzzy subgroup.

Recall that for $x \in G$, $o(x)$ denotes the order of x. Let μ denote an fuzzy
subgroup of G. Let

$$S(\mu) = \{x_a \mid x \in \mu^*, \mu(x) = a\}$$

and

$$\mathcal{FS}(\mu) = \{\chi \mid \chi \text{ is an fuzzy subset of } G \text{ and } \chi \subseteq \mu\}.$$

The following definition corresponds to Definition 6.1.2.

Definition 6.5.1. *Let* $\mu \in \mathcal{F}(G)$ *and let* $\chi \in \mathcal{FS}(\mu)$. *Then* χ *(or* $S(\chi)$*) is*
*said to be a **minimal generating set** for* μ *if* $\mu = \langle \chi \rangle \cup 0_{\mu(0)}$ *and* $\forall$ x_a
$\in S(\chi)$, $\langle \chi - x_a \rangle \cup 0_{\mu(0)} \subset \langle \chi \rangle$, *where* $(\chi - x_a)(y) = (\chi)(y)$ *if* $y \neq x$ *and*
$(\chi - x_a)(y) = 0$ *otherwise.*

Proposition 6.5.2. *Let* $\mu \in \mathcal{F}(G)$. *Suppose that* $\mu^* = \langle x \rangle = \langle y \rangle$. *For all* u
$\in \mu(G)$, *let* n_u, m_u *be the smallest positive integers such that* $\mu_u = \langle n_u x \rangle =$
$\langle m_u y \rangle$. *Then* $n_u = m_u$ $\forall$ $u \in \mu(G)$.

Proof. There exist $q_1, q_2 \in \mathbb{Z}$ such that $x = q_1 y$ and $y = q_2 x$. Suppose that
μ^* has infinite order. Now $x = q_1 q_2 x$ and so $q_1 = q_2 = \pm 1$. Thus $n_u = m_u$ $\forall$
$u \in \mu(G)$. Suppose that μ^* has finite order q. Then q, q_1 are relatively prime
else $o(x) < o(y)$. Thus $\langle n_u y \rangle = \langle n_u q_1 y \rangle = \langle n_u x \rangle$. Similarly, $\langle m_u x \rangle = \langle m_u y \rangle$.
Hence $n_u = m_u$ $\forall$ $u \in \mu(G)$. $\qquad \square$

Let $\mu \in \mathcal{F}(G)$ and $x \in \mu^*$. Let μ^x be the fuzzy subset of G such that $\mu^x(z) = \mu(z) \ \forall \ z \in \langle x \rangle$ and $\mu^x(z) = 0 \ \forall \ z \in G \backslash \langle x \rangle$. Then μ^x is an fuzzy subgroup of G. Let $I_x = \{n_u | \ u \in \mu^x(G) \backslash \{0\}$, n_u the smallest positive integer such that $(\mu^x)_u = \langle n_u x \rangle\}$. Suppose that $\mu^* = \oplus_{j \in \mathcal{A}} \langle x_j \rangle$. Let I_μ denote the ordered pair, $(\cup_{j \in \mathcal{A}} \{(I_{x_j}, o(x_j))\}, \mu(G) \backslash \{0\})$. In the remainder of this section, G' denotes an Abelian group.

Lemma 6.5.3. *Let $\mu \in \mathcal{F}(G)$ and $\nu \in \mathcal{F}(G')$. Suppose that $\mu^* = \langle x \rangle$ and $\nu^* = \langle y \rangle$. Then $I_\mu = I_\nu$ if and only if $\exists$ isomorphism f of μ^* onto ν^* such that $f(\mu) = \nu$ on ν^*. If $I_\mu = I_\nu$, then every isomorphism f of μ^* onto ν^* is such that $f(\mu) = \nu$ on ν^*.*

Proof. Suppose that $I_\mu = I_\nu$. Then clearly $\mu^* \simeq \nu^*$ for some isomorphism f and we may take f so that $f(x) = y$. Then $\forall \ u \in \mu(G)$, $f(n_u x) = n_u y$. Since $\mu(G) \backslash \{0\} = \nu(G') \backslash \{0\}$, $f(\mu_u) = \nu_u \ \forall \ u \in \mu(G) \backslash \{0\}$. Thus since f is one-to-one, $\forall \ z \in \mu^*$, $f(\mu)(f(z)) = \vee \{\mu(w) \ | \ f(w) = f(z)\} = \mu(z) = \nu(f(z))$, where the last equality holds since $f(\mu_u) = \nu_u \ \forall \ u \in \mu(G) \backslash \{0\}$. Hence $f(\mu) = \nu$ on ν^*. Conversely, suppose $\exists$ isomorphism f of μ^* onto ν^* such that $f(\mu) = \nu$. Then $\forall \ z \in \mu^*$, $\nu(f(z)) = f(\mu)(f(z)) = \mu(z)$ as above. Hence $\mu(G) \backslash \{0\} = \text{Im}(\nu) \backslash \{0\}$. Clearly, $o(x) = o(y)$. Now $I_x = I_y$ since $\nu_u = f(\mu)_u = f(\mu_u) \ \forall \ u \in \mu(G) \backslash \{0\}$. Thus $I_\mu = I_\nu$. For any isomorphism f of μ^* onto ν^*, where $I_\mu = I_\nu$, the proof that $f(\mu) = \nu$ follows in a similar manner as above since I_ν is independent of the generator of ν^* by Proposition 6.5.2. $\qquad \square$

Let $\mu \in \mathcal{F}(G)$ and $\nu \in \mathcal{F}(G')$. Suppose that $\mu^* = \oplus_{j \in \mathcal{A}} \langle x_j \rangle$ and $\nu^* = \oplus_{j \in \mathcal{A}} \langle y_j \rangle$. By Theorem 6.1.12,

$$\mu^{x_j} = \langle \{(n_{u_j} x_j)_{u_j} | u_j \in \mu^{x_j}(G) \backslash \{0\}\} \rangle$$

and

$$\nu^{y_j} = \langle \{(m_{v_j} y_j)_{v_j} | v_j \in \nu^{y_j}(G) \backslash \{0\}\} \rangle$$

$\forall \ j \in \mathcal{A}$. We let $\mu^j = \mu^{x_j} \ \forall \ j \in \mathcal{A}$.

The next two results show that I_μ is a complete system of invariants for μ when μ is a direct sum of fuzzy subgroups whose supports are cyclic and where the x_j below have orders ∞ or powers of primes. There is no loss in generality in using the same index set $\mathcal{A}$ for μ and ν.

Theorem 6.5.4. *Let $\mu \in \mathcal{F}(G)$ and $\nu \in \mathcal{F}(G')$. Suppose that $\mu = \oplus_{j \in \mathcal{A}} \mu^j$ and $\nu = \oplus_{j \in \mathcal{A}} \nu^j$, where $\mu^* = \oplus_{j \in \mathcal{A}} \langle x_j \rangle$, $\nu^* = \oplus_{j \in \mathcal{A}} \langle y_j \rangle$. If $I_\mu = I_\nu$, then $\exists$ isomorphism f of μ^* onto ν^* such that $f(\mu) = \nu$.*

Proof. By Lemma 6.5.3, $\exists$ isomorphism f_j of $\langle x_j \rangle$ onto $\langle y_j \rangle$ (assuming the x_j and y_j are suitably arranged) such that $f_j(x_j) = y_j$ and $f_j(\mu^j) = \nu^j \ \forall \ j \in \mathcal{A}$. Let $f = \oplus_{j \in \mathcal{A}} f_j$. Then f is an isomorphism of μ^* onto ν^*. Now $f(\mu)(\sum_j m_j y_j) = f(\mu)(\sum_j m_j f(x_j)) = f(\mu)(f \sum_j m_j x_j) = \mu(\sum_j m_j x_j)$ since f is one-to-one. Also $\mu(\sum_j m_j x_j) = (\oplus_{j \in \mathcal{A}} \mu^j)(\sum_j m_j x_j) = \wedge \{\mu^j(m_j x_j) \ | \ j \in \mathcal{A}\}$ (since μ^* is a direct sum) $= \wedge\{w_j \mid w_j \in \mu^j(G)\backslash\{0\}, j \in \mathcal{A}\}$, where $w_j = \mu(0)$ if $m_j = 0$ and w_j is such that $m_j x_j \in \langle n_{w_j} x_j \rangle \backslash \langle n_{u_j} x_j \rangle \ \forall \ u_j \in \mu^j(G)$ such that $u_j > w_j, n_{w_j}, n_{u_j} \in I_{x_j}$, otherwise. Now $n_{w_j}, n_{u_j} \in I_{y_j}, m_j y_j \in \langle n_{w_j} y_j \rangle \backslash \langle n_{u_j} y_j \rangle$ $\forall \ u_j \in \nu^j(G)$ such that $u_j > w_j$ and so $\nu(\sum m_j y_j) = \wedge\{w_j \mid w_j \in \text{Im}(\nu^j)\backslash\{0\}, j \in \mathcal{A}\}$ in a similar manner. Thus $\mu(\sum_j m_j x_j) = \nu(\sum_j m_j y_j)$. Hence $f(\mu) = \nu$. $\qquad\square$

Theorem 6.5.5. *Let $\mu \in \mathcal{F}(G)$ and $\nu \in \mathcal{F}(G')$. Suppose that f is an isomorphism of μ^* onto ν^* such that $f(\mu) = \nu$. If $\mu^* = \oplus_{j \in \mathcal{A}} \langle x_j \rangle$, then $\nu^* = \oplus_{j \in \mathcal{A}} \langle f(x_j) \rangle$, $I_\mu = I_\nu$, and $f(\sum_{j \in \mathcal{A}} \mu^j) = \sum_{j \in \mathcal{A}} \nu^j$.*

Proof. Clearly, $\nu^* = \oplus_{j \in \mathcal{A}} \langle f(x_j) \rangle$, $f(\mu^j) = \nu^j$, and $o(x_j) = o(f(x_j)) \ \forall \ j \in \mathcal{A}$. By Lemma 6.5.3, $I_{x_j} = I_{f(x_j)} \ \forall \ j \in \mathcal{A}$. Now $\nu(\sum_j m_j f(x_j)) = f(\mu)(f \sum_j m_j x_j) = \mu(\sum_j m_j x_j)$. Thus $\mu(G)\backslash\{0\} = \nu(G)\backslash\{0\}$. Hence $I_\mu = I_\nu$. Now,

$$(f(\sum_j \mu^j))(\sum_j m_j f(x_j)) = (f(\sum_j \mu^j))(f(\sum_j m_j x_j))$$
$$= (\sum_j \mu^j)(\sum_j m_j x_j)$$
$$= \wedge\{\mu^j(m_j x_j) \mid j \in \mathcal{A}\} (\text{since } \mu^* \text{ is a direct sum})$$
$$= \wedge\{f(\mu^j)(f(m_j x_j)) \mid j \in \mathcal{A}\}$$
$$= \wedge\{\nu^j(m_j f(x_j)) \mid j \in \mathcal{A}\}$$
$$= (\sum_j \nu^j)(\sum m_j f(x_j)).$$

Thus $f(\sum_j \mu^j) = \sum_j \nu^j$. $\qquad\square$

Definition 6.5.6. *Let $\mu \in \mathcal{F}(G)$ and $\gamma \in \mathcal{F}(\mu)$ and $x_a \subseteq \mu$. Then the fuzzy subset $x_a + \gamma$ is called the **left fuzzy coset** of γ in μ with representative x_a.*

Using the definition of the sum of two fuzzy subsets, it follows easily that $\forall \ z \in G$, $(x_a + \gamma)(z) = a \wedge \gamma(z - x)$. It also follows that for all a in $[0, \mu(0)]$, $(\mu/\gamma)^{(a)} = \{x_a + \gamma \mid x_a \subseteq \mu, x \in G\}$ is a commutative group under $+$, where $(x_a + \gamma) + (y_a + \gamma) = (x + y)_a + \gamma$.

Theorem 6.5.7. *Let $\mu \in \mathcal{F}(G)$ and $\gamma \in \mathcal{F}(\mu)$. Define $\mu/\gamma = \{ x_a + \gamma \mid x_a \subseteq \mu, x \in G, a \in [0,1]\}$. Then $(\mu/\gamma, +)$ is a commutative semi-group with identity. If $\gamma(0) = \mu(0)$, then μ/γ is completely regular [18] and $\mu/\gamma = \cup_{a \in M}(\mu/\gamma)^{(a)}$, where $M = \{a \mid 0 \le a \le \mu(0)\}$.* $\qquad\square$

Theorem 6.5.8. *Let $\mu \in \mathcal{F}(G)$ and $\gamma \in \mathcal{F}(\mu)$. Then $(\mu/\gamma)^{(a)} \simeq \mu_a/\gamma_a \ \forall \ a$ such that $0 \le a \le \mu(0)$.* $\qquad\square$

Let f be a homomorphism of G into G'. Then $\forall\ x_a \subseteq 1_G$, $f(x_a)(f(x)) = \vee\{x_a(y) \mid f(y) = f(x)\} = \vee\{x_a(x)\} = a = f(x)_a(f(x))$. If $f(z) \neq f(x)$, then $f(x_a)(f(z)) = \vee\{x_a(y) \mid f(y) = f(z)\} = 0 = f(x)_a(f(z))$ since $y \neq x$. If $y \in G'\backslash f(G)$, then $f(x_a) = 0$ by definition. Thus $f(x_a) = f(x)_a$.

Let $\mu, \nu \in \mathcal{F}(G)$. Let $\gamma \in \mathcal{F}(\mu)$, $\delta \in L(\nu)$, and f be an isomorphism of γ^* onto δ^* such that $f(\gamma) = \delta$. Then $(f(x_a + \gamma))(f(z)) = \vee\{(x_a + \gamma)(y) \mid f(y) = f(z)\} = (x_a + \gamma)(z)$ (since f is one-to-one) $= a \wedge \gamma(-x + z) = a \wedge \delta(f(-x + z)) = a \wedge \delta(-f(x) + f(z)) = (f(x)_a + \delta)(f(z))$. That is, $(f(x_a + \gamma))(f(z)) = (x_a + \gamma)(z) = (f(x)_a + \delta)(f(z))$. This discussion is relevant to (3) of the next result.

Theorem 6.5.9. *Let $\mu, \nu \in \mathcal{F}(G)$, $\gamma \in \mathcal{F}(\mu)$, and $\delta \in \mathcal{F}(\nu)$. Suppose that $\mu^* = \gamma^* = \oplus_{j \in A} \langle x_j \rangle$, $\nu^* = \delta^* = \oplus_{j \in A} \langle y_j \rangle$, $\gamma = \oplus_{j \in A} \mu^\alpha$, and $\delta = \oplus_{j \in A} \nu^j$. Suppose further that $I_\gamma = I_\delta$. Then $\exists$ isomorphism f of γ^* onto δ^* such that $f(x_j) = y_j \ \forall\ j \in A$ and $f(\gamma) = \delta$. Also the following conditions are equivalent.*
(1) $f(\mu) = \nu$;
(2) $\forall\ a$ such that $0 \leq a \leq \mu(0)$, $x_a \subseteq \mu$ if and only if $f(x)_a \subseteq \nu$;
(3) $\forall\ a$ such that $0 < a \leq \mu(0)$, f_a defined by $f_a(x_a + \gamma) = f(x)_a + \delta$ is an isomorphism of $(\mu/\gamma)^{(a)}$ onto $(\nu/\delta)^{(a)}$.

Proof. That f exists follows from Theorem 6.5.4.

(1) $\Leftrightarrow$ (2): If (1) holds, then (2) holds immediately. Suppose that (2) holds. Since $x_{\mu(x)} \subseteq \mu$, $f(x)_{\mu(x)} \subseteq \nu$ by hypothesis and since $f(x)_{\nu(f(x))} \subseteq \nu$, $x_{\nu(f(x))} \subseteq \mu$ by hypothesis. Thus $\mu(x) = a$ implies $\nu(f(x)) \geq a$ and $\nu(f(x)) = b$ implies $\mu(x) \geq b$. Hence $\mu(x) = \nu(f(x))$. Thus $\nu = f(\mu)$ since f is one-to-one.

(2) $\Leftrightarrow$ (3): Suppose that (2) holds. Now $x_a + \gamma = y_a + \gamma \Leftrightarrow (-y + x)_a \subseteq \gamma \Leftrightarrow f(-y + x)_a \subseteq f(\gamma) \Leftrightarrow (-f(y) + f(x))_a \subseteq f(\gamma) \Leftrightarrow f(x)_a + f(\gamma) = f(y)_a + f(\gamma) \Leftrightarrow f(x)_a + \delta = f(y)_a + \delta \Leftrightarrow f_a(x_a + \gamma) = f_a(y_a + \gamma)$. Thus f_a is single-valued and one-to-one. Now $x_a + \gamma \in (\mu/\gamma)^{(a)} \Leftrightarrow x_a \subseteq \mu \Leftrightarrow f(x)_a \subseteq \nu \Leftrightarrow f(x)_a + \delta \in (\nu/\delta)^{(a)} \Leftrightarrow f_a(x_a + \gamma) \in (\nu/\delta)^{(a)}$. Thus f_a maps $(\mu/\gamma)^{(a)}$ onto $(\nu/\delta)^{(a)}$ since f maps μ^* onto ν^* and so $f(x)_a + \delta$ is arbitrary in $(\nu/\delta)^{(a)}$. Now $f_a((x_a + \gamma) + (y_a + \gamma)) = f_a((x + y)_a + \gamma)) = f(x + y)_a + \delta = (f(x) + f(y))_a + \delta = (f(x)_a + \delta) + (f(y)_a + \delta)) = f_a(x_a + \gamma) + f_a(y_a + \gamma)$. Hence f_a is an isomorphism. Conversely, suppose (3) holds. Then f_a has domain $(\mu/\gamma)^{(a)}$ and image $(\nu/\delta)^{(a)}$. Hence (2) holds since $x_a \subseteq \mu \Leftrightarrow x_a + \gamma \in (\mu/\gamma)^{(a)} \Leftrightarrow f(x)_a + \delta \in (\nu/\delta)^{(a)} \Leftrightarrow f(x)_a \subseteq \nu$. $\square$

We now illustrate Theorem 6.5.9 and other results.

Example 6.5.10. *Let $G = \langle x \rangle \oplus \langle y \rangle$, where $o(x) = p$ and $o(y) = p^3$, p a prime. Let $H = \langle x + py \rangle$. Define the fuzzy subset μ of G by $\mu(z) = 1$ if $z \in H$ and $\mu(z) = \frac{1}{2}$ otherwise. Then μ is a fuzzy subgroup of G. The fuzzy subgroups μ^x and μ^y are such that $\mu^x(0) = 1$, $\mu^x(z) = \frac{1}{2}$ if $z \in \langle x \rangle \setminus \langle 0 \rangle$, $\mu^x(z) = 0$ otherwise, $\mu^y(z) = 1$ if $z \in \langle p^2 y \rangle$, $\mu^y(z) = \frac{1}{2}$ if $z \in \langle y \rangle \setminus \langle p^2 y \rangle$, and $\mu^y(z) = 0$ otherwise. Now $\mu^x + \mu^y = \mu^x \oplus \mu^y$ by Theorem 1.5.9. Let $\nu = \gamma = \delta = \mu^x \oplus \mu^y$. Then $\mu^* = \nu^* = \gamma^* = \delta^* = G$. Clearly, $I_\gamma = I_\delta$ and so $\exists$ an* *isomorphism f of γ^* onto δ^* such that $f(\gamma) = \delta$ by Theorem 6.5.9. However, $(x + py)_1 \subseteq \mu$, but $f(x + py)_1 \not\subseteq \nu$ else $f(H) \subseteq \langle p^2 y \rangle$ and so $f(H)$ has order $\leq p$. Hence by Theorem 6.5.9, $f(\mu) \neq \nu$. In fact, $\nexists$ any isomorphism f of μ^* onto ν^* such that $f(\mu) = \nu$ for the same reason. We note that $I_\mu = I_\nu$. However, by Example 6.1.17, μ is not a direct sum of fuzzy subgroups of G whose supports are cyclic. Thus Theorem 6.5.4 is not contradicted.*

6.6 Basic and p-Basic Fuzzy Subgroups

The representation of G as a direct sum of cyclic groups yields structure results for G. However, not every group G is a direct sum of cyclic groups. But there exist largest subgroups of G that are direct sums of cyclic groups of order infinity and powers of primes. If these largest subgroups are pure, they are of importance in determining properties of G, [4]. These largest pure subgroups, called basic subgroups, are invariants of p-groups [[4], Theorem 35.2, p. 148]. This yields a system of cardinal numbers as a system of invariants of G, [[4], p. 148]. The theory of p-groups is based to a large extent on basic subgroups. Also the concept of basic subgroups as well as other concepts from Abelian group theory are of importance when carried over to the study of inseparable field extensions.

Let $x \in \mu^*$. Let μ^x be the fuzzy subset of G defined by $\mu^x(y) = \mu(y)$ for all $y \in \langle x \rangle$ and $\mu^x(y) = 0$ for all $y \in G \backslash \langle x \rangle$. Then μ^x is a fuzzy subgroup of G. Let $I_x = \{n_u | u \in \text{Im}(\mu^x) \backslash \{0\}$, n_u the smallest positive integer such that $(\mu^x)_u = \langle n_u x \rangle\}$. Suppose that $\mu^* = \oplus_{j \in \mathcal{A}} \langle x_j \rangle$. Let I_μ denote the ordered pair, $(\cup_{j \in \mathcal{A}} \{I_{x_j}, o(x_j)\}, \text{Im}(\mu) \backslash \{0\})$. In the remainder of this section, G' denotes an Abelian group and ν a fuzzy subgroup of G'.

If f is an isomorphism of G onto G', then $f(\mu)$ is the fuzzy subset of G' defined by $\forall~y \in G'$, $f(\mu)(y) = \mu(x)$, where $y = f(x)$. This definition coincides with the usual definition of a homomorphism applied to μ since f is an isomorphism here. By Theorem 1.2.11, $f(\mu)$ is a fuzzy subgroup of G'.

Suppose that $\mu^* = \oplus_{j \in \mathcal{A}} \langle x_j \rangle$ and $\nu^* = \oplus_{j \in \mathcal{A}} \langle y_j \rangle$. By Theorem 6.1.12,

$$\mu^{x_j} = \left\langle \{(n_{u_j} x_j)_{u_j} | u_j \in \mu^{x_j}(G) \backslash \{0\}) \right\rangle$$

and

$$\nu^{y_j} = \left\langle \{(m_{v_j} y_j)_{v_j} | v_j \in \nu^{y_j}(G') \backslash \{0\}) \right\rangle$$

$\forall~j \in \mathcal{A}$.

We call an fuzzy subgroup γ of G **cyclic** if and only if $\exists~x \in G$ such that $\gamma = \langle x_a \rangle$, where $a = \gamma(x)$.

Definition 6.6.1. *Let $\mu \in \mathcal{F}(G)$ and $\nu \in \mathcal{F}(\mu)$. Then ν is said to be **basic** in μ if*

(1) *ν is a direct sum of cyclic fuzzy subgroups of G whose supports are cyclic of prime power orders or of infinite order;*

(2) ν is pure in μ,

(3) $\forall$ a such that $0 \leq a \leq \mu(0)$, $\forall$ $x_a + \nu \in \mu/\nu$, $\forall$ $n \in \mathbb{N}$, $\exists$ $y_b + \nu \in \mu/\nu$ such that $n(y_b + \nu) = x_a + \nu$.

Proposition 6.6.2. *Let* $\mu \in \mathcal{F}(G)$ *and* $\nu \in \mathcal{F}(\mu)$. *Then* (3) *of Definition* 6.6.1 *implies* μ^*/ν^* *is divisible.*

Proof. Let $x + \nu^* \in \mu^*/\nu^*$ and let $n \in \mathbb{N}$. Then $x_a + \nu \in \mu/\nu$ for all a such that $0 \leq a \leq \mu(0)$. Thus $\exists$ $y_b + \nu \in \mu/\nu$ such that $x_a + \nu = ny_b + \nu$. Hence $a = b \leq \nu(-x + ny)$. Thus $\nu(-x + ny) > 0$ and so $-x + ny \in \nu^*$. Hence $x + \nu^* = ny + \nu^*$ and so μ^*/ν^* is divisible. $\qquad \square$

It follows that (3) of Definition 6.6.1 is equivalent to the exact same statement with b replaced by a.

Proposition 6.6.3. *Let* $\mu \in \mathcal{F}(G)$ *and* $\nu \in \mathcal{F}(\mu)$. *Then* (3) *of Definition* 6.6.1 *holds if and only if* $\forall$ a *such that* $0 \leq a \leq \nu(0)$, μ_a/ν_a *is divisible.*

Proof. Suppose that (3) holds. Let a be such that $0 \leq a \leq \nu(0)$. Let $x + \nu_a \in \mu_a/\nu_a$ and $n \in \mathbb{N}$. Then $\mu(x) \geq a$ and so $x_a \subseteq \mu$. Thus $\exists$ $y_a + \nu$ such that $n(y_a + \nu) = x_a + \nu$ and so $(ny)_a + \nu = x_a + \nu$. Hence $(-x + ny)_a \subseteq \nu$. Thus $\nu(-x + ny) \geq a$. Hence $-x + ny \in \nu_a$. Thus $ny + \nu_a = x + \nu_a$. Hence μ_a/ν_a is divisible. Conversely, suppose that μ_a/ν_a is divisible for all a such that $0 \leq a \leq \nu(0)$. Let a be such that $0 \leq a \leq \nu(0)$. Let $x_a + \nu \in \mu/\nu$ and $n \in \mathbb{N}$. Then $x_a \subseteq \mu$ and so $\mu(x) \geq a$. Thus $x \in \mu_a$. Hence $x + \nu_a \in \mu_a/\nu_a$. Thus $\exists$ $y + \nu_a \in \mu_a/\nu_a$ such that $n(y + \nu_a) = x + \nu_a$. Hence $ny + \nu_a = x + \nu_a$. Thus $-x + ny \in \nu_a$. Hence $\nu(-x + ny) \geq a$. Thus $(-x + ny)_a \subseteq \nu$. Hence $ny_a + \nu = x_a + \nu$. Thus (3) holds. $\qquad \square$

Theorem 6.6.4. *Let* $\mu \in \mathcal{F}(G)$ *and* $\nu \in \mathcal{F}(\mu)$. *If* ν *is basic in* μ, *then* $\forall$ a *such that* $0 \leq a \leq \nu(0)$, ν_a *is basic in* μ_a.

Proof. The desired result holds from the Propositions 6.4.13, 6.6.3 and Theorem 1.5.10. $\qquad \square$

Theorem 6.6.5. *Let* $\mu \in \mathcal{F}(G)$. *Suppose that* μ *has the sup property. Let* $\nu \in \mathcal{F}(\mu)$. *If* ν *is basic in* μ, *then* ν^* *is basic in* μ^*.

Proof. By Proposition 6.6.2, μ^*/ν^* is divisible. Now $\nu = \oplus_{j \in \mathcal{A}} \gamma_j$, where $\gamma_j \in \mathcal{F}(\mu)$ and γ_j is cyclic. Thus $\nu^* = \oplus_{j \in \mathcal{A}} \gamma_j^*$ by Theorem 1.5.9 and γ_j^* is cyclic $\forall$ $j \in \mathcal{A}$. Since ν is pure in μ, we have by Proposition 6.4.15 and Proposition 6.4.16 that $\forall$ $n \in \mathbb{N}$, $n\nu = \nu \cap n\mu$ implies $n\nu^* = (n\nu)^* = (\nu \cap n\mu)^* = \nu^* \cap n\mu^*$, i.e., ν^* is pure in μ^*. $\qquad \square$

Let p denote a fixed prime.

Definition 6.6.6. *Let* $\mu \in \mathcal{F}(G)$ *and* $\nu \in \mathcal{F}(\mu)$. *Then* ν *is said to be* p-**pure** *in* μ *if* $\forall$ *L-singletons* $x_a \subseteq \nu$ *with* $a > 0$, $\forall$ $k \in \mathbb{N}$, $\forall$ $y_a \subseteq \mu$, $p^k(y_a) = x_a$ *implies that* $\exists$ $z_a \subseteq \nu$ *such that* $p^k(z_a) = x_a$.

Definition 6.6.7. Let $\mu \in \mathcal{F}(G)$ and $\nu \in \mathcal{F}(\mu)$. Then ν is said to be p-**basic** in μ if

(1) ν is a direct sum of cyclic fuzzy subgroups whose supports are cyclic of order a power of p or of infinite order;

(2) ν is p-pure in μ;

(3) $\forall$ a such that $0 < a \leq \mu(0)$, $\forall$ $x_a + \nu$, $\forall$ $k \in \mathbb{N}$, $\exists$ $y_b + \nu$ such that $p^k(x_a + \nu) = y_b + \nu$.

Theorem 6.6.8. Let $\mu \in \mathcal{F}(G)$ and $\nu \in \mathcal{F}(\mu)$.

(1) Then (3) of Definition 6.6.7 implies μ^*/ν^* is p-divisible;

(2) ν is p-pure in μ if and only if ν_a is p-pure in μ_a $\forall$ a such that $0 < a \leq \mu(0)$;

(3) Part (3) of Definition 6.6.7 holds if and only if μ_a/ν_a is p-divisible $\forall$ a such that $0 < a \leq \mu(0)$;

(4) If ν is p-basic in μ, then ν_a is p-basic in $\mu_a \forall a$ such that $0 < a \leq \mu(0)$.

(5) Suppose that μ has the sup property. If ν is p-basic in μ, then ν^* is p-basic in μ^*.

Proof. The proofs are similar to those of basic fuzzy subgroups. $\square$

We let $\mathbb{Z}$ denote the additive group of integers and $\mathbb{Z}(p^n)$ a cyclic group of order p^n, $n \in \mathbb{N}$. We recall that for H a p-basic subgroup of G, $H = \oplus_{n=0}^{\infty} K_n$, where $K_0 = \oplus_{m_0} \mathbb{Z}$ and $K_n = \oplus_{m_n} \mathbb{Z}(p^n)$, $n = 1, 2, \ldots$, and $\{m_0, m_1, \ldots, m_n, \ldots\}$ is a system of invariants for G, [[4], p. 148].

Theorem 6.6.9. Let $\mu \in \mathcal{F}(G)$ and $\nu \in \mathcal{F}(\mu)$ be such that ν is p-basic in μ.

(1) $\cup_{a \in A}\{m_{0a}, m_{1a}, \ldots, m_{na}, \ldots\}$ is a system of invariants for μ, where $\{m_{0a}, m_{1a}, \ldots, m_{na}, \ldots\}$ is the system of invariants determined by ν_a in μ_a and $A = \{a \mid 0 \leq a \leq \nu(0)\}$.

(2) Suppose that μ has the sup property. Then $\{m_0, m_1, \ldots, m_n, \ldots\}$ is a system of invariants for μ, where $\{m_0, m_1, \ldots, m_n, \ldots\}$ is the system determined by ν^* in μ^*.

Proof. The proof is immediate from Theorem 6.6.8 (4) and (5) and the preceding comments. $\square$

Definition 6.6.10. Let $\mu \in \mathcal{F}(G)$ and $\chi \in \mathcal{FS}(\mu)$. Then χ is said to be p-**independent** in μ if $\forall$ $x_1, \ldots, x_k \in \chi^*$ and $\forall$ $a_i \in \{a \mid 0 \leq a \leq \chi(x_i)\}$ for $i = 1, \ldots, k$ and $\forall$ $r \in \mathbb{N}$, $n_1(x_1)_{a_1} + \ldots + n_k(x_k)_{a_k} \subseteq p^r\mu$ ($n_i x_i \neq 0$, $i = 1, \ldots, k$) implies $p^r \mid n_i$ for $i = 1, \ldots, k$. χ is called a p-**basis** of μ if χ is p-independent in μ and $\mu = p\mu + \langle \chi \rangle$.

Proposition 6.6.11. Let $\mu \in \mathcal{F}(G)$ and $\chi \in \mathcal{FS}(\mu)$. Then χ is p-independent in μ if and only if χ^* is p-independent in μ^*.

Proof. Suppose that χ is p-independent in μ. Let $x_1, \ldots, x_k \in \chi^*$. Suppose that $n_1 x_1 + \ldots + n_k x_k \in p^r\mu^*$ ($n_i x_i \neq 0, i = 1, \ldots, k$). Then $n_1 x_1 + \ldots +$

$n_k x_k = p^r x$ for some $x \in \mu^*$. Let $a = a_1 \wedge \ldots \wedge a_k \wedge \mu(x)$, where $a_i = \chi(x_i)$ for $i = 1, \ldots, k$. Then $n_1(x_1)_a + \ldots + n_k(x_k)_a = p^r x_a \subseteq p^r \mu$. Thus $p^r \mid n_i$ for $i = 1, \ldots, k$. Conversely, suppose that χ^* is p-independent in μ^*. Let $x_1, \ldots, x_k \in \chi^*$. Suppose that $n_1(x_1)_{a_1} + \ldots + n_k(x_k)_{a_k} \subseteq p^r \mu$ $(n_i x_i \neq 0, i = 1, \ldots, k)$. Then $(n_1 x_1 + \ldots + n_k x_k)_a \subseteq p^r \mu$, where $a = a_1 \wedge \ldots \wedge a_k > 0$. Thus $(p^r \mu)(n_1 x_1 + \ldots + n_k x_k) \geq a > 0$. Hence $\exists\, x \in G$ such that $n_1 x_1 + \ldots + n_k x_k = p^r x$. Thus $(p^r \mu)(p^r x) > 0$ and so $\mu(x) > 0$. Hence $x \in \mu^*$ and so $p^r x \in p^r \mu^*$. Thus $p^r \mid n_i$ for $i = 1, \ldots, k$. $\qquad\square$

Definition 6.6.12. *Let $\mu \in \mathcal{F}(G)$ and $\chi \in \mathcal{FS}(\mu)$. Then χ is said to be* **maximally p-independent** *in μ if χ is p-independent in μ and $\nexists\, \eta \in \mathcal{FS}(\mu)$ such that η is p-independent in μ and $\chi \subset \eta$.*

Proposition 6.6.13. *Let $\mu \in \mathcal{F}(G)$ and $\chi \in \mathcal{FS}(\mu)$. Then χ is maximally p-independent in μ if and only if χ^* is a p-basis for μ^* and $\forall\, x \in \chi^*$, $\chi(x) = \mu(x)$.*

Proof. Suppose that χ is maximally p-independent in μ. Then χ^* is p-independent in μ^* by Proposition 6.6.11. Suppose that χ^* is not a p-basis of μ^*. Then $\exists\, y \in \mu^* \backslash \chi^*$ such that $\chi^* \cup \{y\}$ is p-independent in μ^*. Thus by Proposition 6.6.11, $\eta \in \mathcal{FS}(\mu)$, where $\eta = \mu$ on $\chi^* \cup \{y\}$ and $\eta(z) = 0 \forall\, z \in G \backslash (\chi^* \cup \{y\})$ is p-independent in μ. However, this contradicts the maximality of χ. Conversely, suppose that χ^* is a p-basis of μ^* and $\forall\, x \in \chi^*$, $\chi(x) = \mu(x)$. Suppose that χ is not maximal. Then $\exists\, \eta \in \mathcal{FS}(\mu)$ such that $\chi \subset \eta$ and η is p-independent in μ. Since $\chi = \mu$ on χ^*, $\chi^* \subset \eta^*$. Since η^* is necessarily p-independent in μ, this contradicts the maximality of χ^*. Thus χ is maximal. $\qquad\square$

It follows easily that if χ is a p-basis of μ, then χ is maximally p-independent in μ. However, the following example shows that the converse is not true.

Example 6.6.14. *Let $G = \langle x \rangle$, where x has order p^2, p a prime. Define the fuzzy subset μ of G by $\mu(z) = 1$ if $z \in \langle px \rangle$ and $\mu(z) = \frac{1}{2}$ if $z \in G \backslash \langle px \rangle$. Then μ is a fuzzy subgroup of G and $\mu^* = G$. Now $(p\mu)(px) = \mu(x) = \frac{1}{2}$. Thus $(p\mu)(0) = 1$, $(p\mu)(z) = \frac{1}{2}$ if $z \in \langle px \rangle \backslash \langle 0 \rangle$, and $(p\mu)(z) = 0$ if $z \in G \backslash \langle px \rangle$. If χ is a p-basis of μ, then $\chi^* = \{z\}$ for some $z \in G \backslash \langle px \rangle$. Now $\mu(z) = \frac{1}{2}$. Thus $\mu \supset p\mu + \langle \chi \rangle$ since $\mu(px) = 1$, but $(p\mu + \langle \chi \rangle)(px) = \vee\{(p\mu)(u) \wedge \langle \chi \rangle (v) \mid px = u + v\} = \vee\{(p\mu)(pix) \wedge \langle \chi \rangle (p(1 - i)x) \mid 0 \leq i < p\} \leq \frac{1}{2}$. Thus μ does not have a p-basis even though $p^2 \mu^* = \langle 0 \rangle$. Now χ is maximal p-independent in μ since χ^* is a p-basis for μ^*.*

Lemma 6.6.15. *Let $\mu \in \mathcal{F}(G)$. $\forall\, a \in (0, \mu(0)]$, $p^r(\mu_a) \subseteq (p^r \mu)_a$. If either μ has the sup property or μ^* is torsion-free, then $p^r(\mu_a) = (p^r \mu)_a$.*

Proposition 6.6.16. *Let $\mu \in \mathcal{F}(G)$ and $\chi \in \mathcal{FS}(\mu)$.*
  (1) If χ is p-independent in μ, then χ_a is p-independent in μ_a $\forall\, a$ such that $0 \leq a \leq \mu(0)$.

(2) *Suppose that μ has the sup property or μ^* is torsion-free. If χ_a is p-independent in μ_a $\forall$ a such that $0 < a \leq \mu(0)$, then χ is p-independent in μ.*

Proof. (1) Let $x_1, \ldots, x_k \in \chi_a$. Suppose that $n_1 x_1 + \ldots + n_k x_k \in p^r(\mu_a)$ $(n_i x_i \neq 0, i = 1, \ldots, k)$. Then $x_1, \ldots, x_k \in \chi^*$ and $n_1 x_1 + \ldots + n_k x_k \in p^r \mu^*$. By Proposition 6.6.11, $p^r \mid n_i$ for $i = 1, \ldots, k$.

(2) Let $x_1, \ldots, x_k \in \chi^*$. Suppose that $n_1 (x_1)_{a_1} + \ldots + n_k (x_k)_{a_k} \subseteq p^r \mu$ $(n_i x_i \neq 0, i = 1, \ldots, k)$. Then $(n_1 x_1 + \ldots + n_k x_k)_a \subseteq p^r \mu$, where $a = a_1 \wedge \ldots \wedge a_k$. By Lemma 6.6.15, $(p^r \mu)_a = p^r(\mu_a)$. Thus $n_1 x_1 + \ldots + n_k x_k \in p^r(\mu_a)$. Since also $x_1, \ldots, x_k \in \chi_a, p^r \mid n_i$ for $i = 1, \ldots, k$. $\square$

Proposition 6.6.17. *Let $\mu \in \mathcal{F}(G)$. Suppose that μ has the sup property or μ^* is torsion-free. Let $\chi \in \mathcal{FS}(\mu)$ be such that $\chi(x) = \mu(x)$ $\forall$ $x \in \chi^*$. If χ_a is a p-basis of μ_a $\forall$ a such that $0 < a \leq \mu(0)$, then χ is maximally p-independent in μ.*

Proof. By Proposition 6.6.16(2), χ is p-independent in μ. Suppose that χ is not maximally p-independent in μ. Then there exists $\eta \in \mathcal{FS}(\mu)$ such that $\chi \subset \eta$ and η is p-independent in μ. By Proposition 6.6.16(1), η_a is p-independent in μ_a for all a such that $0 < a \leq \mu(0)$. Since $\chi^* \subset \eta^*$, there exists a such that $\chi_a \subset \eta_a$. However, this contradicts the maximality of χ_a. $\square$

The following example shows that the converse of Proposition 6.6.17 does not hold.

Example 6.6.18. *Let $G = \mathbb{Z}(p^\infty)$ and let $x \in G$ have order p. Define the fuzzy subset μ of G by $\mu(z) = 1$ if $z \in \langle x \rangle$ and $\mu(z) = \frac{1}{2}$ otherwise. Then μ is a fuzzy subgroup of G and $\mu^* = G$. Now 0_G is maximally p-independent in μ since $\emptyset$ is a p-basis of μ^*. However, $\mu_1 = \langle x \rangle$ and $(0_G)_1 = \emptyset$ is not a p-basis of $\langle x \rangle$. Note also that 0_G is not a p-basis of μ.*

Let $\mu \in L(G)$. If χ is maximally p-independent in μ, we determine in the following to what extent $\langle \chi \rangle$ is a p-basic L-subgroup of μ.

Proposition 6.6.19. *Let $\mu \in \mathcal{F}(G)$ and $\chi \in \mathcal{FS}(\mu)$. If χ is p-independent in μ, then χ is independent in μ.*

Proof. χ is p-independent in μ $\Leftrightarrow$ χ^* is p-independent in μ^* by Proposition 6.6.11. If χ^* is p-independent in μ^*, then χ^* is independent in μ^*. Now χ^* is independent in μ^* $\Leftrightarrow$ χ is independent in μ. $\square$

Theorem 6.6.20. *Let $\mu \in \mathcal{F}(G)$ and $\chi \in \mathcal{FS}(\mu)$. Then χ is independent in μ if and only if $\langle \chi \rangle = \oplus_{x_a \in S_\chi} \langle x_a \rangle$, where $S_\chi = \{x_a \mid x \in \chi^*, \chi(x) = a\}$.*

Proof. See Theorem 6.2.5. $\square$

Theorem 6.6.21. *Let $\mu \in \mathcal{F}(G)$ and $\chi \in \mathcal{FS}(\mu)$. If χ is p-independent in μ, then $\langle \chi \rangle = \oplus_{x_a \in S_\chi} \langle x_a \rangle$, where $S_\chi = \{x_a \mid x \in \chi^*, \chi(x) = a\}$.*

Proof. The desired result follows by Proposition 6.6.19 and Theorem 6.6.20.
□

Lemma 6.6.22. *Let $\mu \in \mathcal{F}(G)$ and $\chi \in \mathcal{FS}(\mu)$. If χ is independent in μ, then $\langle \chi_a \rangle = \langle \chi \rangle_a$ for all a such that $0 \leq a \leq \mu(0)$.*

Proof. If $a > \vee \{\chi(x) \mid x \in \chi^*\}$, then the result follows easily. Since $\chi_a \subseteq \langle \chi \rangle_a$, $\langle \chi_a \rangle \subseteq \langle \chi \rangle_a$. Let $x \in \langle \chi \rangle_a$. Then $\langle \chi \rangle (x) \geq a$. Now $\langle \chi \rangle = \oplus_{x_S \in S_\chi} \langle x_S \rangle$ by Theorem 6.6.20. Thus x has an unique representation, $x = n_1 x_1 + \ldots + n_k x_k$, where $x_i \in \chi^*$. Thus $a \leq a_1 \wedge \ldots \wedge a_k$, where $\chi(x_i) = a_i$ for $i = 1, \ldots, k$. Hence $x \in \langle \chi_a \rangle$. Thus $\langle \chi \rangle_a \subseteq \langle \chi_a \rangle$.
□

Theorem 6.6.23. *Let $\mu \in \mathcal{F}(G)$ and $\chi \in \mathcal{FS}(\mu)$. If χ is p-independent in μ, then $\langle \chi \rangle$ is p-pure in μ.*

Proof. Since χ is p-independent in μ, χ_a is p-independent in $\mu_a \; \forall \; a \in \{a | 0 \leq a \leq \mu(0)\}$ by Proposition 6.6.16. Thus $\langle \chi \rangle_a = \langle \chi_a \rangle$ is p-pure in μ_a by Lemma 6.6.22 and [4]. Hence $\langle \chi \rangle$ is p-pure in μ.
□

Let $\mu \in \mathcal{F}(G)$. We also note that if $\chi \in \mathcal{FS}(\mu)$ is maximal p-independent, then $\langle \chi^* \rangle = \langle \chi \rangle^*$ is a p-basic subgroup of μ^* since χ^* is a p-basis of μ^* by Proposition 6.6.13.

Theorem 6.6.24. *Let $\mu \in \mathcal{F}(G)$ and $\chi \in \mathcal{FS}(\mu)$.*
  (1) *Suppose that χ is p-independent in μ. If $\langle \chi \rangle$ is a p-basic in μ, then χ is a p-basis of μ.*
  (2) *Suppose that μ is finite-valued. If χ is a p-basis of μ, then $\langle \chi \rangle$ is p-basic in μ.*

Proof. (1) Let $x_a \subseteq \mu$. Then by (3) of Definition 6.6.1, $\exists \; y_a \subseteq \mu$ such that $x_a + \langle \chi \rangle = p y_a + \langle \chi \rangle$. Thus $\mu = p\mu + \langle \chi \rangle$.
  (2) By Theorem 6.6.21 and Lemma 6.6.22, it suffices to show that $\mu_a / \langle \chi \rangle_a$ is divisible $\forall \; a \in \{a | 0 \leq a \leq \mu(0)\}$. Since μ is finite-valued, $p(\mu_a) = (p\mu)_a$ by Lemma 6.6.15 and $(p\mu + \langle \chi \rangle)_a = (p\mu)_a + \langle \chi \rangle_a$ since $p\mu$ and $\langle \chi \rangle$ are necessarily finite-valued. Thus from $\mu = p\mu + \langle \chi \rangle$, it follows that $\mu_a = p(\mu_a) + \langle \chi_a \rangle$ by Lemma 6.6.22. Hence by Proposition 6.6.16, χ_a is a p-basis of μ_a. Thus $\mu_a / \langle \chi \rangle_a$ is divisible $\forall \; a$ such that $0 \leq a \leq \mu(0)$.
□

Theorem 6.6.25. *Let $\mu \in \mathcal{F}(G)$ and $\nu \in \mathcal{F}(\mu)$. If ν is p-basic in μ, then χ is a p-basis of μ, where $\nu = \oplus_{j \in A} \langle (x_j)_{a_j} \rangle$ and χ is the fuzzy subset of G defined by $\chi(z) = a_j$ if $z = x_j$ and $\chi(z) = 0$ otherwise.*

Proof. Now $\nu^* = \oplus_{j \in A} \langle x_j \rangle$. Thus χ^* is a p-basis of ν^*. Since ν^* is p-pure in μ^*, χ^* is p-independent in μ^*. Since μ^*/ν^* is divisible by Proposition 6.6.2, χ^* is a p-basis of μ^*. Thus χ is maximally p-independent in μ by Proposition 6.6.13. Now $\nu = \langle \chi \rangle$. Hence χ is a p-basis of μ by Theorem 6.6.24(1).
□

The proof of Theorem 6.6.25 also shows that if ν is p-basic in μ, then $\mu = \nu$ on ν^*.

Theorem 6.6.26. *Let* $\mu \in \mathcal{F}(G)$. *If* μ^* *has a unique p-basic subgroup, then* μ *has a unique p-basic subgroup.*

Proof. Let ν and γ be p-basic subgroups of μ. Then ν^* and γ^* are p-basic subgroups of μ^* and so $\nu^* = \gamma^*$. Now $\nu = \langle \chi \rangle$ and $\gamma = \langle \eta \rangle$ for p-bases χ and η of μ by Theorem 6.6.25. By Proposition 6.6.13, $\chi = \mu$ on χ^* and $\eta = \mu$ on η^*. Thus $\nu = \mu$ on ν^* and $\gamma = \mu$ on γ^*. But $\nu^* = \gamma^*$ and so $\nu = \gamma$.   □

Let $\mu \in \mathcal{F}(G)$. Criteria for μ^* to contain exactly one p-basic subgroup can be found in [[4], Theorem 35.3, p. 149].

Theorem 6.6.27. *Let* $\mu \in \mathcal{F}(G)$. *Then the following assertions hold.*
   (1) *Suppose that* μ^* *has a unique p-basic subgroup and that* μ *is finite-valued. If* χ *and* η *are p-bases of* μ, *then* $I_{\langle \chi \rangle} = I_{\langle \eta \rangle}$.
   (2) *Suppose that* $p^e \mu = \theta$ *for some positive integer e. If* χ *and* η *are p-bases of* μ, *then* $I_{\langle \chi \rangle} = I_{\langle \eta \rangle}$.

Proof. For assertion (1), $\langle \chi \rangle$ and $\langle \eta \rangle$ are p-basic subgroups of μ by Theorem 6.6.24(2). Thus $\langle \chi \rangle = \langle \eta \rangle$. For assertion (2), since $p^e \mu = 0_G$, we have that $\mu = \langle \chi \rangle = \langle \eta \rangle$. In either case, let f be the identity map on $\langle \chi \rangle^*$. Then $f(\langle \chi \rangle) = \langle \eta \rangle$. Now $\langle \chi \rangle^* = \langle \chi^* \rangle = \oplus \langle x_j \rangle$ and so $\langle \eta \rangle^* = \langle \eta^* \rangle = \oplus \langle f(x_j) \rangle$ and $I_{\langle \chi \rangle} = I_{\langle \eta \rangle}$ by Theorem 6.5.5.   □

Suppose that G is torsion. Suppose that G has a finite p-basis. Let G' and K be subgroups of G, K' a subgroup of $K \cap G'$, and H a subgroup of G'. Suppose that $K' \subset K$ and $G' \subset G$. We note that it is impossible for K' and H to be p-basic subgroups of G' and for K and H to be p-basic subgroups of G : Suppose otherwise. Then we that have $G' = pG' + K'$, $G' = pG' + H$, $G = pG + K \supset pG + K'$ (since G has a finite p-basis), and $G = pG + H$. Thus $G = pG + G' = pG + pG' + K' = pG + K'$, a contradiction.

Theorem 6.6.28. *Suppose that* G *is torsion. Let* $\mu \in \mathcal{F}(G)$. *Suppose that* μ^* *has a finite p-basis and that* μ *is finite-valued. Let* $\nu, \gamma \in \mathcal{F}(\mu)$. *If* ν *and* γ *are p-basic in* μ, *then* $\nu(G) \backslash \{0\} = \gamma(G) \backslash \{0\}$.

Proof. Let $\mathcal{B} = \nu(G) \backslash \{0\}$ and $\mathcal{C} = \gamma(G) \backslash \{0\}$. Since $\mu = \nu$ on ν^* and $\mu = \gamma$ on γ^*, $\mathcal{B} \subseteq \mu(G)$ and $\mathcal{C} \subseteq \mu(G)$. Since μ is finite-valued, ν and γ are finite-valued. Suppose that $\mathcal{B} \neq \mathcal{C}$. Then either $\exists \, a \in \mathcal{B}$ such that $a \notin \mathcal{C}$ or $\exists \, b \in \mathcal{C}$ such that $b \notin \mathcal{B}$, say a exists. Then $a < \nu(0)$. Since $\mathcal{B}$ and $\mathcal{C}$ are finite sets, either (1) $\exists$ largest $b \in \mathcal{C}$ such that $b < a$ or (2) $a < \wedge \{b \mid b \in \mathcal{C}\}$. Suppose that (1) holds. Then $\exists$ smallest $v \in \mathcal{B}$, $v \notin \mathcal{C}$, such that $b < v \leq a$. Now $\exists$ smallest $u \in \mathcal{C}$ such that $b < v < u$ since $\gamma(0) > v$. Thus $\gamma_v = \gamma_u$ and $\nu_v \supset \nu_u$. By Theorem 6.6.8(4), γ_v and ν_v are p-basic in μ_v and γ_u and ν_u are p-basic in μ_u. However, as noted above, this is impossible since $\mu_v \supset \mu_u$, where the latter inclusion holds since $v, u \in \mu(G)$. Suppose that (2) holds. Let $u = \wedge \{b \mid b \in \mathcal{C}\}$. Then $\gamma_a = \gamma_u$ and $\nu_a \supset \nu_u$ and we have a contradiction as just argued.   □

Theorem 6.6.29. *Suppose that G is torsion. Let $\mu \in \mathcal{F}(G)$. Suppose that μ^* has a finite p-basis and that μ is finite-valued. Let $\nu, \gamma \in \mathcal{F}(\mu)$. If ν and γ are p-basic in μ, then $\exists$ isomorphism f of ν^* onto γ^* such that $f(\nu) = \gamma$ and $I_\nu = I_\gamma$.*

Proof. By Theorem 6.6.25, ν and γ are finitely generated. Now $\nu = \oplus_{j \in \mathcal{A}} \langle (x_j)_{a_j} \rangle$, where $\nu^* = \oplus_{j \in \mathcal{A}} \langle x_j \rangle$ and $\gamma = \oplus_{m \in \mathcal{B}} \langle (y_m)_{b_m} \rangle$, where $\gamma^* = \oplus_{m \in \mathcal{B}} \langle y_m \rangle$ by Theorem 6.6.25. By Theorem 6.6.25, $|\mathcal{A}| = |\mathcal{B}| < \infty$. By Theorem 6.6.28, $\nu(G) \backslash \{0\} = \gamma(G) \backslash \{0\}$ and these sets are finite. Thus $\nu(G) \backslash \{0\} = \{a_1, \ldots, a_n\} = \gamma(G) \backslash \{0\}$, where $a_1 < \ldots < a_n$. Since ν is finite-valued, $\nu_a = \oplus_{j \in \mathcal{A}} \langle (x_j)_{a_j} \rangle_a \ \forall \ a \in \nu(G)$. Now $\mathcal{A} = \mathcal{A}_1 \cup \ldots \cup \mathcal{A}_n$, where $j \in \mathcal{A}_i$ if and only if $a_j = a_i$. Thus $\nu_{a_i} = \oplus_{j \in \mathcal{A}_i} \langle x_j \rangle \oplus \oplus_{j \in \mathcal{A}_{i+1} \cup \ldots \cup \mathcal{A}_n} \langle x_j \rangle$ and $\oplus_{j \in \mathcal{A}_i} \langle x_j \rangle \simeq \nu_{a_i} / \nu_{a_{i+1}}$ and $\oplus_{j \in \mathcal{A}_{i+1} \cup \ldots \cup \mathcal{A}_n} \langle x_j \rangle \simeq \nu_{a_{i+1}}$ for $i = 1, \ldots, n-1$. Thus $\nu_{a_i} \simeq (\nu_{a_i} / \nu_{a_{i+1}}) \oplus \nu_{a_{i+1}}$ and similarly $\gamma_{a_i} \simeq (\gamma_{a_i} / \gamma_{a_{i+1}}) \oplus \gamma_{a_{i+1}}$. Now there exists an isomorphism g_i of ν_{a_i} onto γ_{a_i} for $i = 1, \ldots, n$ by Theorem 6.6.4. Since $\nu_{a_{i+1}}$ is a direct summand of ν_{a_i} and $\gamma_{a_{i+1}}$ is a direct summand of γ_{a_i} and these groups are finitely generated, there exists an isomorphism f_i of $\nu_{a_i} / \nu_{a_{i+1}}$ onto $\gamma_{a_i} / \gamma_{a_{i+1}}$ such that $g_i = f_i \oplus g_{i+1}$, $i = 1, \ldots, n-1$. Let $f_n = g_n$. Then $f = f_1 \oplus \ldots \oplus f_n$ is an isomorphism of $\nu^* = \nu_{a_1} = (\nu_{a_1} / \nu_{a_2}) \oplus \ldots \oplus (\nu_{a_{n-1}} / \nu_{a_n}) \oplus \nu_{a_n}$ onto $\gamma^* = \gamma_{a_1} = (\gamma_{a_1} / \gamma_{a_2}) \oplus \ldots \oplus (\gamma_{a_{n-1}} / \gamma_{a_n}) \oplus \gamma_{a_n}$ such that $f(\nu) = f(\gamma)$. Hence $I_\nu = I_\gamma$ by Theorem 6.5.5. $\square$

References

1. S. C. Cheng and Z. Wang, Divisible TL-subgroups and pure TL-subgroups, *Fuzzy Sets and Systems* 78 (1996) 387-393.
2. P. S. Das, Fuzzy subgroups and level subgroups, *J. Math. Anal. Appl.* 84 (1981) 264-269.
3. J. K. Deveney and J. N. Mordeson, Inseparable extensions and primary abelian groups, *Arch. Math.* 33 (1979) 538-545.
4. L. Fuchs, *Infinite Abelian Groups*, Vol. 36 Academic Press, New York 1970.
5. W. Gu and T. Lu, The properties of fuzzy divisible groups, *Fuzzy Sets and Systems* 56 (1993) 195-198.
6. L. Y. Kulikov, On the theory of abelian groups of arbitrary cardinality, *Mat. Sbornik* 16 (1945) 129-162 (in Russian).
7. T. Lu and W. Gu, Abelian fuzzy group and its properties, *Fuzzy Sets and Systems* 64 (1994) 415-420.
8. D. S. Malik and J. N. Mordeson, Fuzzy subgroups of abelian groups, *Chinese J. Math.* (1991) 129-145.
9. D. S. Malik, J. N. Mordeson, and P. S. Nair, Fuzzy generators and fuzzy direct sums of abelian groups, *Fuzzy Sets and Systems* 50 (1992) 193-199.
10. M. A. A. Mishref, Primary fuzzy subgroups, *Fuzzy Sets and Systems* 112 (2000) 313-318.
11. J. N. Mordeson, Modular extensions and abelian groups, *Arch. Math.* 36 (1981) 13-20.
12. J. N. Mordeson, Generating properties of fuzzy algebraic structures, *Fuzzy Sets and Systems* 55 (1993) 107-120.

13. J. N. Mordeson, Invaraiants of fuzzy subgroups, *Fuzzy Sets and Systems* 63 (1994) 81-85.
14. J. N. Mordeson, Fuzzy group subalgebras, *J. Fuzzy Math.* 3 (1995) 69-81.
15. J. N. Mordeson, Fuzzy group subalgebras II, *J. Fuzzy Math.* 3 (1995) 885-897.
16. J. N. Mordeson and D, S, Malik, *Fuzzy Commutative Algebra*, World Scientific, Inc., 1998.
17. J. N. Mordeson and M. K. Sen, Basic fuzzy subgroups, *Inform. Sci.* 82 (1995) 167-179.
18. M. Petrich, *Introduction to Semigroups*, Charles E. Merrill Publishing Co., Columbus Ohio, 1973.
19. S. Sebastian and S. B. Sandar, Commutative *L*-fuzzy subgroups, *Fuzzy Sets and Systems* 68 (1994) 115-121.
20. F. I. Sidky and M. A. Mishref, Divisible and pure fuzzy subgroups, *Fuzzy Sets and Systems* 34 (1990) 377-382.
21. W. Waterhouse, The structure of insepsarable field extensions, *Trans. Amer. Math. Soc.* 211 (1975) 39-56.

7

Direct Products of Fuzzy Subgroups and Fuzzy Cyclic Subgroups

In Chapter 6, a necessary and sufficient condition for a fuzzy subgroup to be a weak direct sum of fuzzy subgroups was obtained by employing known structure theorems for Abelian groups. In [31] and [23], the problem of expressing a normal fuzzy subgroup of a direct product of groups as a fuzzy direct product of certain subgroups of a group G was examined. This is essential in order to obtain structure theorems involving fuzzy subgroups of a group G and fuzzy subgroups of it subgroups. In this chapter, we first we introduce the notion of the fuzzy direct product of fuzzy subgroups of subgroups of a group and employ these ideas to the problem in group theory of obtaining conditions under which a group G can be expressed as the direct product of its normal subgroups. We extend the definition of weak direct sum introduced in Chapter 6 for Abelian groups to the general case and establish a one-to-one correspondence between these two fuzzy direct products. Thus the relevant results of fuzzy direct products can be applied to fuzzy weak direct products.

We let G denote a group throughout this chapter. Let $\{\mu_i \mid i \in I\}$ be a collection of fuzzy subsets of G. Then the fuzzy subsets $\cap_{i \in I} \mu_i$ and $\prod_{i \in I} \mu_i$ of G are defined by $\forall x \in G, (\cap_{i \in I} \mu_i)(x) = \wedge \{\mu_i(x) \mid i \in I\}$ and $(\prod_{i \in I} \mu_i)(x) = \vee \{\wedge \{\mu_i(x_i) \mid i \in I\} \mid x = \prod_{i \in I} x_i, x_i \in G, i \in I\}$, where all but a finite number of the x_i are the identity in the product $\prod_{i \in I} x_i$. Let μ be a fuzzy subgroup of G. The chain of subgroups $\{\mu_t \mid t \in \text{Im}(\mu)\}$ is called the chain of level subgroups of μ. Recall that $\mu_* = \{x \in G \mid \mu(x) = \mu(e)\}$ is a subgroup of G. Two fuzzy subgroups of G are said to be **equivalent** if they have the same chain of level subgroups. It is easy to check that two fuzzy subgroups of G are equal if and only if they are equivalent and they have the same images. Recall that fuzzy subgroup μ of G is normal if and only if its chain of level subgroups forms a normal chain of G. A fuzzy subgroup of G is said to be **subnormal** if its chain of level subgroups forms a subnormal chain of G. It is clear that every normal fuzzy subgroup is a subnormal fuzzy subgroup. However the converse is not true, [[28], No. 355, p. 142].

7.1 Fuzzy Direct Products

We now consider fuzzy direct products. This work is mainly by Alkhamees [5]. The following definition is a slight modification of the notion of direct product of fuzzy subgroups given in Chapter 1.

Definition 7.1.1. *Let $\{G_i \mid i \in I\}$ be a collection of normal subgroups of G and μ_i a fuzzy subgroup of G_i $\forall i \in I$. Let μ be a fuzzy subgroup of G. Then μ is called the **fuzzy direct product** of the μ_i, denoted by $\mu = \prod_{i \in I}^{\otimes} \mu_i$, if the following conditions hold:*
(1) $G = \prod_{i \in I}^{\otimes} G_i$;
(2) $\mu(e) = \mu_i(e)$ $\forall i \in I$;
(3) if $x \in G$, $x \neq e$, then $\mu(x) = \wedge\{\mu_i(x_i) \mid x_i \neq e, i \in I\}$, where $x = \prod_{i \in I} x_i$ is the unique representation of x as a product of elements of G.

Example 7.1.2. *Let $G = \{e, a, b, c\}$ be the Klein 4-group, where e is the identity of G. Let $H = \langle a \rangle$ and $K = \langle b \rangle$. Let μ, ν, ρ be the fuzzy subgroups of G, H, K, respectively, defined as follows:*

$$\mu(b) = \mu(c) = s_0, \mu(a) = s_1, \mu(e) = s_2,$$

$$\nu(a) = s_1, \nu(e) = s_2,$$

$$\rho(b) = s_0, \rho(e) = s_2,$$

where $0 \leq s_0 < s_1 < s_2 \leq 1$. Clearly, $G = H \otimes K$. Now $\mu(a) = \nu(a), \mu(b) = \rho(b)$, and $\mu(c) = \nu(a) \wedge \rho(b)$. Thus $\mu = \nu \otimes \rho$.

Theorem 7.1.3. *Let $\{G_i \mid i \in I\}$ be a collection of subgroups of G and let μ_i be a fuzzy subgroup of G_i $\forall i \in I$. A fuzzy subgroup μ of G is the fuzzy direct product of the μ_i, $i \in I$, if and only if the following conditions hold:*
(1) μ_i is the restriction of μ to G_i $\forall i \in I$;
(2) $\mu_s = \prod_{i \in I}^{\otimes} (\mu_i)_s$ $\forall s \in \text{Im}(\mu)$.

Proof. Assume conditions (1) and (2) hold. Let $s^* = \wedge\text{Im}(\mu)$ and $s_i = \wedge\text{Im}(\mu_i)$ $\forall i \in I$. Then clearly $G = \mu_{s^*} = \prod_{i \in I}^{\otimes} (\mu_i)_{s_i} = \prod_{i \in I}^{\otimes} G_i$, where the second equality follows from the fact that $s_i \geq s^*$ and so $(\mu_i)_{s^*} = (\mu_i)_{s_i} \forall i \in I$. Let $s \in \text{Im}(\mu)$ and $z \in G$ be such that $\mu(z) = s$. Let $z = z_{k_1} z_{k_2} ... z_{k_n}$ be the unique representation of z as a product of elements of G_i, where $z_{k_i} \neq e$ for all $i = 1, ..., n$. Since $z \in \mu_s$ and $\mu_s = \prod_{i \in I}^{\otimes} (\mu_i)_s, z_{k_i} \in (\mu_{k_i})_s, i = 1, ..., n$. Thus $\mu_{k_i}(z_{k_i}) \geq s$ for $i = 1, ..., n$. Let $t = \wedge\{\mu_{k_i}(z_{k_i}) \mid i = 1, ..., n\}$ and $J = \{i \mid \mu_{k_i}(z_{k_i}) = t, i = 1, ..., n\}$. If $J = \{1, ..., n\}$, let $k = z_{k_i}$ and $h = z z_{k_i}^{-1}$. Then $\mu(k) \leq \mu(z) \leq \mu(zh^{-1}) = \mu(k)$. Thus $s = \mu(z) = \mu(k) = t$. If $J \neq \{1, ..., n\}$, then clearly z can be expressed as $z = hk$, where $h = \prod_{i \in I \setminus J} z_{k_i}$ and $k = \prod_{j \in J} z_{k_j}$.

We next show that $\mu(z) = \mu(k)$. If $\mu(hk) \geq \mu(h)$, then $\mu(k) = \mu((h^{-1}h)k) \geq \mu(h) \wedge \mu(hk) \geq \mu(h)$.

$$\mu(k) = \mu((h^{-1}h)k) \geq \mu(h) \wedge \mu(hk) \geq \mu(h).$$

Thus $\mu(k) = \mu(h)$, a contradiction. Hence $\mu(hk) < \mu(h)$. Now $\mu(hk) \geq \mu(h) \wedge \mu(k)$ and so $\mu(hk) \geq \mu(k)$. Also

$$\mu(k) = \mu((h^{-1}h)k) \geq \mu(h) \wedge \mu(hk) = \mu(hk).$$

Hence we have that $\mu(z) = \mu(hk) = \mu(k) = t$. Therefore, if $z \in G$, $z \neq e$, then $\mu(z) = \wedge\{\mu_i(z_i) \mid z_i \neq e, i \in I\}$,

$$\mu(z) = \wedge\{\mu_i(z_i) \mid z_i \neq e, i \in I\},$$

where $z = \prod_{i \in I} z_i$ is the unique representation of z as a product of elements of G_i. Thus μ is a fuzzy direct product of the μ_i.

Conversely, suppose μ is a fuzzy direct product of the μ_i. Then clearly μ_i is the restriction of μ to G_i $\forall i \in I$. Also, by the definition of fuzzy direct product, it follows easily that $\mu_s = \prod_{i \in I}^{\otimes} (\mu_i)_s$ $\forall s \in \text{Im}(\mu)$. $\square$

Corollary 7.1.4. *Let μ be a fuzzy direct product of $\mu_i, i \in I$. Then $\text{Im}(\mu) = \cup_{i \in I} \text{Im}(\mu_i)$.*

Definition 7.1.5. *Let μ be a fuzzy subgroup of G and let $\{\mu_i \mid i \in I\}$ be a collection of fuzzy subgroups of G. If μ is a fuzzy direct product of the μ_i, we use the notation $P_I^{\mu} = \{\mu_i \mid i \in I\}$. Let ν be a fuzzy subgroup of G and let $\{\nu_i \mid i \in I\}$ be a collection of fuzzy subgroups of G such that ν is a fuzzy direct product of the ν_i. Then we say that P_I^{μ} is **equivalent** to P_I^{ν} if the following conditions are satisfied:*

(1) μ_i is equivalent to ν_i $\forall i \in I$.

(2) $\forall s \in \text{Im}(\mu), (\mu_i)_s = (\nu_i)_t$ $\forall i \in I$, where $t = \wedge\{f_i(s) \mid s \in \text{Im}(\mu_i), i \in I\}$ and f_i is a bijection of $\text{Im}(\mu_i)$ onto $\text{Im}(\nu_i)$ $\forall i \in I$.

Example 7.1.6. *Let G, H, K and μ, ν, ρ be defined as in Example 7.1.2. Let μ', ν', ρ' be the fuzzy subgroups of G, H, K, respectively, defined as follows:*

$$\mu'(a) = \mu'(ab) = s_0, \mu'(b) = s_1, \mu'(e) = s_2$$

$$\nu'(a) = s_0, \nu'(e) = s_2$$

$$\nu'(a) = s_0, \nu'(e) = s_2$$

$$\rho'(b) = s_1, \rho'(e) = s_2,$$

where $0 \leq s_0 < s_1 < s_2 \leq 1$. Then clearly $\mu = \nu \otimes \rho$, $\mu' = \nu' \otimes \rho'$, ν is equivalent to ν' and ρ is equivalent to ρ'. However, μ is not equivalent to μ' since their level sets differ.

Theorem 7.1.7. *Suppose that μ and ν are fuzzy subgroups of G such that μ and ν are fuzzy direct products of the collections of fuzzy subgroups $\{\mu_i \mid i \in I\}$ and $\{\nu_i \mid i \in I\}$, repectively. Then μ is equal (equivalent) to ν if and only if P_I^{μ} is equal (equivalent) to P_I^{ν}.*

Proof. Suppose μ is equivalent to ν. Let f be the bijection of $\text{Im}(\mu)$ onto $\text{Im}(\nu)$ such that $\mu_s = \nu_{f(s)}$ for all $s \in \text{Im}(\mu)$. Let f_i denote the restriction of f to $\text{Im}(\mu_i)$. Suppose $s \in \text{Im}(\mu_j)$, where $j \in I$. Then $s \in \text{Im}(\mu)$. By Theorem 7.1.3,

$$\mu_s = \prod_{i \in I}^{\otimes} (\mu_i)_s, \quad \nu_{f(s)} = \prod_{i \in I}^{\otimes} (\nu_i)_{f(s)}.$$

Since $\mu_s = \nu_{f(s)}$,

$$\prod_{i \in I}^{\otimes} (\mu_i)_s = \prod_{i \in I}^{\otimes} (\nu_i)_{f(s)}.$$

Since $\mu_s \cap G_j = (\mu_j)_s$ and $\mu_s = \nu_{f(s)}$ and so $\mu_s \cap G_j = \nu_{f(s)} \cap G_j = (\nu_j)_{f(s)}$, we have $(\mu_j)_s = (\nu_j)_{f(s)}$ $\forall s \in \text{Im}(\mu_j)$. Hence μ_j is equivalent to ν_j $\forall j \in I$. The other condition of Definition 7.1.5 can easily be arrived at from the fact that $\prod_{i \in I}^{\otimes}(\mu_i)_s = \mu_s = \nu_{f(s)} = \prod_{i \in I}^{\otimes}(\nu_i)_{f(s)}$. Therefore, P_I^μ is equivalent to P_I^ν. Conversely, suppose P_I^μ is equivalent to P_I^ν. Let f_i be the bijection of $\text{Im}(\mu_i)$ onto $\text{Im}(\nu_i)$ such that $(\mu_i)_s = (\nu_i)_{f_i(s)}$ for all $i \in I$. Let f be the function from $\text{Im}(\mu)$ into $\text{Im}(\nu)$ defined as follows:

$$\forall s \in \text{Im}(\mu), f(s) = \wedge\{f_i(s) | s \in \text{Im}(\mu_i), i \in I\}.$$

Then $\forall s \in \text{Im}(\mu)$, $(\mu_i)_s = (\mu_i)_{s_i}$ for some $s_i \in \text{Im}(\mu_i)$ and hence $(\mu_i)_s = (\nu_i)_{f_i(s_i)}$ since μ_i is equivalent to ν_i. By condition (2) of Definition 7.1.5, we have $(\mu_i)_s = (\nu_i)_{f(s)}$. Therefore, $(\nu_i)_{f(s)} = (\nu_i)_{f_i(s_i)}$ $\forall i \in I$. Also, $\forall s \in \text{Im}(\mu)$,

$$\mu_s = \prod_{i \in I}^{\otimes} (\mu_i)_s = \prod^{\otimes} i \in I(\mu_i)_{s_i}$$

$$= \prod_{i \in I}^{\otimes} (\nu_i)_{f_i(s_i)} = \prod^{\otimes} i \in I(\nu_i)_{f(s)}$$

$$= \nu_{f(s)}.$$

Hence if $f(s) = f(t)$, then $\nu_{f(s)} = \nu_{f(t)}$. Since $\mu_s = \nu_{f(s)}$ and $\mu_t = \nu_{f(t)}, \mu_s = \mu_t$. Thus $s = t$. Therefore, f is one-one. Let $t \in \text{Im}(\nu)$. Then $t \in \text{Im}(\nu_i)$ for some $i \in I$. Suppose $s = \vee\{s_i \in \text{Im}(\mu_i) \mid f_i(s_i) = t, i \in I\}$. Then clearly $f(s) = t$ and so f is bijective. Since $\mu_s = \nu_{f(s)}$ $\forall s \in \text{Im}(\mu)$, μ is equivalent to ν. □

Lemma 7.1.8. *Let μ be a fuzzy subgroup of G. Suppose that μ is subnormal (normal) and that μ is a fuzzy direct product of fuzzy subgroups $\mu_i, i \in I$. Then μ_i is subnormal (normal) $\forall i \in I$.*

Proof. Suppose μ is subnormal (normal). Let $s, t \in \text{Im}(\mu_j)$ be such that $s < t$ and such that there is no $k \in \text{Im}(\mu_j)$ such that $s < k < t$. Then $s, t \in \text{Im}(\mu)$. Suppose $s' \in \text{Im}(\mu)$ is such that $s < s'$ and such that there is no $k' \in \text{Im}(\mu)$ such that $s < k' < s'$. This implies $s' \leq t$. Further,

$$\mu_s = \prod_{i \in I}^{\otimes} (\mu_i)_s$$

$$\mu_{s'} = \prod_{i \in I}^{\otimes} (\mu_i)_{s'}$$

$$= \prod_{i \in I, i \neq j}^{\otimes} (\mu_i)_{s'} \otimes (\mu_j)_t.$$

Since μ is subnormal (normal), we have $\mu_{s'} \lhd \mu_s (\mu_{s'} \lhd G)$. Thus $(\mu_j)_t \lhd (\mu_j)_s ((\mu_j)_t \lhd G)$. Therefore, μ_j is subnormal (normal). □

We next consider direct products of fuzzy subgroups of composition (principal) length.

Definition 7.1.9. *Let μ be a fuzzy subgroup of G. If $\mathrm{Im}(\mu)$ is finite, let $l(\mu) = |\mathrm{Im}(\mu)| - 1$. Then $l(\mu)$ is called the **length** of μ.*

Definition 7.1.10. *A normal fuzzy subgroup μ of G is called a fuzzy subgroup of **principal length** if the chain of level subgroups of μ forms a principal (chief) series.*

Definition 7.1.11. *A subnormal fuzzy subgroup μ of G is called a fuzzy subgroup of **composition length** if the chain of level subgroups of μ forms a composition series of G.*

It is known that the composition length and the principal length are unique, [[28], pp. 121-126 and p.142]. Furthermore, if the composition length exists, it is maximal of all lengths of subnormal fuzzy subgroups of G and if the principal length exists, it is maximal of all lengths of normal fuzzy subgroups of G.

It is known that a group has a composition (principal) series if and only if it satisfies the subnormal (normal) chain conditions. By this result and by the fact that all subnormal (normal) chains of such groups have equivalent composition (principal) series as refinements [[28], Theorem 7.9 p. 125 and p. 142], the following theorem holds.

Theorem 7.1.12. *A group G has a composition (principal) series if and only if every subnormal (normal) fuzzy subgroup of G is of finite length.*

We need the following result of group theory [[28], Theorem 7.36, p. 142]. Recall that a principal factor of a group G is a factor group H/K where H and K are consecutive terms of a principal series of G.

Theorem 7.1.13. *Suppose that G has principal series. Let H, K be normal subgroups of G with $K \subset H$. Then H/K is a principal factor of G if and only if H/K is a minimal normal subgroup of G/K.*

Proposition 7.1.14. *Let H, K be normal subgroups of G such that $H \cap K = \{e\}$. Let N be a normal subgroup of H. Then $N \otimes K \lhd H \otimes K$.*

Proof. Let $hk \in H \otimes K$ and $h'k' \in N \otimes K$. Since $K \lhd G$,

$$hk(h'k')(hk)^{-1} = hkh'k'k^{-1}h^{-1} = hh'k_1k'k^{-1}h^{-1} = hh'h^{-1}k_2 \in N \otimes K,$$

where $k_1, k_2 \in K$.   □

Theorem 7.1.15. *Suppose that G has a composition series and μ is a subnormal fuzzy subgroup of G such that μ is a fuzzy direct product of fuzzy subgroups μ_i of subgroups G_i of G, $i = 1, ..., n$. Then μ is of composition length if and only if μ_i is of composition length and $\mathrm{Im}(\mu_i) \cap \mathrm{Im}(\mu_j) = \{\mu(e)\}$ $\forall i \ne j, i, j = 1, ..., n$.*

Proof. We prove the theorem by induction on n. It suffices to prove it for $n = 2$ and then repeat the same argument by taking $H = \prod_{i \in I}^{\otimes} G_i$ ($I = \{1, 2, ..., n-1\}$) and $K = G_n$. Thus suppose that H and K are subgroups of G and that ν and ρ are fuzzy subgroups of H and K, respectively, such that $\mu = \nu \otimes \rho$. Assume ν and ρ are of composition length and $\mathrm{Im}(\nu) \cap \mathrm{Im}(\rho) = \{\mu(e)\}$. Suppose μ is not of composition length. Then the chain of level subgroups of μ can be refined to obtain another subnormal chain. Hence there exists a a subgroup N of G such that $\mu_t \lhd N \lhd \mu_s$, where $s, t \in \mathrm{Im}(\mu)$ with $s < t$ and there is no r in $\mathrm{Im}(\mu)$ with $s < r < t$. Since $\mathrm{Im}(\nu) \cap \mathrm{Im}(\rho) = \{\mu(e)\}, \nu(e) = \mu(e) = \rho(e)$. Since $H \cap K = \{e\}$, it is clear that the only common element in the domains of ν and ρ is e. Let $x \in N \backslash \mu_t$. Then $s \le \mu(x) < t$ and so $\mu(x) = s$. Suppose neither ν nor ρ are defined on x. Then $x = hk$, where $h \in \nu_s \subseteq H$ and $k \in \rho_s \subseteq K$. Now $s = \mu(x) = \nu(h) \wedge \rho(k)$ and $\nu(h) \ne \rho(k)$. Say $s = \mu(x) = \nu(h) < \rho(k)$. Since there is no r in $\mathrm{Im}(\mu)$ such that $s < r < t$, there is no such r in $\mathrm{Im}(\rho)$ by Corollary 7.1.4. Thus $k \in \rho_t \subseteq \mu_t \subseteq N$ and so $k \in K \cap N$. Since $x \in N, h \in N$. Thus $h \in H \cap N$. Suppose ν is defined on x. Then ρ is not defined on x. Hence $\mu(x) = s = \nu(x)$. Thus $x \in \nu_s \cap N \subseteq H \cap N$. Suppose $x \in \mu_t$. Then $x = hk$, where $h \in \nu_t \subseteq H$ and $k \in \rho_t \subseteq K$. Since $\nu_t, \rho_t \subseteq \mu_t \subseteq N, h \in H \cap N$ and $k \in K \cap H$. Hence $N = (H \cap N) \otimes (H \cap N)$. Let $N' = H \cap N$ and $N'' = K \cap N$. Also, $\mu_s = \nu_s \otimes \rho_s, \mu_t = \nu_t \otimes \rho_t$ and $\mathrm{Im}(\nu) \cap \mathrm{Im}(\rho) = \{\mu(e)\}$ imply that either $\nu_t \ne N'$ or $\rho_t \ne N''$. Suppose $\nu_t \ne N'$. Define the fuzzy subset ν' of H as follows: $\forall x \in H$,
$$\nu'(x) = \begin{cases} r' & x \in N' \backslash \nu_t \\ \nu(x) & \text{otherwise,} \end{cases}$$

where $r' \in [0, 1]$ with $t' < r' < s'$ and $s', t' \in \mathrm{Im}(\nu)$ such that $\nu_{s'} = \nu_s$ and $\nu_{t'} = \nu_t$. Then it follows that ν' is subnormal fuzzy subgroup of H. However, this contradicts the assumption that ν is of composition length since $l(\nu') > l(\nu)$. Therefore, μ is of composition length.

Conversely, assume that μ is of composition length, $s, t \in \mathrm{Im}(\nu)$ with $s < t$ and there is no r in $\mathrm{Im}(\nu)$ with $s < r < t$. Also, let $s' \in \mathrm{Im}(\mu)$ with $s < s'$

and suppose there are no r' in $\mathrm{Im}(\mu)$ such that $s < r' < s'$. Then $\mu_s = \nu_s \otimes \rho_s$ and $\mu_{s'} = \nu_{s'} \otimes \rho_{s'}$. If $\rho_s \neq \rho_{s'}$, then using the fact that ρ is a subnormal fuzzy subgroup of K (Lemma 7.1.8) and Proposition 7.1.14, it follows easily that $\nu_s \otimes \rho_{s'}$ is a proper normal subgroup of μ_s. Since $\mu_{s'}$ is a proper normal subgroup of $\nu_s \otimes \rho_{s'}$, $(\nu_s \otimes \rho_{s'})/\mu_{s'}$ is a proper normal subgroup of $\mu_s/\mu_{s'}$. This contradicts the fact $\mu_s/\mu_{s'} \cong \nu_s/\nu_t$ is simple since μ is of composition length. Therefore, $\rho_s = \rho_{s'}$ and hence $\mu_s/\mu_{s'} \cong \nu_s/\nu_t$. Thus ν_s/ν_t is a simple group and hence ν is of composition length. Similarly, it can be shown that ρ is of composition length. Finally, let $s' \in \mathrm{Im}(\nu) \cap \mathrm{Im}(\rho), s' \neq \mu(e)$, and $s \in \mathrm{Im}(\mu)$ with $s < s'$. Suppose there is no r' in $\mathrm{Im}(\mu)$ such that $s < r' < s'$. Then $(\nu_s \otimes \rho_{s'})/\mu_{s'}$ is a nontrivial normal subgroup of $\mu_s/\mu_{s'}$. This contradicts the fact that $\mu_s/\mu_{s'}$ is a simple group. Therefore, $\mathrm{Im}(\nu) \cap \mathrm{Im}(\rho) = \{\mu(e)\}$.   □

Theorem 7.1.16. *Suppose that G has a principal series and μ is a normal fuzzy subgroup of G such that μ is a fuzzy direct product of fuzzy subgroups μ_i of subgroups G_i of G, $i = 1, \cdots, n$. Then μ is of principal length if and only if μ_i is of principal length and $\mathrm{Im}(\mu_i) \cap \mathrm{Im}(\mu_j) = \{\mu(e)\} \; \forall i \neq j, \; i, j = 1, \ldots, n$.*

Proof. As in the previous theorem, it suffices to prove the theorem for $n = 2$. Suppose H, K are subgroups of G and that ν and ρ are fuzzy subgroups of H and K, respectively. If ν and ρ are of principal length, then by a similar argument as used in the proof of the first part of the previous theorem, it can be shown that μ is of principal length.

Conversely, assume that μ is of principal length. Let $s, t \in \mathrm{Im}(\nu)$ with $s < t$ and such that there is no $r \in \mathrm{Im}(\nu)$ such that $s < r < t$. Let $s' \in \mathrm{Im}(\mu)$. Then $\mu_s = \nu_s \otimes \rho_s$ and $\mu_{s'} = \nu_{s'} \otimes \rho_{s'} = \nu_t \otimes \rho_{s'}$. Suppose $\rho_s \neq \rho_{s'}$. Since $\nu_s = H \cap \mu_s$ and $\mu_s, H \lhd G, \nu_s \lhd G$ and since ρ is a normal fuzzy subgroup of K by Lemma 7.1.8, $\rho_{s'} \lhd K$. Thus by using a similar method to that used in Proposition 7.1.14, it follows that $\nu_s \otimes \rho_{s'} \lhd G$. Hence $(\nu_s \otimes \rho_s)/\mu_{s'}$ is a normal subgroup of $G/\mu_{s'}$. However, since $(\nu_s \otimes \rho_{s'})/\mu_{s'}$ is a proper subgroup of $\mu_s/\mu_{s'}$, we have a contradiction of the fact that $\mu_s/\mu_{s'}$ is a minimal normal subgroup of $G/\mu_{s'}$ by Theorem 7.1.13 since μ is of principal length. Therefore, $\rho_s = \rho_{s'}$ and so $\mu_s/\mu_{s'} \cong \nu_s/\nu_t$. Since $(H \otimes \rho_{s'})/\mu_{s'}$ is a subgroup of $G/\mu_{s'}$ and $\mu_s/\mu_{s'}$ is a subgroup of $G/\mu_{s'}$, $\mu_s/\mu_{s'}$ is a minimal normal subgroup of $(H \otimes \rho_{s'})/\mu_{s'} \cong H/\nu_t$ and $\mu_s/\mu_{s'} \cong \nu_s/\nu_t$. Thus it follows that ν_s/ν_t is a minimal normal subgroup of H/ν_t. By Theorem 7.1.13, it follows that ν is fuzzy subgroup of H of principal length. Similarly, it follows that ρ is fuzzy subgroup of K of principal length. Finally, let $s' \in \mathrm{Im}(\nu) \cap \mathrm{Im}(\rho), s' \neq \mu(e)$, and $s \in \mathrm{Im}(\mu)$ with $s < s'$. Suppose there is no r' in $\mathrm{Im}(\mu)$ such that $s < r' < s'$. Then it follows as above that $(\nu_s \otimes \nu_{s'})/\mu_{s'}$ is a non-trivial normal subgroup of $\mu_s/\mu_{s'}$. This contradicts the fact that $\mu_s/\mu_{s'}$ is a minimal normal subgroup of $G/\mu_{s'}$. Therefore, $\mathrm{Im}(\nu) \cap \mathrm{Im}(\rho) = \{\mu(e)\}$.   □

We next obtain necessary and sufficient conditions for a group G to be a direct product of its normal subgroups.

Theorem 7.1.17. *Suppose that G has a principal series. Then G is a direct product of its normal subgroups $G_1, \ldots, G_n$ if and only if there exists a fuzzy subgroup μ of G of principal length such that $l(\mu) = \sum^n i = 1 l(\mu_i)$ and $\mathrm{Im}(\mu_i) \cap \mathrm{Im}(\mu_j) = \{\mu(e)\}, i \neq j$, where μ_i is the restriction of μ to G_i, $i, j = 1, \ldots, n$.*

Proof. We prove the theorem by induction on n. Suppose H and K are normal subgroups of G and μ is a fuzzy subgroup of G of principal length such that $l(\mu) = l(\nu) + l(\rho)$ and $\mathrm{Im}(\nu) \cap \mathrm{Im}(\rho) = \{\mu(e)\}$, where ν, ρ are the restrictions of μ to H, K, respectively. Let $x \in H \cap K$. Then $\mu(x) = \nu(x) = \rho(x)$. Hence $\mu(x) = \mu(e)$ and so $x \in \mu_*$. However, $\mu_* = \{e\}$ since μ is of principal length. Thus $H \cap K = \{e\}$ and so $H \otimes K$ is a normal subgroup of G. Since $l(\mu) = l(\nu) + l(\rho), |(\mathrm{Im}(\nu)\backslash\{\mu(e)\}) \cup (\mathrm{Im}(\rho)\backslash\{\mu(e)\})| = |\mathrm{Im}(\mu)\backslash\{\mu(e)\}|$. Suppose $\mathrm{Im}(\mu) = \{s_0, s_1, \ldots, s_n\}, \mu_{s_i} = \mu_i, \nu_{s_i} = \nu_i$ and $\rho_{s_i} = \rho_i$, where $0 \leq s_0 < s_1 < \ldots < s_n \leq 1$. Define the fuzzy subset δ of $H \otimes K$ as follows: $\forall x \in H \otimes K$,

$$\delta(x) = s_i \ if \ x \in \mu_i \backslash \mu_{i+1}, i = 0, 1, \ldots, n.$$

Then since $\mu_i \lhd G \ \forall i = 1, 2, \ldots, n$, it follows that δ is a normal fuzzy subgroup of $H \otimes C$ with $\delta_{s_i} = \nu_i \otimes \rho_i \ \forall i = 1, 2, \ldots, n$. However, clearly the length of δ is equal to the length of μ. Therefore, $G = H \otimes K$ and $\delta = \mu$, otherwise G has a normal fuzzy subgroup of length greater than the length of μ which is impossible by the assumption that μ is of principal length.

Conversely, suppose that $G = H \otimes K$, and

$$H = H_0 \rhd H_1 \rhd \ldots \rhd H_h = \{e\}$$

$$K = K_0 \rhd K_1 \rhd \ldots \rhd K_k = \{e\}$$

are principal series of H and K, respectively. By a similar argument as used in the proof of Proposition 7.1.14, it follows that $H \otimes K_i \lhd G \ \forall i = 1, \ldots, k$. Clearly, $H_j \lhd G \ \forall j = 1, \ldots, h$. Thus

$$G = (H \otimes K_0) \rhd (H \otimes K_1) \rhd \ldots \rhd (H \otimes K_{k-1}) \rhd H_0 \rhd H_1 \rhd \ldots \rhd H_h = \{e\}$$

is a normal series of G. By Theorem 7.1.13, we have that K_i/K_{i+1} is a minimal normal subgroup of K/K_{i+1}. However, since $(H \otimes K_i)/(H \otimes K_{i+1})$ is isomorphic to K_i/K_{i+1} and $(H \otimes K)/(H \otimes K_{i+1})$ is isomorphic to K/K_{i+1}, we have that $(H \otimes K_i)/(H \otimes K_{i+1})$ is a minimal normal subgroup of $(H \otimes K)/(H \otimes K_{i+1})$ and by Theorem 7.1.13 again, we have that it is a principal factor. Hence the $H \otimes K_i$ are terms of the principal series of G, $i = 0, 1, \ldots, k - 1$. Let μ be a fuzzy subgroup of G which has the above series as its chain of level subgroups. Suppose

$$\mathrm{Im}(\mu) = \{s_0, s_1, \ldots, s_k, s_{k+1}, \ldots, s_{k+h}\},$$

where $0 \leq s_0 < s_1 < \ldots < s_{k+h} \leq 1$. Set $\nu = \mu|_H$ and $\rho = \mu|_K$. Then clearly μ is of principal length, $l(\mu) = l(\nu) + l(\rho)$ and $\mathrm{Im}(\nu) \cap \mathrm{Im}(\rho) = \{\mu(e)\}$. $\square$

The following theorem can be proved in a similar manner as the previous theorem.

Theorem 7.1.18. *Suppose that G has a composition series. Then G is a direct product of its normal subgroups $G_1, ..., G_n$ if and only if there exists a fuzzy subgroup μ of G of composition length such that $l(\mu) = \sum_{i=1}^{n} l(\mu_i)$ and $\mathrm{Im}(\mu_i) \cap \mathrm{Im}(\mu_j) = \{\mu(e)\}$ $\forall i \neq j$, where μ_i are the restrictions of μ to G_i, $i = 1, ..., n$.*

Theorem 7.1.19. *Suppose G has a composition (principal) series and μ is a fuzzy subgroup of G of composition (principal) length. Then μ is fuzzy direct product of its restrictions $\mu_1, ..., \mu_n$ to the normal subgroups $G_1, ..., G_n$ of G if and only if $l(\mu) = \sum_{i=1}^{n} l(\mu_i)$ and $\mathrm{Im}(\mu_i) \cap \mathrm{Im}(\mu_j) = \{\mu(e)\}$ $\forall i \neq j, i, j = 1, \ldots, n$.*

Proof. It is sufficient to prove the theorem for $n = 2$. Let H and K be subgroups of G and let ν and ρ be fuzzy subgroups of H and K, respectively. Assume that $\mu = \nu \otimes \rho$, $s \in \mathrm{Im}(\nu) \cap \mathrm{Im}(\rho), s \neq \mu(e)$, and s', t, r are elements of $\mathrm{Im}(\mu), \mathrm{Im}(\nu), \mathrm{Im}(\rho)$, respectively, with $s', t, r < s$. Assume also that there are no k_1, k_2, k_3 in $\mathrm{Im}(\mu)$, $\mathrm{Im}(\nu), \mathrm{Im}(\rho)$, respectively, such that $s < k_1 < s', s < k_2 < t$ and $s < k_3 < r$. Then as in the proof of Theorem 7.1.15, we obtain that if μ is of composition length or as in the proof of Theorem 7.1.16 that if μ is of principal length, $\nu_s = \nu_t$ or $\rho_s = \rho_r$. However, this contradicts the above assumption. Therefore, $\mathrm{Im}(\nu) \cap \mathrm{Im}(\rho) = \{\mu(e)\}$. Since $\mu = \nu \otimes \rho$, $\mathrm{Im}(\mu) = \mathrm{Im}(\nu) \cup \mathrm{Im}(\rho)$. However, $\mathrm{Im}(\nu) \cap \mathrm{Im}(\rho) = \{\mu(e)\}$. Therefore, $l(\mu) = l(\nu) + l(\rho)$.

The converse follows as in the proof of Theorem 7.1.17 in the case that μ is of principal length. A similar technique for the case that μ is of composition length can be used. $\square$

We now consider the correspondence between fuzzy direct products and fuzzy weak direct products.

Definition 7.1.20. *Let $\{\nu_i \mid i \in I\}$ be a collection of subnormal fuzzy subgroups of G. Then a subnormal fuzzy subgroup ν of G is said to be a **fuzzy weak direct product** of the ν_i if the following conditions hold:*
  (1) $\nu = \prod_{i \in I} \nu_i$,
  (2) $(\nu_j \cap (\prod_{i \in I, i \neq j} \nu_i))(x) = 0$ $\forall x \in G, x \neq e$.

Definition 7.1.20 is a generalization of [[25], Definition 4.1, p. 140] to the noncommutative case. Using a similar technique as in the proof of [[25], Theorem 4.6, p. 142] and [[25], Corollary 4.7, p. 142], this corollary can be generalized to obtain the following result.

Theorem 7.1.21. *Let ν be a subnormal fuzzy subgroup of G and $\{\nu_i \mid i \in I\}$ a collection of subnormal fuzzy subgroups of G such that $\nu_i \subseteq \nu$ $\forall i \in I$. Then ν is a fuzzy weak direct product of ν_i if and only if $\nu_i(e) = \nu(e)$ $\forall i \in I$ and*

$$\nu_s = \prod_{i \in I}^{\otimes} (\nu_i)_s \ \forall s \in (0, \nu(e)].$$

Suppose that $\{G_i \mid i \in I\}$ is a collection of normal subgroups of G and that $G = \prod_{i \in I}^{\otimes} G_i$. Suppose μ and ν are subnormal fuzzy subgroups of G. Let P_I^μ be defined as previously. Let W_I^ν denote the set $\{\nu_i \mid i \in I\}$, where the ν_i are subnormal fuzzy subgroups of G such that ν is a fuzzy weak direct product of the ν_i, $(\nu_i)^*$ is a subgroup of G_i and $\nu_i \subseteq \nu \ \forall i \in I$. Now $\forall i \in I$, define the fuzzy subset μ_i' of G by $\forall x \in G$, $\mu_i'(x) = \mu_i(x)$ if $x \in G_i$ and $\mu_i(x) = 0$ otherwise. Let $(P_I^\mu)'$ denote the set of all the μ_i' and let $(W_I^\nu)'$ denote the set of all the ν_i' such that ν_i' is the restriction of ν_i to $G_i, i \in I$. Let $\mathcal{A} = \{P_I^\mu \mid \mu \in \mathcal{F}(G) \text{ and } \mu \text{ is subnormal}\}$, $\mathcal{B} = \{W_I^\nu \mid \nu \in \mathcal{F}(G) \text{ and } \nu \text{ is subnormal}\}$, $\mathcal{A}' = \{(P_I^\mu)' \mid \mu \in \mathcal{F}(G) \text{ and } \mu \text{ is subnormal}\}$, and $\mathcal{B}' = \{(W_I^\nu)' \mid \nu \in \mathcal{F}(G) \text{ and } \nu \text{ is subnormal}\}$.

The following results show how a fuzzy weak direct product can be produced from a fuzzy direct product and vice versa and how through this process a one-to-one correspondence between $\mathcal{A}$ and $\mathcal{B}$ can be established. Hence the relevant results of this section can be applied to W_I^ν.

Proposition 7.1.22. *Let $P_I^\mu \in \mathcal{A}$ and $W_I^\nu \in \mathcal{B}$. Then $(P_I^\mu)' \in \mathcal{B}, W_I^\nu \in \mathcal{A}, ((P_I^\mu)')' = P_I^\mu$ and $((W_I^\nu)')' = W_I^\nu$.*

Proof. By Lemma 7.1.8, μ_i is a subnormal fuzzy subgroup of $G_i \ \forall i \in I$. Since $(\mu_i')_s = (\mu_i)_s \ \forall s \in (0, \mu(e)]$, it follows easily that μ_i' is a subnormal fuzzy subgroup of $G \ \forall i \in I$. By Theorem 7.1.3, the μ_i are the restrictions of μ to G_i. Also, $\mu_i'(x) = \mu(x) \ \forall x \in G_i$ and so $\mu_i' \subseteq \mu \ \forall i \in I$. By Theorem 7.1.3,

$$\mu_s = \prod_{i \in I}^{\otimes} (\mu_i)_s = \prod_{i \in I}^{\otimes} (\mu_i')_s \ \forall s \in (0, \mu(e)].$$

Thus by Theorem 7.1.21, we have that $(P_I^\mu)' \in \mathcal{B}$. Clearly, μ_i' is a subnormal fuzzy subgroup of G_i and $(\mu_i')' = \mu_i \ \forall i \in I$. Therefore, $((P_I^\mu)')' = P_I^\mu$.

Since the ν_i' are the restrictions of ν_i to G_i and $(\nu_i')s = (\nu_i)s \ \forall s \in (0, \nu(e)]$, clearly the ν_i' are fuzzy subnormal subgroups of G_i. Also, by Theorem 7.1.21, $\nu_i(e) = \nu(e) \ \forall i \in I$. Hence $\nu_i'(e) = \nu(e) \ \forall i \in I$. Let $x \in G_i, x \neq e$. By Theorem 7.1.21, $\nu_s = \prod_{i \in I}^{\otimes} (\nu_i)_s \ \forall \ s \in (0, \nu(e)]$ and so it follows that $\nu^* = \prod_{i \in I}^{\otimes} (\nu_i)^*$ and if $\nu_i'(x) = 0$, then $\nu_i'(x) = \nu_i(x) = \nu(x) \ \forall i \in I$. If $\nu_i'(x) > 0$, then since $\nu_s = \prod_{i \in I}^{\otimes} (\nu_i)_s \ \forall \ s \in (0, \nu(e)]$, it follows that $\nu_i'(x) = \nu_i(x) = \nu(x) \ \forall i \in I$. Therefore, $\nu_i'(x) = \nu(x) \ \forall x \in G_i$. Hence the ν_i' are the restrictions of ν to G_i. Since $\nu^* = \prod_{i \in I}^{\otimes} (\nu_i)^*$ and $\nu_s = \prod^{\otimes} i \in I(\nu_i)_s \ \forall s \in (0, \nu(e)]$, we have that $\nu_s = \prod_{i \in I}^{\otimes} (\nu_i')_s \ \forall s \in \text{Im}(\nu)$. Therefore, $W_I^\nu \in \mathcal{A}$ by Theorem 7.1.3. Finally, since the (ν_i') are subnormal fuzzy subgroups of G and $(\nu_i')' = \nu_i$, it follows that $((W_I^\mu)')' = W_I^\mu$. $\qquad \square$

We now have the following theorem.

Theorem 7.1.23. *Define ϕ from $\mathcal{A}$ into $\mathcal{B}$ and ψ from $\mathcal{B}$ into $\mathcal{A}$ as follows:*

$$\forall P_I^\mu \in \mathcal{A}, \phi(P_I^\mu) = (P_I^\mu)' \text{ and } \forall W_I^\nu \in \mathcal{B}, \ \psi(W_I^\mu) = (W_I^\mu)'.$$

$\forall P_I^\mu \in \mathcal{A}, \phi(P_I^\mu) = (P_I^\mu)'$ *and* $\forall \ W_I^\mu \in \mathcal{B}, \ \psi(W_I^\mu) = (W_I^\mu)'.$ *Then ϕ and ψ are one-one correspondences such that $\phi \circ \psi$ is the identity of $\mathcal{B}$ and $\phi \circ \psi$ is the identity of $\mathcal{A}$.*

7.2 Fuzzy p-groups

In this section, we are interested in the work in [17]. We introduce the notion of a fuzzy p-subgroup and prove that a fuzzy subgroup can be written as the intersection of its minimal fuzzy p-subgroups.

Recall that we denote the identity of a group G by e, the order of G by $o(G)$, the order of x in G by $o(x)$, and the greatest common divisor of integers $n_1, ..., n_t$ by $(n_1, ..., n_t)$.

Definition 7.2.1. *Let μ be a fuzzy subgroup of a group G. If p is prime, then μ is called a **fuzzy p-subgroup** of G if $FO_\mu(x)$ is a power of p for all $x \in G$.*

It follows that the order of a fuzzy p-subgroup is a power of p if the order is finite.

In Chapter 6, we introduced the notion of a p-primary fuzzy subgroup of an Abelian group by using fuzzy singletons. We showed that there is a unique maximal p-primary fuzzy subgroup contained in a fuzzy subgroup for each prime p. However, the notion of a fuzzy p-subgroup as defined in Chapter 6 is different from the one used here.

Proposition 7.2.2. *Let μ be a fuzzy subgroup of a group G such that μ_* is a normal subgroup of G. Then μ is a fuzzy p-subgroup if and only if G/μ_* is a p-group.*

Proof. Suppose μ is a fuzzy p-subgroup. Let $x \in G$. Then $FO_\mu(x) = p^t$ for some nonnegative integer t. Hence $x^{p^t} \in \mu_*$. Thus G/μ_* is a p-group. Conversely, suppose G/μ_* is a p-group. Then for $x \in G$, $x^{p^t} \in \mu_*$ for some nonnegative integer t and so $\mu(x^{p^t}) = \mu(e)$. Hence μ is a fuzzy p-subgroup of G by Proposition 1.6.5. $\qquad\square$

Now we consider the order of the intersection of fuzzy subgroups.

Proposition 7.2.3. *Let μ and ν be fuzzy subgroups of a group G such that $\mu(e) = \nu(e)$. Let $O(\mu) = mp_1^{r_1}...p_t^{r_t}$ and $O(\nu) = np_1^{s_1}...p_t^{s_t}$ for $r_i, s_i \in \mathbb{N}, i = 1, ..., t$. Suppose that $(m, p_i) = (n, p_i) = 1$, where $(m, n) = 1$ and the p_i are distinct primes for $i = 1, ..., t$. Then $O(\mu \cap \nu) = mnp_1^{u_1}...p_t^{u_t}$, where $u_i = r_i \vee s_i$ for all $i = 1, ..., t$.*

Proof. Let $k = mnp_1^{u_1}...p_t^{u_t}$. Then

$$(\mu \cap \nu)(x^k) = \mu(x^k) \wedge \nu(x^k)$$
$$= \mu(e) \wedge \nu(e)$$
$$= (\mu \cap \nu)(e) \text{ for all } x \in G$$

since $O(\mu)$ and $O(\nu)$ both divide k. Thus $O(\mu \cap \nu)|k$. By Theorem 2.3.22, $O(\mu)$ and $O(\nu)$ both divide $O(\mu \cap \nu)$. However, k is the least common multiple of $O(\mu)$ and $O(\nu)$. Hence we have the desired conclusion. $\qquad\square$

Corollary 7.2.4. *The intersection of a finite number of fuzzy p-subgroups of a group G such that their images of e are equal is again a fuzzy p-subgroup of G.*

Corollary 7.2.5. *Let μ be a fuzzy subgroup of a group G. If there exists a minimal fuzzy p-subgroup ν of G containing μ such that $\nu(e) = \mu(e)$, then ν is unique.*

We denote the unique minimal fuzzy p-subgroup in Corollary 7.2.5 by $\mu_{(p)}$ if it exists. However, $\mu_{(p)}$ does not exist in general as can be seen by the following example.

Example 7.2.6. *Let p be prime. Define the fuzzy subgroup μ of $\mathbb{Z}$ by $\forall x \in \mathbb{Z}$,*

$$\mu(x) = \begin{cases} t_0 & \text{if } x = 0, \\ t_1 & \text{otherwise,} \end{cases}$$

where $t_0 > t_1$. For all $n \in \mathbb{N}$, define a fuzzy subgroup ν_n of $\mathbb{Z}$ by $\forall x \in \mathbb{Z}$,

$$\nu_n(x) = \begin{cases} t_0 & \text{if } x \in p^n\mathbb{Z}, \\ t_1 & \text{otherwise.} \end{cases}$$

Then $\forall n \in \mathbb{N}, FO_{\nu_n}(x) = p^j$ if $x \in p^{n-j}\mathbb{Z}\backslash p^{n-j+1}\mathbb{Z}$ for $j = 1,...,n$ and $FO_{\nu_n}(x) = p^0$ if $x \in p^n\mathbb{Z}$. Hence $\nu_1 \supset \nu_2 \supset ...$ is an infinite descending chain of fuzzy p-subgroups of $\mathbb{Z}$ containing μ and such that $\cap_{i=1}^{\infty}\nu_n = \mu$. Thus it follows that $\mu_{(p)}$ does not exist since μ is not a fuzzy p-subgroup.

In [21], a fuzzy subgroup μ of G is defined to be a fuzzy Sylow p-subgroup of G if $|\text{Im}(\mu)| \leq 2$ and μ_* is a Sylow p-subgroup of G. Let μ be a fuzzy Sylow p-subgroup of a finite nilpotent group G and ν a fuzzy subgroup of G with $\nu_* = \{e\}$. Then ν_* is the Sylow p-subgroup G_p of G. However, $(\nu_{(p)})_*$ is the p-complement of G, where $\nu_{(p)}$ is the minimal fuzzy p-subgroup of G containing ν (cf., Proposition 7.2.9 and Lemma 7.2.8) and $G/(\nu_{(p)})_*$ is isomorphic to G_p. Thus the notion of a fuzzy Sylow p-subgroup of [21] is somewhat different from the notion of a minimal fuzzy p-subgroup presented here. However, these two notions are related as follows: (i) $\cap\{\nu_{(q)} \mid q$ is a prime and $q \neq p\}$ is a fuzzy Sylow p-subgroup of G if the number of the images of ν is less than or equal to 2, (ii) the minimal fuzzy p-subgroup $\mu_{(p)}$ of G containing μ is such that $(\mu_{(p)})_* = G$.

In the remainder of the section, we are concerned with the problem of whether or not a fuzzy subgroup μ can be written as the intersection of its minimal p-subgroups.

Lemma 7.2.7. *Let μ be a fuzzy subgroup of a group G. Let $x, y \in G$. If $(FO_\mu(x), FO_\mu(y)) = 1$ and $xy = yx$, then $\mu(xy) = \mu(x) \wedge \mu(y)$.*

Proof. Let $FO_\mu(y) = m$. Then

$$\mu(xy) \le \mu((xy)^m) = \mu(x^m y^m) = \mu(x^m)$$

by Lemma 2.1.2. However, $\mu(x^m) = \mu(x)$ by Proposition 1.6.11. Hence $\mu(xy) \le \mu(x)$. Also, $\mu(xy) \le \mu(y)$ holds similarly. Thus $\mu(xy) = \mu(x) \wedge \mu(y)$. $\square$

Lemma 7.2.8. *Let μ be a fuzzy subgroup of a group G satisfying the conditions $(1), (2)$ and (3) for some prime p.*

(1) $FO_\mu(x)$ is finite for all $x \in G$.

(2) $H_p = \{x \in G \mid (FO_\mu(x), p) = 1\}$ and $K_p = \{x \in G \mid FO_\mu(x)$ is a power of $p\}$ are subgroups of G.

(3) For all $x, y \in G$, there exist expressions $x = x_1 x_2 = x_2 x_1$ and $y = y_1 y_2 = y_2 y_1$ of x and y respectively, as in Theorem 1.6.16, such that $x_1, y_1 \in H_p$, $x_2, y_2 \in K_p$, $x_1 y_2 = y_2 x_1$ and $x_2 y_1 = y_1 x_2$.

Define a fuzzy subset ν_p of G by $\nu_p(x) = \mu(x_2)$, where x_2 is the element in Theorem 1.6.16, i.e., where $x = x_1 x_2 = x_2 x_1$, $FO_\mu(x_1) = m$ and $FO_\mu(x_2) = p^t$ with $FO_\mu(x) = mp^t$ and $(m, p) = 1$. Then ν_p is a fuzzy p-subgroup of G containing μ.

Proof. By Theorem 1.6.16, ν_p is well-defined. For $x, y \in G$, let $x = x_1 x_2 = x_2 x_1$ and $y = y_1 y_2 = y_2 y_1$ be expressions of x and y, respectively, as in condition (3). Let $z = xy$. By conditions (2) and (3), $z = z_1 z_2 = z_2 z_1$ is an expression of $z = xy$, where $z_1 = x_1 y_1$ and $z_2 = x_2 y_2$. Thus

$$\begin{aligned}
\nu_p(xy) &= \mu(z_2) = \mu(x_2 y_2) \\
&\ge \mu(x_2) \wedge \mu(y_2) \\
&= \nu_p(x) \wedge \nu_p(y).
\end{aligned}$$

Clearly, $\nu_p(x^{-1}) \ge \nu_p(x)$ for all $x \in G$. Therefore, ν_p is a fuzzy subgroup of G. Clearly, ν_p is a fuzzy p-subgroup of G containing μ. $\square$

Proposition 7.2.9. *Let ν_p be the fuzzy subgroup defined in Lemma 7.2.8. Then $\nu_p = \mu_{(p)}$. In particular, if $FO_\mu(x)$ is a power of p for $x \in G$, then $\mu(x) = \mu_{(p)}(x)$.*

Proof. Let ξ be a fuzzy p-subgroup of G such that $\mu \subseteq \xi \subseteq \nu_p$. Let $x \in G$. Suppose that $FO_\mu(x) = mp^t$ and $(m, p) = 1$. If $t = 0$ then, $\mu(x^m) = \mu(e)$. Now $\mu \subseteq \xi$. Thus $\xi(x^m) = \xi(e)$ and so $\xi(x) = \mu(e) = \nu_p(x)$ since ν_p is a fuzzy p-subgroup and by Proposition 1.6.5. If $m = 1$, then $\mu(x) \leq \xi(x) \leq \nu_p(x) = \mu(x)$ and so $\xi(x) = \nu_p(x)$. Finally, let $x = x_1 x_2 = x_2 x_1$ be an expression of x as in Lemma 7.2.8. Then $\xi(x_1) = \mu(e)$ and $\xi(x_2) = \nu_p(x_2) = \mu(x_2)$ by the above cases. Hence $\xi(x) = \xi(x_1 x_2) = \mu(x_2) = \nu_p(x)$. $\qquad\square$

Theorem 7.2.10. *Let μ be a fuzzy subgroup of a group G such that the conditions of Lemma 7.2.8 hold for all primes p. Then $\mu = \cap \mu_{(p)}$.*

Proof. Let $x \in G$ and let $FO_\mu(x) = p_1^{r_1}...p_t^{r_t}$, where the p_i are distinct primes and the r_i are positive integers, $i = 1, ..., t$. Let $q_i = FO_\mu(x)/(p_i^{r_i})$, $i = 1, ..., t$. Then $(q_1, ..., q_t) = 1$. Hence there exist $k_1, ..., k_t \in \mathbb{Z}$ such that $q_1 k_1 + ... + q_t k_t = 1$. Now $FO_\mu(x^{q_i k_i})$ divides $p_i^{r_i}$ for all $i = 1, ..., t$ and the $FO_\mu(x^{q_i k_i})$ are pairwise coprime. Therefore,

$$
\begin{aligned}
\mu(x) &= \mu(x^{q_1 k_1} ... x^{q_t k_t}) \\
&= \mu(x^{q_1 k_1}) \wedge ... \wedge \mu(x^{q_t k_t}) \text{ by Lemma 7.2.7} \\
&= \mu_{(p_1)}(x^{q_1 k_1}) \wedge ... \wedge \mu_{(p_t)}(x^{q_t k_t}) \text{ by Proposition 7.2.9} \\
&= \mu_{(p_1)}(x) \wedge ... \wedge \mu_{(p_t)}(x) \\
&\geq (\cap_{(p)} \mu)(x).
\end{aligned}
$$

Thus $\mu \supseteq \cap \mu_{(p)}$. Clearly, $\mu \subseteq \mu_{(p)}$. Hence we have the desired conclusion. $\quad\square$

Lemma 7.2.11. *Let μ be a fuzzy subgroup of a group G. If either (1) or (2) below hold, then the conditions of Lemma 7.2.8 hold for all primes p.*
  (1) G is a finite nilpotent group.
  (2) G is an Abelian group and $FO_\mu(x)$ is finite for all $x \in G$.

Proof. If G is finite and nilpotent, then G is the direct product of its Sylow subgroups. Thus the remainder of the proof follows easily. $\qquad\square$

Corollary 7.2.12. *Let μ be a fuzzy subgroup of a group G. If G is an Abelian group with $FO_\mu(x) < \infty$ for all $x \in G$ or if G is a finite nilpotent group, then $\mu = \cap \mu_{(p)}$.*

That Theorem 7.2.10 does not hold in general can be seen by the following example.

Example 7.2.13. *Let $D_3 = \langle a, b | a^3 = b^2 = e, ba = a^2 b \rangle$ be the dihedral group with 6 elements. Define the fuzzy subgroup μ of D_3 by $\forall x \in D_3$,*

$$
\mu(x) = \begin{cases} t_0 & \text{if } x = e, \\ t_1 & \text{otherwise}, \end{cases}
$$

where $t_0 > t_1$. Then $\mu_{(2)}$ is as follows:

$$\mu_{(2)}(x) = \begin{cases} t_0 & \text{if } x \in \langle a \rangle, \\ t_1 & \text{otherwise,} \end{cases}$$

since $\mu_{(2)}(b^2) = \mu_{(2)}((ab)^2) = \mu_{(2)}((ba)^2) = \mu_{(2)}(e)$. Let p be an odd prime and ν a fuzzy subgroup of G. Then $\nu(b^p) = \nu(e) \Rightarrow \nu(b) = \nu(e)$. Also, $\nu(a^p) = \nu(e) \Rightarrow p = 3 \Rightarrow a^p = e$. Furthermore, $\nu((ab)^3) = \nu(ab)$ and so $\nu((ab)^3) = \nu(e) \Rightarrow \nu(ab) = \nu(e)$. Similarly, $\nu((ba)^3) = \nu(e) \Rightarrow \nu(ba) = \nu(e)$. Thus $\mu_{(3)}(x) = \mu(e) \ \forall x \in D_3$. In fact, it follows that for all odd primes p, $\mu_{(p)}$ is a trivial fuzzy subgroup, i.e., $\mu_{(p)}(x) = \mu(e)$ for all $x \in D_3$. Since $FO_\mu(a)$ is not a power of $2, \mu$ is not a fuzzy 2-subgroup. Thus $\cap \mu_{(p)} = \mu_{(2)} \neq \mu$.

Theorem 7.2.14. *Let μ be a fuzzy subgroup of G. Suppose G is either Abelian with $O(\mu) < \infty$ or G is a finite nilpotent group. If n is a positive integer that divides $O(\mu)$, then there is a fuzzy subgroup ξ of G containing μ such that $O(\xi) = n$ and $\xi(e) = \mu(e)$.*

Proof. Let $O(\mu) = p_1^{r_1}...p_t^{r_t}$, where the p_i are distinct primes and the r_i are positive integers for $i = 1, ..., t$. Then for all $i = 1, ..., t$, $\mu_{(p_i)}$ exists by Lemma 7.2.11, $\mu_{(p_i)}$ equals ν_{p_i} (defined in Lemma 7.2.8) by Proposition 7.2.9, and $O(\nu_{p_i}) = p_i^{r_i}$ by Proposition 2.3.19. We may assume that $n = p_1^{u_1}...p_s^{u_s}$ with $s \leq t$ and $1 \leq u_i \leq r_i, i = 1, ..., s$. For all $i = 1, ..., s$, define a fuzzy subgroup ξ_i of G by $\forall x \in G$,

$$\xi_i(x) = \begin{cases} \mu(e) & \text{if } p_i^{u_i} \text{ divides } FO_{\nu_{p_i}}(x_2), \\ \nu_{p_i}(x) & \text{otherwise,} \end{cases}$$

where $x = x_1 x_2 = x_2 x_1$ is an expression of x as in Lemma 7.2.8. Then $\xi_i(e) = \mu(e)$ and $O(\xi_i) = p_i^{u_i}$ since $p_i^{u_i} | p_i^{r_i}, i = 1, ..., s$. Thus $\cap \xi_i$ is the desired fuzzy subgroup of G. $\qquad\square$

We now apply our results to direct products of fuzzy subgroups.

It was shown in [31] that every fuzzy subgroup of the direct product of certain groups can be written as the direct product of its projections. This result was improved in [3] and [23] (see Corollary 7.2.17 and Corollary 7.2.16). In this section, we extend their results by using the notion of orders of fuzzy subgroups.

Let μ_i be a fuzzy subgroup of a group $G_i, i = 1, ..., n$, where n is a positive integer. The direct product $\mu_1 \otimes ... \otimes \mu_n$ is the fuzzy subgroup of the direct product $G_1 \otimes ... \otimes G_n$ defined by

$$(\mu_1 \otimes ... \otimes \mu_n)(x_1, ..., x_n)$$
$$= \mu_1(x_1) \wedge ... \wedge \mu_n(x_n).$$

Let μ be a fuzzy subgroup of the direct product of $G_1 \otimes ... \otimes G_n$. For all $i = 1, ..., n$, the projection μ_i of μ on G_i is the fuzzy subgroup of G_i defined by

$$\mu_i(x) = \mu(e_1, ..., e_{i-1}, x, e_{i+1}, ..., e_n)$$

for all $x \in G_i$. The following result extends [[3], Theorem 5.3, p. 99] by replacing the assumption $(G_i, G_j) = 1$ by $(O(\mu_i), O(\mu_j)) = 1$ for all distinct $i, j \in \{1, ..., n\}$.

Theorem 7.2.15. *Let G_i be a group for $i = 1, ..., n$, where $n \in \mathbb{N}$. Let μ be a fuzzy subgroup of the direct product of the G_i, $i = 1, ..., n$. Suppose $O(\mu)$ is finite. Let μ_i be the projection of μ on G_i for all $i = 1, ..., n$. If $(O(\mu_i), O(\mu_j)) = 1$ for all distinct $i, j \in \{1, ..., n\}$, then $\mu = \mu_1 \otimes ... \otimes \mu_n$.*

Proof. Let $(x_1, ..., x_n) \in G_1 \otimes ... \otimes G_n$. Let $i, j \in \{1, ..., n\}, i < j$. Then $(FO_{\mu_i}(x_i), FO_{\mu_j}(x_j)) = 1$ since $(O(\mu_i), O(\mu_j)) = 1$. Hence

$$(FO_\mu(e_1, ..., e_{i-1}, x_i, e_{i+1}, ..., e_n), FO_\mu(e_1, ..., e_{j-1}, x_j, e_{j+1}, ..., e_n)) = 1.$$

Thus

$$\mu(e_1, ..., e_{i-1}, x_i, e_{i+1}, ..., e_{j-1}, x_j, e_{j+1}, ..., e_n)$$
$$= \mu(e_1, ..., e_{i-1}, x_i, e_{i+1}, ..., e_n) \wedge \mu(e_1, ..., e_{j-1}, x_j, e_{j+1}, ..., e_n)$$
$$= \mu_i(x_i) \wedge \mu(j)(x_j)$$

by Lemma 7.2.7. Therefore,

$$\mu(x_1, ..., x_n) = \mu_1(x_1) \wedge ... \wedge \mu_n(x_n)$$

by induction on n since the $O(\mu_i)$ are pairwise co-prime, where $i = 1, ..., n$. $\square$

The following two corollaries follow easily from Theorem 7.2.15.

Corollary 7.2.16. *[3, 23]. Let μ be a fuzzy subgroup of the direct product of the groups $G_1, ..., G_n$, where the G_i have pairwise coprime orders for $i = 1, ..., n$. Then μ can be written as the direct product $\mu_1 \otimes ... \otimes \mu_n$, where each μ_i is the projection of μ on G_i.*

Proof. By Theorem 2.3.17, the order of μ_i divides the order of G_i for all $i \in \{1, ..., n\}$. $\square$

Corollary 7.2.17. *[31]. Let μ be a fuzzy subgroup of the direct product of groups $G_1, ..., G_n$, where the G_i are cyclic groups of distinct prime power orders for $i = 1, ..., n$. Then μ can be written as the direct product $\mu_1 \otimes ... \otimes \mu_n$, where every μ_i is the projection of μ on G_i, $i = 1, ..., n$.*

Corollary 7.2.17 can be easily proved by using Theorem 7.2.10. Let p_i be the prime factor of $o(G_i)$ for all $i \in \{1, ..., n\}$. Then the projection of $\mu_{(p_i)}$ can be identified with $\mu_i, i = 1, ..., n$, and it follows easily that $\cap_{i=1}^{n} \mu_{(p_i)} = \mu_1 \otimes ... \otimes \mu_n$.

7.3 Fuzzy Subgroups Having Property *

The notion of a generalized characteristic fuzzy subgroup was introduced in [1], presented in Section 4.2, and a characterization of all finite cyclic groups in terms of generalized characteristic fuzzy subgroups was given. In this section, the notion of the fuzzy order of an element is used to introduce the notion of what is called property (*) [16]. This is a generalized and more fuzzified notion of the notion of a generalized characteristic fuzzy subgroup. We characterize all finite cyclic groups in terms of this generalized notion and give an easy proof for the characterization in [1]. Finally, we suggest an improvement upon a theorem in [8] concerning the normal fuzzy subgroups of a group of square free order. The results of this section are from [16].

Throughout this section, we denote the identity of a group G by e, the order of x in G by $o(x)$, and the greatest common divisor of integers m and n by (m, n).

$FO_\mu(x)$ is always a divisor of $o(x)$. If $\mu_* = \{e\}$, then $FO_\mu(x) = o(x)$. However, the two statements $o(x) = o(y)$ and $FO_\mu(x) = FO_\mu(y)$ are independent of each other, as is illustrated by Example 1.6.4 and the following example.

Example 7.3.1. *Define the fuzzy subset μ of $\mathbb{Z}_2 \oplus \mathbb{Z}_4$ by $\mu(x) = t_0$ if $x \in \langle (0, 2) \rangle$ and $\mu(x) = t_1$ otherwise, where $t_0 > t_1$. Then μ is a fuzzy subgroup of $\mathbb{Z}_2 \oplus \mathbb{Z}_4$ and $FO_\mu((0, 1)) = FO_\mu((1, 0)) = 2$, but $o((0, 1)) = 4$ and $o((1, 0)) = 2$.*

Recall from Section 4.2 that a fuzzy subgroup μ of a finite group G is called a generalized characteristic fuzzy subgroup (briefly, a GCFS) if $o(x) = o(y)$ implies $\mu(x) = \mu(y)$ for all x, y in G. We now replace $o(x)$ by $FO_\mu(x)$ in this definition.

Definition 7.3.2. *A fuzzy subgroup μ of a group G is said to have **property** (*) if $FO_\mu(x) = FO_\mu(y)$ implies $\mu(x) = \mu(y)$ for all x, y in G.*

The fuzzy subgroup μ in Example 1.6.4 has property (*), but μ is not a GCFS.

Proposition 7.3.3. *Let μ be a GCFS of a finite group G and let $x, y \in G$. If $o(x) = o(y)$, then $FO_\mu(x) = FO_\mu(y)$.*

Proof. Suppose that $o(x) = o(y)$. Then $o(x^n) = o(y^n)$ and so $\mu(x^n) = \mu(y^n)$ for all positive integers n. Thus $FO_\mu(x) = FO_\mu(y)$. $\qquad\qquad\square$

Proposition 7.3.4. *Let μ be a GCFS of a finite p-group G and let $x, y \in G$. If $FO_\mu(x) = FO_\mu(y) > 1$, then $o(x) = o(y)$.*

Proof. Let $FO_\mu(x) = FO_\mu(y) = p^t$, where $t \in \mathbb{N}$. Then $o(x) = p^{t+m}$ and $o(y) = p^{t+n}$ for some nonnegative integers m and n by Proposition 1.6.5. We may assume that $m \geq n$. Since $o(x^{p^{m-n}}) = p^{t+n} = o(y)$, $FO_\mu(x^{p^{m-n}}) = FO_\mu(y) = p^t$ by Proposition 7.3.3. However, $FO_\mu(x^{p^{m-n}}) = FO_\mu(x)/(FO_\mu(x), p^{m-n}) = p^t/(p^t, p^{m-n})$ by Theorem 1.6.10. Thus $m = n$ since $t \geq 1$.   □

The following may be considered a corollary to Theorem 1.6.10.

Corollary 7.3.5. *If μ is a GCFS of a finite nilpotent group G, then μ has property (∗).*

Proof. Let $x, y \in G$. Suppose $FO_\mu(x) = FO_\mu(y)$. Then by Proposition 1.6.5 and Theorem 1.6.13, $FO_\mu(x_p) = FO_\mu(y_p)$ for every prime p, where x_p and y_p denote the p-component of x and y, respectively. Thus $\mu(x_p) = \mu(y_p)$ by Proposition 7.3.4. Hence $\mu_p(x_p) = \mu_p(y_p)$ for every prime p, where μ_p denotes the projection of μ on the p-component G_p of G. Therefore, $\mu(x) = \mu(y)$ by Corollary 7.2.16.   □

We thus see that the property (∗) is a generalized and more fuzzified notion of the notion of GCFS for at least nilpotent finite groups. This leads to the following question.

Does a GCFS of a finite group always have the property (∗)?

We point out that a fuzzy subgroup may neither be a GCFS nor have property (∗). See, for example, Lemma 7.3.11.

Lemma 7.3.6. *Let μ be a fuzzy subgroup of a group G. Then μ_* is normal in G if and only if $FO_\mu(x) = FO_\mu(y^{-1}xy)$ for all x, y in G.*

Proof. Suppose μ_* is normal in G. Let $x \in G$. If $FO_\mu(x) = \infty$, then $x^n \notin \mu_*$ for all positive integers n. Thus for all $y \in G$, $(y^{-1}xy)^n = y^{-1}x^ny \notin \mu_*$ and so $FO_\mu(y^{-1}xy) = \infty$. If $FO_\mu(x) = n$ for some $n \in \mathbb{N}$, then $x^n \in \mu_*$ and so $(y^{-1}xy)^n = y^{-1}x^ny \in \mu_*$. Thus $FO_\mu(y^{-1}xy)$ divides $n = FO_\mu(x)$ by Proposition 1.6.5. Hence $FO_\mu(x) = FO_\mu(y^{-1}xy)$ for all $x, y \in G$. Conversely, if $x \in \mu_*$, then $FO_\mu(y^{-1}xy) = FO_\mu(x) = 1$. Thus $y^{-1}xy \in \mu_*$ for all $y \in G$.   □

Proposition 7.3.7. *Let μ be a fuzzy subgroup of a group G. If μ has property (∗) and μ_* is normal in G, then μ is normal in G.*

Proof. The result follows from Lemma 7.3.6.   □

The assumption in Proposition 7.3.7 that μ_* is normal in G cannot be removed. This can be seen by Example 7.3.8. The converse of the proposition is not true in general as can be seen by Example 7.3.9. Furthermore, under the assumption in the proposition, if G/μ_* is finite, then $g(\mu)$ is clearly a

GCFS, where g is the natural homomorphism of G onto G/μ_*. However, a level subgroup of μ is not a characteristic subgroup of G in general. For example, $\mu_{t_0} = \langle ab \rangle$ is not a characteristic subgroup of G in Example 1.6.4.

Example 7.3.8. *Define the fuzzy subset μ of the dihedral group*
$D_3 = \langle a, b | a^3 = b^2 = e, ba = a^2 b \rangle$ *by* $\forall x \in G, \mu(x) = t_0$ *if* $x \in \langle b \rangle$ *and* $\mu(x) = t_1$ *otherwise, where* $t_0 > t_1$, *Then* μ *is a fuzzy subgroup of G and μ has property* $(*)$. *However, μ_* is not normal in G and μ is not normal in G.*

Example 7.3.9. *Define the fuzzy subset μ of the Klein 4-group*
$G = \langle a, b | a^2 = b^2 = (ab)^2 = e \rangle$ *by* $\mu(e) = t_0$, $\mu(a) = t_1$, *and* $\mu(b) = \mu(ab) = t_2$, *where* $t_0 > t_1 > t_2$. *Then* μ *is a normal fuzzy subgroup of G. Since $FO_\mu(a) = 2 = FO_\mu(b)$ and $\mu(a) \neq \mu(b)$, μ does not have property* $(*)$.

Proposition 7.3.10. *Let f be a homomorphism of a group G onto a group H and let μ and ν be fuzzy subgroups of G and H, respectively. Then the following properties hold.*
  (1) *If μ has property $(*)$ and μ is f-invariant, then $f(\mu)$ has property $(*)$.*
  (2) *If ν has property $(*)$, then $f^{-1}(\nu)$ has property $(*)$.*

Proof. (1) Since μ is f-invariant, $(f(\mu))(f(x)^n) = \mu(x^n)$ for all integers n.
  Hence if $FO_{f(\mu)}(f(x)) = FO_{f(\mu)}(f(y))$, then $FO_\mu(x) = FO_\mu(y)$ and so $(f(\mu))(f(x)) = \mu(x) = \mu(y) = (f(\mu))(f(y))$. Thus $f(\mu)$ has the property $(*)$.
  (2) Since $(f^{-1}(\nu))(x^n) = \nu(f(x^n))$ for all integers n, if $FO_{f^{-1}(\nu)}(x) = FO_{f^{-1}(\nu)}(y)$, then $FO_\nu(f(x)) = FO_\nu(f(y))$ and so $(f^{-1}(\nu))(x) = \nu(f(x)) = \nu(f(y)) = (f^{-1}(\nu))(y)$. Thus $f^{-1}(\nu)$ has the property $(*)$. $\square$

Lemma 7.3.11. *Let H and K be subgroups of a group G. If $x \in H \backslash K$ and $y \in K \backslash H$ are such that $o(x) = o(y)$, then there exists a fuzzy subgroup μ of G such that μ is neither a GCFS nor has property $(*)$.*

Proof. Define the fuzzy subset μ of G by $\forall z \in G$, $\mu(e) = t_1$, $\mu(z) = t_2$ if $z \in H \backslash \{e\}$, and $\mu(z) = t_3$ otherwise, where $t_1 > t_2 > t_3$. Then μ is a desired fuzzy subgroup of G by Definition 7.3.2 and Proposition 7.3.3 since $\mu(x) = t_2 \neq t_3 = \mu(y)$. $\square$

Corollary 7.3.12. *Let G be a finite group. If either (1) or (2) holds, then G is cyclic.*
  (1) *Every fuzzy subgroup of G is a GCFS.*
  (2) *Every fuzzy subgroup μ of G has property $(*)$ and μ_* is normal in G.*

Proof. In the quaternion group $Q = \langle a, b | a^4 = e, b^2 = a^2, ba = a^3 b \rangle$, $a \in \langle a \rangle \backslash \langle b \rangle$, $b \in \langle b \rangle \backslash \langle a \rangle$, and $o(a) = o(b) = 4$. Thus if (1) holds, then since every subgroup of G can be realized as a level subgroup of some fuzzy subgroup and since a GCFS of a finite group must be normal, G is Abelian by Theorem 4.1.10 and Lemma 7.3.11. Also, if (2) holds, then since any subgroup of G can be realized as a level subgroup of some fuzzy subgroup, G is Abelian by Proposition 7.3.7, Theorem 4.1.10, and Lemma 7.3.11. Hence G is cyclic by Lemma 7.3.11. $\square$

Corollary 7.3.12 gives an easy proof for the sufficiency of the theorem in [1] which states that if G is a finite group, then G is cyclic if and only if every fuzzy subgroup of G is a GCFS. We now give another characterization of finite cyclic groups which follows immediately from Corollary 7.3.12 and Theorem 1.7.3.

Theorem 7.3.13. *Let G be a finite group. Then G is cyclic if and only if every fuzzy subgroup μ of G has property $(*)$ and μ_* is normal in G.*

The following proposition is an improvement of Theorem 4.1.7.

Proposition 7.3.14. *Let μ be a normal fuzzy subgroup of a group G such that G/μ_* is of square free order. Let $x, y \in G$. Then the following assertions hold.*
 (1) *If $FO_\mu(x)$ divides $FO_\mu(y)$, then $\mu(y) \le \mu(x)$.*
 (2) *If $FO_\mu(x) = FO_\mu(y)$, then $\mu(y) = \mu(x)$, i.e., μ has property $(*)$.*

Proof. Let g be the natural homomorphism of G onto G/μ_*. Then $FO_\mu(x) = o(g(x))$ and $\mu(x) = (g(\mu))(g(x))$ for all $x \in G$. However, $g(\mu)$ is normal in G/μ_* by Theorem 1.3.13 since μ is normal in G. Thus the desired conclusions follow by Theorem 4.1.7. $\square$

7.4 Cyclic Fuzzy Subgroups and Cyclic Fuzzy p-subgroups

In this section, we introduce the notion of a fuzzy cyclic subgroup and a fuzzy cyclic p-subgroup of a group. We use these notions to obtain characterizations for a finite cyclic group and for a finite cyclic p-group. We also obtain another characterization for a finite cyclic p-group using the notion of the order of a fuzzy subgroup introduced in [17] and presented in Chapter 1. Finally, we use the notion of the fuzzy direct product to obtain necessary and sufficient conditions for a finite group to be the direct product of its cyclic p-subgroups or of its cyclic subgroups. The results here mainly from [4].

For a fuzzy subgroup μ of a group G, recall that $\mu_* = \{x \in G \mid \mu(x) = \mu(e)\}$ is a subgroup of G. If $\mu_* = \{e\}$, then $FO_\mu(x) = o(x)$ for all x in G. Recall also that in Section 7.1, the relation $\sim$ on the set of all fuzzy subgroups of a group G was defined as follows: $\forall \mu, \nu \in \mathcal{F}(G), \mu \sim \nu$ if and only if μ and ν have the same chain of level subgroups.

Let μ be a fuzzy subgroup of a group G. Recall that if there exists a positive integer n such that for all $x \in G, \mu(x^n) = \mu(e)$, then the least such positive integer is called the order of μ, written $O(\mu)$. If no such n exists, μ is said to have infinite order.

Let μ be a fuzzy subgroup of a group G. Let p be a prime. Recall that μ is said to be a fuzzy p-subgroup of G if $FO_\mu(x)$ is a power of p for all x in

G. We also recall that the order of a fuzzy p-subgroup is a power of p if the group is finite.

The next result follows from Section 7.1.

Theorem 7.4.1. *Suppose that G is a finite group. Then G is the direct product of its normal subgroups $G_1, ..., G_n$ if and only if there exist fuzzy subgroups $\mu_1, ..., \mu_n$ of maximal chain of $G_1, ..., G_n$, respectively, such that their fuzzy direct product is a fuzzy subgroup of G.*

We introduce the notion of a fuzzy cyclic subgroup of a group G and obtain a characterization for finite cyclic groups.

Definition 7.4.2. *A fuzzy subgroup μ of a group G is said to be a **fuzzy cyclic subgroup** of G if for all $t \in \mathrm{Im}(\mu)$, there exists element $z \in \mu_t$ such that $\nu(z) \leq \nu(x)$ for all fuzzy subgroups ν of G and for all x in μ_t.*

For an example, consider the Klein 4-group $G = \{e, a, b, c\}$ and consider the fuzzy subgroups μ, ν and ρ of G defined by $\mu(e) = r_2, \mu(a) = r_1, \mu(b) = \mu(ab) = r_0, \nu(e) = s_2, \nu(b) = s_1, \nu(a) = \nu(ab) = s_0$ and $\rho(e) = t_2, \rho(ab) = t_1, \rho(a) = \rho(b) = t_0$, where $r_0, r_1, r_2, s_0, s_1, s_2$, and t_0, t_1, t_2 are elements of [0,1] such that $r_0 < r_1 < r_2, s_0 < s_1 < s_2$, and $t_0 < t_1 < t_2$. Then clearly $G = \mu_{r_0}, r_0 = \mu(ab) = \mu(b) < \mu(a)$ while $\nu(b) > \nu(a)$ and $\rho(ab) > \rho(a)$. Thus μ is not fuzzy cyclic.

Theorem 7.4.3. *Let G be a finite group. Then G is cyclic if and only if every fuzzy subgroup of G is fuzzy cyclic.*

Proof. Suppose every fuzzy subgroup of G is fuzzy cyclic. Let μ be a fuzzy subgroup of G with $\mathrm{Im}(\mu) = \{t_0, ..., t_n\}, t_0 < t_1 < ... < t_n$. Let z be an element of $G = \mu_{t_0}$ such that $\nu(z) \leq \nu(x)$ for all $x \in G$ and for all fuzzy subgroups ν of G. Suppose $G \neq \langle z \rangle$. Let $x \in G \backslash \langle z \rangle$. Then $\chi_{\langle z \rangle}$ is a fuzzy subgroup of G and $\chi_{\langle z \rangle}(z) > \chi_{\langle z \rangle}(x)$, a contradiction. Thus $G = \langle z \rangle$. Conversely, suppose G is cyclic. Then $G = \langle z \rangle$ for some $z \in G$. Let μ be a fuzzy subgroup of G. Suppose $t \in \mathrm{Im}(\mu)$. Then μ_t is cyclic and so $\mu_t = \langle z^k \rangle$ for some $k \in \mathbb{N}$. Let ν be a fuzzy subgroup of G. Let $x \in \mu_t$. Then $x = z^{km}$ for some $m \in \mathbb{Z}$. Thus $\nu(x) = \nu(z^{km}) \geq \nu(z^k) \wedge ... \wedge \nu(z^k) = \nu(z^k)$. Hence μ is fuzzy cyclic. □

Theorem 7.4.4. *Let G be a finite group and μ be a fuzzy subgroup of G. Then the following conditions are equivalent.*
   (1)   *μ is fuzzy cyclic and $\mu_* = \{e\}$.*
   (2)   *$FO_\mu(z) = o(G)$ for some z in G.*
   (3)   *G is Abelian and $O(\mu) = o(G)$.*

Proof. (1) $\Rightarrow$ (2) Since μ is fuzzy cyclic, an argument as in the proof of the last theorem shows that G is cyclic. Suppose that $G = \langle z \rangle$ for some z in G.

Then the condition that $\mu_* = \{e\}$ is equivalent to $FO_\mu(x) = o(x)$ for all x in G. Hence $FO_\mu(z) = o(z) = o(G)$.

(2) $\Rightarrow$ (1) Suppose that $FO_\mu(z) = o(G)$ for some z in G. Since $FO_\mu(z)$ divides $o(z)$, $o(z) = o(G)$. Thus $G = \langle z \rangle$. Hence by Theorem 7.4.3, every fuzzy subgroup of G is fuzzy cyclic and so μ is fuzzy cyclic. Furthermore, if $o(\mu_*) = n$ and $m = o(G)/n$, then $\mu_* = \langle z^m \rangle$ and hence $FO_\mu(z) = m$. However, we have $FO_\mu(z) = o(G)$ and so $\mu_* = \{e\}$.

(2) $\Rightarrow$ (3) Suppose $FO_\mu(z) = o(G)$ for some z in G. Then from the proof of (1) $\Leftrightarrow$ (2), it follows that G is cyclic and hence Abelian. This together with Proposition 1.6.5 and Corollary 2.3.20 (2) imply that $O(\mu) = FO_\mu(z)$. Thus $O(\mu) = o(G)$.

(3) $\Rightarrow$ (2) Suppose that G is Abelian and $O(\mu) = o(G)$. Then by Corollary 2.3.20 (2), there exists z in G such that $FO_\mu(z) = O(\mu)$. Hence $FO_\mu(z) = o(G)$. $\qquad\square$

We next introduce the notion of a fuzzy cyclic p-subgroup of a group G and use it to obtain a characterization of finite cyclic p-groups.

Definition 7.4.5. *A fuzzy subgroup μ of a group G is said to be a **fuzzy cyclic p-subgroup** of G if for all $x, y \in G$,*
(1) $\mu(x) = \mu(y) \Leftrightarrow \langle x \rangle = \langle y \rangle$,
(2) $\mu(x) < \mu(y) \Leftrightarrow \langle x \rangle \supset \langle y \rangle$.

It follows easily that the condition $\langle x \rangle = \langle y \rangle (\langle x \rangle \supset \langle y \rangle)$ implies $\nu(x) = \nu(y)$ $(\nu(x) \le \nu(y))$ for all fuzzy subgroups ν of G and for all $x, y \in G$.

It follows from Theorem 7.4.8 that if G is finite and μ satisfies the conditions in the previous definition, then μ is fuzzy cyclic and there exists a fixed prime number p such that $O(\mu)$ is a power of p. This is the reason μ is called a fuzzy cyclic p-subgroup of G.

For an example, consider a cyclic group $G = \langle z \rangle$ of order 6 and the fuzzy subgroup μ of G defined by

$$\mu(e) = t_2, \mu(z^2) = \mu(z^4) = t_1, \mu(z) = \mu(z^3) = \mu(z^5) = t_0,$$

where t_0, t_1, t_2 are elements of $[0, 1]$ such that $t_0 < t_1 < t_2$. Then clearly μ is a fuzzy cyclic group of maximal chain and $\mu(z) = \mu(z^3)$. However $\langle z \rangle \ne \langle z^3 \rangle$ and thus μ is not a fuzzy cyclic p-subgroup of G.

Theorem 7.4.6. *Let G be a finite group. Then G is a cyclic p-subgroup if and only if G has a fuzzy cyclic p-subgroup.*

Proof. Suppose G is a cyclic p-group. Then $|G| = p^n$ for some $n \in \mathbb{N}$. Consider the following chain of subgroups of G,

$$G = G_0 \supset G_1 \supset \ldots \supset G_n = \{e\},$$

where G_i is the subgroup of G generated by an element of order p^{n-i}, $i = 0, 1, ..., n$.

Let μ be the fuzzy subgroup of G which has the above chain as its chain of level subgroups and $\text{Im}(\mu) = \{t_0, ..., t_n\}$ such that $t_0 < t_1 < ... < t_n$. If $\mu(x) = \mu(y)$ ($\mu(x) < \mu(y)$) for x, y in G, then $\langle x \rangle = \mu_{t_i}$ and $\langle y \rangle = \mu_{t_j}$ for some $i, j \in \{0, 1, \cdots, n\}$. It follows easily that $\mu(x) = t_i$ and $\mu(y) = t_j$. This together with $\mu(x) = \mu(y)$ ($\mu(x) < \mu(y)$) yields $i = j (i < j)$ and consequently that $\langle x \rangle = \langle y \rangle$ ($\langle x \rangle \supset \langle y \rangle$). Hence μ is a fuzzy cyclic p-subgroup of G.

Conversely, let μ be a fuzzy cyclic p-subgroup of G with $\text{Im}(\mu) = \{t_0, t_1, ..., t_n\}$

$$\text{Im}(\mu) = \{t_0, t_1, ..., t_n\}$$

such that $t_0 < t_1 < ... < t_n$. Let $z \in G$ be such that $\mu(z) = t_0$. Let $x \in G$. Then $\mu(x) \geq \mu(z)$ and so $\langle z \rangle \supseteq \langle x \rangle$. Hence $x \in \langle z \rangle$ and so $G = \langle z \rangle$. Let $H = \langle h \rangle$ and $K = \langle k \rangle$ be subgroups of G. Then either $\mu(h) \leq \mu(k)$ or $\mu(h) \leq \mu(k)$. Thus we have either $H \supseteq K$ or $H \subseteq K$. Hence either $o(K)$ divides $o(H)$ or $o(H)$ divides $o(K)$. Thus G is a cyclic p-subgroup. $\square$

Corollary 7.4.7. *Let G be a finite p-group. Then G is cyclic if and only if G has a fuzzy cyclic p-subgroup.*

The following result gives a characterization for a fuzzy cyclic p-subgroup of a group.

Theorem 7.4.8. *Let μ be a fuzzy subgroup of a finite group G. Then the following conditions are equivalent.*
  (1) μ is a fuzzy cyclic p-subgroup.
  (2) All subgroups of G are chained as level subgroups of μ.
  (3) μ is a fuzzy cyclic and a fuzzy p-subgroup of maximal chain.

Proof. (1) $\Rightarrow$ (2) Let μ be a fuzzy cyclic p-subgroup of G. Then G is a cyclic p-group by Theorem 7.4.6. Let H be a subgroup of G. Then $H = \langle h \rangle$ for some $h \in H$. Now $\mu_{\mu(h)} = \langle h' \rangle$, where h' is an element of G. Clearly, $\mu(h') = \mu(h)$. However, since μ is a fuzzy cyclic p-subgroup, we have that $\langle h' \rangle = \langle h \rangle$ and hence $\mu_{\mu(h)} = H$. Therefore, all subgroups of G are chained as level subgroups of μ.

(2) $\Rightarrow$ (1) Suppose

$$G = G_0 \supset G_1 \supset ... \supset G_n = \{e\}$$

is the chain of all subgroups of G and are chained as level subgroups of μ. If $x \in G_0/G_1$, then $\langle x \rangle = G_i$ for some $i \in \{0, 1, \cdots, n\}$. Since $x \notin G_1$, we have $G = \langle x \rangle$. Also it is clear that G is a cyclic p-group. By a similar technique that was used in the first part of the proof of Theorem 7.4.6, it follows that μ is a fuzzy cyclic p-subgroup of G.

(2) $\Rightarrow$ (3) Let μ be a fuzzy cyclic p-subgroup of G. By Theorem 7.4.6, G is a cyclic p-group. Since G is a cyclic p-group, μ is fuzzy cyclic by Theorem

7.4.3 and μ is a fuzzy p-subgroup by Theorem 2.3.17. Also, as in the proof of
(1) $\Leftrightarrow$ (2), it follows from condition (2) that μ is of maximal chain.

(3) $\Rightarrow$ (2) Since μ is of maximal chain, μ_* is trivial and hence G is a p-
group by Proposition 7.2.2. Also, since μ is fuzzy cyclic, it follows as in the
proof of Theorem 7.4.3 that G is cyclic. Thus since G is a cyclic p-group, G
has a unique maximal series which is the chain of all its subgroups. However,
since μ is of maximal chain, it has the unique maximal series of G as its chain
of level subgroups. $\square$

By the previous theorem, we have the following corollary which gives another
characterization of a finite cyclic p-group.

Corollary 7.4.9. *Let G be a finite group. Then the following properties are
equivalent.*
  (1) *G is a cyclic p-group.*
  (2) *All fuzzy subgroups of maximal chain are equivalent.*
  (3) *Every fuzzy subgroup of maximal chain is a fuzzy cyclic p-subgroup.*

Combining the above theorem with Theorem 7.4.4, we obtain the following
result.

Corollary 7.4.10. *Let G be a finite p-group and μ a fuzzy subgroup of G.
Then the following properties are equivalent.*
  (1) *μ is a fuzzy cyclic p-subgroup.*
  (2) *μ is fuzzy cyclic and of maximal chain.*
  (3) *μ is of maximal chain and $FO_\mu(z) = o(G)$ for some z in G.*
  (4) *G is abelian, $O(\mu) = o(G)$, and μ is of maximal chain.*

Remark 7.4.11. It follows easily that the equivalence relation $\sim$ preserves
the following properties concerning fuzzy subgroups μ and ν of G :
  (1) Fuzzy order of an element of a group G with respect to a fuzzy subgroup
μ, (that is, $\mu \sim \nu \Rightarrow FO_\mu(x) = FO_\nu(x)$ for all x in G) and hence it preserves
the order of a fuzzy subgroup, (that is, $\mu \sim \nu \Rightarrow O(\mu) = O(\nu)$);
  (2) Fuzzy p-subgroup (that is, $\mu \sim \nu$ and μ is fuzzy p-subgroup $\Rightarrow \nu$ is
fuzzy p-subgroup);
  (3) Fuzzy cyclic subgroup;
  (4) Fuzzy cyclic p-subgroup.

We now consider direct products. The following result is a direct consequence of Theorems 7.4.3 and 7.4.1.

Theorem 7.4.12. *Let G be a finite group and $G_1, ..., G_n$ be cyclic subgroups
of G, where $n \in \mathbb{N}$. Then G is direct product of $G_1, ..., G_n$ if and only if there
exist fuzzy cyclic subgroups of $G_1, ..., G_n$ of maximal chain such that their
fuzzy direct product is a fuzzy subgroup of G.*

Theorem 7.4.13. *Let G be a finite group and let $p_1, ..., p_n$ be the prime factors of $o(G)$. Then G is cyclic if and only if there exist fuzzy cyclic p_i-subgroups of p_i-primary components of $G, i = 1, ..., n$, such that their fuzzy direct product is a fuzzy subgroup of G.*

Proof. Suppose G is cyclic. Let $I = \{1, ..., n\}$. Suppose $G = \prod_{i \in I}^{\otimes} G_i$, where the G_i are the primary p_i-components of G. Then by Theorem 7.4.1, there exist fuzzy subgroups μ_i of G_i of maximal chain such that their direct product is a fuzzy subgroup of $G, i = 1, ..., n$. Since G_i is a cyclic p_i-subgroup of G, it follows from Corollary 7.4.10 that μ_i is a fuzzy cyclic p_i-subgroup of G_i for all $i = 1, ..., n$.

Conversely, let $\mu = \prod_{i \in I}^{\otimes} \mu_i$ be a fuzzy subgroup of G, where the μ_i are fuzzy cyclic p_i-subgroups of p_i-primary components of G. By Theorem 7.4.6, it follows that the G_i are cyclic p_i-subgroups of G. By Theorem 7.4.1, we have that G is the direct product of cyclic p_i-subgroups of G. Since the p_i are distinct, G is cyclic. $\square$

The next result follows from Theorem 7.4.13, Corollary 7.4.10 and Theorem 7.1.16.

Corollary 7.4.14. *Let G be a finite Abelian group. Then a fuzzy subgroup of G is fuzzy cyclic of maximal chain if and only if it is a fuzzy direct product of fuzzy cyclic p_i-subgroups of the p_i-primary components of G.*

Let $I = \{1, 2, ..., n\}$. Suppose that G is a finite cyclic group and $G = \prod_{i \in I}^{\otimes} G_i$, where the G_i are the p_i-primary components of G and $p_1, ..., p_n$ are the prime factors of $o(G)$. Let $\mu, \mu_1, ..., \mu_n$ be fuzzy subgroup of $G, G_1, ..., G_n$, respectively, such that μ is the fuzzy direct product of the μ_i. Then $\mu_1, ..., \mu_n$ are called **fuzzy p_i-primary components** of μ. For such fuzzy subgroups μ of G, by preserving order, we mean, first fix the ordering of $p_1, ..., p_n$ and then that of the fuzzy p_i-primary components of μ.

If P_I^μ and $P_I^\nu \in \mathcal{A}$ for $I = \{1, ..., n\}$, we say that μ is **associated** to ν if whenever μ_i is equal (equivalent) to $\nu_i \ \forall i = 1, ..., n$, then μ is equal (equivalent to ν.

Theorem 7.4.15. *Let G be a finite cyclic group. Then all associated fuzzy subgroups of G which are the fuzzy direct product of their fuzzy primary components and preserving order are equivalent.*

Proof. Let $G = \prod_{i \in I}^{\otimes} G_i$, where the G_i are the p_i-primary components of G and $I = \{1, 2, ..., n\}$. Then it is known that this decomposition is unique. Suppose $\mathcal{A}$ is the set of all P_I^μ, where μ is a fuzzy subgroup of G and the μ_i are the fuzzy p_i-primary components of μ. Suppose that P_I^μ, P_I^ν are elements of $\mathcal{A}$. Since the μ_i, ν_i are fuzzy cyclic p_i-subgroups, they are of maximal chain by Theorem 7.4.8. Thus by Corollary 7.4.9, we have that μ_i is equivalent to ν_i for all $i = 1, ..., n$. By Theorem 7.1.7, it follows that μ is equivalent to ν. $\square$

7.5 Fuzzy p^*-subgroups

In this section, we introduce the notion of a fuzzy p^*-subgroup and characterize fuzzy subgroups of torsion groups and cyclic groups by their minimal
fuzzy p-subgroups and minimal fuzzy p^*-subgroups. We extend the notion of
the direct product of finitely many fuzzy subgroups to the direct sum of an
arbitrary family of fuzzy subgroups of a commutative group. We also give a
condition for a fuzzy subgroup of the direct sum of groups to be the direct
sum of its projections. We show that a fuzzy subgroup of a torsion group is
completely characterized by its minimal fuzzy p-subgroups. We generalize the
notion of a fuzzy p-subgroup by presenting the notion of a fuzzy p^*-subgroup.
We show that a fuzzy subgroup of a cyclic group is completely characterized
by its minimal fuzzy p^*-subgroups. The results of this section are mainly from
[18].

Let f be a homomorphism of a group G into a group H. Let μ be a fuzzy
subgroup of G. Recall that μ is called f-invariant if $f(x) = f(y)$ implies
$\mu(x) = \mu(y)$ for all $x, y \in G$.

Lemma 7.5.1. *Let f be a homomorphism of a group G onto a group H.
Let μ and ν be fuzzy subgroups of G and H, respectively. Then the following
assertions hold.*

(1) *If $FO_\mu(x) < \infty$ for some $x \in G$, then $FO_{f(\mu)}(f(x)) < \infty$ and
$FO_{f(\mu)}(f(x))$ divides $FO_\mu(x)$.*

(2) *If μ is f-invariant, then $FO_{f(\mu)}(f(x)) = FO_\mu(x)$ for all $x \in G$.*

(3) *$FO_\nu(f(x)) = FO_{f^{-1}(\nu)}(x)$ for all $x \in G$.*

Proof. (1) Let $FO_\mu(x) = n$. Then $(f(\mu))(f(x)^n) \geq \mu(x^n) = \mu(e)$ and so
$FO_{f(\mu)}(f(x))$ divides $FO_\mu(x)$ by Proposition 1.6.5.

(2) For every integer n, $(f(\mu))(f(x)^n) = (f(\mu))(f(x^n)) = \vee\{\mu(z) \mid f(z) = f(x^n), z \in G\} = \mu(x^n)$.

(3) For every integer n, $(f^{-1}(\nu))(x^n) = \nu(f(x^n)) = \nu((f(x))^n)$.   □

Proposition 7.5.2. *Let f be a homomorphism of a group G onto a group H
and let μ and ν be fuzzy subgroups of G and H, respectively. Then the following
assertions hold.*

(1) *If $O(\mu) < \infty$, then $O(f(\mu)) < \infty$ and $O(f(\mu))$ divides $O(\mu)$.*

(2) *If μ is f-invariant, then $O(f(\mu)) = O(\mu)$.*

(3) *$O(f^{-1}(\nu)) = O(\nu)$.*

Proof. The proof follows easily from Lemma 7.5.1.   □

The following result now follows.

Proposition 7.5.3. *Let μ be a fuzzy subgroup of a group G such that μ_* a
normal subgroup of G. Then μ is a fuzzy p-group if and only if G/μ_* is a
p-group.*

Note that the order of a fuzzy p-subgroup of G is a power of p if this order is finite. Also, the homomorphic image and the homomorphic preimage of a fuzzy p-subgroup are again fuzzy p-subgroups.

Relevant classical group-theoretical results and definitions can be found in [30].

Definition 7.5.4. *Let μ be a fuzzy subgroup of an Abelian group G. Then μ is called **fuzzy torsion** if $FO_\mu(x)$ is finite for all $x \in G$.*

Proposition 7.5.5. *Let μ be a fuzzy subgroup of an Abelian group G. Then μ is fuzzy torsion if and only if G/μ_* is torsion.*

We denote the minimal fuzzy p-subgroup of G containing the fuzzy subgroup μ of G by $\mu_{(p)}$ when it exists. That $\mu_{(p)}$ does not exist in general was shown in Section 7.3.

The following result is immediate from Corollary 7.2.12. However the converse does not hold as is shown in the next example.

Proposition 7.5.6. *Every fuzzy torsion subgroup of an Abelian group can be expressed as the intersection of the $\mu_{(p)}$.*

Example 7.5.7. *Let $p_1 < p_2 < ...$ denote the primes. Define a fuzzy subgroup μ of the additive group $\mathbb{Z}$ of all integers as follows: $\forall x \in \mathbb{Z}$,*

$$\mu(x) = \begin{cases} 1 & \text{if } x = 0, \\ 1 - 1/(i+1) & \text{if } x \in p_1...p_i\mathbb{Z} \text{ and } x \notin p_1...p_{i+1}\mathbb{Z} \text{ for some } i \geq 1, \\ 0 & \text{otherwise.} \end{cases}$$

For each prime p_i, define the fuzzy subgroup μ_i of $\mathbb{Z}$ as follows: $\forall x \in \mathbb{Z}$,

$$\mu_i(x) = \begin{cases} 1 & \text{if } x \in p_i\mathbb{Z}, \\ 1 - 1/i & \text{otherwise.} \end{cases}$$

Let $x \in G$. Then

$$\mu(x) = 1 - 1/(i+1) \Leftrightarrow x \in p_1\mathbb{Z} \cap ... \cap p_i\mathbb{Z}, x \notin p_{i+1}\mathbb{Z}$$
$$\Leftrightarrow (\cap \mu_i)(x) = 1 \wedge ... \wedge 1 \wedge (1 - 1/(i+1)) \wedge \{t_j \mid j = i+1, i+2, ...\}$$
$$= 1 - 1/(i+1),$$

where t_j is either 1 or $1 - 1/(j+1)$ for $j = i+1, i+2, ...$ Thus it follows that $\mu = \cap \mu_i$. Also, $\mu_i = \mu_{(p_i)}$ for each i, but $FO_\mu(1)$ is infinite.

Definition 7.5.8. *Let G_i be an Abelian group and let μ_i be a fuzzy subgroup of $G_i \forall i \in I$. The **direct sum** $\oplus_{i \in I} \mu_i$ is the fuzzy subgroup of the direct sum $\oplus_{i \in I} G_i$ defined by $\oplus_{i \in I} \mu_i(\sum x_i) = \wedge\{\mu_i(x_i) \mid i \in I\}$.*

Theorem 7.5.9. *Let μ be a fuzzy subgroup of the direct sum of the Abelian groups G_i, where $i \in I$. Let μ_i be the projection of μ on G_i for all $i \in I$. Suppose $(FO_{\mu_j}(x), FO_{\mu_k}(y)) = 1$ for all $j, k \in I$ such that $j \neq k$ and for all $x \in G_j$ and $y \in G_k$. Then $\mu = \oplus_{i\in I}\mu_i$.*

Proof. Since all but finitely many coordinates of each of the elements of $\oplus_{i\in I}G_i$ are the identity, the proof is same as in the case of a finite direct product, Theorem 7.2.15. $\square$

Corollary 7.5.10. *Let μ be a fuzzy subgroup of a torsion group G. Then $\mu = \oplus\mu_p$, where μ_p is the projection of μ on the p-primary subgroup G_p of G for each prime p.*

Proof. The proof follows by Proposition 1.6.5 and Theorem 7.5.9. $\square$

Lemma 7.5.11. *Let μ be a fuzzy subgroup of a torsion group G. For each prime p, the projection of $\mu_{(p)}$ on the p-primary subgroup G_p of G is equal to the projection μ_p of μ on G_p. Furthermore, the projection of $\mu_{(p)}$ on the p-complement subgroup G'_p of G has the constant image $\mu(0)$. In other words, $\mu_{(p)}$ may be identified with the projection of μ_p on G.*

Proof. Let $x \in G_p$. Then $o(x)$ is a power of p and so $\mu_{(q)}(x) = \mu(0)$ for all primes $q \neq p$ by Proposition 1.6.5. Thus $\mu_{(p)}(x) = \mu(x)$ by Proposition 7.5.6. On the other hand, if $x \in G'_p$, then $(o(x), p) = 1$ and so $\mu_{(p)}(x) = \mu(0)$ by Proposition 1.6.5. $\square$

By Lemma 7.5.11, Corollary 7.5.10 can also be derived from Proposition 7.5.6.

Let μ and ν be fuzzy subgroups of the groups G and H, respectively. Recall that μ and ν are said to be isomorphic if there exists a isomorphism f of G onto H such that $f(\mu) = \nu$.

Theorem 7.5.12. *Let μ and ν be fuzzy subgroups of the torsion groups G and H, respectively. Then μ and ν are isomorphic if and only if $\mu_{(p)}$ and $\nu_{(p)}$ are isomorphic for every prime p.*

Proof. By Proposition 7.5.6, $\mu_{(p)}$ and $\nu_{(p)}$ exist for every prime p. The necessity follows by Proposition 7.5.2. We now consider the sufficiency. For each prime p, let $f_p : G \to H$ be the isomorphism such that $f_p(\mu_{(p)}) = \nu_{(p)}$. Since G and H are torsion, $g_p = f_p|_{G_p}$ is an isomorphism from G_p onto H_p and $g_p(\mu_p) = \nu_p$ by Lemma 7.5.11. Define $f : G \to H$ by $f(\sum x_p) = \sum(g_p(x_p))$, where x_p belongs to the component G_p of G. Then f is clearly an isomorphism, and $f(\mu) = \nu$ by Corollary 7.5.10. $\square$

Due to Proposition 7.5.6 and Theorem 7.5.12, the study of fuzzy subgroups of a torsion group is reduced to the study of its fuzzy p-subgroups. The following example shows that Theorem 7.5.12 does not hold if G and H are not torsion groups.

Example 7.5.13. *For all $n \in \mathbb{N}$, define the fuzzy subgroup ξ_n of $\mathbb{Z}$ by for all $k \in \mathbb{Z}$,*

$$\xi_n(k) = \begin{cases} 1 & \text{if } k \in n\mathbb{Z}, \\ 0 & \text{otherwise}, \end{cases}$$

and let $\mu = \xi_5 \oplus \xi_6$ and $\nu = \xi_3 \oplus \xi_{10}$. Then μ and ν are fuzzy torsion, and $\mu_{(p)}$ and $\nu_{(p)}$ are isomorphic for every prime p. However, μ and ν are not isomorphic.

We now introduce the notion of a fuzzy p^*-subgroup.

Definition 7.5.14. *Let μ be a fuzzy subgroup of a group G and p a prime. Then μ is said to be a **fuzzy p^*-subgroup** if for all $x \in G$, $\wedge\{n \in \mathbb{N} \mid \mu(x) < \mu(x^n)\}$ is a power of p whenever this minimum exists.*

Lemma 7.5.15. *Let ν be a fuzzy p^*-subgroup of a group G and let n, m, t be positive integers such that $n = p^t m$ and $(p, m) = 1$. Then $\nu(x^n) = \nu(x^{p^t})$ for all $x \in G$.*

Proof. Let $x \in G$. Since $n = p^t m$, $\nu(x^{p^t}) \leq \nu(x^n)$. Assume that $\nu(x^{p^t}) < \nu(x^n)$. Set $y = x^{p^t}$. Then $\nu(y) < \nu(y^m)$. Since ν is a fuzzy p^*-subgroup, there exists $s \in \mathbb{N}$ such that $p^s = \wedge\{k \in \mathbb{N} \mid \nu(y) < \nu(y^k)\} \leq m$. Thus $\nu(y^j) = \nu(y)$ for $j = 1, ..., p^s - 1$. By the Euclidean algorithm, there are integers q and r such that $m = p^s q + r$ and $0 \leq r < p^s$. Now $r \neq 0$ since $(p, m) = 1$. Thus $\nu(y) < \nu(y^{p^s}) \leq \nu(y^{p^s q})$ and $\nu(y) = \nu(y^r)$. Since $\nu(y^{p^s q}) > \nu(y) = \nu(y^r)$, $\nu(y^m) = \nu(y^{p^s q} y^r) = \nu(y^{p^s q}) \wedge \nu(y^r) = \nu(y^r) = \nu(y)$ by comments following Lemma 1.2.5. However, this is impossible. Thus $\nu(x^{p^t}) = \nu(x^n)$. $\square$

Proposition 7.5.16. *Let μ be a fuzzy subgroup of a group G. Let p be a prime. Then there exists a unique minimal fuzzy p^*-subgroup, written $\mu_{(p)^*}$, of G containing μ.*

Proof. Let $\mathcal{I} = \{\nu_i \mid i \in I\}$ be the set of all fuzzy p^*-subgroups of G containing μ. Then $\mathcal{I}$ is nonempty since a constant fuzzy subgroup is a fuzzy p^*-subgroup. Set $\nu = \cap_{i \in I} \nu_i$. For $x \in G$, let $\nu(x) < \nu(x^n)$ for some positive integer n, where $n = p^t m$ and $(p, m) = 1$. Then for all $i \in I$, $\nu_i(x^n) = \nu_i(x^{p^t})$ since ν_i is a fuzzy p^*-subgroup. Hence $\nu(x^n) = \nu(x^{p^t})$. Thus ν is the desired $\mu_{(p)^*}$. $\square$

Lemma 7.5.17. *Let μ be a fuzzy subgroup of a group G and let $FO_\mu(x) = n$, where $x \in G$ and $n \in \mathbb{N}$. If m is an integer such that $(n, m) = 1$, then $\mu(x^m) = \mu(x)$.*

Proof. Since $(n, m) = 1$, there exist integers s and t such that $ns + mt = 1$. Thus $\mu(x) = \mu(x^{ns+mt}) = \mu((x^n)^s (x^m)^t) \geq \mu(x^n) \wedge \mu(x^m) = \mu(x^m) \geq \mu(x)$. $\square$

Theorem 7.5.18. *Every fuzzy p-subgroup of a group G is a fuzzy p^*-subgroup of G.*

Proof. Let $x \in G$ and let $n = p^t m$ be a positive integer such that $\mu(x) < \mu(x^n)$ and $(p, m) = 1$. Set $y = x^{p^t}$. Then $FO_\mu(y)$ is a power of p. Hence $(FO_\mu(y), m) = 1$. Thus $\mu(x^{p^t}) = \mu(y) = \mu(y^m) = \mu(x^n)$ by Lemma 7.5.17.   □

Thus the notion of fuzzy p^*-subgroups is an extension of the notion of fuzzy p-subgroups. Let p be a prime p. Then $\mu_{(p)}$ does not exist in general, but $\mu_{(p)^*}$ always exists.

Example 7.5.19. *Define the fuzzy subset μ of $\mathbb{Z}$ as follows: $\forall n \in \mathbb{Z}$,*

$$\mu(n) = \begin{cases} 1 \text{ if } n = 0, \\ 0 \text{ otherwise.} \end{cases}$$

Then μ is not a fuzzy p-subgroup of $\mathbb{Z}$, but μ is vacuously a fuzzy p^*-subgroup of $\mathbb{Z}$.

Proposition 7.5.20. *Let f be a homomorphism of G onto H. Let μ and ν be fuzzy subgroups of G and H, respectively. Then the following assertions hold.*
(1) Suppose that μ is f-invariant. If μ is a fuzzy p^*-subgroup of G, then $f(\mu)$ is a fuzzy p^*-subgroup of H.
(2) If ν is a fuzzy p^*-subgroup of H, then $f^{-1}(\nu)$ is a fuzzy p^*-subgroup of G.

Proof. (1) Let $n \in \mathbb{N}$ and $y \in H$. Then $(f(\mu))(y^n) = \vee\{\mu(z) \mid f(z) = y^n, z \in G\} = \mu(z) \forall z \in G$ such that $f(z) = y^n$. Hence $(f(\mu))(y^n) = \mu(w^n)$, where $w \in G$ is such that $f(w) = y$.
(2) Let $n \in \mathbb{N}$ and $x \in G$. Then $(f^{-1}(\nu))(x^n) = \nu(f(x^n)) = \nu(f(x)^n)$, where $y \in H$ and $x \in G$.   □

Proposition 7.5.20(1) does not hold in general if μ is not f-invariant. This can be seen from the following example.

Example 7.5.21. *Define the fuzzy subset μ of $\mathbb{Z}$ as follows: $\forall n \in \mathbb{Z}$,*

$$\mu(n) = \begin{cases} 1 & \text{if } n = 0, \\ \frac{1}{2} & \text{if } n \in 3\mathbb{Z} \setminus \{0\}, \\ 0 & \text{otherwise.} \end{cases}$$

Let f be a homomorphism of $\mathbb{Z}$ onto $\mathbb{Z}_2$. Then μ is a fuzzy 3^*-subgroup, but $f(\mu)$ is a fuzzy 2^*-subgroup. Note that $f(1) = f(3)$ while $\mu(1) \neq \mu(3)$ and so μ is not f-invariant.

We now consider fuzzy subgroups of a cyclic group. Corresponding to a fuzzy subgroup μ of a cyclic group $\langle x \rangle$, there exists a (finite or infinite) strictly increasing sequence $\{n_i\}$ of positive integers as follows:

$$n_1 = \wedge\{k \in \mathbb{N} \mid \mu(x^k) = \wedge\{\mu(y) \mid y \in \langle x \rangle\}\} = 1,$$
$$n_2 = \wedge\{k \in \mathbb{N} \mid \mu(x^k) = \wedge(\{\mu(y) \mid y \in \langle x \rangle\} \backslash \{\mu(x)\})\},$$
$$n_3 = \wedge\{k \in \mathbb{N} \mid \mu(x^k) = \wedge(\{\mu(y) \mid y \in \langle x \rangle\} \backslash \{\mu(x), \mu(x^{n_2})\})\},$$

and so on.

The sequence $\{n_i\}$ is independent of any specific choice of a generator of the cyclic group $\langle x \rangle$. In fact, if y is another generator of $\langle x \rangle$, we have $\mu(x^r) = \mu(y^r)$ for all integers r. Also, it follows that $n_i | n_{i+1}$ for all $i = 1, 2, \dots$.

Let p be a prime p. Let $\{m_i\}$ be a sequence of nonnegative integers such that $p^{m_i} | n_i$ and $p^{m_i+1} \nmid n_i, i = 1, 2, \dots$. Consider the strictly increasing maximal subsequence $\{m_{i_j}\}$ of $\{m_i\}$ such that each index i_j is minimal. The sequence $\{m_i\}$ depends on the particular choice of a prime p.

Theorem 7.5.22. *Let μ be a fuzzy subgroup of a cyclic group $\langle x \rangle$. Let n_i and m_i be defined as above. Let p be a prime. Then $\mu_{(p)^*}$ is determined as follows:*
(1) *If $FO_\mu(x)$ is finite, then*

$$\mu_{(p)^*}(x^t) = \begin{cases} \mu(x^{n_{i_j-1}}) & \text{if } t \in (p^{m_{i_j-1}})\mathbb{Z} \backslash (p^{m_{i_j}})\mathbb{Z}, \\ \mu(e) & \text{otherwise, i.e., } t \in (p^{m_{i_j}}\mathbb{Z}) \text{ and } m_{i_j} \text{ is the last element of} \\ & \text{the subsequence,} \end{cases}$$

(2) *If $FO_\mu(x)$ is infinite, then*

$$\mu_{(p)^*}(x^t) = \begin{cases} \mu(x^{n_{i_j-1}}) & \text{if } t \in (p^{m_{i_j-1}})\mathbb{Z} \backslash (p^{m_{i_j}})\mathbb{Z}, \\ \vee\{\mu(y) \mid y \in \langle x \rangle, y \neq e\} & \text{if } t \in (p^{m_{i_j}})\mathbb{Z} \backslash \{0\} \text{ and } m_{i_j} \text{ is the last} \\ & \text{element of the subsequence,} \\ \mu(e) & \text{if } t = 0. \end{cases}$$

Proof. (1) Since $\mu_{(p)^*}$ is a fuzzy p^*-subgroup, the $\mu_{(p)^*}(x^t)$ are equal for all $t \in (p^{m_{i_j-1}})\mathbb{Z} \backslash (p^{m_{i_j}})\mathbb{Z}$. Also, $\mu(x^{n_{i_j-1}})$ is the maximal value of the $\mu(x^t)$ for these t. However, $\mu_{(p)^*}$ contains μ. Thus $\mu_{(p)^*}(x^t) = \mu(x^{n_{i_j-1}})$ for all such t since $\mu_{(p)^*}$ is minimal. Next, let $t \in (p^{m_{i_j}})\mathbb{Z}$, where m_{i_j} is the last element of the subsequence. If $(FO_\mu(x), p) \neq 1$, then $FO_\mu(x) \in (p^{m_{i_j}})\mathbb{Z}$ and similarly $\mu_{(p)^*}(x^t) = \mu(e)$. If $(FO_\mu(x), p) = 1$, then $m_{i_j} = 1$ and we have $\mu_{(p)^*}(x) = \mu_{(p)^*}(x^{FO_\mu(x)}) = \mu(e)$.

(2) The first part, even if it involves infinitely many steps, is quite similar to the first part of (1). For the second part, the $\mu_{(p)^*}(x^t)$ are equal for all $t \in (p^{m_{i_j}})\mathbb{Z} \backslash \{0\}$ since m_{i_j} is the last element of the subsequence and $\mu_{(p)^*}$ is a fuzzy p^*-subgroup. Thus for all such t, $\mu_{(p)^*}(x^t) \geq \vee\{\mu(x^t) \mid t \in (p^{m_{i_j}})\mathbb{Z} \backslash \{0\}\} = \vee\{\mu(y) \mid y \in \langle x \rangle, y \neq e\}$ since $\mu_{(p)^*}$ contains μ. However, $\mu_{(p)^*}$ is minimal. Hence $\mu_{(p)^*}(x^t) = \vee\{\mu(y) \mid y \in \langle x \rangle, y \neq e\}$ for all $t \in (p^{m_{i_j}})\mathbb{Z} \backslash \{0\}$. $\square$

Corollary 7.5.23. *Assume the hypothesis of Theorem 7.5.22. If $FO_\mu(x)$ is finite, then $\mu_{(p)^*} = \mu_{(p)}$. If $FO_\mu(x)$ is infinite, $\vee\{\mu(y) \mid y \in \langle x \rangle, y \neq e\} = \mu(e)$,* and if $\{m_{i_j}\}$ is a finite sequence, then $\mu_{(p)^*} = \mu_{(p)}$. Otherwise, $\mu_{(p)}$ does not exist.

Corollary 7.5.24. Let μ be a fuzzy subgroup of a cyclic group $G = \langle x \rangle$. Then μ is a fuzzy p^*-subgroup if and only if either G/μ_* is a p-group or given $t \in [0, \mu(e)]$, the t-level subgroup of μ is $\langle x^{p^n} \rangle$ for some nonnegative integer n.

Proof. The proof follows by applying Proposition 7.5.3 to Corollary 7.5.23.

□

Let p be a prime. Then Theorem 7.5.22 and Corollary 7.5.23 characterize $\mu_{(p)}$ and $\mu_{(p)^*}$ for a fuzzy subgroup μ of a cyclic group. On the other hand, the following theorem and corollary show that a fuzzy subgroup of a cyclic group is completely determined by its fuzzy p^*-subgroups.

Theorem 7.5.25. Let μ be a fuzzy subgroup of cyclic group $\langle x \rangle$. Then $\mu = \cap \mu_{(p_i)^*}$, where the intersection is over the set of all distinct primes p_i.

Proof. Since $\mu_{(p_i)^*}$ contains μ for each p_i, it suffices to show that $\mu \supseteq \cap \mu_{(p_i)^*}$. Assume the hypothesis of Theorem 7.5.22. Let $t \in \mathbb{N}$. Then there exists $n_k \in \{n_i\}$ such that $\mu(x^{n_k}) = \mu(x^t)$. If n_k is the last element of the sequence $\{n_i\}$, then $\mu(x^{n_k}) = \vee \{\mu(x^s)|s \in \mathbb{N}\}$. Let p be a prime such that p is not a factor of n_k. Then the subsequence $\{m_{i_j}\}$, which depends on the p_i, is precisely $\{0\}$. Thus $\mu_{(p)^*}(x^t) = \mu(x^{n_k}) = \mu(x^t)$ by Theorem 7.5.22. On the other hand, if n_k is not the last element of the sequence $\{n_i\}$, then there exists a prime p such that $p^{m_k} < p^{m_{k+1}}$ and $p^{m_{k+1}} \nmid t$ since n_{k+1} is a proper multiple of n_k, where $\{m_i\}$ is the sequence which depends on the p. However, $p^{m_k}|t$ since $n_k|t$. Hence $\mu_{(p)^*}(x^t) = \mu(x^{n_k}) = \mu(x^t)$ by Theorem 7.5.22. Thus there exists a prime p such that $\mu_{(p)^*}(x^t) = \mu(x^t) \ \forall t \in \mathbb{N}$. Therefore, $\mu \supseteq \cap \mu_{(p_i)^*}$.   □

Corollary 7.5.26. Let μ and ν be fuzzy subgroups of the cyclic groups G and H, respectively. Then μ and ν are isomorphic if and only if $\mu_{(p)^*}$ and $\nu_{(p)^*}$ are isomorphic for every prime p.

Proof. If $|G| = |H|$ is finite, then the proof follows by Theorem 7.5.12 and Corollary 7.5.23. Thus let G and H be infinite cyclic groups. The necessity is easily obtained by Proposition 7.5.20. We now consider the sufficiency. For each prime p, let $f_p : G \rightarrow H$ be an isomorphism such that $f_p(\mu_{(p)^*}) = \nu_{(p)^*}$. Since G and H are infinite cyclic groups, they are isomorphic under exactly two isomorphisms g and h, say, such that $g(x) = y$ and $h(x) = y^{-1}$, where x and y are generators of G and H, respectively. Let f be one of these isomorphisms. Then $f(\mu) = \nu$ by Theorem 7.5.25.   □

References

1. S. Abou-Zaid, On generalized characteristic fuzzy subgroups of a finite group, *Fuzzy Sets and Systems* 43 (1991) 234-241.
2. S. Abou-Zaid, On fuzzy subnormal and pronormal subgroups of finite group, *Fuzzy Sets and Systems* 47 (1992) 346-349.
3. M. Akgül, Some properties of fuzzy groups, *J. Math. Anal. Appl.* 133 (1988) 93-100.
4. Y. Alkhamees, Fuzzy cyclic subgroups and fuzzy cyclic *p*-subgroups, *J. Fuzzy Math.* 3 (1995) 911-919.
5. Y. Alkhamees, Fuzzy direct product of fuzzy subgroups of subgroups, *J. Fuzzy Math.* 6 (1998) 307 - 318.
6. J. M. Anthony and H. Sherwood, A characterization of fuzzy subgroups, *Fuzzy Sets and Systems* 7 (1982) 297-305.
7. J. M. Anthony and H. Sherwood, Fuzzy subgroups redefined, *J. Math. Anal. Appl.* 69 (1979)124-130.
8. M. Asaad, Groups and fuzzy subgroups, *Fuzzy Sets and Systems* 39 (1991) 323-328.
9. P. Bhattacharya, Fuzzy subgroups: Some characterizations, *J.Math Anal. Appl.* 128(1987), 241-252.
10. P.S. Das, Fuzzy groups and level subgroups, *J.Math. Anal. Appl.* 84(1981) 246-269.
11. V. N. Dixit, R.Kumar and N. Ajmal, Level subgroups and union of fuzzy subgroups, *Fuzzy Sets and Systems* 37 (1990) 359-371.
12. B. Huppert, *Endliche Gruppen I*, Springer-Verlag, Berlin, 1967.
13. J. G. Kim, On fuzzy orders of elements of fuzzy subgroups, *J. Kyungsung Univ.* 13 (1992) 251-257.
14. J. G. Kim, Fuzzy subgroups and minimal fuzzy *p*-subgroups, *J. Fuzzy Math.* 2 (1994) 913-921.
15. J. G. Kim, Fuzzy orders relative to fuzzy subgroups, *Inform. Sci.* 80 (1994) 341-348.
16. J. G. Kim, Fuzzy subgroups having the property (∗), *Inform. Sci.* 80 (1994) 235-241.
17. J. G. Kim, Orders of fuzzy subgroups and fuzzy *p*-subgroups, *Fuzzy Sets and Systems* 61(1994), 225-230.
18. J. G. Kim and H. D. Kim, A characterization of fuzzy subgroups of some abelian groups, *Inform. Sci.* 80 (1994) 243-252.
19. J. G. Kim and H. D. Kim, Some characterizations of fuzzy subgroups: via fuzzy p^*-subsets and fuzzy p^*-subgroups, *Fuzzy Sets and Systems* 102 (1999) 327-332
20. J. G. Kim, Some characterizations of fuzzy subgroups, *Fuzzy Sets and Systems* 87 (1997) 243-249.
21. R. Kumar, Fuzzy Sylow subgroups, *Fuzzy Sets and Systems* 46 (1992) 267-271.
22. R. Kumar, Fuzzy subgroups, fuzzy ideals and fuzzy cosets, some properties, *Fuzzy Sets and Systems* 48(1992)267-274.
23. I. J. Kumar, R.K. Saxena and P. Yadav, Fuzzy normal subgroups and fuzzy quotients, *Fuzzy Sets and Systems*, 46 (1992) 121-132.
24. Wang-jin Liu, Fuzzy invariants subgroups and fuzzy ideals, *Fuzzy Sets and Systems* 8 (1982) 133-139.
25. D. S. Malik and J. N. Mordeson, Fuzzy subgroups of abelian groups, *Chinese J. of Math.* 19, 2 (1991) 129-145.

26. N. P. Mukherjee and P. Bhattacharya, Fuzzy normal subgroups and fuzzy cosets, *Inform. Sci.* 34 (1984) 225-239.
27. S. Ray, Isomorphic fuzzy groups, *Fuzzy Sets and Systems* 50 (1992) 201-207.
28. J. S. Rose, A course on group theory, Cambridge University Press, Cambridge, 1985.
29. A. Rosenfeld, Fuzzy groups, *J. Math Anal. Appl.* 35 (1971) 512-517.
30. J. J. Rotman, *An Introduction to the Theory of Groups*, 3rd ed., Allyn and Bacon, Boston, MA, 1984.
31. H. Sherwood, Products of fuzzy subgroups, *Fuzzy Sets and Systems* 11 (1983) 79-89.
32. B. W. Wetherilt, Semidirect products of fuzzy subgroups, *Fuzzy Sets and systems* 16 (1985) 237-242.
33. L.A. Zadeh, Fuzzy sets, *Inform. and Control,* 8 (1965) 338-353.

Equivalence of Fuzzy Subgroups of Finite Abelian Groups

In this chapter, we determine the number of fuzzy subgroups of certain finite Abelian groups with respect to a suitable equivalence relation. This is motivated by the realization that in a theoretical study of fuzzy groups, fuzzy subgroups are distinguished by their level sets and not by their images in $[0, 1]$. We first present some results on the equivalence of fuzzy subsets of a given set. The results of this chapter are mainly from [4, 9, 10, 11, 12].

8.1 A Relation on the Set of Fuzzy Subsets of a Set

Let X be a set. Define the relation $\sim$ on $\mathcal{FP}(X)$ as follows: $\forall \mu, \nu \in \mathcal{FP}(X)$, $\mu \sim \nu$ if and only if for all $x, y \in X$, $\mu(x) > \mu(y)$ if and only if $\nu(x) > \nu(y)$ and $\mu(x) = 0$ if and only if $\nu(x) = 0$. Then it follows easily that $\sim$ is an equivalence relation on $\mathcal{FP}(X)$. If $\sim$ is restricted to the crisp subsets of X, $\mathcal{C}(X) = \{f \mid f : X \to \{0, 1\}\}$, then $\sim$ reduces to equality of subsets. The condition $\mu(x) = 0$ if and only if $\nu(x) = 0$ states that the supports of μ and ν are equal. This condition does not follow from the condition that $\mu(x) > \mu(y)$ if and only if $\nu(x) > \nu(y)$. Example 8.1.1 illustrates this fact. In this section, we consider $\sim$ on $\mathcal{FP}(X)$, where X is a set.

Example 8.1.1. *Let $G = <a>$ be a cyclic group of order 2. Define the fuzzy subsets μ and ν of G as follows : $\forall x \in G$,*

$$\mu(x) = \begin{cases} 1 & \text{if } x = e, \\ t & \text{if } x = a; \end{cases}$$

$$\nu(x) = \begin{cases} 1 & \text{if } x = e, \\ 0 & \text{if } x = a, \end{cases}$$

where $0 < t < 1$. Clearly, $\mu(x) > \mu(y)$ if and only if $\nu(x) > \nu(y)$ for all $x, y \in G$. However, $\mu^* \neq \nu^*$ and therefore μ is not equivalent to ν.

Proposition 8.1.2. *Let μ and ν be fuzzy subsets of a set X. If $\mu \sim \nu$, then $|Im(\mu)| = |Im(\nu)|$.*

Proof. Define f from $\text{Im}(\mu)$ into $\text{Im}(\nu)$ by $f(\mu(x)) = \nu(x)$ for all $x \in X$. Then it follows easily that f is well-defined, one-to-one, and onto. $\qquad \square$

Example 8.1.3. *Consider the symmetric group on three letters,*
$S_3 = \{e, a, a^2, b, ab, a^2b\}$. *Define the fuzzy subsets μ and ν of S_3 as follows:*
$\forall x \in S_3$,

$$\mu(x) = \begin{cases} 1 & \text{if } x = e, \\ 1/2 & \text{if } x = b, \\ 1/3 & \text{otherwise;} \end{cases}$$

$$\nu(x) = \begin{cases} 1 & \text{if } x = e, \\ 1/2 & \text{if } x = ab, \\ 1/3 & \text{otherwise.} \end{cases}$$

Then $\text{Im}(\mu) = \text{Im}(\nu)$ and $\mu^* = \nu^*$. However, $\mu(b) > \mu(ab)$, but $\nu(b) \not> \nu(ab)$. Therefore, μ is not equivalent to ν.

Proposition 8.1.4. *Let μ and ν be fuzzy subsets of X. Suppose that for all $t \in (0, 1]$, there exists $s \in (0, 1]$ such that $\mu_t = \nu_s$ and for all $t \in (0, 1]$, there exists $s' \in (0, 1]$ such that $\nu_t = \mu_{s'}$. Then $\mu \sim \nu$.*

Proof. If $\mu^* = \emptyset$, then clearly $\nu^* = \emptyset$. Thus $\nu(x) = 0 = \mu(x)$ for all $x \in X$. Hence $\mu \sim \nu$ trivially. Suppose $\mu^* \neq \emptyset$. Let $x \in \mu^*$. Then $x \in \mu_t$ for some $t \in (0, 1]$. By hypothesis, there exists $s \in (0, 1]$ such that $\mu_t = \nu_s$. Hence $\nu(x) \geq s > 0$. Thus $x \in \nu^*$. Hence $\mu^* \subseteq \nu^*$. Similarly, one can show that $\nu^* \subseteq \mu^*$. Therefore, $\mu^* = \nu^*$. Now let $t = \mu(x) > \mu(y)$ for $x, y \in G$. By hypothesis, $x \in \nu_s = \mu_t$ for some $s \in (0, 1]$. If $\nu(x) \leq \nu(y)$, then $\nu(y) \geq s$ and so $y \in \nu_s = \mu_t$, a contradiction. Hence $\nu(x) > \nu(y)$. Similarly, $\nu(x) > \nu(y)$ implies $\mu(x) > \mu(y)$. $\qquad \square$

We now show a partial converse to the above proposition. By the notation $(1, 1]$, we mean the empty set.

Proposition 8.1.5. *Suppose μ and ν are fuzzy subsets of X such that $\mu \sim \nu$. Then for all $t \in [0, 1]$, there is an $s \in [0, 1]$ such that $\mu_t = \nu_s$ or $\mu_t = \nu^{-1}(s, 1]$.*

Proof. Let $a = \vee\{\mu(x) \mid x \in X\}$ and $b = \vee\{\nu(x) \mid x \in X\}$. For $t = a$, choose $s = b$. Then clearly $\mu_t = \nu_s$. For $t < a$, consider $t_1 = \wedge\{\mu(x) \mid x \in \mu_t\}$. Suppose $t_1 = \mu(y)$ for some $y \in X$. Then s can be chosen to be $\nu(y)$. To show that μ_t is contained in ν_s, let $x \in \mu_t$. Then $\mu(x) \geq t$ and so $\mu(x) \geq t_1 = \mu(y)$. Since $\mu \sim \nu$, $\nu(x) \geq \nu(y) = s$. Thus $x \in \nu_s$. Hence $\mu_t \subseteq \nu_s$. The reverse inclusion can be shown similarly.

Suppose $t_1 \neq \mu(y)$ for all $y \in X$. Then $\mu_t = \mu_{t_1} = \mu^{-1}(t_1, 1] = \mu^{-1}(t, 1]$. Let $s = \wedge\{\nu(x) \mid x \in \mu_t\}$. Let $x \in \mu_t$. Then $\mu(x) > t_1$ and so $\mu(x) > \mu(y)$ for some $y \in \mu_t$. Since $\mu \sim \nu$, $\nu(x) > \nu(y) \geq s$. Thus $x \in \nu^{-1}(s, 1]$. Hence $\mu_t \subseteq \nu^{-1}(s, 1]$. Now let $x \in \nu^{-1}(s, 1]$. Then $\nu(x) > s$ and so $\nu(x) > \nu(y)$ for some $y \in \mu_t$. Since $\mu \sim \nu$, $\mu(x) > \mu(y) > t$. Hence $x \in \mu_t$. Thus $\nu^{-1}(s, 1] \subseteq \mu_t$ and so $\mu_t = \nu^{-1}(s, 1]$. Now consider the case where $a < 1$ and $t > a$. Then $\mu_t = \emptyset$. Suppose $b < 1$. Then $\nu_1 = \emptyset = \mu_t$. If for all $x \in X, \nu(x) < b = 1$, then

$\nu_b = \emptyset = \mu_t$. If $\nu(x) = 1 = b$ for some $x \in X$, then for any $s \in [0,1], \nu_s \neq \emptyset$. In this case, choose $s = 1$. Then $\nu^{-1}(s,1] = \emptyset = \mu_t$.   □

The following examples show that the converse of Proposition 8.1.4 does not hold.

Example 8.1.6. *Let* $X = \{x\}$. *Define the fuzzy subsets* μ *and* ν *of* X *as follows:* $\mu(x) = 1/2$ *and* $\nu(x) = 1$. *Then* $\mu \sim \nu$. *Let* $t = 3/4$. *Then* $\mu_t = \emptyset$, *but* $\nu_s = X$ *for all* $s \in [0,1]$. *However, Proposition 8.1.5 assures that* $\mu_{3/4} = \nu^{(-1)}(1,1]$.

Example 8.1.7. *Let* $X = [0,1]$. *Define the fuzzy subset* μ *of* X *by* $\mu(x) = 1 - \frac{1}{2}x$ *if* $x \in [0,1)$ *and* $\mu(1) = 0$. *Define the fuzzy subset* ν *of* X *by* $\nu(x) = 1 - x$ *for all* $x \in [0,1]$. *Then* $\mu \sim \nu$ *and* $1 \in \mathrm{Im}(\mu), 1 \in \mathrm{Im}(\nu)$. *Let* $t = 1/2$. *Then* $\mu_t = [0,1)$. *However, there is no* $s \in [0,1]$ *such that* $\nu_s = [0,1)$. *Note,* $\mu_{1/2} = \nu^{(-1)}(0,1]$.

In the next example, we show that the conditions $t > 0$ and $s > 0$ in Proposition 8.1.4 are needed. This example also shows that the converse of Proposition 8.1.5 doesn't hold.

Example 8.1.8. *Let* $X = \{x\}$. *Define the fuzzy subsets* μ *and* ν *of* X *as follows:* $\mu(x) = 1/2$ *and* $\nu(x) = 0$. *If* $t \in [0,1/2]$, *let* $s = 0$. *Then* $\mu_t = X = \nu_s$. *If* $t \in (1/2,1]$, *let* $s \in (0,1]$. *Then* $\mu_t = \emptyset = \nu_s$. *However,* μ *and* ν *are not equivalent.*

We now consider an equivalence of fuzzy subsets for which the converse of Proposition 8.1.4 holds.

Definition 8.1.9. *Let* X *be a set and* μ *a fuzzy subset of* X. *Let* $x \in X$. *Suppose* $\forall \epsilon > 0$, *there exists* $y \in X$ *such that* $\mu(x) + \epsilon > \mu(y) > \mu(x)$. *Then* $\mu(x)$ *is called a **right** μ**-limit point** of* $\mathrm{Im}(\mu)$. *Let* $\mu^R = \{x \in X \mid \mu(x)$ *is a right* μ*-limit point of* $\mathrm{Im}(\mu)\}$.

Definition 8.1.10. *Let* X *be a set and let* μ *and* ν *be fuzzy subsets of* X *such that* $\mu \sim \nu$. *If* $\mu^R = \nu^R$, *then* μ *and* ν *are said to be **strongly equivalent** written* $\mu \approx \nu$.

It is clear that the notion of strong equivalence is an equivalence relation. Also, if μ is a fuzzy subset of a set X such that $\mathrm{Im}(\mu)$ is finite, then $\mu^R = \emptyset$. Hence when considering the equivalence of fuzzy subgroups of finite groups, the notions of equivalence and strong equivalence are the same. The notion of strong equivalence will be used in the consideration of fuzzy subgroups of infinite groups.

Theorem 8.1.11. *Let* μ *and* ν *be fuzzy subsets of a set* X. *Then* $\mu \approx \nu$ *if and only if* $\mu^* = \nu^*$ *and for all* $t \in [0,1], \mu_t \neq \emptyset$ *implies there is an* $s \in [0,1]$ *such that* $\mu_t = \nu_s$ *and for all* $s \in [0,1], \nu_s \neq \emptyset$ *implies there is a* $t \in [0,1]$ *such that* $\nu_s = \mu_t$.

Proof. Suppose that the conditions hold. Since $\mu^* = \nu^*, \mu(x) = 0 \Leftrightarrow \nu(x) = 0$ for all $x \in X$. Suppose $\mu(x) > \mu(y)$. Let $t = \mu(x)$. Then $x \in \mu_t$ and $\mu_t \neq \emptyset$. Thus there is an $s \in [0,1]$ such that $\mu_t = \nu_s$. Suppose $\nu(y) \geq \nu(x)$. Then $\nu(y) \geq \nu(x) \geq s$ since $x \in \mu_t = \nu_s$. Thus $y \in \nu_s = \mu_t$, a contradiction. Hence $\nu(x) > \nu(y)$. Similarly, $\nu(x) > \nu(y)$ implies $\mu(x) > \mu(y)$. Hence $\mu \sim \nu$. In order to prove $\mu \approx \nu$, it remains to be shown that $\mu^R = \nu^R$. Since $\mu \sim \nu$, it follows for all $x, y \in X$ that $\mu(x) \geq \mu(y)$ if and only if $\nu(x) \geq \nu(y)$ and that $\mu(x) = 0$ if and only if $\nu(x) = 0$. Thus $\mu_{\mu(y)} = \nu_{\nu(y)}$ and $\{y \in X \mid \mu(y) = \mu(x)\} = \{y \in X \mid \nu(y) = \nu(x)\}$ for all $x \in X$. Let $z \in \mu^R$. Then for all $t > \mu(z)$, there exists $y \in X$ such that $\mu(y) > \mu(z)$ and $\mu(y) < t$. Thus $\mu_t \neq \mu_{\mu(z)} \backslash \{y \in X \mid \mu(y) = \mu(z)\}$. Suppose $t \leq \mu(z)$. Then $\mu_t \neq \mu_{\mu(z)} \backslash \{y \in X \mid \mu(y) = \mu(z)\}$ since $\{y \in X \mid \mu(y) = \mu(z)\} \neq \emptyset$. Suppose $\nu(z) \notin \nu^R$. Then there exists $s \in [0,1]$ such that $s > \nu(z)$ and for all $x \in X, \nu(x) > \nu(z) \Leftrightarrow \nu(x) \geq s$. Hence $\nu_s = \nu_{\nu(z)} \backslash \{y \in X \mid \nu(y) = \nu(z)\} = \mu_{\mu(z)} \backslash \{y \in X \mid \mu(y) = \mu(z)\}$. However there is no $t \in [0,1]$ such that $\mu_t = \nu_s$, a contradiction. Thus $z \in \nu^R$. Therefore, $\mu^R \subseteq \nu^R$. By symmetry, $\nu^R \subseteq \mu^R$. Consequently, $\mu \approx \nu$.

Conversely, suppose $\mu \approx \nu$. Then $\mu \sim \nu$ and $\mu^* = \nu^*$. Let $a = \vee\{\mu(x) \mid x \in X\}$ and $b = \vee\{\nu(x) \mid x \in X\}$. If $t > a$, then $\mu_t = \emptyset$. Thus we only need to consider the case $t \leq a$. Since $\mu \sim \nu$, we have that $\mu(x) = a \Leftrightarrow \nu(x) = b$. If $t = a$ and $\mu_t \neq \emptyset$, then there exists $x_0 \in X$ such that $\mu(x_0) = a = t$. Let $s = b$. Then clearly $\mu_t = \nu_s$ since $\mu(x) = a \Leftrightarrow \nu(x) = b$. For $t < a$, we consider $t_1 = \wedge\{\mu(x) \mid x \in \mu_t\}$. The case when $\mu(z) = t_1$ for some $z \in X$ has been considered in Proposition 8.1.5. Thus suppose $t_1 \neq \mu(x)$ for all $x \in X$. Then $\mu_t = \mu_{t_1} = \mu^{-1}(t_1, 1] = \mu^{-1}(t, 1]$. Let $s = \wedge\{\nu(x) \mid x \in \mu_t\}$. Then it is shown in Proposition 8.1.5 that $\mu_t = \nu^{-1}(s, 1] \subseteq \nu^s$. If for any $x \in X, \nu(x) \neq s$, then $\nu^{-1}(s, 1] = \nu_s$ and so $\mu_t = \nu_s$. If there exists $z \in X$ such that $\nu(z) = s$ and $z \in \mu_t$, then $\{x \in X \mid \nu(x) = s\} \subseteq \mu_t$ since $\mu \sim \nu$. Thus $\mu_t = \nu_s$. Suppose $z \notin \mu_t$. If $s = \nu(z) \notin \nu^R$, then $s \in \{\nu(x) \mid x \in \mu_t\}$ since $s = \wedge\{\nu(x) \mid x \in \mu_t\}$. Thus there exists $x_0 \in \mu_t$ such that $\nu(x_0) = s = \nu(z)$ and so $\mu(z) = \mu(x_0) \geq t$ since $\mu \sim \nu$. Hence $z \in \mu_t$, a contradiction. Thus $s = \nu(z) \in \nu^R$ and so $\mu(z) \in \mu^R$. For all $x \in \mu_t, \nu(x) > s$ and so $\mu(x) > \mu(z)$ since $t_1 \neq \mu(z)$. Hence $\mu(z) < t_1$. Since $\mu(z) \in \mu^R$, there exists $y \in X$ such that $\mu(z) < \mu(y) < t_1$ and $y \notin \mu_t = \mu^{-1}(s, 1]$. However, $\nu(y) > \nu(z) = s$ and so $y \in \nu^{-1}(s, 1]$. However, this is impossible. Thus $z \notin \mu_t$ doesn't hold. Therefore, $\mu_t = \nu_s$.   □

8.2 Fuzzy Subgroups of *p*-groups

Let G denote a finite Abelian group. If μ is a fuzzy subgroup of G, we assume that $\mu(0) = 1$, where 0 is the identity of G. Then the only allowable fuzzy subgroup μ of a trivial group is such that $\mu(0) = 1$. For a prime p, a finite group G is called a p-group if the order of G is a power of p. A cyclic p-group is isomorphic to $\mathbb{Z}_{p^n}$ for some positive integer n. Let k be a positive integer. For the cyclic group $\mathbb{Z}_{p^n}$, we write $\mathbb{Z}_{p^n} \supset \mathbb{Z}_{p^k}$, $k < n$, to mean that $\mathbb{Z}_{p^n}$ contains the cyclic subgroup $\langle p^{n-k} \rangle$ of order p^k. For example, by $\mathbb{Z}_{3^3} \supset \mathbb{Z}_{3^2}$ we mean

$\mathbb{Z}_{3^3} \supset \langle 3 \rangle$. We say two fuzzy subgroups are **distinct** if they are not equivalent. A maximal chain $\mathbb{Z}_{p^n} \supset \mathbb{Z}_{p^{n-1}} \supset ... \supset \mathbb{Z}_p \supset 0$ defines a fuzzy subgroup μ as follows: μ assumes t_n on $\mathbb{Z}_{p^n}\backslash\mathbf{Z}_{p^{n-1}}$, t_{n-1} on $\mathbb{Z}_{p^{n-1}}\backslash\mathbb{Z}_{p^{n-2}}, ..., t_1$ on $\mathbb{Z}_p\backslash\{0\}$ and 1 on $\{0\}$, where $1 \geq t_1 \geq t_2 \geq ... \geq t_{n-1} \geq t_n \geq 0$. The fuzzy subgroup μ just defined is simply denoted by $1t_1t_2...t_{n-1}t_n$. We refer to the heights 1, $t_1, ..., t_n$ as symbols. If $n = 1$, then the group is $\mathbb{Z}_p$. Any fuzzy subgroup of $\mathbb{Z}_p$ is equivalent to one of the following three: $11, 1t, 10$, where $1 > t \geq 0$ and the 11 represents the crisp trivial subgroup $\mathbb{Z}_p$ and 10 represents the trivial subgroup $\{0\}$ of $\mathbb{Z}_p$. $1t$ represents the fuzzy subgroup $\mu(x) = 1$ when $x = 0$ and $\mu(x) = t$ when $x \neq 0$. Clearly, these are the only distinct equivalence classes of fuzzy subgroups of $\mathbb{Z}_p$ since the only crisp subgroups of $\mathbb{Z}_p$ are $\{0\}$ and $\mathbb{Z}_p$ itself.

Example 8.2.1. *Let $n = 2$. Then there are seven distinct equivalence classes of fuzzy subgroups of Z_{p^2} corresponding to the maximal chain, $\mathbb{Z}_{p^2} \supset \mathbb{Z}_p \supset \{0\}$. These are given by the symbols*

$$111, 11t, 110, 1tt, 1ts, 1t0, 100,$$

where by $1ts$ we mean the fuzzy subgroup μ defined by $\mu(x) = 1$ when $x = 0, \mu(x) = t$ when $x \in Z_p\backslash\{0\}$ and $\mu(x) = s$ otherwise, with $1 > t > s \geq 0$. It is easily seen that the number of fuzzy subgroups whose support is Z_{p^2} is one more than the number of fuzzy subgroups whose supports are properly contained in Z_{p^2}. Clearly, $7 = \sum_{k=0}^{2} 2^k = 2^{2+1} - 1$. Now 2^2 is the number of fuzzy subgroups whose support is $Z_{p^2}, 2^1$ is the number of fuzzy subgroups whose support is Z_p, and 2^0 is the number of fuzzy subgroups whose support is $\{0\}$.

Example 8.2.2. *Let $n = 3$. We consider the number of distinct equivalence classes of fuzzy subgroups of Z_{p^3}. Consider the list of symbols in the previous example. We attach symbols $1, t, 0$ to $111; t, s, 0$ to $11t; 0$ to $110; t, s, 0$ to $1tt; s, 0, r$ to $1ts; 0$ to $1t0; 0$ to 100. This yields*

$$1111, 111t, 1110, 11tt, 11ts, 11t0, 1100,$$
$$1ttt, 1tts, 1tt0, 1tss, 1ts0, 1tsr, 1t00, 1000.$$

Thus there are 15 distinct fuzzy subgroups of $\mathbb{Z}_{p^3}$. Clearly, $15 = \sum_{k=0}^{3} 2^k = 2^{3+1} - 1$. Each term in the sum in LHS represents the number of fuzzy subgroups of $\mathbb{Z}_{p^3}$ with specific support as described as follows: For $k = 0, 1, 2, 3$, it follows that 2^k is the number of fuzzy subgroups whose support is $\mathbb{Z}_{p^k}$.(Recall that for the cyclic group $\mathbb{Z}_{p^n}$, we write $\mathbb{Z}_{p^n} \supset \mathbb{Z}_{p^k}$, $k < n$, to mean that $\mathbb{Z}_{p^n}$ contains the cyclic subgroup $\langle p^{n-k} \rangle$ of order p^k.)

Proposition 8.2.3. *Let $n \in \mathbb{N}$. Then there are $2^{n+1} - 1$ distinct equivalence classes of fuzzy subgroups of $\mathbb{Z}_{p^n}$.*

Proof. We first note that there is a one-to-one correspondence between the distinct fuzzy subgroups whose support is $\mathbb{Z}_{p^n}$ and distinct fuzzy subgroups whose support is properly contained in $\mathbb{Z}_{p^n}$. In this correspondence, the fuzzy subgroup 111...1 is excluded. We now prove by induction that the number of not equivalent fuzzy subroups of $\mathbb{Z}_{p^n}$ is $\sum_{k=0}^{n} 2^k = 2^{n+1} - 1$. The statement is clearly true for $n = 1, 2$, and 3 as shown in the above examples. Assume the statement is true for $n = k$, i.e., the number of fuzzy subgroups of $\mathbb{Z}_{p^k}$ is $2^{k+1} - 1 = \frac{2^{k+1}}{2} + \frac{2^{k+1}}{2} - 1$, the induction hypothesis. There are $\frac{2^{k+1}}{2}$ not equivalent fuzzy subgroups of $\mathbb{Z}_{p^k}$ whose support is $\mathbb{Z}_{p^k}$. Each gives rise to two fuzzy subgroups of $\mathbb{Z}_{p^{k+1}}$ whose support is $\mathbb{Z}_{p^{k+1}}$ and one fuzzy subgroup whose support is $\mathbb{Z}_{p^k}$. Thus the $\frac{2^{k+1}}{2}$ fuzzy subgroups of $\mathbb{Z}_{p^k}$ give rise to $2(\frac{2^{k+1}}{2}) + (\frac{2^{k+1}}{2})$ fuzzy subgroups of $\mathbb{Z}_{p^{k+1}}$. The remaining $\frac{2^{k+1}}{2} - 1$ fuzzy subgroups of $\mathbb{Z}_{p^k}$ have supports that are properly contained in $\mathbb{Z}_{p^k}$ and thus yield $\frac{2^{k+1}}{2} - 1$ fuzzy subgroups of $\mathbb{Z}_{p^{k+1}}$ by simply attaching zero to each. Thus the number of not equivalent fuzzy subgroups of $\mathbb{Z}_{p^{k+1}}$ is $2(\frac{2^{k+1}}{2}) + \frac{2^{k+1}}{2} + \frac{2^{k+1}}{2} - 1 = 2^{k+1} + 2^{k+1} - 1 = 2 \cdot 2^{k+1} - 1 = 2^{(k+1)+1} - 1$.   □

We now consider $\mathbb{Z}_p \oplus \mathbb{Z}_q$, where p and q are distinct primes. We characterize all equivalent fuzzy subgroups of $\mathbb{Z}_{p^n} \oplus \mathbb{Z}_q$. Let $n = 1$. We see that $\mathbb{Z}_p \oplus \mathbb{Z}_q$ has the following maximal chains each of which can be identified with the chain $\mathbb{Z}_{p^2} \supset \mathbb{Z}_p \supset \{0\}$, namely, $\mathbb{Z}_p \oplus \mathbb{Z}_q \supset \mathbb{Z}_p \oplus \{0\} \supset \{0\}$ and $\mathbb{Z}_p \oplus \mathbb{Z}_q \supset \{0\} \oplus \mathbb{Z}_q \supset \{0\}$. Each of these chains yield 7 distinct fuzzy subgroups according to Example 8.2.1. Of these 7, three yield identical fuzzy subgroups viz. $111, 1tt, 100$. The other 4 viz. $11t, 110, 1ts, 1t0$ yield distinct fuzzy subgroups as can easily be checked by writing it out fully. Therefore $\mathbb{Z}_p \oplus \mathbb{Z}_q$ has 11 nonequivalent fuzzy subgroups. Hence each one of $\mathbb{Z}_6, \mathbb{Z}_{15}, \mathbb{Z}_{35}$ has 11 distinct fuzzy subgroups.

Similarly, we can determine distinct fuzzy subgroups of $\mathbb{Z}_{p^2} \oplus \mathbb{Z}_q$ by considering the three maximal chains each of which is equivalent to the chain $\mathbb{Z}_{p^3} \supset \mathbb{Z}_{p^2} \supset \mathbb{Z}_p \supset \{0\}$. Each chain yields $2^{3+1} - 1 = 15$ fuzzy subgroups according to Example 8.2.2. Of these 15, three yield identical fuzzy subgroups as above and 8 yield two distinct fuzzy subgroups and 4 yield three distinct fuzzy subgroups. Thus we have $3 \cdot 1 + 8 \cdot 2 + 4 \cdot 3 = 31$ distinct fuzzy subgroups of $\mathbb{Z}_{p^2} \oplus \mathbb{Z}_q$. For example, $\mathbb{Z}_{12}, \mathbb{Z}_{18}, \mathbb{Z}_{20}$ each have 31 distinct fuzzy subgroups.
More generally, we have the following result.

Theorem 8.2.4. $\mathbb{Z}_{p^n} \oplus \mathbb{Z}_q$ *has* $2^{n+1}(n + 2) - 1$ *distinct fuzzy subgroups.*

Proof. We prove the result by induction on n. We have seen that the statement is true for the first few n. Assume now that the statement is true for $n = k$, that is, $\mathbb{Z}_{p^k} \oplus \mathbb{Z}_q$ has $2^{k+1}(k+2) - 1$ distinct fuzzy subgroups. We want to show that $\mathbb{Z}_{p^{k+1}} \oplus \mathbb{Z}_q$ has $2^{k+2}(k + 3) - 1$ distinct fuzzy subgroups. The number of distinct fuzzy subgroups with $\mathbb{Z}_{p^k} \oplus \mathbb{Z}_q$ as support is $2^k(k + 2)$. Each of these fuzzy subgroups yield three further distinct fuzzy subgroups and the rest of the fuzzy subgroups have support strictly contained in $\mathbb{Z}_{p^k} \oplus \mathbb{Z}_q$ and thus yield $2^k(k+2) - 1$ distinct fuzzy subgroups. Now $\mathbb{Z}_{p^{k+1}} \oplus \mathbb{Z}_q$ has one maximal chain more than $\mathbb{Z}_{p^k} \oplus \mathbb{Z}_q$. This chain is $\mathbb{Z}_{p^{k+1}} \oplus \mathbb{Z}_q \supset \mathbb{Z}_{p^{k+1}} \oplus \{0\} \supset \cdots \supset \{0\}$. The only fuzzy subgroups from this chain that are not listed from the previous chains are those that have distinct symbols on $\mathbb{Z}_{p^{k+1}} \oplus \{0\}$ and $\mathbb{Z}_{p^{k+1}} \oplus \mathbb{Z}_q$. This set of fuzzy subgroups will come from the set of fuzzy subgroups corresponding to the maximal chain of length $k+2$. Now the number of such fuzzy subgroups is $[(2^{k+3} - 1) + 1]/2 = 2^{k+2}$ since the total number of distinct fuzzy subgroups of a maximal chain of length $k+2$ is $2^{k+3} - 1$. Combining, we have the number of distinct fuzzy subgoups in total to be $3 \times 2^k(k+2) + 2^k(k+2) - 1 + 2^{k+2} = 2^{k+2}(k+3) - 1$. Thus the induction is complete. $\qquad\square$

The next case that we are interested in, is the number of possible distinct fuzzy subgroups of $\mathbb{Z}_p \oplus \mathbb{Z}_p$, where p is prime. However, first we need to determine the number of maximal chains of $\mathbb{Z}_p \oplus \mathbb{Z}_p$.

Theorem 8.2.5. *Let p be a prime. Then the number of maximal chains of $G = \mathbb{Z}_p \oplus \mathbb{Z}_p$ is $p + 1$.*

Proof. Since the order of G is p^2, every nontrivial subgroup of G is cyclic and of order p. Thus every maximal chain of subgroups in G is of the form $G \supset <a> \supset \{0\}$, where a is of the form (i, j) for $1 \leq i, j < p$. The distinct subgroups generated by a are all given by

$$(1, p), (p, 1), (1, 1), (1, 2), (1, 3), ..., (1, p - 2), (i, p - i)$$

for any fixed i strictly between 1 and $p - 1$. The subgroup generated by (i, j) with $i \neq j$ and $i + j \neq p$ contains one of $(1, 2), (1, 3), ..., (1, p - 2)$. The case $i = j$ is covered by $(1, 1)$ and the case $i + j = p$ is covered by $(i, p - i)$. The only two other cases that are not covered by these cases are $(1, p)$ and $(p, 1)$ which are listed above. $\qquad\square$

Proposition 8.2.6. *For p a prime, the number of distinct fuzzy subgroups of $G = \mathbb{Z}_p \oplus \mathbb{Z}_p$ is $4p + 7$.*

Proof. Since there are three levels in every maximal chain of G, we have a similar situation as in Example 8.2.1. Of the 7 possible symbols, there are 4 that will give rise to distinct fuzzy subgroups and these are of the form $11t, 110, 1ts$, and $1t0$. Thus by Theorem 8.2.5, there are $4(p + 1)$ distinct fuzzy subgroups for this situation. The rest of the symbols each give rise to one fuzzy subgroup. Hence there are $4(p + 1) + 3 = 4p + 7$ distinct fuzzy subgroups. $\qquad\square$

The notion of equivalence can be viewed as a special case of the following notion of fuzzy isomorphism: Let G be a group and μ and ν be fuzzy subgroups of G. If there exists an isomorphism $f : \mu^* \to \nu^*$ such that $\mu(a) > \mu(b) \Leftrightarrow \nu(f(a)) > \nu(f(b))$ where $a, b \in \mu^*$, then μ is said to be **fuzzy isomorphic** to ν and we write $\mu \simeq \nu$. It is clear that the notion of fuzzy isomorphism is an equivalence relation on the family of all fuzzy subgroups of G. The notion of equivalence of fuzzy subgroups as we have defined is finer than the usual notion of fuzzy isomorphism of two fuzzy subgroups as defined above. That is, if two fuzzy subgroups are equivalent then they are fuzzy isomorphic, but not vice versa as illustrated by the following example.

Example 8.2.7. *Let* $G = \mathbb{Z}_2 \oplus \mathbb{Z}_2$. *Define fuzzy subgroups* μ *and* ν *of* G *as follows:* $\forall x \in G$,
$$\mu(x) = \begin{cases} 1 & \text{if } x = (0,0), \\ t & \text{if } x \in \mathbf{Z}_2 \oplus \{0\} \backslash \{(0,0)\}, \\ s & \text{otherwise;} \end{cases}$$
$$\nu(x) = \begin{cases} 1 & \text{if } x = (0,0), \\ t & \text{if } x \in \{0\} \oplus \mathbf{Z}_2 \backslash \{(0,0)\}, \\ s & \text{otherwise,} \end{cases}$$
where $t > s$. *The function on* G *given by* $(a,b) \rightarrow (b,a)$ *is a fuzzy isomorphism. Thus* $\mu \simeq \nu$, *but* μ *is not equivalent to* ν.

We also note that if μ and ν are fuzzy subgroups of G such that ν is defined as in Example 8.2.7, $\mu(0,0) = \mu(1,0) = 1$ and $\mu(x) = t$ otherwise, then μ is not isomorphic to ν although their supports are isomorphic. To see this, suppose μ and ν are isomorphic and let $f : \mu^* \rightarrow \nu^*$ be an isomorphism. Then $\mu(1,0) = 1 = \mu(0,0)$ implies $\nu(f(1,0)) = 1 = \nu(0,0)$. Thus $f(1,0) \neq (1,1)$ or $(1,0)$. Thus we have $f(1,0) = (0,1)$ which implies $\nu(0,1) = 1$, a contradiction.

If $\mu \simeq \nu$, then $|\text{Im}(\mu)\backslash\{1\}| = |\text{Im}(\nu)\backslash\{1\}|$. We have shown previously that $\mathbb{Z}_{2^2}$ has 7 distinct (nonequivalent) fuzzy subgroups, namely,

$$111, 11t, 110, 1tt, 1ts, 1t0, 100.$$ The three maximal chains of $\mathbb{Z}_2 \oplus \mathbb{Z}_2$ yield 15 distinct fuzzy subgroups of $\mathbb{Z}_2 \oplus \mathbb{Z}_2$. However, $\mathbb{Z}_2 \oplus \mathbb{Z}_2$ has 7 non-isomorphic fuzzy subgroups. For example, the μ and ν defined in the previous example are fuzzy isomorphic, implying that the fuzzy subgroup $1ts$ counts only once in the case of isomorphism. However, it counts three times in the case of equivalence.

8.3 Pad Keychains

We now consider the problem of determining the number of distinct equivalence classes of fuzzy subgroups of $G = \mathbb{Z}_{p_1} \oplus ... \oplus \mathbb{Z}_{p_n}$, where $p_1, p_2, ..., p_n$ are distinct primes. We introduce the notion of a keychain of a chain of length $n+1$ and the index of a keychain in order to determine the number of fuzzy subgroups of G. We use the terminology as introduced in [10].

In the previous section, we examined $\mathbb{Z}_{p^n}$, $\mathbb{Z}_p \oplus \mathbb{Z}_p$, and $\mathbb{Z}_{p^n} \oplus \mathbb{Z}_q$ where p and q are distinct primes. We now investigate the number of fuzzy subgroups of G of the form $\mathbb{Z}_{p_1} \oplus ... \oplus \mathbb{Z}_{p_n}$ for distinct primes $p_1, p_2, ..., p_n$. Since the membership values of fuzzy subgroups of finite groups form a finite chain, we consider chains of real numbers in $[0, 1]$ each of length $n + 1$.

For the equivalence relation $\sim$, we denote the equivalence class containing μ by $[\mu]$. A **finite n-chain** is a collection of numbers in $[0, 1]$ of the form $1 > t_1 > t_2 > ... > t_{n-1} > t_n$, where t_n may or may not be zero. This is written simply as $1t_1t_2...t_n$ in descending order. The numbers $1, t_1, ..., t_{n-1}, t_n$ are called **pins** . We say that 1 occupies the first position and t_i occupies the $(i + 1)$-th position for $i = 1, ..., n$. The length of an n-chain is $n + 1$. Thus the number of positions available in an n-chain is equal to the length of the chain which is $n + 1$. The positions play a crucial role in the investigation. Unless otherwise stated, we fix n.

An n-chain is called a **keychain** if $1 \geq t_1 \geq t_2 \geq ... \geq t_{n-1} \geq t_n$. A consecutive occurrence of equality signs is said to be an **interlocking position** of pins. Interlocked pins are called **components**. However, a 1 standing alone in the first position not interlocked with any other t's is not considered to be a component of any keychain. A k-pad ($1 \leq k \leq n$) is a keychain that contains k distinct components. For example, $1 > t_1 = t_2 > t_3 = t_4 = t_5 > t_6 > 0$ is a 4-pad keychain of a 7-chain, whereas $1 = t_1 = t_2 = t_3 > t_4 > t_5 = t_6 = t_7$ is a 3-pad keychain of a 7-chain. The number of pins found in the interlocked position forming a component is called the **padidity** of the component. Thus in our example of a 4-pad keychain, the padidities of the components are respectively $2, 3$ and 1. The **index** of a k-pad keychain is defined to be the set of padidities of various components of the keychain in which singleton components are ignored for the sake of simplicity. Thus the index of a keychain forms a 'partition' of the number n. In our example, for $n = 7$ and $k = 4$, the partition of 7 given by $2 + 3 + 1 + 1$ corresponds to a 4-pad keychain whose index is $(2, 3)$. Another partition of 7 given by $3 + 1 + 3$ corresponds to a 3-pad keychain whose index is $(3, 3)$. More generally, by an index of a k-pad keychain we mean a finite set (unordered) of positive integers $(l_1, l_2, ..., l_k)$, where $l_i \geq 2$ for $i = 1, ..., k$ and $l_1 + l_2 + ... + l_k \leq n$. From these examples, it we see that the index and the set of padidities determine each other. We next determine the number of keychains of a chain of length n. When we write $*$-pad, the $*$ denotes a natural number between 1 and $n - 1$.

For any $n \geq 2$, there are three 1-pad keychains. They are of the form $111...1, 1tt...t, 100...0$, with the padidity being $n - 1$.

The following propositions illustrate the inductive steps needed for proving the main result. Accordingly, in the following, we assume $2 \leq k \leq n - 2$.

Proposition 8.3.1. *The number of $(n - k)$-pad keychains of length n with index k is $4(n - k)$.*

Proof. Since the index of any $(n - k)$-pad keychain is given to be k, it follows that one pin is repeated k times and the other $(n - k - 1)$-pins are distinct. Consider the $(n - k)$-pad keychain given as follows:

$$1 \geq t_1 = t_2 = ... = t_k > t_{k+1} > ... > t_{n-1}.$$

Corresponding to this chain, there are 4 distinct $(n - k)$-pad keychains:
$1 = t_1 = ... = t_k > t_{k+1} > ... > t_{n-1}$ and $1 > t_1 = t_2 = ... = t_k > t_{k+1} > ... > t_{n-1}$ with $t_{n-1} > 0$ and two more keychains obtained by letting $t_{n-1} = 0$.
That is, there are $4(n - k)$-pad keychains corresponding to t_1. Similarly, we obtain 4 distinct $(n - k)$-pad keychains corresponding to $t_2, ..., t_{n-k}$ on up to $t_{n-k}, ..., t_{n-1}$. Clearly, these are the only possible $(n - k)$-pad keychains of index k, yielding a total of $4(n - k)$. From the above argument, it follows that there are four $(n - 1)$-pad keychains for any $n \geq 3$. These are

$$1t_1 t_2 ... t_{n-1} > 0,$$
$$11t_2 ... t_{n-1} > 0,$$
$$1t_1 t_2 ... t_{n-2} 0,$$
$$11t_2 ... t_{n-2} 0.$$

$\square$

The next proposition considers the number of keychains when the index consists of two natural numbers of the form (k_1, k_2).

Proposition 8.3.2. *The number of* $(n - k_1 - k_2 + 1)$-*pad keychains of length* n *with index* (k_1, k_2) *is*

$$\frac{4(n - k_1 - k_2 + 1)!}{(n - k_1 - k_2 - 1)!} \ \text{if } k_1 \neq k_2,$$
$$\frac{4(n - k_1 - k_2 + 1)!}{2!(n - k_1 - k_2 - 1)!} \ \text{if } k_1 = k_2,$$

$k_i > 1$ *for* $i = 1, 2$.

Proof. Assume $k_1 \neq k_2$. Since the index of any $(n - k_1 - k_2 + 1)$-pad keychain is given to be (k_1, k_2), we have that one pin is repeated k_1 times and the other one is repeated k_2 times, while the remaining $n - k_1 - k_2 - 1$ pins are all distinct. Consider the $(n - k_1 - k_2 + 1)$-pad keychain given as follows:

$$1 \geq t_1 = t_2 = ... = t_{k_1} > t_{k_1+1} = t_{k_1+2}$$
$$= ... = t_{k_1+k_2} > t_{k_1+k_2+1} > ... > t_{n-1}....$$

We collapse the first $k_1 + 1$ positions into one so that t_1 is the leading pin as follows:

$$t_1 > t_{k_1+1} = t_{k_1+2} = ... = t_{k_1+k_2} > t_{k_1+k_2+1} > ... > t_{n-1}....$$

The length of this chain is $n - k_1$. By Proposition 8.3.1, it follows that the number of $(n - k_1 - k_2)$-pad keychains of length $n - k_1$ with index k_2 is $4(n - k_1 - k_2)$. By considering the collapsed positions, it follows that the number of $(n - k_1 - k_2 + 1)$-pad keychains of length n with index (k_1, k_2) corresponding to t_1 being repeated k_1 times is $4(n - k_1 - k_2)$. Another $4(n - k_1 - k_2)$ keychains are obtained by interchanging the roles of k_1 and k_2 in the above discussion. Thus the number of $(n - k_1 - k_2 + 1)$-pad keychains of length n with index (k_1, k_2) corresponding to the two repetitions of t_1 is $8(n - k_1 - k_2)$. We repeat the above process with t_1 replaced by t_2. Thus if t_2 is repeated k_1 times, then we may have the following chain

$$1 \geq t_1 > t_2 = t_3 = \ldots = t_{k_1+1} > t_{k_1+2} \geq \ldots \geq t \ldots .$$

If, in addition, one other pin is repeated k_2 times, then one such keychain is

$$1 \geq t_1 > t_2 = \ldots = t_{k_1+1} > t_{k_1+2}$$
$$= \ldots = t_{k_1+k_2+1} > t_{k_1+k_2+2} > \ldots > t_{n-1} \ldots .$$

Collapsing the first $k_1 + 2$ positions into one, we have the chain of length $n - 1 - k_1$:

$$t_2 > t_{k_1+2} = \ldots = t_{k_1+k_2+1} > t_{k_1+k_2+2} > \ldots > t_{n-1} \ldots .$$

By Proposition 8.3.1, the number of $(n - k_1 - 1 - k_2)$-pad keychains of length $n - k_1 - 1$ with index k_2 is $4(n - k_1 - k_2 - 1)$. By considering the collapsed positions, it follows that the number of $(n-k_1-k_2+1)$-pad keychains of length n with index (k_1, k_2) corresponding to the two repetitions of t_2 is $8(n-k_1-1-k_2)$. Proceeding inductively, we obtain $8 \cdot 1$ of $(n - k_1 - k_2 + 1)$-pad keychains of index (k_1, k_2) corresponding to the k_1 and k_2 repetitions of $t_{n-k_1-k_2}$. (The inductive step must stop here because the subscript of t cannot be strictly less than $n - k_1 - k_2$ for any keychain of index (k_1, k_2). Thus the total number of $(n - k_1 - k_2 + 1)$-pad keychains of index (k_1, k_2) is

$$8[(n - k_1 - k_2) + (n - k_1 - k_2 - 1) + \ldots + 2 + 1].$$

Using the formula, $\sum_{i=1}^{n} i = (n + 1)n/2$, it follows that this number is equal to

$$\frac{4(n - k_1 - k_2 + 1)!}{(n - k_1 - k_2 - 1)!} .$$

We now consider the case, where $k_1 = k_2$. Since $k_1 = k_2$, interchanging the roles of k_1 and k_2 yields identical $(n - k_1 - k_2 + 1)$-pad keychains, unlike the case, where $k_1 \neq k_2$. Thus the number of $(n - k_1 - k_2 + 1)$-pad keychains in this case is half the number in the case for distinct k_1, k_2. Hence we have

$$\tfrac{1}{2}(8[(n - k_1 - k_2) + (n - k_1 - k_2 - 1) + \ldots + 2 + 1])$$

$(n - k_1 - k_2 + 1)$-pad keychains. This latter number is equal to

$$\frac{1}{2} \frac{4(n - k_1 - k_2 + 1)!}{(n - k_1 - k_2 - 1)!} ,$$

the desired number. ☐

Proposition 8.3.3. *The number of* $(n - k_1 - k_2 - k_3 + 2)$*-pad keychains of length* n *with index* (k_1, k_2, k_3) *is* $\frac{4(n-k_1-k_2-k_3+2)!}{m!(n-k_1-k_2-k_3-1)!}$*, where* m *is the number of identical* k_i*'s,* $k_i > 1$ *for* $i = 1, 2, 3$*. If* k_1, k_2, k_3 *are all distinct, then the number of* $(n - k_1 - k_2 - k_3 + 2)$*-pad keychains of length* n *with index* (k_1, k_2, k_3) *is*

$$\frac{4(n-k_1-k_2-k_3+2)!}{(n-k_1-k_2-k_3-1)!}.$$

Proof. We first consider the case where all the k_i's are distinct. Since the index of any $(n - k_1 - k_2 - k_3 + 2)$-pad keychain is given to be (k_1, k_2, k_3), we know that one pin is repeated k_1 times, another one is repeated k_2 times, and a third one is repeated k_3 times while the remaining $n - k_1 - k_2 - k_3 - 1$ pins are all distinct. Consider the $(n - k_1 - k_2 - k_3 + 2)$-pad keychain given as follows:

$$1 \geq t_1 = t_2 = ... = t_{k_1} > t_{k_1+1} = t_{k_1+2} = ... = t_{k_1+k_2}$$
$$> t_{k_1+k_2+1} = ... = t_{k_1+k_2+k_3} > t_{k_1+k_2+k_3+1} > ... > t_{n-1}....$$

We collapse the first $k_1 + 1$ positions into one so that t_1 is the leading pin as follows:

$$t_1 > t_{k_1+1} = ... = t_{k_1+k_2} > t_{k_1+k_2+1} = t_{k_1+k_2+2}$$
$$= ... = t_{k_1+k_2+k_3} > t_{k_1+k_2+k_3+1} > ... > t_{n-1}....$$

The length of this chain is $n - k_1$.

By Proposition 8.3.2, it follows that the number of $(n - k_1 - k_2 - k_3 + 1)$-pad keychains of length $n - k_1$ with index (k_2, k_3) is $4(n - k_1 - k_2 - k_3 + 1)(n - k_1 - k_2 - k_3)$. By considering the collapsed positions, it follows that the number of $(n - k_1 - k_2 - k_3 + 2)$-pad keychains of length n with index (k_1, k_2, k_3) corresponding to t_1 being repeated k_1 times is $(4n - k_1 - k_2 - k_3 + 1)(n - k_1 - k_2 - k_3)$. An additional $8(n - k_1 - k_2 - k_3 + 1)(n - k_1 - k_2 - k_3)$ keychains are obtained by interchanging the roles of k_1, k_2 and k_3 in the above discussion. Thus the number of $(n - k_1 - k_2 - k_3 + 2)$-pad keychains of length n with index (k_1, k_2, k_3) corresponding to the three repetitions of t_1 is $12(n - k_1 - k_2 - k_3 + 1)(n - k_1 - k_2 - k_3)$. We repeat the above process with t_1 replaced by t_2. Thus if t_2 is repeated k_1 times, then we may have the following chain

$$1 \geq t_1 > t_2 = t_3 = ... = t_{k_1+1} > t_{k_1+2} > ... > t_{n-1}....$$

If, in addition, one other pin is repeated k_2 times, and a third one is repeated k_3 times, then one such keychain is as follows:

$$1 \geq t_1 > t_2 = ... = t_{k_1+1} > t_{k_1+2} = ... = t_{k_1+k_2+1}$$
$$> t_{k_1+k_2+2} = ... = t_{k_1+k_2+k_3+1} > t_{k_1+k_2+k_3+2} > ... > t_{n-1}....$$

Collapsing the first $k_1 + 2$ positions into one, we have the chain of length $n - 1 - k_1$:

$$t_2 > t_{k_1+2} = ... = t_{k_1+k_2+1} > t_{k_1+k_2+2}$$
$$= ... = t_{k_1+k_2+k_3+1} > t_{k_1+k_2+k_3+2} > ... > t_{n-1}....$$

By Proposition 8.3.2, the number of $(n-k_1-1-k_2-k_3+1)$-pad keychains of length $n-k_1-1$ with index (k_2, k_3) is $4(n-k_1-k_2-k_3)(n-k_1-k_2-k_3-1)$. By considering the collapsed positions, it follows that the number of $(n-k_1-k_2-k_3+2)$-pad keychains of length n with index (k_1, k_2, k_3) corresponding to the three repetitions of t_2 is $12(n-k_1-k_2-k_3)(n-k_1-k_2-k_3-1)$.

Proceeding inductively, we get $12(3{\cdot}2)$ of the $(n-k_1-k_2-k_3+2)$-pad keychains of index (k_1, k_2, k_3) corresponding to the three repetitions of $t_{n-k_2-k_2-k_3-1}$ being k_1, k_2, k_3, respectively. We also get $12(2{\cdot}1)$ of the $(n-k_1-k_2-k_3+2)$-pad keychains of index (k_1, k_2, k_3) corresponding to the three repetitions of $t_{n-k_2-k_2-k_3}$ being k_1, k_2, k_3, respectively. (The inductive step must stop here since the subscript of t cannot be strictly less than $n-k_1-k_2-k_3$ for any keychain of index (k_1, k_2, k_3).) Thus the total number of $(n-k_1-k_2-k_3+2)$-pad keychains of index (k_1, k_2, k_3) is

$$12[(n-k_1-k_2-k_3+1)(n-k_1-k_2-k_3)$$
$$+(n-k_1-k_2-k_3)(n-k_1-k_2-k_3-1+\ldots+3\cdot2+2\cdot1)].$$

By using the formula $n(n+1)+(n-1)n+\ldots+2.3+1.2 = \frac{1}{3}n(n+1)(n+2)$, it follows that this number is equal to

$$\frac{4(n-k_1-k_2-k_3+2)!}{(n-k_1-k_2-k_3-1)!} \ .$$

We now consider the case where exactly two of k_1, k_2, k_3 are equal. It follows as in Proposition 8.3.2 that dividing this number by 2! yields the desired number. If all three of the k_i's are equal, then the desired number is obtained by dividing by 3!.   □

We use the following notation for the general case. Suppose $S = (k_1, k_2, ..., k_m)$ is the given index of a keychain, where not all k_i's need to be distinct. Then S can be partitioned by a set $\{F_1, F_2..., F_s\}$ of equivalence classes induced by an equivalence relation $\sim$ on $\{k_1, k_2, ..., k_m\}$ given by $k_i \sim k_j$ if and only if $k_i = k_j, 1 \le i, j \le m$. Let l_i denote the number of elements in the class F_i for each $i = 1, 2, ..., s$.

The next theorem deals with the most general case.

Theorem 8.3.4. *The number of $(n-k_1-k_2-...-k_m+m-1)$-pad keychains of length n with index $(k_1, k_2, ..., k_m)$ is*

$$\frac{4(n-k_1-k_2-...-k_m+m-1)!}{(n-k_1-k_2-...-k_m-1)!}$$

if all the padidities k_i are distinct, otherwise the number is

$$\frac{4(n-k_1-k_2-...-k_m+m-1)!}{l_1!l_2!...l_s!(n-k_1-k_2-...-k_m-1)!}.$$

Proof. Suppose all the k_i's are distinct. Then the result is true for $m = 1, 2$ and 3, respectively. Assume the result is true when the number of k_i's is $m-1$, the induction hypothesis. Consider the $(n-1)$-chain

$$1 \geq t_1 \geq t_2 \geq ... \geq t_{n-1}.$$

Suppose t_1 is repeated k_1 times. Then, we have the chain

$$1 \geq t_1 = t_2 = ... = t_{k_1} > t_{k_1+1} \geq ... \geq t_{n-1}.$$

Now collapse the first $k_1 + 1$ positions so that t_1 is a leading pin as follows:

$$t_1 > t_{k_1+1} \geq ... \geq t_{n-1}.$$

This is a keychain of length $(n - k_1)$. From this chain, it follows that the number of $(n - k_1 - k_2 - ... - k_m + m - 2)$-pad keychains of length $n - k_1$ with index $(k_2, k_3, ..., k_m)$ is

$$\frac{4(n-k_1-k_2-...-k_m+m-2)!}{(n-k_1-k_2-...-k_m-1)!}.$$

By considering the collapsed positions, it follows that the number of $(n - k_1 - k_2 - ... - k_m + m - 1)$-pad keychains of length n with index $(k_1, k_2, .., k_m)$ corresponding to t_1 being repeated k_1 times is

$$\frac{4(n-k_1-k_2-...-k_m+m-2)!}{(n-k_1-k_2-...-k_m-1)!}.$$

We obtain an additional $4(m-1)\frac{(n-k_1-k_2-...-k_m+m-2)!}{(n-k_1-k_2-...-k_m-1)!}$ keychains by interchanging the roles of $k_1, k_2, ...,$ and k_m in the above discussion. Thus the number of $(n - k_1 - k_2 - ... - k_m + m - 1)$-pad keychains of length n with index $(k_1, k_2, ..., k_m)$ corresponding to the m repetitions of t_1 is

$$4m\frac{(n-k_1-k_2-...-k_m+m-2)!}{(n-k_1-k_2-...-k_m-1)!}.$$

We repeat the above process with t_1 replaced by t_2. Thus if t_2 is repeated k_1 times, then the first chain becomes

$$1 \geq t_1 > t_2 = t_3 = ... = t_{k_1+1} > t_{k_1+2} \geq ... \geq t_{n-1}.$$

The keychain $t_2 > t_{k_1+2} \geq ... \geq t_{n-1}$ is of length $(n - k_1 - 1)$. From this chain, the induction hypothesis assures us that the number of $(n - k_1 - 1 - k_2 - ... - k_m + m - 2)$-pad keychains of index $(k_2, k_3, \cdots, k_m)$ is

$$4\frac{(n-k_1-1-k_2-...-k_m+m-2)!}{(n-k_1-k_2-...-k_m-2)!}.$$

Combining this with the chain

$$1 \geq t_1 > t_2 = t_3 = ... = t_{k_1+1} > t_{k_1+2} \geq ... \geq t_{n-1}$$

and using the previous discussion, it follows that the number of $(n - k_1 - k_2 - ... - k_m + m - 1)$-pad keychains of length n with index $(k_1, k_2, ..., k_m)$ corresponding to the m repetitions of t_2 is

$$4m\frac{(n-k_1-k_2-...-k_m+m-3)!}{(n-k_1-k_2-...-k_m-2)!}.$$

Proceeding inductively, corresponding to $t_{n-k_1-k_2-...-k_m-1}$ being repeated $k_1, k_2, ..., k_m$ times respectively, the number of $(n - k_1 - k_2 - ... - k_m + m - 1)$-pad keychains of index $(k_1, k_2, ..., k_m)$ is $4m(m!)$. Also, there are $4m(m - 1)!$ similar keychains corresponding to the repetitions of $t_{n-k_1-k_2-...-k_m}$.

(The inductive step must stop here since the subscript of t cannot be strictly less than $n-k_1-k_2-...-k_m$ for any keychain of index $(k_1, k_2, ..., k_m)$.) Thus the total number of $(n - k_1 - k_2 - ... - k_m + m - 1)$-pad keychains of index $(k_1, k_2, ..., k_m)$ is $4m[(n - k_1 - k_2 - ... - k_m + m - 2)(n - k_1 - k_2 - ... - k_m + m - 3) \cdots (n - k_1 - k_2 - ... - k_m) +$
$(n - k_1 - k_2 - ... - k_m + m - 3)...(n - k_1 - k_2 - ... - k_m - 1) + ... + m(m - 1)...2 + (m - 1)(m - 2)...2 \cdot 1]$ which can be easily seen to be equal to $\frac{4(n-k_1-k_2-...-k_m+m-1)!}{(n-k_1-k_2-...-k_m-1)!}$, where we have used the fact that $n(n + 1)...(n + k - 1) + ... + 2 \cdot 3...(k + 1) + 1 \cdot 2...k = \frac{1}{k+1}n(n + 1)(n + 2)...(n + k)$.

Now consider the case, where some k_i's are equal. Then as in Proposition 8.3.3, the desired number is obtained by dividing the expression above by the factorials of the numbers of identical k_i's. $\quad\square$

Corollary 8.3.5. *If* $n = k_1 + k_2 + \cdots + k_m + s + 1$, *then the number of* $(n-k_1-k_2-...-k_m+m-1)$-*pad keychains of length* n *with index* $(k_1, k_2, ..., k_m)$ *is*

$$\frac{4(s+m)!}{s!}$$

if all the k_i *are distinct, where* s *is the number of nonrepeating pins from the second position on. Otherwise, the number is*

$$\frac{4(s+m)!}{s!l_1!l_2!\cdots l_t!},$$

where $l_1, l_2, \ldots, l_t$ *are the numbers of identical* k_i's, $k_i > 1$, *for* $i = 1, 2, ..., m$.

Proof. Suppose all the k_i's are distinct. By Theorem 8.3.4, the desired number is

$$4\frac{(n-k_1-k_2-...-k_m+m-1)!}{(n-k_1-k_2-...-k_m-1)!}.$$

Clearly, $s + 1 = n - k_1 - k_2 - ... - k_m$. Thus the desired number is

$$4\frac{(s+1+m-1)!}{(s+1-1)!} = 4\frac{(s+m)!}{s!}.$$

The case when some k_i's are identical follows by arguments similar to those previously given. $\quad\square$

Example 8.3.6. (1) *We determine the number of 4-pad keychains of length 8 with index* $(2, 2, 2)$. *By Corollary 8.3.5, we have* $8 = 2 + 2 + 2 + 1 + 1$ *so that* $s = 1$, $m = 3$, $l_1 = 3$. *Hence the number of* $(8 - 2 - 2 - 2 + 3 - 1)$-*pad keychains of the given index is* $4\frac{(3+1)!}{3!} = 16$.

The chain $1 \geq t_1 = t_2 > t_3 = t_4 > t_5 = t_6 > t_7$ is a representative 4-pad keychain of length 8 with index $(2, 2, 2)$. Note that $s = 1$ since t_7 is the only nonrepeating pin from position 2 on.

(2) We now determine the number of 4-pad keychains of length 10 with index $(2, 2, 2, 3)$. By Corollary 8.3.5, we have $10 = 2 + 2 + 2 + 3 + 0 + 1$ and $4 = 10 - 2 - 2 - 2 - 3 + 4 - 1$. Hence $m = 4$, $l_1 = 3$ and $s = 0$. Thus the desired number is $4\frac{(4!)}{3!} = 16$. Note that all pins (from the second position) repeat, hence $s = 0$.

(3) We now find the number of 5-pad keychains of length 11 with index $(2, 2, 3)$. As before, we have $11 = 2 + 2 + 3 + 1 + 1 + 1 + 1$. Hence $m = 3$, $l_1 = 2$ and $s = 3$. Thus the desired number is $4\frac{(3+3)!}{3!2!} = 240$.

Recall that any fuzzy subgroup of $G = \mathbb{Z}_p$ is equivalent to one of the following three: $11, 1t, 10$ where the 11 represents the crisp trivial subgroup $\mathbb{Z}_p$ and 10 represents the trivial subgroup $\{0\}$ of $\mathbb{Z}_p$. $1t$ represents the fuzzy subgroup $\mu(x) = 1$ if $x = 0$ and $\mu(x) = t$ if $x \neq 0 \forall x \in G$. Clearly, these are the only distinct equivalence classes of fuzzy subgroups of $\mathbb{Z}_p$ since the only crisp subgroups of $\mathbb{Z}_p$ are $\{0\}$ and $\mathbb{Z}_p$ itself. Here we always assume $1 > t \geq 0$.

Proposition 8.3.7. *Let $G = \mathbb{Z}_{p_1} \oplus ... \oplus \mathbb{Z}_{p_{n-1}}$, where the p_i are distinct primes and $n \in \mathbb{N}$ with $n \geq 3$. Then the number of distinct fuzzy subgroups of G represented by one $(n-1)$-chain of length n with index k is $\frac{(n-1)!}{k!}$.*

Proof. Let $G_i = \mathbb{Z}_{p_1} \oplus ... \oplus \mathbb{Z}_{p_i} \oplus \{0\} \oplus ... \oplus \{0\}$. Consider a fuzzy subgroup μ of G defined by $\mu(x) = t_i$ for $x \in G_i \backslash G_{i-1}$ with all the t_i distinct except for $i = j$ in which case j appears k times and $t_0 = 1$. We may represent μ as $1t_1 t_2 ... t_{j-1} t_j ... t_j t_{j+k} ... t_{n-1}$. We may also write,

$$
\mu(x) = \begin{cases}
1 & \text{if } x = 0 \\
t_1 & \text{if } x \in G_1 \backslash \{0\} \\
t_2 & \text{if } x \in G_2 \backslash G_1 \\
... & \\
t_{j-1} & \text{if } x \in G_{j-1} \backslash G_{j-2} \\
t_j & \text{if } G_{j+k-1} \backslash G_{j-1} \\
t_{j+k} & \text{if } x \in G_{j+k} \backslash G_{j+k-1} \\
... & \\
t_{n-1} & \text{otherwise.}
\end{cases}
$$

If we replace G_1 by a subgroup $G_{1k} = \{0\} \oplus ... \oplus \mathbb{Z}_k \oplus \{0\} \oplus ... \oplus \{0\}$ isomorphic to G_1, then another fuzzy subgroup μ not equivalent to the previous one is defined. Since there are $n - 2$ distinct subgroups isomorphic to G_1, there are $n - 1$ distinct fuzzy subgroups of the form μ. Similarly, with G_1 fixed, we may replace G_2 by $G_{2k} = \mathbb{Z}_1 \oplus \{0\} \oplus ... \oplus \{0\} \oplus \mathbb{Z}_k \oplus \{0\} \oplus ... \oplus \{0\}$. This yields an additional $n - 2$ nonequivalent fuzzy subgroups. Repeating the process with G_{j-1}, an additional $n - j + 1$ distinct fuzzy subgroups arise. Next we replace G_j with G_{jk}. This gives $\binom{n-j}{k}$ more fuzzy subgroups. Finally, G_{n-1} yields 1 fuzzy subgroup. By the Fundamental Principle of Counting, we have $\frac{(n-1)!}{k!}$ distinct fuzzy subgroups. $\qquad\square$

Corollary 8.3.8. *Let* $n = k + m + 1$. *Let* $G = \mathbb{Z}_{p_1} \oplus ... \oplus \mathbb{Z}_{p_{n-1}}$. *Then the number of distinct fuzzy subgroups of G represented by all $(n-1)$-chains each of length n with index k, is*

$$4(n-k)\frac{(n-1)!}{k!} = 4\frac{(m+1)!(n-1)!}{m!k!}.$$

Proof. The desired result follows immediately from Proposition 8.3.7, Proposition 8.3.1 and Corollary 8.3.5. $\qquad\square$

Proposition 8.3.9. *Let* $G = \mathbb{Z}_{p_1} \oplus ... \oplus \mathbb{Z}_{p_{n-1}}$, *where the p_i are distinct primes and $n \in \mathbb{N}$ with $n \geq 3$. Then the number of distinct fuzzy subgroups represented by one $(n-1)$-chain of length n with index (k_1, k_2) is*

$$\frac{(n-1)!}{k_1!k_2!}$$

if $k_1 \neq k_2$. If $k_1 = k_2$, then $\frac{(n-1)!}{2k_1!k_2!}$ is the desired number.

Proof. The proof follows as in Proposition 8.3.7 except that here there are two factors of the form $\binom{n-j}{k_s}$. Using the Fundamental Principle of Counting as in Proposition 8.3.7, the desired result follows. $\qquad\square$

Proposition 8.3.9 is easily extended to the general case when the index of the fuzzy subgroup is $(k_1, k_2, ..., k_m)$.

Corollary 8.3.10. *Let* $n = k_1 + k_2 + ... + k_m + s + 1$. *Let* $G = \mathbb{Z}_{p_1} \oplus ... \oplus \mathbb{Z}_{p_{n-1}}$. *Then the number of distinct fuzzy subgroups represented by all $(n-1)$-chains each of length n with index $(k_1, k_2, ... k_m)$ is*

$$4\frac{(s+m)!(n-1)!}{s!k_1!k_2!...k_m!}$$

if all the k_i are distinct, where s is the number of nonrepeating pins from the 2nd position. If some k_i's are identical, divide the first expression by the factorials of the numbers of the identical k_i's.

Example 8.3.11. *We find the number of distinct fuzzy subgroups of $\mathbb{Z}_{p_1} \oplus ... \oplus \mathbb{Z}_{p_6}$ represented by all 6-chains of length 7 with index*
(1) 3, (2) $2 \cdot 2$, (3) $2 \cdot 2 \cdot 2$, (4) $2 \cdot 3$.
By Corollary 8.3.10, we have (1) $n = 7 = 3 + 3 + 1$ and $s = 3$, $k_1 = 3$, $m = 1$ and $3 < 7 - 1$. Thus the desired number is

$$4\frac{(3+1)!(7-1)!}{3!3!} = (4 \cdot 4!6!)/(6 \cdot 6) = 1920.$$

(2) $n = 7 = 2 + 2 + 2 + 1$ *and* $s = 2, k k_2 = 2, l = 2 = $ *number of identical* k_i's, $m = 2$ *and* $2 < 7 - 1$. *Hence the desired number is*

$$4 \frac{(2+2)!(7-1)!}{2!2!2!2!} = 4(24)(6)(120)/16 = 4320.$$

(3) $n = 7 = 2 + 2 + 2 + 0 + 1$ *and* $s = 0, m = 3, k_1 = k_2 = k_3 = 2, l = 3$. *Thus the desired number is*

$$4 \frac{(0+3)!(7-1)!}{0!2!2!2!3!} = 360.$$

(4) $n = 7 = 2 + 3 + 1 + 1$ *and* $s = 1, m = 2$. *Hence the required number is*

$$\frac{(1+2)!(7-1)!}{2!3!} = 1440.$$

8.4 $\mathbb{Z}_{p^n} \oplus \mathbb{Z}_{q^m}$

In this section, we determine the number of distinct equivalence classes of fuzzy subgroups of $G = \mathbb{Z}_{p^n} \oplus \mathbb{Z}_{q^m}$ where p and q are distinct primes and $n, m \in \mathbb{N}$. We first derive a formula for the number of maximal chains of G. We then derive a combinatorial formula for the number of distinct fuzzy subgroups of G. We utilize the complete characterization of finite Abelian groups in the crisp case [3]. Throughout this section, we denote G simply by $p^n q^m$ and their subgroups $H = \mathbb{Z}_{p^i} \oplus \mathbb{Z}_{q^j}$ in a similar fashion $p^i q^j$ for $1 \leq i \leq n$ and $1 \leq j \leq m$. (Recall that for the cyclic group $\mathbb{Z}_{p^n}$, we write $\mathbb{Z}_{p^n} \supset \mathbb{Z}_{p^k}$, $k < n$, to mean that $\mathbb{Z}_{p^n}$ contains the cyclic subgroup of order p^k generated by $p^{n-k} \cdot 1$.)

For any $n \in \mathbb{N}$ and $m = 0$, it is clear that G has only one maximal chain, namely $\mathbb{Z}_{p^n} \supset \mathbb{Z}_{p^{n-1}} \supset \ldots \supset \mathbb{Z}_p \supset \{0\}$.

Proposition 8.4.1. *Let* $G = \mathbb{Z}_{p^n} \oplus \mathbb{Z}_q$ *where* p *and* q *are distinct primes. Then the number of maximal chains of* G *is* $n + 1$.

Proof. The proof is by induction on n. Let $n = 1$. Then G has two maximal chains, namely $G = \mathbb{Z}_p \oplus \mathbb{Z}_q \supset \{0\} \oplus \mathbb{Z}_q \supset \{0\}$ and $G = \mathbb{Z}_p \oplus \mathbb{Z}_q \supset \mathbb{Z}_p \oplus \{0\} \supset \{0\}$. Assume that $\mathbb{Z}_{p^k} \oplus \mathbb{Z}_q$ has $k + 1$ maximal chains. Now $\mathbb{Z}_{p^{k+1}} \oplus \mathbb{Z}_q$ has two maximal subgroups namely $\mathbb{Z}_{p^k} \oplus \mathbb{Z}_q$ and $\mathbb{Z}_{p^{k+1}} \oplus \{0\}$ has only one maximal chain. Thus $\mathbb{Z}_{p^{k+1}} \oplus \mathbb{Z}_q$ has $k + 2$ maximal chains. The desired result follows by induction. $\qquad \square$

By symmetry, it follows that $\mathbb{Z}_p \oplus \mathbb{Z}_{q^m}$ has $m + 1$ maximal chains.

The factorial expressions in the next proposition are written in a form for ease of use in inductive steps later. Also, $r_i = 1$ for all i and $m \geq 2$ in the next result, where $m \geq 2$ since the case for $m = 1$ has been considered in Proposition 8.4.1.

Proposition 8.4.2. *Let* $G = \mathbb{Z}_{p^2} \oplus \mathbb{Z}_{q^m}$, *where* p *and* q *are distinct primes and* $m \geq 2$. *Then the number of maximal chains of* G *is* $\sum_{i=-1}^{m-1} r_i\,(m-i)$, *where* $r_i = \frac{(2+i-1)!}{(2-2)!(1+i)!}$.

Proof. The proof is by induction on m. Let $m = 2$. Then $G = \mathbb{Z}_{p^2} \oplus \mathbb{Z}_{q^2}$ has 6 maximal chains, namely

$$p^2 q^2 \supset p^2 q \supset pq \supset p \supset 0$$
$$p^2 q^2 \supset p^2 q \supset pq \supset q \supset 0$$
$$p^2 q^2 \supset p^2 q \supset p^2 \supset p \supset 0$$
$$p^2 q^2 \supset pq^2 \supset pq \supset p \supset 0$$
$$p^2 q^2 \supset pq^2 \supset pq \supset q \supset 0$$
$$p^2 q^2 \supset pq^2 \supset q^2 \supset q \supset 0.$$

Now

$$\sum_{i=-1}^{2-1} r_i\,(2-i) = 6.$$

Assume that the result is true for $k \in \mathbb{N}, k \geq 2$, the inductive hypothesis. We show that $\mathbb{Z}_{p^2} \oplus \mathbb{Z}_{q^{k+1}}$ has $\sum_{i=-1}^{k} r_i\,(k+1-i)$ maximal chains. The number of maximal chains of $\mathbb{Z}_{p^2} \oplus \mathbb{Z}_{q^{k+1}}$ can be determined from $\mathbb{Z}_{p^2} \oplus \mathbb{Z}_{q^k}$.

Now $p^2 q^k$ has $\sum_{i=-1}^{k-1} r_i\,(k-i)$ maximal chains, pq^k has $k+1$ maximal chains, and q^{k+1} has one maximal chain. Thus $\mathbb{Z}_{p^2} \oplus \mathbb{Z}_{q^{k+1}}$ has $\sum_{i=-1}^{k-1} r_i\,(k-i)$ plus $k+1+1$ maximal chains which is equal to $\sum_{i=-1}^{k} r_i\,(k+1-i)$. Thus the desired result follows by induction. □

By symmetry, it follows that $\mathbb{Z}_{p^n} \oplus \mathbb{Z}_{q^2}$ has $\sum_{i=-1}^{n-1} r_i\,(n-i)$ maximal chains. In order to help the reader understand the inductive process of finding the maximal chains of $G = \mathbb{Z}_{p^n} \oplus \mathbb{Z}_{q^m}$ in general, we present an additional case.

Proposition 8.4.3. *Let* $G = \mathbb{Z}_{p^3} \oplus \mathbb{Z}_{q^m}$, *where* p *and* q *are distinct primes. Then the number of maximal chains of* G *is* $\sum_{i=-1}^{m-1} r_i\,(m-i)$, *where* $r_i = \frac{(3+i-1)!}{(3-2)!(1+i)!}$, $m \geq 2$.

Proof. Let $m = 2$. There are 10 maximal chains of $p^3 q^2$. This can be seen as follows:

From Proposition 8.4.2, six of them are in the form

$$p^3q^2 \supset p^2q^2 \supset p^2q \supset pq \supset p \supset 0$$
$$p^3q^2 \supset p^2q^2 \supset p^2q \supset pq \supset q \supset 0$$
$$p^3q^2 \supset p^2q^2 \supset p^2q \supset p^2 \supset p \supset 0$$
$$p^3q^2 \supset p^2q^2 \supset pq^2 \supset pq \supset p \supset 0$$
$$p^3q^2 \supset p^2q^2 \supset pq^2 \supset pq \supset q \supset 0$$
$$p^3q^2 \supset p^2q^2 \supset pq^2 \supset q^2 \supset q \supset 0.$$

another three in the form

$$p^3q^2 \supset p^3q \supset p^2q \supset pq \supset p \supset 0$$
$$p^3q^2 \supset p^3q \supset p^2q \supset pq \supset q \supset 0$$
$$p^3q^2 \supset p^3q \supset p^2q \supset p^2 \supset p \supset 0$$

and one more in the form

$$p^3q^2 \supset p^3q \supset p^3 \ldots \qquad .$$

It is easily verified that the number 10 is obtained upon substitution of $m = 2$ in the formula.

The inductive step can be proved in a manner similar as in the proof of Proposition 8.4.2.

By symmetry, we can interchange p and q in G without affecting the number of maximal chains. The above propositions illustrate how the inductive proofs can be employed effectively using extended diagrams from the k-level to the $(k + 1)$-level. □

Theorem 8.4.4. *Let* $G = \mathbb{Z}_{p^n} \oplus \mathbb{Z}_{q^m}$, *where* p *and* q *are distinct primes. Then the number of maximal chains of* G *is*

$$\sum_{i=-1}^{m-1} r_i (m - i), \ \ where \ r_i = \frac{(n + i - 1)!}{(n - 2)! (1 + i)!}, \ n \ge 2.$$

Proof. The proof is ny induction on m with n fixed. For $m = 2, 3$, the theorem holds by Propositions 8.4.2 and 8.4.3. Assume that $\mathbb{Z}_{p^n} + \mathbb{Z}_{q^k}$ has $\sum_{i=-1}^{k-1} r_i (k - i)$ maximal chains, where $r_i = \frac{(n+i-1)!}{(n-2)!(1+i)!}$, the induction hypothesis. We show that $\mathbb{Z}_{p^n} + \mathbb{Z}_{q^{k+1}}$ has $\sum_{i=-1}^{k} r_i (k + 1 - i)$ maximal chains where $r_i = \frac{(n+1-1)!}{(n-2)!(1+i)!}$.

As in Propositions 8.4.2 and 8.4.3, it follows that (n, k) yields $\sum_{i=-1}^{k-1} r_{ni} (k - i)$ maximal chains, where $r_{ni} = \frac{(n+i-1)!}{(n-2)!(1+i)!}$. The cases for $((n - 1), k), ..., (0, (k + 1))$ are similar. Thus the total number of maximal chains is

$$\sum_{j=0}^{n-2} \sum_{i=-1}^{k-1} r_{(n-j)i} (k-i) + k + 1 + 1$$

$$= \sum_{j=0}^{n-2} \sum_{i=-1}^{k-1} \frac{(n-j+i-1)! (k-i)}{(n-j-2)! (1+i)!} + k + 1 + 1$$

$$= \sum_{j=0}^{n-2} [k+1 + (n-j-1)k + \frac{(n-j)(n-j-1)(k-1)}{2!}$$

$$+ \frac{(n+k-j-2) \dots (n-j-1)}{k!}] + (k+1) + 1$$

$$= (n-1)(k+1) + k[(n-1) + (n-2) + \dots + 1]$$

$$+ \frac{(k-1)}{2!} \sum_{j=0}^{n-2} (n-j)(n-j-1) + \dots$$

$$+ \frac{(k-2)}{3!} \sum_{j=0}^{n-2} (n-j+1)(n-j)(n-j-1) + \dots$$

$$+ k + 1 + 1.$$

For all $n, k \in \mathbb{N}$,

$$n(n+1) \dots (n+k-1) + \dots + 2.3 \dots (k+1) + 1.2 \dots k$$

$$= \frac{1}{k+1} n(n+1)(n+2) \dots (n+k).$$

Thus with n replaced by $n-1$ and for $k = 2, 3, \dots$, we have

$$n(n-1) + (n-1)(n-2) + \dots + 2.1 = \frac{(n-1)n(n+1)}{3};$$

$$(n+1)n(n-1) + \dots + 3.2.1 = \frac{(n-1)n(n+1)(n+2)}{4};$$

and so forth. Thus the total number of maximal chains is

$$(n-1)(k+1) + k\frac{n(n-1)}{2} + \frac{(k-1)}{2!} \frac{(n-1)n(n+1)}{3}$$

$$+ \frac{(k-2)}{3!} \frac{(n-1)n(n+1)(n+2)}{4}$$

$$+ \frac{(k-3)}{4!} \frac{(n-1)n(n+1)(n+2)(n+3)}{5} + \dots + k + 1 + 1$$

$$= (n-1)(k+1) + \frac{k}{2!} n(n-1)$$

$$+ \frac{(k-1)}{3!} (n-1)n(n+1) + \frac{(k-2)}{4!} (n-1)n(n+1)(n+2)$$

$$+ \dots + k + 1 + 1.$$

Now

$$\sum_{i=-1}^{k} r_i \left(k+1-i\right) = \sum_{i=-1}^{k} \frac{(n+i-1)!}{(n-2)!\,(1+i)!}\left(k+1-i\right)$$

$$= (k+2) + (n-1)\left(k+1\right) + \frac{n\left(n-1\right)}{2!} k$$

$$+ \frac{(n+1)\,n\left(n-1\right)}{3!}\left(k-1\right)$$

$$+ \frac{(n+2)\left(n+1\right)n\left(n-1\right)}{4!}\left(k-2\right) + \ldots$$

$$+ \frac{(n+k-1)\left(n+k-2\right)\ldots n\left(n-1\right)}{(k+1)!} 1.$$

Clearly the expressions in the right hand side of the immediately preceding equations are equal. Thus we have the desired result.   □

Given a maximal chain of length $n+1$ of subgroups of G, there are 2^{n+1} distinct equivalence classes of fuzzy subgroups. However, for two such maximal chains, it is not necessarily true that the number of distinct equivalence classes of fuzzy subgroups is $2 \times 2^{n+1}$. This is because some fuzzy subgroups on distinct maximal chains determine the same equivalence class of fuzzy subgroups such as $1ttsr$ on the following two maximal chains

$$p^2 q^2 \supset p^2 q \supset pq \supset p \supset 0,$$
$$p^2 q^2 \supset p^2 q \supset pq \supset q \supset 0.$$

Therefore, the number of distinct fuzzy subgroups of $G = \mathbb{Z}_{p^n} \oplus \mathbb{Z}_{q^m}$ is fewer than $2^{n+m+1} \times \sum_{i=-1}^{m-1} r_i \left(m-i\right)$, where the sum is the number of maximal chains. It is worth pointing out that the length of each maximal chain is $n+m+1$. We now present a formula for the precise number of distinct fuzzy subgroups of $G = \mathbb{Z}_{p^n} \oplus \mathbb{Z}_{q^m}$.

Theorem 8.4.5. *The number of distinct fuzzy subgroups of $G = \mathbb{Z}_{p^n} \oplus \mathbb{Z}_{q^m}$ is*

$$2^{n+m+1} \sum_{r=0}^{m} 2^{-r} \binom{n}{n-r}\binom{m}{r} - 1, \;\; where \; m \leq n.$$

We prove the theorem by using a series of propositions, fixing m and inducting on n.

The case $m = 0$ is trivial, so we do not state it as a proposition but offer the following brief explanation:

It follows that the above formula reduces to $2^{n+1}2^0 \binom{n}{n}\binom{m}{0} - 1$. In this case, there is only one maximal chain of length $n+1$. By Proposition 8.2.3, the number of fuzzy subgroups is $2^{n+1} - 1$. There is nothing to prove in this case.

For the case $m = 1$, we need the following lemma. For $i \in \mathbb{N}$, $0 \leq i \leq n-1$, we denote the maximal chain $0 \subset p \subset p^2 \subset ... \subset p^i \subset p^i q \subset p^{i+1}q \subset ... \subset p^{n-1}q \subset p^n q$ by C_{i+1}. For $i = 0$, we mean the maximal chain C_1 given by $0 \subset q \subset pq \subset ... \subset p^{n-1}q \subset p^n q$. By C_0, we mean the maximal chain $0 \subset p \subset p^2 \subset ... \subset p^n \subset p^n q$.

Lemma 8.4.6. *The number of fuzzy subgroups on the maximal chain C_{i+1}, $0 \leq i \leq n-1$,*

$$0 \subset p \subset p^2 \subset ... \subset p^i \subset p^i q \subset p^{i+1}q \subset ... \subset p^{n-1}q \subset p^n q$$

distinct from any fuzzy subgroup on any other maximal chain is 2^{n+1}.

Proof. Consider the fuzzy subgroups on C_{i+1} given by the set S_{i+1} of key-chains $1 \geq t_1 \geq ... \geq t_i \geq t_{i+1} \geq t_{i+2} \geq ... \geq t_{n+1}$. It follows that there are $\frac{2^{n+2}}{2}$ distinct keychains in S_{i+1}. Any fuzzy subgroup μ given by an element of S_{i+1} on C_{i+1} for any fixed $0 \leq i_0 \leq n-1$ is distinct from any fuzzy subgroup on C_0 since μ is represented by $... \subset (p^{i_0}q)^{t_{i_0}+1} \subset ...$ which cannot appear in any fuzzy subgroup on C_0. For the same reason, μ is distinct from any fuzzy subgroup on C_0. For the same reason, μ is distinct from any fuzzy subgroup on C_j for $i_0 + 1 < j \leq n-1$. If $1 \leq j \leq i_0 + 1$, then μ is distinct from any fuzzy subgroup ν on C_j since ν would be represented by $... \subset (p^{j-1}q)^{t_j} \subset ...$ which cannot appear in μ. $\square$

Proposition 8.4.7. *The number of distinct fuzzy subgroups of $G = \mathbb{Z}_{p^n} \oplus \mathbb{Z}_{q^1}$ is*

$$2^{n+1+1} \sum_{r=0}^{1} 2^{-r} \binom{n}{n-r} \binom{1}{r} - 1, \text{ where } n \geq 1.$$

Proof. By Proposition 8.4.2 with $m = 1$, it follows that the number of maximal chains of G is equal to $n + 1$. Only one such maximal chain C_0 goes through p^n and is given by

$$0 \subset p \subset p^2 \subset ... \subset p^i \subset ... \subset p^n \subset p^n q.$$

The length of this maximal chain is $n + 2$ and so the number of fuzzy subgroups on C_0 is

$$2^{n+2} - 1.$$

All other n maximal chains $C_1, C_2, ..., C_n$ go through $p^{n-1}q$. These maximal chains can be distinguished from each other by writing $p^{n-1}q$ as qp^{n-1}, pqp^{n-2}, $p^2qp^{n-3}, ..., p^{n-1}q$. By $p^i qp^{n-i-1}$, we mean the maximal chain C_{i+1}

$$0 \subset p \subset p^2 \subset ... \subset p^i \subset p^i q \subset p^{i+1}q \subset ... \subset p^{n-1}q.$$

By Lemma 8.4.6, the number of fuzzy subgroups on C_{i+1} is 2^{n+1}. Therefore, the total number of fuzzy subgroups on these n distinct maximal chains is

$$n2^{n+1}.$$

For $m = 1$, the formula in Theorem 8.4.5 gives $2^{n+2}(1 + \frac{n}{2}) - 1$ which agrees with the sum of the numbers $2^{n+2} - 1$ and $n2^{n+1}$ above.

Next we consider the case $m = 2$. First we determine the number of fuzzy subgroups on specific maximal chains. For $0 \le i \le j \le n - 1$, let C_{ij} denote the maximal chain, $0 \subset p \subset \ldots \subset p^i \subset p^i q \subset p^{i+1} q \subset \ldots \subset p^j q \subset p^j q^2 \subset p^{j+1} q^2 \subset \ldots \subset p^{n-1} q^2 \subset p^n q^2$. For $i = 0$, C_{0j} denotes the maximal chain $0 \subset q \subset pq \subset \ldots \subset p^j q \subset p^j q^2 \subset \ldots \subset p^n q^2$. For $j = n - 1$, $C_{i(n-1)}$ denotes the maximal chain $C_{i(n-1)} 0 \subset q \subset q^2 \subset \ldots \subset p^i q^2 \subset \ldots \subset p^{n-1} q^2 \subset p^n q^2$.   □

Lemma 8.4.8. *The number of fuzzy subgroups on the maximal chain C_{ij}*

$$0 \subset p \subset \ldots \subset p^i \subset p^i q \subset p^{i+1} q \subset \ldots \subset p^j q$$
$$\subset p^j q^2 \subset p^{j+1} q^2 \subset \ldots \subset p^{n-1} q^2 \subset p^n q^2$$

distinct from any fuzzy subgroup on any other maximal chain is 2^{n+1}, for $0 \le i \le j \le n - 1$.

Proof. Consider the fuzzy subgroups on C_{ij} given by the set S_{ij} of keychains $1 \ge t_1 \ge \ldots \ge t_i \ge t_{i+1} \ge t_{i+2} \ge \ldots \ge t_j \ge t_{j+1} \ge \ldots \ge t_{n+2}$. It follows that there are $\frac{2^{n+3}}{4}$ distinct keychains in S_{ij}. Any fuzzy subgroup μ given by an element of S_{ij} on C_{ij} for any fixed $0 \le i_0 \le j_0 \le n - 1$ is distinct from any fuzzy subgroup on C_0

$$0 \subset p \subset p^2 \subset \ldots \subset p^n \subset p^n q$$

since μ is represented by $\ldots \subset (p^{i_0} q)^{t_{i_0}+1} \subset \ldots$ which cannot appear in any fuzzy subgroup on C_0. For the same reason, μ is distinct from any fuzzy subgroup on C_k

$$0 \subset p \subset p^2 \subset \ldots \subset p^k \subset p^k q \subset \ldots \subset p^{n-1} q \subset p^r q^s \subset p^n q^2,$$

where either $r = n$, $s = 1$ or $r = n - 1$, $s = 2$ and for $i_0 + 1 < k \le n - 1$. Now if $1 \le k \le i_0 + 1$, then μ is distinct from any fuzzy subgroup ν on C_k since ν would be represented by $\ldots \subset (p^{k-1} q)^{t_k} \subset \ldots$ which cannot appear in μ. We now compare μ with any fuzzy subgroup on C_{kl} for any $i_0 < k < l \le n - 1$. It follows that μ is distinct from any fuzzy subgroup on C_{kl} since μ is represented by $\ldots \subset (p^{i_0} q)^{t_{i_0}+1} \subset \ldots$ which cannot appear in any fuzzy subgroup on C_{kl}. If $0 \le k < l \le i_0$, then μ is distinct from any fuzzy subgroup ν on C_{kl} since ν has in its representation $\ldots \subset (p^{k-1} q)^{t_k} \subset \ldots$ which cannot appear in μ.   □

Proposition 8.4.9. *The number of distinct fuzzy subgroups of $G = \mathbb{Z}_{p^n} \oplus \mathbb{Z}_{q^2}$ is*

$$2^{n+2+1} \sum_{r=0}^{2} 2^{-r} \binom{n}{n-r} \binom{2}{r} - 1, \text{ where } n \ge 2.$$

Proof. By the formula in Theorem 8.4.4 with $m = 2$, the number of maximal chains of G is equal to

$$\sum_{i=-1}^{1} r_i (2 - i) = 3 + 2 (n - 2) + \frac{n (n - 1)}{2},$$

where $r_{-1} = 1$; $r_0 = n - 1$ and $r_1 = \frac{n(n-1)}{2}$. This sum reduces to $1 + 2n + \frac{n(n-1)}{2}$.

Now only one such maximal chain C_0 goes through p^n and it is given by

$$0 \subset p \subset p^2 \subset \ldots \subset p^n \subset p^n q \subset p^n q^2.$$

The length of this maximal chain is $n + 3$. Thus by Proposition 8.2.3, the number of fuzzy subgroups on C_0 is

$$2^{n+3} - 1.$$

There are $2n$ maximal chains passing through $p^{n-1}q$. Following the notation of representing maximal chains as in the previous Proposition 8.4.7, it follows that each such maximal chain can be uniquely written in the following manner:

$$0 \subset p \subset p^2 \subset \ldots \subset p^i \subset p^i q \subset p^{i+1} q \subset \ldots \subset p^{n-1} q \subset p^n q \subset p^n q^2$$

or

$$0 \subset p \subset p^2 \subset \ldots \subset p^i \subset p^i q \subset p^{i+1} q \subset \ldots \subset p^{n-1} q \subset p^{n-1} q^2 \subset p^n q^2.$$

By Lemma 8.4.6, the number of fuzzy subgroups on anyone of the above maximal chains is 2^{n+2}. Therefore, the total number of fuzzy subgroups on these $2n$ distinct maximal chains is

$$2n 2^{n+2}.$$

There are $\frac{n(n-1)}{2}$ maximal chains passing through $p^{n-1}q^2$. By Lemma 8.4.6, the number of fuzzy subgroups on anyone of these maximal chains is 2^{n+1}. Thus the total number of fuzzy subgroups on these $\frac{n(n-1)}{2}$ distinct maximal chains is

$$\frac{n (n - 1)}{2} 2^{n+1}.$$

For $m = 2$, the formula in Theorem 8.4.5 gives $2^{n+3}(1 + \frac{2n}{2} + \frac{n(n-1)}{2 \times 4})$ which agrees with the sum of numbers $2^{n+3} - 1$, $2n2^{n+2}$, and $\frac{n(n-1)}{2}2^{n+1}$ found immediately above. $\qquad \square$

Proof of Theorem 8.4.5: The formula in Theorem 8.4.4 gives the number of maximal chains of G to be equal to $\sum_{i=-1}^{m-1} r_i (m - i)$ $\forall m \in \mathbb{N}$. It follows that this sum reduces to

$$\sum_{r=0}^{m} \binom{n}{n-r}\binom{m}{r}, \text{ where } m \leq n.$$

For any $r = k$, $0 \leq k \leq m$, in the above sum, the term $\binom{n}{n-k}\binom{m}{k}$ represents the total number of elements in the set $S_{p^{n-k}q^k}$ of all maximal chains passing through $p^{n-k}q^k$. First we notice that the length of any maximal chain is $n+m+1$ and therefore the number of fuzzy subgroups on any of the maximal chain is

$$2^{n+m+1} - 1.$$

By an argument similar to the one used in the above two lemmas, it follows that

$$\frac{2^{n+m+1} - 1 + 1}{2^k}$$

many fuzzy subgroups on each of the element of $S_{p^{n-k}q^k}$ are distinct from any other fuzzy subgroup on any other maximal chain. Similarly, the number of fuzzy subgroups on other maximal chains passing through $p^{n-r}q^r$, $0 \leq r \leq k-1$ can be calculated. As r varies over 0 to m, we get the total number of distinct fuzzy subgroups on G. This total is equal to the sum found in the formula of the theorem.

8.5 Sums, Unions, and Intersections

In this section, we study the operations sum, intersection and union, and their behavior with respect to the equivalence $\sim$ on a group. We examine the extent to which a homomorphism preserves the equivalence. By a **flag** C on a group G, we mean a chain of subgroups $0 \subset G_1 \subset G_2 \subset ... \subset G_n = G$. The G_i's are called **components** of C. We recall that a keychain means $1t_1t_2...t_n$, where the t_i's are not all necessarily distinct. The t_i's are called **pins**. The number of distinct t_i's is called the length of the chain. Two keychains are called **distinct** if either their lengths differ or one of them contains at least one pin distinct from the other. By a **pinned-flag** on G, we mean a pair (C, l) of a flag C and a keychain l. Every keychain gives rise to a fuzzy subgroup μ corresponding to a pinned-flag on G given by $\{0\}_1 \subset (G_1)_{t_1} \subset (G_2)_{t_2} \subset ... \subset (G_n)_{t_n}$ in the following way:

$$\mu(x) = \begin{cases} 1 & \text{if } x = 0, \\ t_1 & \text{if } x \in G_1 \backslash \{0\}, \\ t_2 & \text{if } x \in G_2 \backslash G_1, \\ ... \\ t_n & \text{if } x \in G_n \backslash G_{n-1}, \end{cases}$$

where the component $G_n = G$ and $1 \geq t_1 \geq t_2 \geq ... \geq t_n \geq 0$. We denote this simply by $(G_n)_{t_n} = G_{t_n}$. In this case, we say μ is represented by the pinned-flag $\{0\}_1 \subset (G_1)_{t_1} \subset (G_2)_{t_2} \subset ... \subset (G_n)_{t_n}$. Given a homomorphism between two groups, we examine the equivalence classes of homomorphic images and pre-images of fuzzy subgroups.

Proposition 8.5.1. *Let $f : G \to H$ be a homomorphism of a group G into a group H. If $\mu \sim \nu$, then $f(\mu) \sim f(\nu)$.*

Proof. Clearly $\mu^* = \nu^*$ implies $f(\mu^*) = f(\nu^*)$ which in turn implies $f(\mu)^* = f(\nu)^*$. Let $x, y \in H$ be such that $f(\mu)(x) > f(\mu)(y)$. It suffices to show that $f(\nu)(x) > f(\nu)(y)$. Now there exists $x' \in G$ such that $f(x') = x$ and $\mu(x') > \mu(y')$ for all $y' \in G$ such that $f(y') = y$. Since $\mu \sim \nu$, $\nu(x') > \nu(y')$ for all $y' \in G$. Thus $f(\nu)(x) = \vee\{\mu(x')|f(x') = x\} > \vee\{\mu(y')|f(y') = y\} = f(\nu)(y)$. Hence $f(\mu) \sim f(\nu)$. $\qquad\qquad\square$

Let $f : G \to H$ be a homomorphism of a group G into a group H. It follows easily that if $\mu \sim \nu$ in H, then $f^{-1}(\mu) \sim f^{-1}(\nu)$. Also one could consider the behavior of non equivalent fuzzy subgroups under a homomorphism. The following example illustrates that two not equivalent fuzzy subgroups may have equivalent images under a homomorphism.

Example 8.5.2. *Consider the group $\mathbb{Z}_6 = \{0, 1, 2, 3, 4, 5\}$. Define $f : \mathbb{Z}_6 \to \mathbb{Z}_6$ by $f(n) = 2n \ \forall n \in \mathbb{Z}_6$. Then f is a homomorphism of $\mathbb{Z}_6$ into itself. Define the fuzzy subgroups μ and ν of $\mathbb{Z}_6$ as follows: $\forall x \in \mathbb{Z}_6$,*

$$\mu(x) = \begin{cases} 1 & \text{if } x = 0. \\ 1/2 & \text{if } x \in \{2, 4\}, \\ 1/4 & \text{otherwise}; \end{cases}$$

$$\nu(x) = \begin{cases} 1 & \text{if } x = 0, \\ 1/2 & \text{if } x \in \{3\}, \\ 1/4 & \text{otherwise}. \end{cases}$$

Clearly, μ and ν are not equivalent. Now $f(\mu)(2) = \vee\{\mu(x')|f(x') = 2\} = \mu(1) \vee \mu(4) = 1/2$ and $f(\mu)(4) = \vee\{\mu(x')|f(x') = 4\} = \mu(2) \vee \mu(5) = 1/2$. Also, $f(\nu)(2) = \vee\{\nu(x')|f(x') = 2\} = \nu(1) \vee \nu(4) = 1/4$ and $f(\nu)(4) = \vee\{\mu(x')|f(x') = 4\} = \mu(2) \vee \mu(5) = 1/4$. Hence it follows that

$$f(\mu)(x) = \begin{cases} 1 & \text{if } x = 0 \\ 1/2 & \text{if } x \in \{2, 4\} \\ 0 & \text{if } x \in \{1, 3, 5\} \end{cases}$$

$$f(\nu)(x) = \begin{cases} 1 & \text{if } x = 0, \\ 1/4 & \text{if } x \in \{2, 4\} \\ 0 & \text{if } x \in \{1, 3, 5\} \end{cases}$$

Clearly, $f(\mu)$ and $f(\nu)$ are equivalent.

Similarly, we may have non equivalent fuzzy subgroups giving rise to equivalent pre-images under a homomorphism. This can be seen by the following example.

Example 8.5.3. *Let Z_6 and f be as in the previous example. Define the fuzzy subgroups μ and ν of Z_6 as follows:* $\forall x \in Z_6$,

$$\mu(x) = \begin{cases} 1 & \text{if } x = 0. \\ 1/2 & \text{if } x \in \{2,4\}, \\ 1/3 & \text{otherwise;} \end{cases}$$

$$\nu(x) = \begin{cases} 1 & \text{if } x = 0, \\ 1/2 & \text{if } x \in \{3\}, \\ 1/3 & \text{otherwise.} \end{cases}$$

Clearly μ is not equivalent to ν. Using the equations $f^{-1}(\mu)(x) = \mu(f(x))$ and $f^{-1}(\nu)(x) = \nu(f(x))$, it follows that

$$f^{-1}(\mu)(x) = \begin{cases} 1 & \text{if } x \in \{0,3\}, \\ 1/2 & \text{otherwise;} \end{cases}$$

$$f^{-1}(\nu)(x) = \begin{cases} 1 & \text{if } x \in \{0,3\}, \\ 1/3 & \text{otherwise.} \end{cases}$$

Thus, $f^{-1}(\mu)$ and $f^{-1}(\nu)$ are equivalent.

If $f : G \to H$ is an epimorphism, then $f(f^{-1}(\mu)) = \mu$. Thus if μ and ν are not equivalent fuzzy subgroups of H, then $f^{-1}(\mu)$ and $f^{-1}(\nu)$ are not equivalent fuzzy subgroups of G. However, if $f : G \to H$ is a monomorphism, a similar conclusion cannot be drawn. For example, let $G = \mathbb{Z}_2 = \{0,1\}$ and $H = \{0'\}$. Define $f : G \to H$ by $f(0) = 0' = f(1)$. Define $\mu : G \to [0,1]$ by $\mu(0) = 1$ and $\mu(1) = 1/2$. Then $f^{-1}(f(\mu))(1) = f(\mu)(f(1)) = f(\mu)(0') = \mu(0) \vee \mu(1) = 1 \neq 1/2 = \mu(1)$. Thus $f^{-1}(f(\mu)) \neq \mu$. Define $\nu : G \to [0,1]$ by $\nu(0) = 1 = \nu(1)$. Then $\mu \nsim \nu$. However, $f(\mu)(0') = \mu(0) \vee \mu(1) = 1 = \nu(0) \vee \nu(1) = f(\nu)(0')$. thus $f(\mu) \sim f(\nu)$.

In general the operations of intersection, union and sum of fuzzy subgroups do not preserve the equivalence classes of fuzzy subgroups. This is shown in the following example.

Example 8.5.4. *Consider the group of integers $\mathbb{Z}$ under addition. Define the fuzzy subgroups $\mu, \mu', \nu,$ and ν' of $\mathbb{Z}$ as follows:* $\forall x \in \mathbb{Z}$,

$$\mu(x) = \begin{cases} 1 & \text{if } x \in 2\mathbb{Z}, \\ 1/3 & \text{otherwise} \end{cases} \qquad \mu'(x) = \begin{cases} 1 & \text{if } x \in 2\mathbb{Z}, \\ 1/4 & \text{otherwise} \end{cases}$$

$$\nu(x) = \begin{cases} 1 & \text{if } x \in 3\mathbb{Z}, \\ 1/2 & \text{otherwise} \end{cases} \qquad \nu'(x) = \begin{cases} 1 & \text{if } x \in 3\mathbb{Z}, \\ 1/5 & \text{otherwise.} \end{cases}$$

Then,

$$(\mu \cap \nu)(x) = \begin{cases} 1 & \text{if } x \in 6\mathbb{Z}, \\ 1/2 & \text{if } x \in 2\mathbb{Z} \backslash 6\mathbb{Z}, \\ 1/3 & \text{otherwise;} \end{cases}$$

$$(\mu' \cap \nu')(x) = \begin{cases} 1 & \text{if } x \in 6\mathbb{Z}, \\ 1/4 & \text{if } x \in 3\mathbb{Z} \backslash 6\mathbb{Z}, \\ 1/5 & \text{otherwise.} \end{cases}$$

Clearly, $\mu \sim \mu'$ and $\nu \sim \nu'$. However, it follows that $(\mu \cap \nu) \nsim (\mu' \cap \nu')$ since $(\mu \cap \nu)(3) = \frac{1}{3} < (\mu \cap \nu)(2) = \frac{1}{2}$, but $(\mu' \cap \nu')(3) = \frac{1}{4} > (\mu' \cap \nu')(2) = \frac{1}{5}$.

Similarly, it can be shown that $\mu \sim \mu'$ and $\nu \sim \nu'$ does not imply $\mu \cup \nu \sim \mu' \cup \nu'$.

The next example deals with the operation, sum.

Example 8.5.5. *Consider the group of integers $\mathbb{Z}$ under addition. Define the fuzzy subgroups μ, μ', ν, and ν' of $\mathbb{Z}$ as follows: $\forall x \in \mathbb{Z}$,*

$$\mu(x) = \begin{cases} 1 & \text{if } x \in 6\mathbb{Z}, \\ 1/2 & \text{if } x \in 2\mathbb{Z}\backslash 6\mathbb{Z}, \\ 1/4 & \text{otherwise}, \end{cases} \qquad \mu'(x) = \begin{cases} 1 & \text{if } x \in 6\mathbb{Z}, \\ 3/4 & \text{if } x \in 2\mathbb{Z}\backslash 6\mathbb{Z}, \\ 1/8 & \text{otherwise} \end{cases}$$

$$\nu(x) = \begin{cases} 1 & \text{if } x \in 6\mathbb{Z}, \\ 2/3 & \text{if } x \in 3\mathbb{Z}\backslash 6\mathbb{Z}, \\ 1/6 & \text{otherwise}, \end{cases} \qquad \nu'(x) = \begin{cases} 1 & \text{if } x \in 6\mathbb{Z}, \\ 1/2 & \text{if } x \in 3\mathbb{Z}\backslash 6\mathbb{Z}, \\ 1/6 & \text{otherwise}. \end{cases}$$

Now $(\mu + \nu)(x) = \vee\{\mu(y) \wedge \nu(z) | x = y + z\}$ for all $x \in \mathbb{Z}$ and similarly for $\mu' + \nu'$. Thus,

$$(\mu + \nu)(x) = \begin{cases} 1 & \text{if } x \in 6\mathbb{Z}, \\ 2/3 & \text{if } x \in 3\mathbb{Z}\backslash 6\mathbb{Z}, \\ 1/2 & \text{otherwise}, \end{cases}$$

$$(\mu' + \nu')(x) = \begin{cases} 1 & \text{if } x \in 6\mathbb{Z}, \\ 3/4 & \text{if } x \in 2\mathbb{Z}\backslash 6\mathbb{Z}, \\ 1/2 & \text{otherwise}. \end{cases}$$

Clearly, $\mu \sim \mu'$ and $\nu \sim \nu'$. However, $(\mu + \nu) \nsim (\mu' + \nu')$ since $(\mu + \nu)(3) = \frac{2}{3} > (\mu + \nu)(2) = \frac{1}{2}$, but $(\mu' + \nu')(3) = \frac{1}{2} < (\mu' + \nu')(2) = \frac{3}{4}$.

In the previous example, $\mu + \nu \sim \nu$. However, this need not be true in general. For example if $\frac{1}{6}$ is replaced by 0 in the definition of ν above, then $\mu + \nu \nsim \nu$ since the support of $\mu + \nu$ is now different than that of ν. Also we note, $\mu + \nu \nsim \mu$. Thus $\mu + \nu$ is not equivalent to either μ or ν. In Proposition 8.5.8, we show that if $\mu \sim \nu$, then $\mu + \nu \sim \mu$ and so $\mu + \nu \sim \nu$.

If we take two fuzzy subgroups μ and ν from the same equivalence class C determined by μ and ν, then the inf, sup and sum of μ and ν determine the same equivalence class C.

Proposition 8.5.6. *Let μ and ν be fuzzy subgroups of a group G. If $\mu \sim \nu$, then $\mu \cap \nu \sim \mu$.*

Proof. Since $\mu \sim \nu$, $\mu^* = \nu^*$. Since also $(\mu \cap v)^* = \mu^* \cap \nu^*$, we have $(\mu \cap \nu)^* = \mu^*$. Let $x, y \in G$. If $\mu(x) > \mu(y)$, then $\nu(x) > \nu(y)$ and so $(\mu \cap \nu)(x) > (\mu \cap \nu)(y)$. Now suppose $(\mu \cap \nu)(x) > (\mu \cap \nu)(y)$. We have four cases.

(1) $\mu(x) = (\mu \cap \nu)(x)$ and $\mu(y) = (\mu \cap \nu)(y)$: In this case, clearly $\mu(x) > \mu(y)$.

(2) $\mu(x) = (\mu \cap \nu)(x)$ and $\nu(y) = (\mu \cap \nu)(y)$: Here $\nu(x) \geq \mu(x) > \nu(y)$. Thus $\mu(x) > \mu(y)$ since $\mu \sim \nu$.

(3) $\nu(x) = (\mu \cap \nu)(x)$ and $\nu(y) = (\mu \cap \nu)(y)$: Here $\nu(x) > \nu(y)$. Thus $\mu(x) > \mu(y)$ since $\mu \sim \nu$.

(4) $\nu(x) = (\mu \cap \nu)(x)$ and $\mu(y) = (\mu \cap \nu)(y)$: In this case, $\nu(x) > \mu(y)$. Since $\mu(x) \geq \nu(x), \mu(x) > \mu(y)$. $\qquad\square$

Proposition 8.5.7. *Let μ and ν be fuzzy subgroups of a group G. If $\mu \sim \nu$ then, $\mu \cup \nu \sim \mu$.*

Proof. $(\mu \cup \nu)^* = \mu^* \cup \nu^* = \mu^*$. Let $x, y \in G$. Suppose $(\mu \cup \nu)(x) > (\mu \cup \nu)(y)$. Suppose $\mu(x) = (\mu \cup \nu)(x)$. Then $\mu(x) > \mu(y) \vee \nu(y)$. Thus $\mu(x) > \mu(y)$. Suppose $\nu(x) = (\mu \cup \nu)(x)$. Then $\nu(x) > \mu(y) \vee \nu(y)$ and so $\nu(x) > \nu(y)$. Since $\mu \sim \nu, \mu(x) > \mu(y)$. Suppose $\mu(x) > \mu(y)$. Then $\nu(x) > \nu(y)$ since $\mu \sim \nu$. Thus $\mu(x) \vee \nu(x) > \mu(y) \vee \nu(y)$. Hence $(\mu \cup \nu)(x) > (\mu \cup \nu)(y)$. $\qquad\square$

Proposition 8.5.8. *Let μ and ν be fuzzy subgroups of a finite group G. If $\mu \sim \nu$, then $\mu + \nu \sim \mu$.*

Proof. $(\mu + \nu)^* = \mu^* + \nu^* = \mu^*$. Suppose $(\mu + \nu)(x) > (\mu + \nu)(y)$. Then there are x_1 and x_2 with $x = x_1 + x_2$ such that $\mu(x_1) \wedge \nu(x_2) > (\mu + \nu)(y)$. Now $(\mu + \nu)(y) \geq \mu(y) \wedge \nu(e) = \mu(y)$. Suppose $\mu(x_1) = \mu(x_1) \wedge \nu(x_2)$. Then $\mu(x_1) > \mu(y)$ and $\nu(x_2) \geq \mu(x_1) > \nu(y)$. Thus $\mu(x_2) > \mu(y)$ since $\mu \sim \nu$. Therefore, $\mu(x) \geq \mu(x_1) \wedge \mu(x_2) > \mu(y)$. Suppose $\nu(x_2) = \mu(x_1) \wedge \nu(x_2)$. Then $\mu(x_1) \geq \nu(x_2) > \nu(y)$. Thus $\mu(x_2) > \mu(y)$ since $\mu \sim \nu$. Also $\mu(x_1) \geq \nu(x_2) > (\mu + \nu)(y) \geq \mu(y) \wedge \nu(e) = \mu(y)$. Hence $\mu(x) > \mu(y)$. Conversely, suppose $\mu(x) > \mu(y)$. Then we show by contradiction that $(\mu + \nu)(x) > (\mu + \nu)(y)$. Suppose $(\mu + \nu)(x) < (\mu + \nu)(y)$. Then by the above argument, we have $\mu(y) > \mu(x)$, a contradiction. Suppose $(\mu + \nu)(x) = (\mu + \nu)(y)$. Then since G is finite, there exists $y_1, y_2 \in G$ such that $(\mu + \nu)(x) = \mu(y_1) \wedge \nu(y_2)$ and $y = y_1 + y_2$. Hence $\mu(y_1) \geq (\mu + \nu)(x) \geq \mu(x)$. Also, $\nu(y_2) \geq \nu(x)$ and so $\mu(y_2) \geq \mu(x)$ since $\mu \sim \nu$. Thus $\mu(y) \geq \mu(y_1) \wedge \mu(y_2) \geq \mu(x)$, a contradiction. Consequently, $(\mu + \nu)(x) > (\mu + \nu)(y)$. $\qquad\square$

Proposition 8.5.9. *Let μ and ν be fuzzy subgroups of a group G. If $\mu \sim \nu$, then $1 - \mu \sim 1 - \nu$.*

Proof. Now $x \in (1 - \mu)^* \Leftrightarrow (1 - \mu)(x) > 0 \Leftrightarrow 0 \leq \mu(x) < 1 \Leftrightarrow 0 \leq \nu(x) < 1 \Leftrightarrow (1 - \nu)(x) > 0 \Leftrightarrow x \in (1 - \nu)^*$. Thus $(1 - \mu)^* = (1 - \nu)^*$. Let $x, y \in G$. Suppose $(1 - \mu)(x) > (1 - \mu)(y)$. Then $\mu(x) < \mu(y)$. Thus $\nu(x) < \nu(y)$. Hence $(1 - \nu)(x) > (1 - \nu)(y)$. Similarly, $(1 - \nu)(x) > (1 - \nu)(y)$ implies $(1 - \mu)(x) > (1 - \mu)(y)$. $\qquad\square$

We now determine the equivalence class of fuzzy subgroups corresponding to the intersection and sum of two fuzzy subgroups in terms of pinned-flags associated with given fuzzy subgroups. Throughout this section, we require the number of components in a pinned-flag to be at least 3, otherwise the discussion becomes trivial. Therefore, we assume $n \geq 2$.

In the next proposition, we look at a special case, namely the characterization of intersection and sum of two equivalence classes of fuzzy subgroups with the same flag of subgroups as their t-cuts.

Proposition 8.5.10. *Let G be a group with a flag C of subgroups $\{0\} = G_0 \subset G_1 \subset G_2 \subset ... \subset G_n = G$. Suppose $\mu = (C_\mu, l_\mu)$ and $\nu = (C_\nu, l_\nu)$ are two fuzzy subgroups whose representative keychains are of the form $l_\mu = 1t_1 t_2 ... t_n$ and $l_\nu = 1s_1 s_2 \cdots s_n$, respectively. Then the following assertions hold:*

    (1) $C_{\mu \cap \nu} = C$ $u \cap \nu \sim 1t_1 \wedge s_1 ... t_n \wedge s_n$,

    (2) $C_{\mu + \nu} = C$ $u + \nu \sim 1t_1 \vee s_1 ... t_n \vee s_n$.

Proof. Let $x \in G$. Then there is an index i such that $1 \leq i \leq n$ with $x \in G_i$, but $x \notin G_{i-1}$. Thus $\mu(x) = t_i$ and $\nu(x) = s_i$.

(1) It suffices to prove that $(\mu \cap \nu)(x) = t_i \wedge s_i$. However, this is clearly true.

(2) We have that $(\mu + \nu)(x) \geq t_i \vee s_i$. Suppose $(\mu + \nu)(x) > t_i \vee s_i$. Then there exist $x_1, x_2 \in G$ such that $x = x_1 + x_2$ and $\mu(x_1) \wedge \nu(x_2) > t_i \vee s_i$. Hence $\mu(x_1) > t_i$ and $\nu(x_2) > s_i$ and so $x_1 \in G_{i-1}$ and $x_2 \in G_{i-1}$. Thus $x \in G_{i-1}$, a contradiction. $\qquad\qquad\square$

In the next proposition, we consider two flags differing in one component.

Proposition 8.5.11. *Suppose μ and ν are fuzzy subgroups of G whose representative keychains are of the form $1t_1 t_2 \cdots t_n$ and whose underlying flags are C_μ and C_ν described as follows:*

    $C_\mu : \{0\} = G_0 \subset G_1 \subset G_2 \subset ... \subset G_{k-1} \subset G_k \subset G_{k+1} \subset ... \subset G_n = G$

    and

    $C_\nu : \{0\} = G_0 \subset G_1 \subset G_2 \subset ... \subset G_{k-1} \subset H_k \subset G_{k+1} \subset ... \subset G_n = G$

    for a fixed k such that $1 \leq k \leq n-1, G_k \neq H_k$.

Then the following assertions hold:

    (1) $\mu \cap \nu = (C_{\mu \cap \nu}, l_{\mu \cap \nu}) : 1t_1 t_2 ... t_{k-1} t_{k+1} t_{k+1} ... t_n$ on C_μ or on C_ν.

and

    (2) $\mu + \nu = (C_{\mu + \nu}, l_{\mu + \nu}) : 1t_1 t_2 ... t_{k-1} t_k t_k t_{k+2} ... t_n$ on C_μ or on C_ν,

where $C_{\mu \cap \nu} = C_\mu = C_\nu$ and $C_{\mu + \nu} = C_\mu = C_\nu$.

Proof. (1) By Proposition 8.5.10, it follows that $(\mu \cap \nu)(x)$ has the same keychain pins as μ and ν on all G_i's, for $i = 1, 2, ..., k-1, k+1, ..., n$. It suffices to prove the case for $i = k$. Let $x \in G_{k+1} \backslash G_{k-1}$. Then $(\mu \cap \nu)(x) \geq t_{k+1}$. Suppose $(\mu \cap \nu)(x) > t_{k+1}$. Then either $(\mu \cap \nu)(x) = t_s$ for some $1 \leq s \leq k$ or $(\mu \cap \nu)(x) = 1$. The latter case implies $\mu(x) = 1 = \nu(x)$. However, $x \notin G_{k-1}$ implies $\mu(x) < t_{k-1}$. This in turn implies $\mu(x) \neq 1$, a contradiction. For the former case, $t_s \geq t_k$. Thus $\mu(x) \geq t_k$. Hence $x \in G_k$. Similarly, $x \in H_k$. Thus $x \in G_k \cap H_k$. By the maximality of the chains involved, $G_k \cap H_k = G_{k-1}$. Therefore, $x \in G_{k-1}$, which is a contradiction of the choice of x. Thus $(\mu \cap \nu)(x) = t_{k+1}$ for all $x \in G_{k+1} \backslash G_{k-1}$.

(2) It follows that $G_{k+1} = G_k + H_k$ by the maximality of chains. Similar to the proof of case (1), for the sum, it suffices to consider $x \in G_{k+1} \backslash G_{k-1}$. Then $(\mu \cap \nu)(x) \geq \mu(x_1) \wedge \nu(x_2) \geq t_k$ for $x = x_1 + x_2$ with $x_1 \in G_k$ and $x_2 \in H_k$. Suppose $\mu(x_1) \wedge \nu(x_2) > t_k$. Then either $\mu(x_1) \wedge \nu(x_2) = t_s$ for some $1 \leq s \leq k - 1$ or $\mu(x_1) \wedge \nu(x_2) = 1$. The latter case implies $x = x_1 + x_2 \in G_{k-1} + G_{k-1} = G_{k-1}$, and in the former case $x = x_1 + x_2 \in G_s + G_s = G_s$. By the chain property, we conclude that $x \in G_{k-1}$, which is a contradiction of the choice of x. Thus $(\mu \cap \nu)(x) = t_k$ for all $x \in G_{k+1} \backslash G_{k-1}$. $\square$

Let μ and ν be fuzzy subgroups of a group whose pinned-flags differ in two nonconsecutive components and are given by $\mu = (C_\mu, l_\mu)$ and $\nu = (C_\nu, l_\nu)$ as follows:

$C_\mu : \{0\} = G_0 \subset G_1 \subset ... \subset G_{i_1} \subset ... \subset G_{i_2} \subset ... \subset G_{i_k} \subset ... \subset G_n = G$
and

$C_\nu : \{0\} = G_0 \subset G_1 \subset ... \subset H_{i_1} \subset ... \subset H_{i_2} \subset ... \subset H_{i_k} \subset ... \subset G_n = G,$

where $1 \leq i_1 < i_1 + 2 \leq i_2 < i_2 + 2 \leq ... < i_{k-1} < i_{k-1} + 2 \leq i_k \leq n - 1, G_{i_k} \neq H_{k_i}$ for $k = 1, 2$.

Then, using a similar argument as in Proposition 8.5.11 inductively, we have the following result.

Corollary 8.5.12. *Let $\mu = (C_\mu, l_\mu)$ and $\nu = (C_\nu, l_\nu)$ be as above. Then the following assertions hold:*
(1) $\mu \cap \nu$ is represented by the keychain

$$l_{\mu \cap \nu} = 1t_1...t_{i_1 - 1} t_{i_1 + 1} t_{i_1 + 1}...t_{i_k - 1} t_{i_k + 1} t_{i_k + 1}...t_n \ on \ C_{\mu \cap \nu} = C_\mu = C_\nu$$

(2) $\mu + \nu$ is represented by the keychain

$$l_{\mu + \nu} = 1t_1...t_{i_1 - 1} t_{i_1} t_{i_1}...t_{i_k - 1} t_{i_k} t_{i_k}...t_n \ on \ C_{\mu + \nu} = C_\mu = C_\nu.$$

We now consider two flags differing in two or more components consecutively, with all other basic assumptions taken to be true on the flags as in the previous propositions:

$C_\mu : ... \subset G_{i-1} \subset G_i \subset G_{i+1} \subset ... \subset G_{i+k-1} \subset G_{i+k} \subset G_{i+k+1} \subset ...$
and

$C_\nu : ... \subset G_{i-1} \subset H_i \subset H_{i+1} \subset ... \subset H_{i+k-1} \subset H_{i+k} \subset G_{i+k+1} \subset ...$
for $1 \leq i < ... < i + k \leq n - 1$, with $k \geq 1$ and $G_{i+k} \neq H_{i+k}$.

In the above, we have only indicated the corresponding distinct components in C_μ and C_ν as the suppressed corresponding components are assumed to be identical in the two flags.

$C_{\mu \cap \nu} : ... \subset G_{i-1} \subset G_{i+1} \cap H_{i+1} \subset ... \subset G_{i+k} \cap H_{i+k} \subset F \subset G_{i+k+1} \subset ...,$
where F can be either G_{i+k} or H_{i+k}.

$C_{\mu + \nu} : ... \subset G_{i-1} \subset E \subset G_i + H_i \subset ... \subset G_{i+k-1} + H_{i+k-1} \subset G_{i+k+1} \subset ...,$
where E can be either G_i or H_i.

In the above, we have only indicated the corresponding distinct components in $C_\mu, C_\nu, C_{\mu \cap \nu}$ and $C_{\mu + \nu}$, and as such the suppressed corresponding components are assumed to be identical in the two flags.

Proposition 8.5.13. *Suppose μ and ν are fuzzy subgroups of a group G whose representative keychains are of the form $1t_1t_2...t_n$ and whose underlying flags are C_μ and C_ν, respectively, as described above. Then the following assertions hold:*

(1) *$\mu \cap \nu$ is represented by the keychain*

$$1t_1t_2...t_{i-1}t_{i+1}t_{i+2}...t_{i+k}t_{i+k+1}t_{i+k+1}t_{i+k+2}...t_n$$

on

$$C_{\mu\cap\nu} : ... \subset G_{i-1} \subset G_{i+1} \cap H_{i+1} \subset ... \subset G_{i+k} \cap H_{i+k} \subset F \subset G_{i+k+1} \subset ...,$$

where F can be either G_{i+k} or H_{i+k} and

(2) *$\mu + \nu$ is represented by the keychain*

$$1t_1t_2...t_{i-1}t_it_it_{i+1}...t_{i+k}t_{i+k+2}...t_n$$

on

$$C_{\mu+\nu} : ... \subset G_{i-1} \subset E \subset G_i + H_i \subset ... \subset G_{i+k-1} + H_{i+k-1} \subset G_{i+k+1} \subset ...,$$

where E can be either G_i or H_i.

Proof. As in Proposition 8.5.11, it suffices to consider only indices $i, i+1, ..., i+k$.

(1) Let $x \in G_{i+1} \cap H_{i+1} \setminus G_{i-1}$. Then clearly $t_{i-1} > (\mu \cap \nu)(x) \geq t_{i+1}$. Now suppose $(\mu \cap \nu)(x) > t_{i+1}$. Then $(\mu \cap \nu)(x) = t_i$. Thus $x \in G_i \cap H_i = G_{i-1}$ by the maximality of C_μ and C_ν. This is a contradiction of our choice of x. Hence $(\mu \cap \nu)(x) = t_{i+1}$. Now for other cases, let $x \in G_{i+j} \cap H_{i+j} \setminus G_{i+j-1} \cap H_{i+j-1}$ for $j = 2, 3, ..., k$. Then $t_{i+j-1} > (\mu \cap \nu)(x) \geq t_{i+j}$. Suppose $(\mu\cap\nu)(x) > t_{i+j}$. Then there exists a pin t_s representing the value of $(\mu\cap\nu)(x)$ and is such that $t_{i+j-1} > t_s > t_{i+j}$, However, this is a contradiction since t_{i+j-1} and t_{i+j} are two consecutive pins. Thus $(\mu \cap \nu)(x) = t_{i+j}$. Let $x \in F \setminus G_{i+k} \cap H_{i+k}$, say $F = H_{i+k}$. Then $\nu(x) \geq t_{i+k}$ and $\mu(x) < t_{i+k}$ which implies $(\mu \cap \nu)(x) = \mu(x) < t_{i+k}$. However, $(\mu \cap \nu)(x) \geq t_{i+k+1}$. Suppose $(\mu \cap \nu)(x) > t_{i+k+1}$. Then this leads to a contradiction as in the previous case. Thus $(\mu \cap \nu)(x) = t_{i+k+1}$. The argument is similar for $F = G_{i+k}$. Finally, let $x \in G_{i+k+1} \setminus F$, say $F = H_{i+k}$. Then $(\mu \cap \nu)(x) \geq t_{i+k+1}$ and $\nu(x) < t_{i+k}$ which implies $(\mu \cap \nu)(x) < t_{i+k}$. As in the previous cases, it follows that $(\mu\cap\nu)(x) = t_{i+k+1}$. The argument is similar for $F = G_{i+k}$ Thus (1) is proved.

(2) Let $x \in E \setminus G_{i-1}$, say $E = H_i$. Then $\nu(x) \geq t_i$. Hence $(\mu \cap \nu)(x) \geq t_i$. Suppose $(\mu \cap \nu)(x) \geq t_{i-1}$. Then $x \in G_{i-1}$, a contradiction. Thus $t_i \leq (\mu \cap \nu)(x) < t_{i-1}$. As in case (1) above, we conclude that $(\mu \cap \nu)(x) = t_i$. Next let $x \in G_i + H_i \setminus E$, say $E = H_i$. Then $(\mu \cap \nu)(x) \geq t_i$ and $\nu(x) < t_i$. Suppose $(\mu \cap \nu)(x) \geq t_{i-1}$. Then $x \in G_{i-1} \subseteq H_i$, a contradiction. Thus $t_i \leq (\mu \cap \nu)(x) < t_{i-1}$. Hence, as before $(\mu \cap \nu)(x) = t_i$. The argument is

similar for $E = G_i$. For other cases, let $x \in G_{i+j} + H_{i+j} \setminus G_{i+j-1} + H_{i+j-1}$, for $j = 1, 2, ..., k - 1$. Then $t_{i+j-1} > (\mu \cap \nu)(x) \geq t_{i+j}$. As in other parts of this proof, it is clear that $(\mu \cap \nu)(x) = t_{i+j}$. Finally, let $x \in G_{i+k+1} \setminus G_{i+k-1} + H_{i+k-1}$. Then $(\mu \cap \nu)(x) \geq t_{i+k+1}$. However, $G_{i+k+1} = G_{i+k} + H_{i+k}$, by the maximality of the chains C_μ and C_ν. Hence $t_{i+k-1} > (\mu \cap \nu)(x) \geq t_{i+k}$. Thus $(\mu \cap \nu)(x) = t_{i+k}$. $\qquad\square$

The determination of the pinned-flags of the intersection and the sum of two fuzzy subgroups μ and ν, where the pins as well as the flags of the pinned-flags C_μ and C_v representing μ and ν are distinct, in general, does not seem to follow any particular pattern as we have derived above. This is illustrated by the following example.

Example 8.5.14. Let $G = \mathbb{Z}_{72}$. Let C_μ and C_v be the pinned-flags of μ and ν on G given by

$$(C_\mu, l_\mu) : 0_1 \subset (\mathbb{Z}_3)_{\frac{1}{2}} \subset (\mathbb{Z}_9)_{\frac{1}{5}} \subset (\mathbb{Z}_{18})_{\frac{1}{6}} \subset (\mathbb{Z}_{36})_{\frac{1}{9}} \subset (\mathbb{Z}_{72})_{\frac{1}{10}}$$

and

$$(C_\nu, l_\nu) : 0_1 \subset (\mathbb{Z}_3)_{\frac{1}{3}} \subset (\mathbb{Z}_6)_{\frac{1}{4}} \subset (\mathbb{Z}_{12})_{\frac{1}{7}} \subset (\mathbb{Z}_{36})_{\frac{1}{8}} \subset (\mathbb{Z}_{72})_{\frac{1}{11}}$$

respectively. It follows that the pinned-flags for $\mu \cap \nu$ and $\mu + \nu$ are

$$(C_{\mu \cap \nu}, l_{\mu \cap \nu}) : 0_1 \subset (\mathbb{Z}_3)_{\frac{1}{3}} \subset (\mathbb{Z}_6)_{\frac{1}{6}} \subset (\mathbb{Z}_{18})_{\frac{1}{8}} \subset (\mathbb{Z}_{36})_{\frac{1}{9}} \subset (\mathbb{Z}_{72})_{\frac{1}{11}}$$

and

$$(C_{\mu + \nu}, l_{\mu + \nu}) : 0_1 \subset (\mathbb{Z}_3)_{\frac{1}{2}} \subset (\mathbb{Z}_6)_{\frac{1}{4}} \subset (\mathbb{Z}_{18})_{\frac{1}{5}} \subset (\mathbb{Z}_{36})_{\frac{1}{7}} \subset (\mathbb{Z}_{72})_{\frac{1}{10}},$$

respectively. In the above calculation, the roles played by the pins and the components of the flags are equally important in a way in which they are tied to each other. Suppose we retain the flags, but the pins for μ and ν are changed to $1\frac{1}{3}\frac{1}{9}\frac{1}{18}\frac{1}{36}\frac{1}{72}$ and $1\frac{1}{72}\frac{1}{90}\frac{1}{100}\frac{1}{110}\frac{1}{120}$ respectively. Then $\mu \cap \nu$ and $\mu + \nu$ have the pinned-flags given by

$$(C_{\mu' \cap \nu'}, l_{\mu' \cap \nu'}) : 0_1 \subset (\mathbb{Z}_3)_{\frac{1}{72}} \subset (\mathbb{Z}_6)_{\frac{1}{90}} \subset (\mathbb{Z}_{12})_{\frac{1}{100}} \subset (\mathbb{Z}_{36})_{\frac{1}{110}} \subset (\mathbb{Z}_{72})_{\frac{1}{120}}$$
$$(C_{\mu' + \nu'}, l_{\mu' + \nu'}) : 0_1 \subset (\mathbb{Z}_3)_{\frac{1}{3}} \subset (\mathbb{Z}_9)_{\frac{1}{9}} \subset (\mathbb{Z}_{18})_{\frac{1}{18}} \subset (\mathbb{Z}_{36})_{\frac{1}{36}} \subset (\mathbb{Z}_{72})_{\frac{1}{72}},$$

respectively. Similarly, we could retain the pins, but change the flags of μ and ν, for example in (1) and (2) above we retain the same pins, but swap the underlying flags. Then it follows that we arrive at different (from (3) and (4)) pinned-flags for $\mu \cap \nu$ and $\mu + \nu$.

8.6 Fuzzy Subgroups of Infinite Cyclic Groups

One of the main purposes of crisp group theory is to determine the number of distinct subgroups a given group possesses. The number of subgroups of an infinite group is infinite and so to characterize distinct subgroups we use the concepts of homomorphism and isomorphism. The determination of how many distinct fuzzy subgroups a given group can possess is an interesting area of research within fuzzy group theory. Several authors have studied the issue of how many distinct fuzzy subgroups a group may have. They have used various concepts including those of fuzzy isomorphism [6, 7] and the cardinality of the set of truth-values [18]. The concept of fuzzy isomorphism incorporates the following notions: Let G be a group and μ and ν be fuzzy subgroups of G. If there exists an isomorphism $f : \mu^* \to \nu^*$ such that $\mu(x) > \mu(y)$ implies $\nu(f(x)) > \nu(f(y))\forall x, y \in \mu^*$, then μ is said to be **isomorphic** to ν. In Section 8.1, the concepts of fuzzy isomorphism [6, 7] and fuzzy equivalence were compared and it was observed that equivalence is finer than isomorphism. Hence the number of fuzzy subgroups of certain finite abelian groups was studied using the equivalence of fuzzy subsets. However, the equivalence approach pays less attention to the key point in group theory that two groups are considered to be the same group if their algebraic properties are same, i.e., they are isomorphic. Since the number of crisp subgroups of a finite group G is also finite, it is possible from the point of view of equivalence to count the number of fuzzy subgroups of G. Thus it seems that in some special cases, equivalence does work. However, for infinite groups, for example an infinite cyclic group, we can construct infinite fuzzy subgroups from the equivalence viewpoint as indicated in Example 8.6.2 below. Thus it seems that in the case of infinite groups, the equivalence relation is not suitable.

Similar problems are encountered when using fuzzy isomorphisms. For example, let $G = \langle a \rangle$ be an infinite cyclic group and μ and ν be fuzzy subgroups of G such that $\mu^* = \nu^* = G$. Then an isomorphism f of μ^* onto ν^* is either the identity or is such that $f(x) = -x \forall x \in G$. In this case, μ and ν are equivalent if and only if they are isomorphic. Hence a new approach is needed to study the classification of fuzzy subgroups of an infinite group. We use the concept of group isomorphism to characterize the similarity of fuzzy subgroups.

Definition 8.6.1. *Let μ and ν be fuzzy subgroups of a group G. If $|\mathrm{Im}(\mu)| = |\mathrm{Im}(\nu)|$, $\mu^* \cong \nu^*$, and for all $t \in [0, 1]$ such that $\mu_t \neq \emptyset$ there exists $s \in [0, 1]$ such that $\mu_t \cong \nu_s$ and for all $s \in [0, 1]$ such that $\nu_s \neq \emptyset$ there exists $t \in [0, 1]$ such that $\mu_t \cong \nu_s$, then μ and ν are called S^*-equivalent, written $\mu \cong \nu$.*

If in Definition 8.6.1, $\mu^* \cong \nu^*$ and $\mu_t \cong \nu_s$ is replaced by $\mu^* = \nu^*$ and $\mu_t = \nu_s$, respectively, then the notion of S^*-equivalence becomes that of strong equivalence given in Definition 8.1.10. If two fuzzy subgroups of a finite group are equivalent, then they are S^*-equivalent. However the converse is not necessarily true since two subgroups of a group may be isomorphic without being equal. Thus the notion of equivalence is a special case of S^*-equivalence when dealing with fuzzy subgroups of finite groups. The initial idea of Definition 8.6.1 first appeared in [16].

Example 8.6.2. *Let G be an infinite cyclic group generated by a. Let μ and ν be fuzzy subsets of G defined as follows. $\forall x \in G, \mu(x) = 1$ if $x \in \langle a^2 \rangle, \mu(x) = 1/2$ otherwise and $\nu(x) = 1$ if $x \in \langle a^3 \rangle$ and $\nu(x) = 1/2$ otherwise. Then μ and ν are fuzzy subgroups of G. Clearly μ and ν are neither equivalent nor isomorphic, but they are S^*-equivalent.*

Example 8.6.3. *Let G be an infinite cyclic group generated by a. Let $\{t_n\}_{n-1}^{\infty}$ be a strictly increasing sequence of numbers in $[0, 1]$. Let μ and ν be the fuzzy subsets of G defined as follows: $\forall x \in G, \mu(x) = t_1$ if $x \in G \backslash \langle a^2 \rangle, \mu(x) = t_n$ if $x \in \langle a^{2n-2} \rangle \backslash \langle a^{2n} \rangle$ for $n = 2, 3, ..., \mu(e) = 1$ and $\nu(x) = t_1$ if $x \in G \backslash \langle a^3 \rangle, \nu(x) = t_n$ if $x \in \langle a^{3n-3} \rangle \backslash \langle a^{3n} \rangle$ for $n = 2, 3, ..., \nu(e) = 1$. Then μ and ν are fuzzy subgroups of G which are neither equivalent nor isomorphic. It is shown in Theorem 8.6.6 that μ and ν are S^*-equivalent.*

Proposition 8.6.4. *Let G be an infinite cyclic group generated by a. Let μ and ν be finite-valued fuzzy subgroups of G. Then $\mu \cong \nu$ if and only if $|\text{Im}(\mu)| = |\text{Im}(\nu)|, \mu^* \cong \nu^*$, and $\mu_* \cong \nu_*$.*

Proof. Suppose that the conditions hold. Clearly, if $t = 0$, then $\mu_0 = \nu_0 = G$. Suppose $t \in (0, 1]$ and $\mu_t \neq \emptyset$. Suppose $\mu_t = \{e\}$. Let $s = \nu(e)$. Then $\mu_t = \mu_* \cong \nu_* = \nu_s$. Suppose $\mu_t \neq \{e\}$. Then $\mu_t = \langle a^k \rangle$ and $\mu(a^k) \geq t > 0$ for some positive integer k. Since $\mu^* \cong \nu^*$ and μ and ν are finite-valued, $\nu^* = \langle a^j \rangle$ for some positive integer j and $s = \wedge \{\nu(x) \mid \nu(x) > 0\} > 0$. Then $\nu_s = \langle a^j \rangle$ and $\mu_t \cong \nu_s$.

Conversely, suppose $\mu \cong \nu$. Suppose $\mu_* = \{e\}$. There is an $s \in [0, 1]$ such that $\mu_* \cong \nu_s$. Thus $\nu_s = \{e\}$. Since $\nu_* \subseteq \nu_s$ and $s \leq \nu(e)$, it follows that $\nu_* = \nu_s$. Thus $\mu_* \cong \nu_*$. Suppose $\mu_* \neq \{e\}$. Then $\mu_* = \langle a^k \rangle$ for some positive integer k. there exists $t \in [0, 1]$ such that $\mu_t \cong \nu_{\nu(e)} = \nu_*$. Thus $\nu_* \neq \{e\}$. Hence $\nu_* = \langle a^j \rangle$ for some positive integer j. Thus $\mu_* \cong \nu_*$. □

Lemma 8.6.5. *Let G be an infinite cyclic group generated by a. Let μ be a fuzzy subgroup of G. If $Im(\mu)$ is an infinite set, then $\mu_* = \{e\}$.*

Proof. Suppose $\mu_* \neq \{e\}$. Then $\mu_* = \langle a^k \rangle$ for some positive integer k. Since $\forall t \leq \mu(e), \mu_* \subseteq \mu_t$, there is a positive integer j such that $\mu_t = \langle a^j \rangle$ and $1 \leq j \leq k$. Thus the number of level subgroups is finite and so $\text{Im}(\mu)$ is finite, a contradiction. Hence $\mu_* = \{e\}$. □

Theorem 8.6.6. *Let G be an infinite cyclic group generated by a. Let μ and ν be fuzzy subgroups of G. If both $Im(\mu)$ and $Im(\nu)$ are infinite, then $\mu \cong \nu$.*

Proof. Since both $\text{Im}(\mu)$ and $\text{Im}(\nu)$ are infinite, both $\text{Im}(\mu)$ and $\text{Im}(\nu)$ are countable. Thus $|\text{Im}(\mu)| = |\text{Im}(\nu)|$. Furthermore, there exists $x \in G$ such that

$\mu(e) > \mu(x) > 0$. Hence $\mu^* \neq \{e\}$. Similarly, $\nu^* \neq \{e\}$. Thus $\mu^* \cong \nu^*$. For all $t \in [0,1]$ such that $\mu_t \neq \emptyset$, either $\mu_t = \{e\}$ or $\mu_t \cong G$. By Lemma 8.6.5, we have in either case that there is an $s \in [0,1]$ such that $\mu_t \cong \nu_s$. Hence $\mu \cong \nu$. $\qquad\square$

From Theorem 8.6.6, we see that, with respect to S^*-equivalence, infinite-valued fuzzy subgroups of an infinite cyclic group are unique. With respect to equivalence, the number of infinite-valued fuzzy subgroups of an infinite cyclic group is infinite. Clearly, the notion of S^*-equivalence can be used to reduce the number of distinct fuzzy subgroups of a finite group compared with the equivalence approach since two fuzzy subgroups which are not equivalent may be S^*-equivalent.

Remark 8.6.7. The number of fuzzy subgroups of a finite Abelian group G with respect to a suitable equivalence relation were determined for special cases of G. However, it remains an open problem to determine the number for an arbitrary finite Abelian group.

References

1. Y. Alkhamees, Fuzzy cyclic subgroups and fuzzy cyclic p-subgroups, *J. Fuzzy Math.* 3(4) (1995) 911-919.
2. P. S. Das, Fuzzy groups and level subgroups, *J. Math. Anal. Appl.* 84 (1981) 264-269.
3. J. B. Fraleigh, *A First Course in Abstract Algebra*, Addison-Wesley, London (1982).
4. J. Jiashang, C. Degang,Wu Congxin, and E. C. C. Tsang, Some notes on the equivalence of fuzzy sets and fuzzy subgroups, *Fuzzy Sets and Systems,* to appear.
5. B.B Makamba, Studies in Fuzzy Groups, Thesis, Rhodes University, Grahamstown (1992).
6. M. Mashinchi and M. Mukaidono, A classification of fuzzy subgroups, Ninth Fuzzy System Symposium, Sapporo, Japan, 1992, 649-652.
7. M. Mashinchi and M. Mukaidono, On fuzzy subgroup classification, Research Report of Meiji University, Japan, vol. 9 (65) (1993) 31-36.
8. M. Mashinchi and Sh. Salili, On fuzzy isomorphism theorems, *J. Fuzzy Math.* 4 (1996) 39-49.
9. V. Murali and B.B. Makamba, On an Equivalence of Fuzzy Subgroups I, *Fuzzy Sets and Systems* 123 (2001) 259-264.
10. V. Murali and B.B. Makamba, On an Equivalence of Fuzzy Subgroups II, *Fuzzy Sets and Systems* 136 (2003) 93-104.
11. V. Murali and B. B. Makamba, Operations on equivalent fuzzy subgroups, *J. Fuzzy Math.,* to appear.
12. V. Murali and B. B. Makamba, Counting the number of fuzzy subgroups of an Abelian group of order $p^n q^m$, *Fuzzy Sets and Systems* 144 (2004) 459-470.
13. N. P. Mukherjee and P. Bhattacharya, Fuzzy Groups: Some Group Theoretic Analogs, *Inform. Sci.* 39 (1986) 247-268.

14. S. Ray, Isomorphic fuzzy groups, *Fuzzy Sets and Systems* 50 (2) (1992) 201-207.
15.   A. Rosenfeld, Fuzzy groups, *J. Math. Anal. Appl.* 35 (1971) 512-517.
16. Su-Yun Li, De-Gang Chen,Wen-Xiang Gu, and Hui Wang, Fuzzy homomorphisms, *Fuzzy Sets and Systems* 79 (1996) 235-238.
17. S. Tung and Y. Zhang, Fuzzy subgroups of the same type, *Fuzzy Sets and Systems* 87 (1997) 377 - 381.
18. S. A. Ziaie, H. V. Kumbhojkar, and M. Mashinchi, On cardinality of truth-value set of a fuzzy subgroup, *Int. J. Scientia Iranica* 3 (1996) 97-103.

9

Lattices of Fuzzy Subgroups

Many results concerning relationships between classes of crisp subsets can be carried over to similar relationships between classes of fuzzy subsets. The purpose of the next few sections is to provide a framework for transferring crisp results into fuzzy results when such a transference is possible. This is accomplished by the presentation of a suitable metatheorem. The results of the first four sections of this chapter are mainly from [15, 16].

9.1 Embedding of Fuzzy Power Sets

Let X be a non-empty set. Recall that $\mathcal{C}(X) = \{f \mid f : X \to \{0,1\}\}$ and that $\mathcal{P}(X)$ denotes the power set of X. We call $\mathcal{C}(X)$ the **crisp power set** of X. Then $\mathcal{P}(X)$ and $\mathcal{C}(X)$ are in a natural one-to-one correspondence and $\mathcal{C}(X) \subseteq \mathcal{FP}(X)$.

In Section 9.4, we present a metatheorem and two closely associated subdirect product representation theorems. Let $\mathcal{C}(X)^J$ denote $\{f \mid f : J \to \mathcal{C}(X)\}$, where J is the half open interval $J = [0,1)$. These theorems arise from first establishing a representation function $\mathfrak{R} : \mathcal{FP}(X) \to \mathcal{C}(X)^J$. The idea behind $\mathfrak{R}$ is very closely related to the concepts of level sets and strong cuts. The results here are generated by using the details of the definition of $\mathfrak{R}$.

We also develop the first subdirect product theorem and the metatheorem. The metatheorem in [15] was used to obtain results of other papers in a unified way. The second subdirect product theorem is applied to extend the modularity results of [3].

Define the function $\mathfrak{C} : \mathcal{P}(X) \to \mathcal{C}(X)$ by $\forall Y \in \mathcal{P}(X)$, $\mathfrak{C}(Y)(x) = 1$ if and only if $x \in Y$. Then $\mathfrak{C}$ establishes a bijection of $\mathcal{P}(X)$ with $\mathcal{C}(X)$. For all $Y \in \mathcal{P}(X)$, $\mathfrak{C}(Y)$ is the characteristic function of Y. Every crisp subset is also a fuzzy subset. Recall that $\mathcal{FP}(X)$ has the partial order $\subseteq$ defined by $\forall \mu, \nu \in \mathcal{FP}(X)$, $\mu \subseteq \nu$ if and only if $\mu(x) \leq \nu(x) \ \forall x \in X$. Thus $\mathcal{FP}(X)$ is a complete lattice with $\mathcal{C}(X)$ a complete sublattice. $\mathcal{P}(X)$ is a complete lattice with respect to the partial order of containment and the function $\mathfrak{C} : \mathcal{P}(X) \to$

$\mathcal{C}(X)$ is an isomorphism of complete lattices. The function $\mathfrak{S} : \mathcal{C}(X) \to \mathcal{P}(X)$ defined by $\mathfrak{S}(\mu) = \{x \in X \mid \mu(x) = 1\}$ is an isomorphism and is the inverse of $\mathfrak{C}$. By use of the functions $\mathfrak{C}$ and $\mathfrak{S}$, it follows that every assertion about crisp subsets is also an assertion about (ordinary) subsets and conversely.

We next define the representation function $\mathfrak{R}$. Recall that $J = [0,1)$.

Definition 9.1.1. *Let X be a nonempty set. Define the function $\mathfrak{R} : \mathcal{FP}(X) \to \mathcal{C}(X)^J$ be defined by $\forall \mu \in \mathcal{FP}(X)$, $\forall r \in J$, and $\forall x \in X$,*

$$\mathfrak{R}(\mu)(r)(x) = \begin{cases} 0 \ \text{if } \mu(x) \leq r \\ 1 \ \text{otherwise.} \end{cases}$$

Proposition 9.1.2. *The function $\mathfrak{R}$ is injective.*

Proof. Suppose $\mathfrak{R}(\mu) = \mathfrak{R}(\nu)$. Let $x \in X$. Since

$$\mathfrak{R}(\mu)(\nu(x))(x) = \mathfrak{R}(\nu)(\nu(x))(x) = 0,$$

it follows from the definition of $\mathfrak{R}$ that $\mu(x) \leq \nu(x)$. Similarly, since

$$\mathfrak{R}(\nu)(\mu(x))(x) = \mathfrak{R}(\mu)(\mu(x))(x) = 0,$$

$\nu(x) \leq \mu(x)$. Thus $\mu(x) = \nu(x)$ for all $x \in X$. Hence $\mu = \nu$. $\square$

Define the relation $\leq$ on $\mathcal{C}(X)^J$ as follows: $\forall \, G, H \in \mathcal{C}(X)^J$, $G \leq H$ if and only if $\forall r \in J$, $G(r) \subseteq H(r)$. Since $\subseteq$ is a partial order of $\mathcal{C}(X)$, $\leq$ is a partial order of $\mathcal{C}(X)^J$. It follows that the inequality, $G(r) \subseteq H(r)$, holds if and only if $\forall x \in X$, $G(r)(x) = 0$ whenever $H(r)(x) = 0$. Equivalently, $G(r) \subseteq H(r)$ holds if and only if $H(r)(x) = 1$ whenever $G(r)(x) = 1$. Further, observe that $\mathcal{C}(X)^J$ has the least and the largest element K and T respectively given by: $K(r)(x) = 0$, $T(r)(x) = 1 \ \forall r \in J$ and $\forall x \in X$. $\mathcal{C}(X)^J$ is also closed under suprema and infima and so $\mathcal{C}(X)^J$ is indeed a bounded lattice.

Proposition 9.1.3. $Im(\mathfrak{R}) = \{G : J \to \mathcal{C}(X) \mid \forall r \in J, \ G(r) = \cup \{G(q) \mid q > r\}\}$.

Proof. Suppose that $G = \mathfrak{R}(\mu)$ for some $\mu \in F(X)$. Let $r \in J$ and $x \in X$. Then $G(r)(x) = \mathfrak{R}(\mu)(r)(x) = 0$ if and only if $\mu(x) \leq r$. However, $\mu(x) \leq r$ if and only if $\mu(x) \leq q$ for all $q > r$. Therefore, $G(r)(x) = 0$ if and only if $G(q)(x) = 0$ for all $q > r$. Thus $G(r)(x) = \vee \{G(q)(x) \mid q > r\}$. Since this equation holds $\forall \, x \in X$, $G(r) = \cup \{G(q) \mid q > r\}$ as required.

Suppose that $G : J \to \mathcal{C}(X)$ is such that $\forall r \in J$, $G(r) = \cup \{G(q) \mid q > r\}$. Define the function $\mu : X \to I$ by $\forall x \in X, \mu(x) = \wedge \{p \in J \mid G(p)(x) = 0\}$. Then it follows that $\mathfrak{R}(\mu) = G$ since $\forall r \in J$ and $\forall x \in X$, the following five assertions are equivalent:

(1) $\mathfrak{R}(\mu)(r)(x) = 0$.
(2) $\mu(x) \leq r$.
(3) For all $q > r$, $\mu(x) \leq q$.
(4) For all $q > r$, $G(q)(x) = 0$.
(5) $G(r)(x) = 0$. $\square$

It can be seen from Proposition 9.1.3 that $\forall \mu \in \mathcal{FP}(X)$, $\mathfrak{R}(\mu)$ is an order reversing function from J into $\mathcal{C}(X)$, i.e., $q \geq r$ implies $\forall x \in X$, $\mathfrak{R}(\mu)(q)(x) \leq \mathfrak{R}(\mu)(r)(x)$.

Proposition 9.1.4. $\mathfrak{R}$ *commutes with infima of finite sets F of fuzzy subsets, i.e.,* $\mathfrak{R}(\cap_{\mu \in F} \mu) = \wedge \{\mathfrak{R}(\mu) \mid \mu \in F\}$ *and with suprema of arbitrary sets S of fuzzy subsets, i.e.,* $\mathfrak{R}(\cup_{\mu \in S} \mu) = \vee \{\mathfrak{R}(\mu) \mid \mu \in S\}$.

Proof. Let F be a finite collection of fuzzy subsets of X. That $\mathfrak{R}(\cap_{\mu \in F} \mu) = \wedge \{\mathfrak{R}(\mu) \mid \mu \in F\}$ can be seen from the fact that $\forall r \in J$ and $\forall x \in X$, the following five statements are equivalent:

(1) $\mathfrak{R}(\cap_{\mu \in F} \mu)(r)(x) = 0$.
(2) $(\cap_{\mu \in F} \mu)(x) \leq r$.
(3) There exists $\mu \in F$ such that $\mu(x) \leq r$.
(4) There exists $\mu \in F$ such that $\mathfrak{R}(\mu)(r)(x) = 0$.
(5) $\wedge \{\mathfrak{R}(\mu) \mid \mu \in F\}(r)(x) = 0$.

Let S be any collection of fuzzy subsets of X. That $\mathfrak{R}(\cup_{\mu \in S} \mu) = \vee \{\mathfrak{R}(\mu) \mid \mu \in S\}$ can be seen from the fact that $\forall r \in J$ and $\forall x \in X$, the following five statements are equivalent:

(6) $\mathfrak{R}(\cup_{\mu \in S} \mu)(r)(x) = 0$.
(7) $(\cup_{\mu \in S} \mu)(x) \leq r$.
(8) For all $\mu \in S$, $\mu(x) \leq r$.
(9) For all $\mu \in S$, $\mathfrak{R}(\mu)(r)(x) = 0$.
(10) $\vee \{\mathfrak{R}(\mu) \mid \mu \in S\}(r)(x) = 0$. □

In the proof above, (2)$\Rightarrow$(3) fails when F is infinite. It can be easily shown that $\mathfrak{R}$ does not, in general, commute with infima of a descending sequences of fuzzy subsets even when X is a singleton. As an example, let $X = \{a\}$ be a set with one element. Define a countable family $\{\mu_i\}_{i \in \mathbb{N}}$ of fuzzy subsets of X as $\mu_i(a) = \frac{1}{i} \, \forall i \in \mathbb{N}$. Then $\cap_{i \in \mathbb{N}} \mu_i(a) = 0$, but $\mu_i(a) \neq 0 \, \forall i \in \mathbb{N}$.

Let $I(X)$ be the image of the function $\mathfrak{R} : \mathcal{FP}(X) \to \mathcal{C}(X)^J$. By Proposition 9.1.4, $I(X)$ is a sublattice and also a complete upper subsemilattice of $\mathcal{C}(X)^J$. By Propositions 9.1.2 and 9.1.4, $\mathfrak{R}$ is an order isomorphism of $\mathcal{FP}(X)$ onto $I(X)$. Note also that $\mathfrak{R}_{0_X} = K$ and $\mathfrak{R}_{1_X} = T$ where $0_X(x) = 0$ and $1_X(x) = 1 \, \forall x \in X$.

9.2 Representation of the Fuzzy Power Algebra

In this section, we present the Subdirect Product Theorem. It plays an essential role in transferring crisp results to fuzzy ones.

Definition 9.2.1. *Let $* : X^n \to X$ be an n-ary operation on X, where $n \in \mathbb{N}$. We define an n-ary operation $*$ on $\mathcal{FP}(X)^n$, $* : \mathcal{FP}(X)^n \to \mathcal{FP}(X)$ by the standard convolution method as follows:* $\forall \mu_1, ..., \mu_n \in \mathcal{FP}(X)$ *and* $\forall x \in X$,

$$*(\mu_1, \ldots, \mu_n)(x) = \vee \{\mu_1(x_1) \wedge \ldots \wedge \mu_n(x_n) \mid x = *(x_1, \ldots, x_n)\}.$$

The operation $*$ on X also yields an n-ary operation $*$ on $\mathcal{P}(X)$ defined by for all subsets $X_1, \ldots, X_n$ of X,

$$*(X_1, \ldots, X_n) = \{*(x_1, \ldots, x_n) \mid x_i \in X_i, \; i = 1, 2, \ldots, n\}.$$

Proposition 9.2.2. *For every n-ary operation $*$ on X with $n \in \mathbb{N}, \mathcal{C}(X)$ is closed with respect to the convolution extension of $*$ to $\mathcal{FP}(X)$. The bijection $\mathfrak{C} : \mathcal{P}(X) \to \mathcal{C}(X)$ commutes with the $*$ operation on $\mathcal{P}(X)$ and $\mathcal{C}(X)$.*

A convolution extension of $*$ from X to $\mathcal{FP}(X)$ induces an operation $*$ on $\mathcal{C}(X)$ which then provides a pointwise operation $*$ for $\mathcal{C}(X)^J$. Hence for $G_1, \ldots, G_n \in \mathcal{C}(X)^J$ and $r \in J$,

$$*(G_1, \ldots, G_n)(r) = *(G_1(r), \ldots, G_n(r)).$$

Proposition 9.2.3. *For every n-ary operation $*$ on X, with $n \in \mathbb{N}$, the representation function $\mathfrak{R} : \mathcal{FP}(X) \to \mathcal{C}(X)^J$ commutes with the convolution extension of $*$, i.e., $\forall \mu_1, \ldots, \mu_n \in \mathcal{FP}(X), \mathfrak{R}(*(\mu_1, \ldots, \mu_n)) = *(\mathfrak{R}(\mu_1), \ldots, \mathfrak{R}(\mu_n))$.*

Proof. The desired result follows from the fact that $\forall \mu_1, \ldots, \mu_n \in \mathcal{FP}(X)$, $r \in J$, and $x \in X$ the following eight assertions are equivalent:

(1) $\mathfrak{R}(*(\mu_1, \ldots, \mu_n))(r)(x) = 1$.

(2) $*(\mu_1, \ldots, \mu_n)(x) > r$.

(3) $\vee\{\mu_1(x_1) \wedge \ldots \wedge \mu_n(x_n) \mid x = *(x_1, \ldots, x_n)\} > r$.

(4) There exists $x_1, \ldots, x_n \in X$ for which $x = *(x_1, \ldots, x_n)$ and $\mu_1(x_1) > r, \ldots, \mu_n(x_n) > r$.

(5) There exists $x_1, \ldots, x_n \in X$ for which $x = *(x_1, \ldots, x_n)$ and

$$\mathfrak{R}(\mu_1)(r)(x_1) = 1, \ldots, \mathfrak{R}(\mu_n)(r)(x_n) = 1.$$

(6) $\vee\{\mathfrak{R}(\mu_1)(r)(x_1) \wedge \ldots \wedge \mathfrak{R}(\mu_n)(r)(x_n) \mid x = *(x_1, \ldots, x_n)\} = 1$.

(7) $*(\mathfrak{R}(\mu_1)(r), \ldots, \mathfrak{R}(\mu_n)(r))(x) = 1$.

(8) $*(\mathfrak{R}(\mu_1), \ldots, \mathfrak{R}(\mu_n))(r)(x) = 1$. $\square$

Now let X be an algebra of arbitrary structure. That is, X is a nonempty set provided with n-ary operations $*_1, \ldots, *_k$ for various values of $n \in \mathbb{N}$.

Example 9.2.4. (1) *X may be a semigroup having one binary operation, usually denoted by ".".*

(2) *X may be a group having one unary operation, usually denoted by $^{-1}$, and one binary operation, usually denoted by ".".*

(3) *X may be a ring having one unary operation, usually denoted by $-$, and two binary operations usually denoted by "+" and ".".*

(4) *X may be a lattice with two binary operations, "$\vee$" and "$\wedge$".*

We have seen that the n-ary operations $*_1, \ldots, *_k$, $n \geq 1$, extends to operations on $\mathcal{P}(X)$, $\mathcal{C}(X)$ and $\mathcal{FP}(X)$. Now $\mathcal{P}(X), \mathcal{C}(X)$, and $\mathcal{FP}(X)$ have two additional binary operations, $\cap$ and $\cup$, as discussed previously. Thus $\mathcal{P}(X), \mathcal{C}(X)$, and $\mathcal{FP}(X)$ become algebras having the corresponding operations $\cap$, $\cup$, $*_1, \ldots, *_k$.

By Propositions 9.1.2, 9.1.4 and 9.2.3, it follows that $\mathfrak{R}$ is an injective homomorphism of the algebra $\mathcal{FP}(X)$ into the product algebra $\mathcal{C}(X)^J$ that establishes an algebraic and order-theoretic isomorphism of $\mathcal{FP}(X)$ with its image $I(X)$. For all $r \in J$, the projection of $\mathcal{FP}(X)$ into the rth coordinate space of $\mathcal{C}(X)^J$ is a surjection. In fact, $\forall r \in J$, $\mathfrak{R}(\mu)(r)$ maps the subset $\mathcal{C}(X)$ of $\mathcal{FP}(X)$ bijectively onto the rth coordinate space $\mathcal{C}(X)$: Let $\mu \in \mathcal{C}(X)$. Then $\forall r \in J$, $x \in X$, $\mathfrak{R}(\mu)(r)(x) = \mu(x)$, i.e., $\mathfrak{R}(\mu)(r) = \mu$.

Recall that an algebra A is said to be a subdirect product of a family of algebras $\{A_i \mid i \in I\}$, where I is an arbitrary index set, if A is isomorphic to a subalgebra of the product algebra $\Pi\{A_i \mid i \in I\}$ with the property that its projection into each coordinate space A_i is a surjection. Now $\mathfrak{R}$ defines an isomorphism of $\mathcal{FP}(X)$ onto $I(X)$ and the projection of $I(X)$ into each coordinate space of $\mathcal{C}(X)^J$ is a surjection. Thus we have the following result.

Theorem 9.2.5. *(The Subdirect Product Theorem)* Let X be an algebra *having n-ary operations $*_1, \ldots, *_k$ for various values of $n \geq 1$. Then $\mathfrak{R}$: $\mathcal{FP}(X) \to \mathcal{C}(X)^J$ is a representation of the $(\cap, \cup, *_1, \ldots, *_k)$-algebra $\mathcal{FP}(X)$ as a subdirect product of copies of the $(\cap, \cup, *_1, \ldots, *_k)$-algebra $\mathcal{C}(X)$.*

9.3 The Metatheorem

We review the recursive definition of an expression over a set Var of *variables*, a set Op of *operations*, and the three *auxiliary symbols* left paren, right paren, and comma in order to eastablish our notation. Each operation symbol $*$ in Op has an integer $n(*) \geq 1$ as its *arity*. The following two sentences constitute the definition of an *expression* over Var and Op : For all $v \in Var$, v is an expression. For all $* \in Op$ and every list of expressions $E_1, \ldots, E_{n(*)}$, where $n(*)$ is the arity of $*$, $*(E_1, \ldots, E_{n(*)})$ is an expression.

Every expression is a finite string of symbols having a syntax rigidly specified by the recursive definition. Let E be an expression. Then the number of distinct symbols in Var that occur in E is a positive integer and the number of distinct symbols in Op that occur in E is a nonnegative integer. Every symbol occurring in E may occur more than once. If no variable other than those in the set $\{v_1, \ldots, v_m\}$ occurs in E, E may be denoted by writing $E = E(v_1, \ldots, v_m)$. The order in which the variable symbols are listed is immaterial and a variable symbol may occur in the list even though it does not occur in the expression E. The convenience of this notation is demonstrated in the next paragraph.

Let X be a non-empty set provided with n-ary operations $*_1, \ldots, *_k$ for various values of $n \geq 1$. Then $\mathcal{P}(X), \mathcal{C}(X)$, and $\mathcal{FP}(X)$ are algebras with operations $\cap, \cup, *_1, \ldots, *_k$. Let $E = E(v_1, \ldots, v_m)$ be an expression over the set of variables $Var = \{v_1, \ldots, v_m\}$ and the set of operations $Op = \{\cap, \cup, *_1, \ldots, *_k\}$. Then for any m elements $\mu_1, \ldots, \mu_m \in \mathcal{FP}(X)$, $E(\mu_1, \ldots, \mu_m)$ is the element of $\mathcal{FP}(X)$ which results when, in E, each occurrence of each v_i is replaced by μ_i and the result is evaluated in $\mathcal{FP}(X)$.

We say that a class C of fuzzy subsets of X is **closed under projections** if $\forall \mu \in C$, $\forall r \in J$, the crisp set $\mathfrak{R}(\mu)(r)$ is also in C.

Theorem 9.3.1 (The Metatheorem). *Let X be an algebra with n-ary operations $*_1, \ldots, *_k$ for various values of $n \geq 1$. Let $\mathcal{FP}(X)$ have the operations $\cap$, $\cup$, $*_1, \ldots, *_k$. Let $D(v_1, \ldots, v_m)$ and $E(v_1, \ldots, v_m)$ be expressions over the variable set $\{v_1, \ldots, v_m\}$ and the operation set $\{\cap, \cup, *_1, \ldots, *_k\}$. Let $C_1, \ldots, C_m$ be classes of fuzzy subsets of X that are closed under projections. The inequality*

$$D(\mu_1, \ldots, \mu_m) \; REL \; E(\mu_1, \ldots, \mu_m)$$

holds for all fuzzy sets $\mu_i \in C_i$, $i = 1, 2, \ldots, m$ if and only if it holds for all crisp sets $\mu_i \in C_i$, $i = 1, 2, \ldots, m$, where REL is any one of the three relations $\subseteq, =, or \supseteq$.

Proof. We consider only the case for which REL is $\subseteq$ since the case for $\supseteq$ follows by symmetry and then the case for $=$ follows by combining $\subseteq$ and $\supseteq$. Suppose the inequality $D(\mu_1, \ldots, \mu_m) \subseteq E(\mu_1, \ldots, \mu_m)$ does not hold. Then $\exists x \in X$ such that

$$D(\mu_1, \ldots, \mu_m)(x) > E(\mu_1, \ldots, \mu_m)(x).$$

Let $r = E(\mu_1, \ldots, \mu_m)(x)$. Then

$$\begin{aligned}
D(\mathfrak{R}(\mu_1)(r), \ldots, \mathfrak{R}(\mu_m)(r))(x) &= D(\mathfrak{R}(\mu_1), \ldots, \mathfrak{R}(\mu_m))(r)(x) \\
&= \mathfrak{R}(D(\mu_1, \ldots, \mu_m))(r))(x) \\
&= 1.
\end{aligned}$$

However,

$$\begin{aligned}
E(\mathfrak{R}(\mu_1)(r), \ldots, \mathfrak{R}(\mu_m)(r))(x) &= E(\mathfrak{R}(\mu_1), \ldots, \mathfrak{R}(\mu_m))(r)(x) \\
&= \mathfrak{R}(E(\mu_1, \ldots, \mu_m))(r))(x) \\
&= 0.
\end{aligned}$$

Let A be an $(\wedge, \vee, *_1, \ldots, *_k)$-algebra. For expressions $D(v_1, \ldots, v_m)$ and $E(v_1, \ldots, v_m)$ over $Var = \{v_1, \ldots, v_m\}$ and $Op = \{\wedge, \vee, *_1, \ldots, *_k\}$, we say that $D(v_1, \ldots, v_m) = E(v_1, \ldots, v_m)$ is an $(\wedge, \vee, *_1, \ldots, *_k)$-identity. The algebra A satisfies the identity, $D(v_1, \ldots, v_m) = E(v_1, \ldots, v_m)$ if $\forall \mu_1, \ldots, \mu_m \in A$, $D(\mu_1, \ldots, \mu_m) = E(\mu_1, \ldots, \mu_m)$. We obtain the following result when either the metatheorem or the subdirect product theorem is combined with the isomorphism of $\mathcal{P}(X)$ with $\mathcal{C}(X)$.

Corollary 9.3.2. *For every algebra X, with n-ary operations $*_1, \ldots, *_k$ for various $n \geq 1$, the three $(\cap, \cup, *_1, \ldots, *_k)$-algebras $\mathcal{P}(X)$, $\mathcal{C}(X)$, and $\mathcal{FP}(X)$ all satisfy precisely the same $(\cap, \cup, *_1, \ldots, *_k)$-identities.*

Let X be an algebra as in Corollary 9.3.2. Note that $v_1 \vee v_1' = 1$, where the prime denotes the appropriate complementation, may be regarded as an identity that is satisfied by $\mathcal{P}(X)$ and $\mathcal{C}(X)$, but not by $\mathcal{FP}(X)$. However, $v_1 \vee v_1' = 1$ is not an $(\cap, \cup, *_1, \ldots, *_k)$-identity: Since 1 does not contain a variable, 1 is not an expression over $Var = \{v_1\}$ and $Op = \{\cap, \cup, *_1, \ldots, *_k\}$. For this reason alone, $v_1 \vee v_1' = 1$ does not satisfy the definition of an $(\cap, \cup, *_1, \ldots, *_k)$-identity. However, even when $X = \mathcal{P}(Y)$, $X = \mathcal{C}(Y)$, or $X = \mathcal{FP}(Y)$ for some set Y, it follows that the prime required in the identity is not the convolution extension of the complement appropriate to such an X.

Corollary 9.3.2 provides two elementary known facts that are fundamental to the theory of fuzzy algebraic structures. One refers only to the information concerning $\cap$ and $\cup$ [34]: *For every nonempty set X, the lattice of fuzzy subsets, $\mathcal{FP}(X)$, is distributive.* Another refers to the operations $*_1, \ldots, *_k$ and the identities they may satisfy. In some cases, identities satisfied by the $(*_1, \ldots, *_k)$-algebra X are also satisfied by $\mathcal{P}(X)$. For example, this is true for the associativity identity and for the commutativity identity. Thus the metatheorem yields the following results: *If $(X, *)$ is a semigroup, then $(\mathcal{FP}(X), *)$ is a semigroup. If $(X, *)$ is commutative, then so is $(\mathcal{FP}(X), *)$.*

9.4 Unifications

In this section, the classes $\mathcal{C}_i$ in the metatheorem are required to satisfy restrictions that are closely related to the structure of the $(*_1, \ldots, *_k)$-algebra X. That these conditions are easily dealt with is a consequence of the following property of $\mathfrak{R}$ which follows immediately from its definition.

Proposition 9.4.1. *For all $x, y \in X$ and $\forall \mu \in \mathcal{FP}(X)$, $\mu(x)$ REL $\mu(y)$ holds if and only if $\forall r \in J$, $\mathfrak{R}(\mu)(r)(x)$ REL $\mathfrak{R}(\mu)(r)(y)$ holds, where REL is any one of the three relations $\leq$, $=$, or $\geq$.*

Several fundamental results in the literature can be obtained by noting that they are consequences of Proposition 9.4.1. Generic patterns for defining fuzzy versions of crisp concepts will allow many results to be treated simultaneously. For each class $\mathcal{C}$ of crisp subsets of X, a fuzzy subset $\mu \in \mathcal{FP}(X)$ is called a **fuzzy $\mathcal{C}$ subset** if and only if $\forall r \in J$, $\mathfrak{R}(\mu)(r) \in \mathcal{C}$. This generic definition is consistent with many definitions from semigroup theory as pointed out in [15]. We do not discuss them here since they do not fall within the framework of group theory. The convenience of this general concept of a fuzzy $\mathcal{C}$ class is illustrated by deriving generalizations of results in a unified manner by reference to the following principle which is immediate from the definition of a fuzzy $\mathcal{C}$ class: *For classes $\mathcal{C}$ and $\mathcal{D}$ of crisp subsets, the class of fuzzy $\mathcal{C}$*

subsets is contained in the class of fuzzy $\mathcal{D}$ subsets if and only if $\mathcal{C}$ is contained in $\mathcal{D}$. From this principle and the observations made above, numerous propositions hold [15].

We now give an application to lattice modularity.

Let $(X,^{-1},*)$ be a group. Recall that $\mathcal{NF}(X)$ denotes the set of all normal fuzzy subgroups in $\mathcal{FP}(X)$. With respect to the partial order, $\subseteq$, of $\mathcal{FP}(X)$, $\mathcal{NF}(X)$ is a lattice for which the INF coincides with the inf of the lattice $\mathcal{FP}(X)$, but for which the SUP does not. By normality, this SUP is very well behaved: $SUP(\mu,\nu) = \mu * \nu \; \forall \mu, \nu \in \mathcal{NF}(X)$. In the present exposition, the fact that the SUP of $\{\mu, \nu\}$ is $\mu * \nu$, is a consequence of Proposition 9.4.1 and the definition of a normal fuzzy subgroup: Since μ and ν are normal fuzzy subgroups, $\mathfrak{R}(\mu)(r)$ and $\mathfrak{R}(\nu)(r)$ are (crisp) normal subgroups $\forall r \in J$. Since $\mathfrak{R}$ commutes with $*$, $\mathfrak{R}(\mu * \nu)(r) = \mathfrak{R}(\mu)(r) * \mathfrak{R}(\nu)(r)$ holds $\forall r \in J$ and is the SUP of its two displayed crisp factors. Since $\mathfrak{R}(\mu * \nu)(r)$ is normal $\forall r \in J$, $\mu * \nu$ is normal. Since $\mathfrak{R}(\mu * \nu)(r) = SUP\{\mathfrak{R}(\mu)(r), \mathfrak{R}(\nu)(r)\} \; \forall r$, $\mu * \nu = SUP\{\mu, \nu\}$ if and only if $\mu(e) = \nu(e)$. It is shown in Proposition 9.1.4 that $\mathfrak{R}$ commutes with SUP.

Corollary 9.4.2. *Let $(X,^{-1},*)$ be a group. Let $\mathcal{NF}(X)$ be the lattice of normal fuzzy subgroups in $\mathcal{FP}(X)$. Let $L\mathfrak{R}$ be the restriction of $\mathfrak{R}: \mathcal{FP}(X) \to \mathcal{C}(X)^J$ to $\mathcal{NF}(X)$. Let $LI(X)$ denote the image of $L\mathfrak{R}$ and let LC consist of the crisp subsets in $\mathcal{NF}(X)$. Then $L\mathfrak{R}$ is an isomorphism of the lattice $\mathcal{NF}(X)$ onto $LI(X)$ that represents $\mathcal{NF}(X)$ as a subdirect product in the lattice LC^J.*

Proof. This result follows from the subdirect product theorem since from that theorem we know that $\mathfrak{R}$ is an injection that commutes with $\cap$ and $*$. Since $\mathcal{NF}(X)$ is closed under $\cap$ and $*$, $L\mathfrak{R}$ is also an injection that commutes with $\cap$ and $*$. Since $*$ provides the SUP operation of $\mathcal{NF}(X)$, the desired result follows. $\qquad\square$

From Corollary 9.4.2, we obtain the following result.

Corollary 9.4.3. *For each group $(X,^{-1},*)$ the lattice of normal fuzzy subgroups and the lattice of crisp normal subgroups satisfy precisely the same $(\cap = INF, * = SUP)$-identities.*

A lattice is modular if and only if satisfies the identity:

$$v_1 \vee (v_2 \wedge (v_1 \vee v_3)) = (v_1 \vee v_2) \wedge (v_1 \vee v_3)$$

Since the lattice of normal subgroups of a group is known to be modular, we have from Corollary 9.4.3 that for every group $(X,^{-1},*)$ the lattice of normal fuzzy subgroups is modular.

Let LF consist of the fuzzy normal subgroups $\mu \in \mathcal{FP}(X)$ for which the set $\{\mu(x) \mid x \in X\}$ is finite. Let the identity element of the group X be e and let $t \in [0,1]$. Let $LFt = \{\mu \in LF \mid \mu(e) = t\}$. Both $\wedge$ and $*$ preserve the property of having a finite image set and the property of assuming a given value at e. Consequently, LF and all of the LFt are sublattices of the lattice of normal fuzzy subgroups in $\mathcal{FP}(X)$. This observation gives the following result [[6], Theorem 3.10]: For each group $(X,^{-1},*)$, the lattice LF is modular and as is LFt, $\forall t \in [0,1]$.

When the concept of a fuzzy characteristic subgroup and a fuzzy fully invariant subgroup are defined generically, we have immediately that for each group $(X,^{-1},*)$, the fuzzy characteristic subgroups (respectively, the fuzzy fully invariant subgroups) form a sublattice of the lattice of normal fuzzy subgroups and so must also be modular. The metatheorem and the subdirect product theorems given here apply to the study of other fuzzy algebraic structures.

9.5 Lattices of Fuzzy Congruences

Most of the results in this and the next section are from [10]. Let $L(G)$ denote $\mathcal{F}(G)$ and $L_n(G)$ denote $\mathcal{NF}(G)$ in the remainder of the chapter, where G is a group. In this section, we discuss certain properties of fuzzy congruences and fuzzy normal subgroups of a group in relation to strong level subsets. We provide generating techniques for fuzzy congruences and fuzzy subgroups of a group. With these techniques, we prove that the lattices of fuzzy congruences and fuzzy normal subgroups of a group are isomorphic. This is a generalization of a result that the lattices of t-fuzzy congruences and fuzzy normal subgroups with tip "t" are isomorphic. We establish this isomorphism by following the metatheorem approach presented earlier in the chapter. We use a function that represents the algebra of fuzzy subsets of a set as a subdirect product of copies of the algebra of crisp subsets.

In [2], the use of the notion of level subsets was shifted to that of strong level subsets. Strong level subsets characterizations are used more effectively to obtain results in fuzzy group theory and other fuzzy algebraic structures. This approach provides a method for fuzzification of results and ideas from classical algebra to fuzzy algebraic structures.

In the present section, we follow this approach to investigate the properties of fuzzy congruences and fuzzy normal subgroups of a group. Fuzzy equivalence relations and fuzzy congruences have been studied in [26, 27]. In [31], the foundation of lattice theoretic studies in the area of fuzzy group theory was laid. It was considered further in the papers [3, 6, 8]. Lattice theoretic results of fuzzy subgroups in a more general setting were presented in [23, 24].

The purpose of [6] was to demonstrate that fuzzy algebraic structures are important when viewed from a lattice theoretic point of view. For this reason, the sublattices $L_t, L_f, L_{ft}, L_{fn_t}$ and L_{n_t} of the lattice $L = L(G)$ of all fuzzy subgroups of a given group G were examined. Also, in [8] various types of sublattices of the lattice E of all fuzzy equivalence relations on a group were constructed. In particular, the sublattices C_t, C_f, C_{ft} and C_{s_t} of the lattice C

of all fuzzy congruence relations were examined. Furthermore, it was shown that the lattice C_t of all t-fuzzy congruences and the lattice L_{n_t} of all fuzzy normal subgroups with tip "t" are isomorphic for all $t \in (0, 1]$. Moreover, in [3] it was established that the lattice L_n of all fuzzy normal subgroups is a sublattice of the lattice L of all fuzzy subgroups of a given group. As in classical group theory, an isomorphism between lattices C and L_n of a given group is presented. The isomorphism between these lattices is established by using the approach of strong level subsets. There is a close link between the approach used here and that of the metatheorem approach given in Sections 9.1-9.4.

Let G be a group. For $\mu \in \mathcal{FP}(G)$ and $\alpha \in \mathcal{FP}(G \times G)$, we let $\vee \mu$ denote $\vee \{\mu(x) \mid x \in G\}$ and $\vee \alpha$ denote $\vee \{\alpha(x, y) \mid (x, y) \in G \times G)\}$. Let $\mu \in \mathcal{FP}(G)$. For a fuzzy subgroup μ, we call $\mu(e)$ the **tip** of the fuzzy subgroup μ. Let $L_t(G)$ denote the set of all fuzzy subgroups of G with tip "t". Let $\nu \in L(G)$. Recall that ν is called normal if and only if $\nu(xy) = \nu(yx) \; \forall x, y \in G$. We let $L_n(G)$ denote the set of all normal fuzzy subgroups of G and $L_{n_t}(G)$ the set of all normal fuzzy subgroups in $L_t(G)$.

Let $\alpha \in \mathcal{FP}(G \times G)$. For a fixed $t \in [0, 1]$, α is called a t-**fuzzy relation** if $\vee \alpha = t$. A t-fuzzy relation α is called t-**reflexive** if $\alpha(x, x) = t, \forall x \in G$. A t-reflexive relation α is called a t-**fuzzy equivalence relation** if α is symmetric, that is, $\alpha(x, y) = \alpha(y, x) \forall x, y \in G$ and α is sup-min transitive, that is, $\alpha(x, y) \geq \vee \{\alpha(x, z) \wedge \alpha(z, y) \mid z \in G\} \forall x, y \in G$. Let $E_t(G)$ denote the set of all t-fuzzy equivalence relations on G and let $E(G)$ denote the set of all fuzzy equivalence relations on G, i.e., $E(G) = \cup_{t \in [0,1]} E_t(G)$. Let $\beta \in E_t(G)$. Then β is called a t-**fuzzy congruence relation** if $\beta(ac, bd) \geq \beta(a, b) \wedge \beta(c, d), \; \forall a, b, c, d \in G$. Let $C_t(G)$ denote the set of all t-fuzzy congruence relations in $E_t(G)$ and $C(G)$ denote the set of all fuzzy congruence relations in $E(G)$, i.e., $C(G) = \cup_{t \in [0,1]} C_t(G)$.

For $\mu \in \mathcal{FP}(G)$, the fuzzy subgroup generated by μ is defined to be the least member of $L(G)$ containing μ and is denoted by $\langle \mu \rangle$. We use the same notation for an ordinary subgroup generated by the level subset (strong level subset) of μ. Similarly for $\alpha \in \mathcal{FP}(G \times G)$ a fuzzy equivalence relation generated by α is denoted by $[\alpha]$ and a fuzzy congruence relation generated by α is denoted by $\lceil \alpha \rceil$. Also, the same notations are used for ordinary equivalence relations and congruence relations generated by level subsets (strong level subsets) of α.

Let $\mu \in \mathcal{FP}(G)$. We assume in the following that $\vee \mu > 0$.

Proposition 9.5.1. *Let $\mu \in \mathcal{FP}(G)$ and $t_0 = \vee \mu$. Then the following conditions are equivalent:*

(1) $\mu \in L(G)$.
(2) $\mu_t^>$ *is a subgroup of G $\forall t \in [0, t_0)$.*
(3) *Every nonempty strong level subset of μ is a subgroup of G.*

Proof. (1) $\Rightarrow$ (2) : Let $t \in [0, t_0)$. Then $\mu_t^> \neq \emptyset$. Let $x, y \in \mu_t^>$. Then $\mu(x) > t$ and $\mu(y) > t$. Since μ is a fuzzy subgroup of $G, \mu(xy^{-1}) \geq \mu(x) \wedge \mu(y) > t$. Thus $xy^{-1} \in \mu_t^>$. Hence $\mu_t^>$ is a subgroup of G.

(2) $\Rightarrow$ (3) : Let $\mu_t^>$ be a nonempty strong level subset of G, where $t \in [0, 1]$. Then there exists $x \in \mu_t^>$ and for any such $x, t < \mu(x) \leq t_0$. Hence by (2), $\mu_t^>$ is a subgroup of G.

(3) $\Rightarrow$ (1) : Let $x, y \in G$ and $\mu(x) = t_1$ and $\mu(y) = t_2$. If either $t_1 = 0$ or $t_2 = 0$, then $\mu(xy^{-1}) \geq 0 = \mu(x) \wedge \mu(y)$. Suppose both $t_1 > 0$ and $t_2 > 0$. Let $0 \leq t < t_1 \wedge t_2$. Then $x, y \in \mu_t^>$. Since $\mu_t^>$ is a subgroup of $G, xy^{-1} \in \mu_t^>$ and so $\mu(xy^{-1}) > t$. Thus $\mu(xy^{-1}) \geq t_1 \wedge t_2 = \mu(x) \wedge \mu(y)$. Hence μ is a fuzzy subgroup of G. $\square$

Proposition 9.5.2. *Let $\mu \in L(G)$. Then the following conditions are equivalent:*

(1) $\mu \in L_n(G)$.
(2) $\mu_t^>$ *is a normal subgroup of $G, \forall t \in [0, t_0)$, where $t_0 = \mu(e)$.*
(3) *Every nonempty strong level subgroup is a normal subgroup of G.*

Proof. (1) $\Rightarrow$ (2) : Let $x \in G$ and $y \in \mu_t^>$, where $t \in [0, t_0)$. Since μ is normal, $\mu(xyx^{-1}) = \mu(y) > t$. Thus $\mu xyx^{-1} \in \mu_t^>$ and so $\mu_t^>$ is normal.

(2) $\Rightarrow$ (3) : Let $\mu_t^>$ be a nonempty strong level subset of G, where $t \in [0, 1]$. Then there exists $x \in \mu_t^>$ and for any such $x, t < \mu(x) \leq t_0$. Hence by (2), $\mu_t^>$ is a normal subgroup of G.

(3) $\Rightarrow$ (1) : Let $x, y \in G$ and $\mu(y) = t_2$. Let $0 \leq t < t_2$. Then $y \in \mu_t^>$. Since $\mu_t^>$ is a normal subgroup of $G, xyx^{-1} \in \mu_t^>$. Thus $\mu(xyx^{-1}) > t$. Hence $\mu(xyx^{-1}) \geq \mu(y)$. Thus μ is normal. $\square$

Theorem 9.5.3. *Let $\alpha \in \mathcal{FP}(G \times G)$. Then the following conditions are equivalent:*

(1) $\alpha \in C(G)$.
(2) $\alpha_t^>$ *is a congruence relation on $G \forall t \in [0, t_0)$, where $t_0 = \vee \alpha$.*
(3) *Every nonempty strong level subset is a congruence relation on G.*

Proof. (1) $\Rightarrow$ (2) : Let $t \in [0, t_0)$. For all $x \in G, \alpha(x, x) = t_0 > t$. Thus $(x, x) \in \alpha_t^>$. Suppose $(x, y) \in \alpha_t^>$. Then $\alpha(y, x) = \alpha(x, y) > t$ and so $(y, x) \in \alpha_t^>$. Suppose $(x, z), (z, y) \in \alpha_t^>$. Then $\alpha(x, y) \geq \vee\{\alpha(x, z') \wedge \alpha(z', y) \mid z' \in G\} \geq \alpha(x, z) \wedge \alpha(z, y) > t$. Thus $(x, y) \in \alpha_t^>$. Hence $\alpha_t^>$ is an equivalence relation. Let $(a, b), (c, d) \in \alpha_t^>$. Then $\alpha(ac, bd) \geq \alpha(a, b) \wedge \alpha(c, d) > t$. Thus $(ac, bd) \in \alpha_t^>$. Therefore, $\alpha_t^>$ is a congruence relation on G.

(2) $\Rightarrow$ (3) : Suppose $\alpha_t^> \neq \emptyset$, where $t \in [0, 1]$. Then there exists $(x, y) \in \alpha_t^>$ and for all such $(x, y), t < \alpha(x, y) \leq t_0$. Hence by (2), $\alpha_t^>$ is an ordinary congruence relation on G.

(3) $\Rightarrow$ (1) : For all $t \in [0, 1]$ such that $\alpha_t^> \neq \emptyset, (x, x) \in \alpha_t^>$ for all $x \in G$. Thus it follows that $\alpha(x, x) = t_0$ for all $x \in G$, where $t_0 = \vee \alpha$. Hence $\alpha_t^> \neq \emptyset$ for all t such that $t \in [0, t_0)$. Let $t \in [0, t_0)$. Then $\alpha_t^>$ is a congruence relation on G. Let $x, y \in G$. Since $(x, y) \in \alpha_t^>$ if and only if $(y, x) \in \alpha_t^>$

for all $t \in [0, t_0)$, it follows that $\alpha(x, y) = \alpha(y, x)$. Let $(x, z), (z, y) \in G$. Then $\alpha(x, z) > t$ and $\alpha(z, y) > t \Rightarrow (x, z), (z, y) \in \alpha_t^> \Rightarrow (x, y) \in \alpha_t^> \Rightarrow \alpha(x, y) > t \Rightarrow \alpha(x, y) \geq \vee\{\alpha(x, z) \wedge \alpha(z, y) \mid z \in G\}$. Hence $\alpha \in E(G)$. Suppose $(a, b), (c, d) \in G \times G$. Then $\alpha(a, b) > t, \alpha(c, d) > t \Rightarrow (a, b), (c, d) \in \alpha_t^> \Rightarrow (ac, bd) \in \alpha_t^> \Rightarrow \alpha(ac, bd) > t$. Therefore, $\alpha(ac, bd) \geq \alpha(a, b) \wedge \alpha(c, d)$. Thus $\alpha \in C(G)$. $\qquad \square$

Theorem 9.5.4. *Let* $\alpha \in \mathcal{FP}(G \times G)$. *Define a fuzzy relation* α^* *on* G *by*
$$\alpha^*(x, y) = \vee\{r \mid (x, y) \in [\alpha_r], r < \vee\alpha\}, \ \forall(x, y) \in G \times G.$$
Then $\alpha^* = [\alpha]$.

Proof. We first show that α^* is a fuzzy equivalence relation. Let $x \in G$. Then $\alpha^*(x, x) = t_0$ since $(x, x) \in [\alpha_r] \ \forall r < t_0$, where $t_0 = \vee\alpha$. *For all* $x, y \in G, \alpha^*(x, y) = \alpha^*(y, x)$ since $(x, y) \in [\alpha_r] \Leftrightarrow (y, x) \in [\alpha_r]$. Let $x, y, z \in G$ and let $t_1 = \alpha^*(x, z)$ and $t_2 = \alpha^*(z, y)$. Then $(x, z) \in [\alpha_{r_1}] \ \forall r_1 < t_1$ and $(z, y) \in [\alpha_{r_2}] \ \forall r_2 < t_2$. Suppose $t_1 \geq t_2$. (A similar argument can be used if $t_1 \leq t_2$.) Hence $(x, z), (z, y) \in [\alpha_{r_2}] \ \forall r_2 < t_2$. Thus $(x, y) \in [\alpha_{r_2}] \ \forall r_2, t_2$. Thus $\alpha^*(x, y) \geq t_2 = \alpha^*(x, z) \wedge \alpha^*(z, y)$. Thus it follows that α^* is a fuzzy equivalence relation. Let $x, y \in G$ and $\alpha(x, y) = r$. Then $(x, y) \in \alpha_r \subseteq [\alpha_r]$. Thus $\alpha^*(x, y) \geq r$. Hence $\alpha \subseteq \alpha^*$. Thus $[\alpha] \subseteq \alpha^*$. We show that $[\alpha] = \alpha^*$ by showing that very fuzzy equivalence relation which contains α contains α^*. Let θ be a fuzzy equivalence relation on G such that $\theta \supseteq \alpha$. Then $\theta_r \supseteq \alpha_r$ and so $\theta_r \supseteq [\alpha_r] \ \forall r \in [0, t_0)$. Let $\alpha^*(x, y) = r$. Then $(x, y) \in [\alpha_{r-\epsilon}] \subseteq \theta_{r-\epsilon} \forall\epsilon$ such that $0 < \epsilon < r$. Hence $\theta(x, y) \geq r - \epsilon\forall\epsilon$ such that $0 < \epsilon < r$. Thus $\theta(x, y) \geq r$. Hence $\alpha^* \subseteq \theta$. Therefore, $\alpha^* = [\alpha]$. $\qquad \square$

Theorem 9.5.5. *Let* $\alpha \in \mathcal{FP}(G \times G)$. *Define fuzzy relations* α^* *and* α^{**} *on* G *by* $\forall(x, y) \in G \times G$,
$$\alpha^*(x, y) = \vee\{r \mid (x, y) \in \lceil\alpha_r\rceil, r < \vee\alpha\},$$
$$\alpha^{**}(x, y) = \vee\{r \mid (x, y) \in \lceil\alpha_r^>\rceil, r < \vee\alpha\}.$$
Then $\alpha^* = \alpha^{**} = \lceil\alpha\rceil$.

Proof. First, we show that $\alpha^{**} \in C(G)$. Let $t_0 = \vee\alpha$. Then by the definition of α^{**}, for all $t \in [t_0, 1)$, the strong level subset $(\alpha^{**})_t^>$ of α^{**} is empty.

We show that
$$(\alpha^{**})_t^> = \lceil\alpha_t^>\rceil \ \forall t \in [0, t_0).$$

Let $(x, y) \in (\alpha^{**})_t^>$. Then $\alpha^{**}(x, y) > t$. Thus $\vee\{r \mid (x, y) \in \lceil\alpha_r^>\rceil, r < \vee\alpha\} > t$. Hence $\exists s \in [0, t_0)$ such that $s > t$ and $(x, y) \in \lceil\alpha_s^>\rceil$. However, $\lceil\alpha_s^>\rceil \subseteq \lceil\alpha_t^>\rceil$ and so $(x, y) \in \lceil\alpha_t^>\rceil$. Thus $(\alpha^{**})_t^> \subseteq \lceil\alpha_t^>\rceil$. Let $(x, y) \in \lceil\alpha_t^>\rceil$. Now an ordinary congruence relation $\lceil\alpha_t^>\rceil$ generated by the subset $\alpha_t^>$ in a semigroup (or group) can be written as
$$\lceil\alpha_t^>\rceil = [(\alpha_t^>)^c] = [\alpha_t^>]^c, \qquad (9.5.1)$$
where $(\alpha_t^>)^c$ and $[\alpha_t^>]^c$ are the smallest compatible relations containing $\alpha_t^>$ and $[\alpha_t^>]$, respectively. Therefore, $(x, y) \in [\alpha_t^>]^c$. Hence $\exists(x_1, y_1) \in G \times G$ and $a, b \in G$ such that

$$(x_1, y_1) \in [\alpha_t^>] \text{ and } (ax_1 b, ay_1 b) = (x, y) \qquad (9.5.2)$$

Thus $\exists x_1 = z_0, z_1, z_2, ..., z_{n-1}, z_n = y_1$ in G such that $\alpha(z_{i-1}, z_i) > t, \forall i = 1, 2, ..., n$. Let $t_1 = \wedge \{\alpha(z_{i-1}, z_i) \mid 1 \le i \le n\}$. Then $t_1 > t$. Let t' such be that $t_1 > t' > t$. Then $(z_{i-1}, z_i) \in \alpha_t^>$, $\forall i = 1, 2, ..., n$. Thus $(x_1, y_1) \in [\alpha_{t'}^>] \subseteq [\alpha_{t'}^>]^c$. Therefore, by the above Expressions 9.5.1 and 9.5.2 it follows that $(x, y) \in \lceil \alpha_{t'}^> \rceil$. Hence

$$\vee \{r \mid (x, y) \in \lceil \alpha_r^> \rceil, r < \vee \alpha\} \ge t' > t.$$

Thus by definition of α^{**}, we have $(x, y) \in (\alpha^{**})_t^>$. Hence

$$(\alpha^{**})_t^> = \lceil \alpha_t^> \rceil, \forall t \in [0, t_0) \qquad (9.5.3)$$

By Theorem 9.5.3(1), $\alpha^{**} \in C(G)$ since each nonempty strong level subset of α^{**} is an ordinary congruence relation generated by the relation $\alpha_t^>$.

Now $\forall t \in [0, t_0), \alpha_t^> \subseteq \lceil \alpha_t^> \rceil$ and $\forall t \in [t_0, 1), \alpha_t^>$ is empty. However, $\lceil \alpha_t^> \rceil = \{(x, x) \mid x \in G\}$. Hence $\alpha_t^> \subseteq \lceil \alpha_t^> \rceil, \forall t \in [0, 1)$. Therefore, by Expression 9.5.3 above, we also have $\alpha \subseteq \alpha^{**}$.

Let $\theta \in C(G)$ be such that $\alpha \subseteq \theta$. Then $\alpha_t^> \subseteq \theta_t^> \forall t \in [0, 1)$. Thus $\lceil \alpha_t^> \rceil \subseteq \theta_t^>$. Therefore, by Expression 9.5.3, we have $(\alpha^{**})_t^> \subseteq \theta_t^> \forall t \in [0, 1)$. Hence $\alpha^{**} \subseteq \theta$ and so $\alpha^{**} = \lceil \alpha \rceil$.

Finally, we show that the fuzzy subsets α^* and α^{**} are identical. By their definitions, it is clear that $\alpha^{**} \subseteq \alpha^*$. To show the reverse inclusion, let $t \in [0, 1)$. Now if $t \in [t_0, 1)$, then $(\alpha^{**})_t^>$ and $(\alpha^*)_t^>$ are empty. On the other hand, for $t \in [0, t_0)$, we let $(x, y) \in (\alpha^*)_t^>$. Then by definition of α^*, $\exists k > t$ such that $(x, y) \in \lceil \alpha_k \rceil$. However, $\lceil \alpha_k \rceil \subseteq \lceil \alpha_t^> \rceil$ and so $(x, y) \in \lceil \alpha_t^> \rceil$. Hence $(\alpha^*)_t^> \subseteq \lceil \alpha_t^> \rceil$. Therefore, by Expression 9.5.3, $(\alpha^*)_t^> \subseteq \lceil \alpha_t^> \rceil = (\alpha^{**})_t^>$, $\forall t \in [0, t_0)$. Hence $(\alpha^*)_t^> \subseteq (\alpha^{**})_t^>$, $\forall t \in [0, 1)$. Thus $\alpha^* \subseteq \alpha^{**}$. $\square$

The technique used in Theorem 9.5.4 and Theorem 9.5.5 can also be used to obtain a similar construction of a fuzzy subgroup generated by an arbitrary fuzzy subset as can be seen in the following result.

Theorem 9.5.6. Let $\mu \in \mathcal{FP}(G)$. Define fuzzy subsets μ^* and μ^{**} of G by $\forall x \in G$,
  (1) $\mu^*(x) = \vee \{r \mid x \in \langle \mu_r \rangle, r < \vee \mu\}$,
  (2) $\mu^{**}(x) = \vee \{r \mid x \in \langle \mu_r^> \rangle, r < \vee \mu\}$.
  Then $\mu^* = \mu^{**} = \langle \mu \rangle$.

Proof. We have that (1) holds since (1) is a restatement of Proposition 5.3.4. For all $r < \vee \mu, \mu_r^> \subseteq \mu_r$ and so $\langle \mu_r^> \rangle \subseteq \langle \mu_r \rangle$. Thus $\mu^{**} \subseteq \mu^* = \langle \mu \rangle$. That $\mu \subseteq \mu^{**}$ and μ^{**} is a fuzzy subgroup of G follows as in the proof of Proposition 5.3.4. Hence (2) holds. $\square$

Lemma 9.5.7. Let $\alpha \in \mathcal{FP}(G \times G)$ and $\mu \in \mathcal{FP}(G)$. Then the following assertions hold.
  (1) $\lceil \alpha \rceil (x, x) = \vee \{\alpha(y, z) \mid (y, z) \in G \times G\} \forall x \in G$.
  (2) $\langle \mu \rangle (e) = \vee \{\mu(x) \mid x \in G\}$.

Lemma 9.5.8. *Let $\alpha \in \mathcal{FP}(G \times G)$ and $\mu \in \mathcal{FP}(G)$. Then*
 (1) $\lceil \alpha \rceil_t^{>} = \lceil \alpha_t^{>} \rceil$, $\forall t \in [0, t_0)$, *where* $t_0 = \langle \alpha \rangle (x, x)$,
 (2) $\langle \mu \rangle_t^{>} = \langle \mu_t^{>} \rangle$, $\forall t \in [0, s_0)$, *where* $s_0 = (\mu)(e)$.

Lemma 9.5.9. *Let α, $\beta \in C(G)$ and $\mu, \nu \in L(G)$. Then*
 (1) $\lceil \alpha \cup \beta \rceil_t^{>} = \lceil \alpha_t^{>} \cup \beta_t^{>} \rceil$ $\forall t \in [0, t_0)$, *where* $t_0 = \alpha(x, x) \vee \beta(x, x)$,
 (2) $\langle \mu \cup \nu \rangle_t^{>} = \langle \mu_t^{>} \cup \nu_t^{>} \rangle$ $\forall t \in [0, s_0)$, *where* $s_0 = \mu(e) \vee \nu(e)$.

Theorem 9.5.10. *Let μ and ν be normal fuzzy subgroups of G. Then the fuzzy subgroup $\langle \mu \cup \nu \rangle$ is normal.*

Proof. By Lemma 9.5.7, $\langle \mu \cup \nu \rangle(e) = \vee(\mu \cup \nu(x) \mid x \in G\} = \vee\{\mu(x) \vee \nu(x) \mid x \in G\} = (\vee\{\mu(x) \mid x \in G\}) \vee (\vee\{\nu(x) \mid x \in G\}) = \mu(e) \vee \nu(e)$. By Lemma 9.5.9, $\langle \mu \cup \nu \rangle_t^{>} = \langle \mu_t^{>} \cup \nu_t^{>} \rangle$ $\forall t \in [0, s_0)$, where $s_0 = \mu(e) \vee \nu(e)$. Let $p = \mu(e) \wedge \nu(e)$ and $q = \mu(e) \vee \nu(e)$. Suppose $t \in [0, p)$. Then by Lemma 9.5.9, $\langle \mu \cup \nu \rangle_t^{>} = \langle \mu_t^{>} \cup \nu_t^{>} \rangle$, $\forall t \in [0, p)$. Since μ and ν are normal, $\mu_t^{>}$ and $\nu_t^{>}$ are normal subgroups of G $\forall t \in [0, p)$ by Proposition 9.5.2. Hence the product $\mu_t^{>} \nu_t^{>}$ is a normal subgroup of G $\forall t \in [0, p)$. Suppose $t \in [p, q)$. Then $\langle \mu \cup \nu \rangle_t^{>} = \langle \mu_t^{>} \rangle$ or $\langle \nu_t^{>} \rangle$ $(= \mu_t^{>}$ or $\nu_t^{>})$, respectively, $\forall t \in [0, s_0)$. Combining the two cases, we have $\langle \mu \cup \nu \rangle_t^{>}$ is a normal subgroup of $G \forall t \in [0, s_0)$. Hence by Proposition 9.5.2, $\langle \mu \cup \nu \rangle$ is a normal fuzzy subgroup of G. □

In [2, 3, 6, 8] various sublattices of the lattices of fuzzy subgroups and fuzzy equivalence relations for a given group were discussed. We recall the following result.

Theorem 9.5.11. *The set $L_n(G)$ forms a complete sublattice of the lattice $L(G)$.*

Proof. Let $\mu, \nu \in L_n(G)$. Then it follows that the join of μ and $\nu, \mu \vee \nu$, is $\langle \mu \cup \nu \rangle$. By the previous theorem, $\mu \vee \nu \in L_n(G)$. The meet of μ and $\nu, \mu \wedge \nu$, is $\mu \cap \nu$. Thus $\mu \wedge \nu \in L_n(G)$. In fact, the intersection of any collection of normal fuzzy subgroups of G is a normal fuzzy subgroup of G. □

The following result can also be easily verified.

Theorem 9.5.12. *The set $C(G)$ forms a complete sublattice of the lattice $E(G)$.*

It is known that there exists a one-to-one correspondence between the set of all congruence relations on a group and the set of all normal subgroups of that group. For a congruence relation R on a group G, a normal subgroup $N(R)$ can be defined as follows:

$$N(R) = \{x \in G | (x, e) \in R\}, \qquad (9.5.4)$$

where e is the identity of G. For a normal subgroup H of G, a congruence relation $C(H)$ can defined as follows:

$$C(H) = \{(x,y) \in G \times G | xy^{-1} \in H\}. \tag{9.5.5}$$

It is well known that the function ϕ from the set of all congruence relations on G onto the set of all normal subgroups of G given by $\varphi(R) = N(R)$ not only establishes a one-to-one correspondence, but also provides an isomorphism between the lattices of congruence relations on G and its normal subgroups. The inverse of this isomorphism is given by $\phi^{-1}(H) = C(H)$. For all congruence relations R_1, R_2 on G,

$$N(\lceil R_1 \cup R_2 \rceil) = (N(R_1) \cup N(R_2)) \tag{9.5.6}$$

and

$$N(R_1 \cap R_2) = N(R_1) \cap N(R_2). \tag{9.5.7}$$

For a fuzzy congruence relation β on a group G, define a fuzzy subset $N(\beta)$ of G as follows:

$$N(\beta)(x) = \beta(x,e), \forall x \in G. \tag{9.5.8}$$

Also, for a normal fuzzy subgroup θ of G, define a fuzzy relation $C(\theta)$ in G as follows:

$$C(\theta)(x,y) = \theta(xy^{-1}), \forall (x,y) \in G \times G. \tag{9.5.9}$$

In order to show that the fuzzy sets $N(\beta)$ and $C(\theta)$ play the role of subsets given in Expressions 9.5.4 and 9.5.5 respectively, we prove the following result.

Lemma 9.5.13. *Let $\beta \in C(G)$ and $\theta \in L_n(G)$ and let $N(\beta)$ and $C(\theta)$ be as defined in the above Expressions 9.5.8 and 9.5.9, respectively. Then the following assertions hold.*
(1) $(N(\beta))_t^> = N(\beta_t^>)$, $\forall t \in [0,1)$.
(2) $(C(\theta))_t^> = C(\theta_t^>)$, $\forall t \in [0,1)$.

Proof. (1) Let $t \in [0,1)$. Then for $x \in G$, we have that $x \in (N(\beta))_t^> \Leftrightarrow N(\beta)(x) > t \Leftrightarrow \beta(x,e) > t$ (by definition) $\Leftrightarrow (x,e) \in \beta_t^> \Leftrightarrow x \in N(\beta_t^>)$ (by Expression 9.5.4). Thus (1) holds.

(2) Let $t \in [0,1)$. Then for $(x,y) \in G \times G$, we have that $(x,y) \in (C(\theta))_t^> \Leftrightarrow C(\theta)(x,y) > t \Leftrightarrow \theta(xy^{-1}) > t$ (by definition) $\Leftrightarrow xy^{-1} \in \theta_t^> \Leftrightarrow (x,y) \in C(\theta_t^>)$ (by Expression 9.5.5). Thus (2) holds. $\square$

Theorem 9.5.14. *[8, 19, 22, 32] Let $\beta \in C(G)$. Then $N(\beta) \in L_n(G)$.*

Proof. Suppose $\beta \in C(G)$. Then $\beta(e,e) = \vee\beta$. Let $t_0 = \vee\beta$. Then $N(\beta)(e) = t_0$. Thus $\beta \in C(G) \Leftrightarrow \beta_t^>$ is a congruence relation on G for all $t \in [0,t_0)$ (by Theorem 9.5.3) $\Rightarrow N(\beta_t^>)$ is a normal subgroup of G for all $t \in [0,t_0)$ $\Leftrightarrow (N(\beta))_t^>$ is a normal subgroup of G for all $t \in [0,t_0) \Leftrightarrow \beta$ is a normal fuzzy subgroup of G (by Proposition 9.5.2). $\square$

Theorem 9.5.15. *[8, 19, 22, 32] Let $\theta \in L_n(G)$. Then $C(\theta) \in C(G)$.*

Proof. Suppose $\theta \in L_n(G)$. Then for all $x \in G$, $C(\theta)(x, x) = \theta(xx^{-1}) = \theta(e)$. Let $t_0 = \theta(e)$. Then $\vee C(\theta) = t_0$. Thus $\theta \in L_n(G) \Leftrightarrow \theta_t^>$ is a normal subgroup of G for all $t \in [0, t_0)$ (by Proposition 9.5.2) $\Rightarrow C(\theta_t^>)$ is a congruence relation on G for all $t \in [0, t_0) \Leftrightarrow (C(\theta))_t^>$ is a congruence relation on G for all $t \in [0, t_0)$ (since $C(\theta_t^>) = \{(x, y) | xy^{-1} \in \theta_t^>\} = \{(x, y) | \theta(x, y^{-1}) > t\} = \{(x, y) | C(\theta)(x, y) > t\} = C(\theta)_t^>) \Leftrightarrow C(\theta)$ is a fuzzy congruence relation on G (by Theorem 9.5.3). $\square$

Theorem 9.5.16. *Let $\beta \in C(G)$ and $\theta \in L_n(G)$. Let $N(\beta)$ and $C(\theta)$ be defined as in Expressions 9.5.8 and 9.5.9, respectively. Then the following assertions hold:*
  (1) $C(N(\beta)) = \beta$.
  (2) $N(C(\theta)) = \theta$.

Proof. (1) Let $x, y \in G$. Then

$$C(N(\beta))(x, y) = N(\beta)(xy^{-1}) = \beta(xy^{-1}, yy^{-1})$$
$$\geq \beta(x, y) \wedge \beta(y^{-1}, y^{-1}) = \beta(x, y).$$

Now

$$\beta(x, y) = \beta(xy^{-1}y, ey) \geq \beta(xy^{-1}, e) \wedge \beta(y, y)$$
$$= \beta(xy^{-1}, e) = N(\beta)(xy^{-1}) = C(N(\beta))(x, y).$$

Hence $C(N(\beta))(x, y) = \beta(x, y)$.
  (2) Let $x \in G$. Then $N(C(\theta))(x) = C(\theta)(x, e) = \theta(xe^{-1}) = \theta(x)$. $\square$

Theorem 9.5.17. *The lattices of fuzzy congruences and fuzzy normal subgroups of group G are isomorphic. That is,*
$$C(G) \cong L_n(G).$$

Proof. Define the function $f : C(G) \to L_n(G)$ by $f(\alpha) = N(\alpha)$, $\forall \alpha \in C(G)$. Let $\theta \in L_n(G)$. Then by Theorem 9.5.15, $C(\theta) \in C(G)$ and so by Theorem 9.5.16 (2), $N(C(\theta)) = \theta$. Thus f is onto. Let $\beta, \gamma \in C(G)$ be such that $f(\beta) = f(\gamma)$. Then $N(\beta) = N(\gamma)$. It follows easily that $\vee \beta = \vee N(\beta)$ and $\vee \gamma = \vee N(\gamma)$. Hence $\vee \gamma = \vee \beta$. Thus by Lemma 9.5.13 (1), $N(\beta_t^>) = N(\gamma_t^>)$ $\forall t \in [0, 1)$. Therefore, $C(N(\beta_t^>)) = C(N(\gamma_t^>))$ and hence $\beta_t^> = \gamma_t^>$, $\forall t \in [0, 1)$. Consequently, $\beta = \gamma$. Thus f is one-to-one. Finally, we show that the function f preserves $\vee$ (join) and $\wedge$ (meet). Let $\beta, \gamma \in C(G)$. We prove that
$$N(\lceil \beta \cup \gamma \rceil) = N(\beta) \cup N(\gamma)$$
  and
$$N(\beta \cap \gamma) = N(\beta) \cap N(\gamma).$$
  Before doing so, we note that $\vee \lceil \beta \cup \gamma \rceil = (\vee \beta) \vee (\vee \gamma)$ and so $\vee N(\lceil \beta \cup \gamma \rceil) = (\vee \beta) \vee (\vee \gamma)$. Also, $\vee \lceil N(\beta) \cup N(\gamma) \rceil = (\vee N(\beta)) \vee (\vee N(\gamma)) = (\vee \beta) \vee (\vee \gamma)$. Since $N(\lceil \beta \cup \gamma \rceil)$ and $(N(\beta) \cup N(\gamma))$ have the same supremum, say t_0, it follows

that $N(\lceil \beta \cup \gamma \rceil)_t^> = \phi = (N(\beta) \cup N(\gamma))_t^> \; \forall t \in [t_0, 1)$. On the other hand, $\forall t \in [0, t_0)$, we have that

$$
\begin{aligned}
N(\lceil \beta \cup \gamma \rceil)_t^> &= N(\lceil \beta \cup \gamma \rceil_t^>) && \text{by Lemma 9.5.8} \\
&= N(\lceil \beta_t^> \cup \gamma_t^> \rceil) && \text{by Lemma 9.5.9} \\
&= (N(\beta_t^>) \cup N(\gamma_t^>)) && \text{using Expression 9.5.6} \\
&= ((N(\beta))_t^> \cup (N(\gamma))_t^>) && \text{by Lemma 9.5.13 (1)} \\
&= (N(\beta) \cup N(\gamma))_t^> && \text{by Lemma 9.5.9 (2).}
\end{aligned}
$$

Therefore, $N(\lceil \beta \cup \gamma \rceil) = (N(\beta) \cup N(\gamma))$. By a similar argument, it follows that

$$
(N(\beta \cap \gamma))_t^> = (N(\beta) \cap N(\gamma))_t^>, \; \forall t \in [0, 1).
$$

Thus $N(\beta \cap \gamma) = N(\beta) \cap N(\gamma)$. $\qquad\qquad\square$

9.6 The Metatheorem Approach

We now consider the metatheorem approach. In the following, we use the subdirect product theorem developed in Section 9.4 to establish our main result. The technique used here and also applied in [15] demonstrates the ease in obtaining the fuzzy versions of the results from their crisp counterparts. In fact, our main Isomorphism Theorem is established here without using the join formula of the lattice of fuzzy normal subgroups and that of fuzzy congruences in a group.

We have the following corollary of Proposition 9.2.3. Let X be a semigroup with operation $*$. Then the systems $(\mathcal{P}(X), *)$ and $(\mathcal{C}(X), *)$ are isomorphic semigroups.

Let $(X, {}^{-1}, *)$ be a group. The set L_n of all normal fuzzy subgroups of X is a lattice under the ordering $\subseteq$ of $\mathcal{FP}(X)$, the inf in L_n coincides with the inf of the lattice $\mathcal{FP}(X)$, whereas the SUP is defined as in Theorem 9.5.6. Moreover, we have $\mu \vee \nu = (\mu \cup \nu) = \mu' * \nu'$, where μ' and ν' are the tip extended pair of fuzzy subgroups obtained by μ and ν respectively. That is, $\mu'(e) = \nu'(e) = \mu(e) \vee \nu(e)$ and $\mu'(x) = \mu(x)$ and $\nu'(x) = \nu(x)$ of all $x \in G, x \neq e$. Let $\mu, \nu \in L_n$. Then $\mu', \nu' \in L_n$. Hence it is immediate that $\mu' * \nu'$ is a normal fuzzy subgroup.

Next, let $\mathfrak{R}|_{L_n}$ be the restriction of the function $\mathfrak{R} : \mathcal{FP}(X) \to \mathcal{C}(X)^J$ to L_n and let $LI(X)$ be the image of $\mathfrak{R}|_{L_n}$. By $L_{\mathcal{C}}$, let us denote the set of all crisp normal subgroups in $\mathcal{C}(X)$. Then $L_{\mathcal{C}}$ is contained in L_n and so we state the following subdirect product theorem for the lattice of fuzzy normal subgroups.

Theorem 9.6.1. *Let* $(X, {}^{-1}, *)$ *be a group. Then* $\mathfrak{R}|_{L_n} : L_n \to L_{\mathcal{C}}^J$ *is an isomorphism of the lattice* L_n *onto* $LI(X)$ *that represents* L_n *as a subdirect product of the copies of* $L_{\mathcal{C}}$ *in the lattice* $L_{\mathcal{C}}^J$.

Proof. Since $\mathfrak{R}$ is a bijection from $\mathcal{FP}(X)$ onto $I(X)$, the function $\mathfrak{R}|_{L_n}$ is also a bijection from L_n onto $LI(X)$. As in the subdirect theorem, the projection in each coordinate space $L_{\mathcal{C}}$ is a surjection since for all $r \in J$ and $f \in L_{\mathcal{C}}$, $\mathfrak{R}(f)(r) = f$. Also, L_n is closed under INF and SUP since for $\mu, \nu \in L_n$, $\mu \vee \nu = \mu' * \nu'$ is a normal fuzzy subgroup. To show that $\mathfrak{R}|_{L_n}$ is an isomorphism, it remains to show that $\mathfrak{R}|_{L_n}$ commutes with INF and SUP. Since $\mathfrak{R}$ commutes with inf by Proposition 9.1.4, $\mathfrak{R}|_{L_n}$ commutes with INF. That $\mathfrak{R}|_{L_n}$ commutes with SUP can be shown as follows: for $\mu, \nu \in L_n$ we have

$$\mu \vee \nu = \mu' * \nu' = \mu' \vee \nu'.$$

Thus for all $r \in J$,

$$\begin{aligned}
\mathfrak{R}(\vee\nu)(r) &= \mathfrak{R}(\mu' \vee \nu')(r) \\
&= \mathfrak{R}(\mu' * \nu')(r) \\
&= \mathfrak{R}(\mu')(r) * \mathfrak{R}(\nu')(r) \\
&= \mathfrak{R}(\mu)(r)' * \mathfrak{R}(\nu)(r)' \\
&= \mathfrak{R}(\mu)(r) \vee \mathfrak{R}(\nu)(r) \\
&= \mathfrak{R}(\mu) \vee \mathfrak{R}(\nu)(r),
\end{aligned}$$

where $\mathfrak{R}(\mu)(r)'$ and $\mathfrak{R}(\nu)(r)'$ are the tip extended pair of fuzzy subgroups associated with $\mathfrak{R}(\mu)(r)$ and $\mathfrak{R}(\nu)(r)$ respectively (for details see [15, 16]). □

Let C denote the set of all fuzzy congruences in the set of all fuzzy subsets $\mathcal{FP}(X \times X)$ and let $\mathcal{C}(X \times X)$ be the set of all crisp subsets of $X \times X$. Denote by C_C the set of all crisp congruences in $\mathcal{C}(X \times X)$. Then C_C is a subset of C by [[8], Theorem 3.6]. Now consider the function $\mathfrak{R}$ from the lattice $\mathcal{FP}(X \times X)$ into the product lattice $\mathcal{C}(X \times X)^J$. In view of the subdirect product theorem, the function $\mathfrak{R} : \mathcal{FP}(X \times X) \to \mathcal{C}(X \times X)^J$ is a representation of the lattice $\mathcal{FP}(X \times X)$ as a subdirect product of the copies of $\mathcal{C}(X \times X)$ in the lattice $\mathcal{C}(X \times X)^J$.

In order to formulate the subdirect product theorem for the lattice C of fuzzy congruences in X, we denote by $\mathfrak{R}|_C$ the restriction of the function $\mathfrak{R} : \mathcal{FP}(X \times X) \to \mathcal{C}(X \times X)^J$ to the lattice C and denote by $CI(X \times X)$ the image of $\mathfrak{R}|_C$. Then we have the following subdirect product theorem for the lattice of fuzzy congruences of a group.

Theorem 9.6.2. *Let* $(X, ^{-1}, *)$ *be a group. Then* $\mathfrak{R}|_C : C \to C_C^J$ *is an isomorphism of the lattice* C *onto* $CI(X \times X)$ *that represents* C *as a subdirect product of the copies of* C_C *in the lattice* C_C^J.

As a consequence of the above subdirect product theorems for the lattices of fuzzy normal subgroups and fuzzy congruences of a group, we next prove the isomorphism theorem.

Theorem 9.6.3. *Let* G *be a group. Then the lattice* L_n *of normal fuzzy subgroups of* G *is isomorphic with the lattice* C *of its fuzzy congruences.*

Proof. In view of subdirect product theorems for the lattices of fuzzy normal subgroups and fuzzy congruences, it suffices to show that the lattices $LI(G)$ and $CI(G \times G)$ are isomorphic. Since the lattice $L_{\mathcal{C}}$ of crisp normal subgroups is isomorphic to the lattice of crisp congruences $C_{\mathcal{C}}$, the product lattices $L_{\mathcal{C}}^J$ and $C_{\mathcal{C}}^J$ are also isomorphic. The isomorphism $\theta : C_{\mathcal{C}}^J \to L_{\mathcal{C}}^J$ is given by

$$\theta(A)(r) = N(A(r)), \forall A \in G_{\mathcal{C}}^J, \ \forall r \in J,$$

where $N(A(r))$ is the fuzzy normal subgroup as defined previously in expression 9.5.8. Since $A(r) \in C_{\mathcal{C}'}$, it follows that $N(A(r)) \in L_{\mathcal{C}}, \ \forall r \in J$. It can be shown that under the restriction of θ to $CI(G \times G)$, we obtain the image $LI(G)$. That is, the restriction of θ to $CI(G \times G)$ is a bijection onto $LI(G)$ and hence is an isomorphism.

□

9.7 Fuzzy Subgroups With The Sup Property

In this section, we examine fuzzy subgroups with the sup property. The results are essentially from [4]. We characterize this type of fuzzy subgroup in terms of their level subsets. We show that the property of being a fuzzy subgroup with the sup property is invariant under a homomorphism and under the homomorphic preimage of a fuzzy subgroup with the sup property. We consider the class L_{n_t} of fuzzy normal subgroups of a group G, each of which assumes the same value "t" at the identity of G. It is proved that the subclass L_{ns_t} of fuzzy subgroups with sup property of L_{n_t} constitutes a sublattice of L_{n_t}. The modularity of L_{ns_t} follows as a consequence of the modularity of L_{n_t}.

It has been shown in [6] that the class of all fuzzy subgroups L_t, each of which assumes the same value t at the identity element of the given group, forms a complete sublattice of the lattice L of all fuzzy subgroups of that group. This class of fuzzy subgroups is known as the class of fuzzy subgroups with tip "t" and is shown to be of value in the study of fuzzy subgroups. For example, in [2], a fuzzy version of the correspondence theorem is obtained for this class of fuzzy subgroups. Moreover, in [1] and [6], it is established that in this class of fuzzy subgroups, the notion of set product due to Liu [20] can be used in the study of fuzzy subgroups as successfully as the notion of product of complexes in classical group theory.

In [1], it was shown that in the class of fuzzy subgroups with tip "t" of a given group, if the set product of two fuzzy subgroups is a fuzzy subgroup, then it is the fuzzy subgroup generated by their union. This result leads us to the formation of lattices of fuzzy normal subgroups. In fact, it is proved in [6] that the class L_{fn_t} of all fuzzy normal subgroups of a group, each of which has finite range and tip "t" is a modular sublattice of the lattice L of all fuzzy subgroups. On the other hand, it is shown in [8] that the condition of finiteness of the range set of the fuzzy subgroups can be dropped. A different technique is used here to establish this result. We prove that the class of all fuzzy normal subgroups with sup property and tip "t" of a group constitute a sublattice of L, which is modular.

We recall some definitions and results from previous chapters. Let μ and ν be fuzzy subsets of a group G. Then the product $\mu \circ \nu$ is the fuzzy subset of G, defined by for all z in G, $\mu \circ \nu(z) = \vee\{\mu(x) \wedge \nu(x) \mid z = xy, x, y \in G\}$. A fuzzy subset μ in a set S is said to have sup property if for each nonempty subset $A \subseteq S$, there exists $x_0 \in A$ such that $\vee\{\mu(x) \mid x \in A\} = \mu(x_0)$. Let f be a homomorphism from a group G onto a group G', and let μ and ν be fuzzy subgroups of G and G', respectively. Then $f(\mu)$ and $f^{-1}(\nu)$ are fuzzy subgroups of G' and G, respectively. The intersection of an arbitrary collection of fuzzy subgroups (normal fuzzy subgroups) of a group G is a fuzzy subgroup (normal fuzzy subgroup) of G.

It is clear from the definition that any fuzzy subgroup of a finite group has the sup property. Moreover, a fuzzy subgroup with a finite image also possesses the sup property.

We now provide an example of a fuzzy subgroup with the sup property whose image is countably infinite.

Example 9.7.1. *Let G be the group of nonnegative real numbers less than 1, under the operation of addition modulo 1. Define the fuzzy subset μ of G as follows:*

$$\mu(x) = \begin{cases} 1 & \text{if } x \in \langle 1/2 \rangle, \\ \frac{1}{2}(1 + \frac{1}{2^n}) & \text{if } x \in \langle 1/2^{n+1} \rangle \setminus \langle 1/2^n \rangle; n = 1, 2, ..., \\ 0 & \text{if } x \in G \setminus \overset{\infty}{\underset{n=1}{\cup}} \langle 1/2^n \rangle. \end{cases}$$

Then μ is a fuzzy subgroup of G and its chain of level subgroups is given by:

$$\langle 1/2 \rangle \subseteq \langle 1/2^2 \rangle \subseteq \langle 1/2^3 \rangle \subseteq \cdots \subseteq G.$$

Let A be a nonempty subset of G. Then either $A \subseteq G \setminus \overset{\infty}{\underset{n=1}{\cup}} \langle 1/2^n \rangle$ or $A \cap \langle 1/2^n \rangle \neq \emptyset$ for some positive integer n. Thus $\vee\{\mu(x) \mid x \in A\} = 0 = \mu(x_0) \forall x_0 \in A$ if $A \subseteq G \setminus \overset{\infty}{\underset{n=1}{\cup}} \langle 1/2^n \rangle$ and otherwise $\vee\{\mu(x) \mid x \in A\} = \frac{1}{2}(1 + \frac{1}{2^n}) = \mu(x_0)$ for the smallest such positive integer n such that $A \cap \langle 1/2^n \rangle \neq \emptyset$ and where $x_0 \in A \cap \langle 1/2^n \rangle$. Hence μ has the sup property.

We say that a fuzzy subgroup of μ of G attains its supremum everywhere on G if for each nonempty subset $A \subseteq G$, there exists $x_0 \in G$ such that $\vee_{x \in A} \mu(x) = \mu(x_0)$.

Example 9.7.2. *Let $\mathbb{Z}$ be the group of integers under the binary operation of usual addition. Define the fuzzy subset μ of $\mathbb{Z}$ as follows:*

$$\mu(x) = \begin{cases} 0 & \text{if } x \in \mathbb{Z} \setminus \langle 2 \rangle, \\ \frac{1}{2}(1 - \frac{1}{2^n}) & \text{if } x \in (2^n) \setminus (2^{n+1}), n = 1, 2, 3..., \\ \frac{1}{2} & \text{if } x = 0. \end{cases}$$

Then μ is a fuzzy subgroup of $\mathbb{Z}$. Moreover, μ attains its supremum everywhere on $\mathbb{Z}$. However, μ does not possess the sup property since μ fails to attain its supremum $1/2$ on the subset $\mathbb{Z} \setminus \langle 0 \rangle$.

Example 9.7.3. *In Example 9.7.2, define the fuzzy subset μ as given by*

$$\mu(x) = \begin{cases} 0 & \text{if } x \in \mathbb{Z}\backslash\langle 2\rangle, \\ \frac{1}{2}(1 - \frac{1}{2^n}) & \text{if } x \in \langle 2^n\rangle\backslash\langle 2^{n+1}\rangle, n = 1, 2, 3..., \\ 1 & \text{if } x = 0. \end{cases}$$

Then μ is a fuzzy subgroup of Z which neither attains its supremum everywhere nor has the sup property.

The following theorem characterizes the notion of the sup property in terms of level subsets.

Theorem 9.7.4. *Let μ be a fuzzy subgroup of G. Then μ has the sup property if and only if $\mu_t \subset \cap_{t_i < t, t_i \in \text{Im}(\mu)} \mu_{t_i}$ for all $t \in [0, 1]$.*

Proof. Suppose μ has the sup property. Let $t \in [0, 1]$. Define the subset A_t of G as follows:
$$A_t = \{x \in G \mid \mu(x) < t\}.$$
Let
$$t_1 = \vee\{\mu(x) \mid x \in A_t\}.$$
Since μ has the sup property, there exists $x_0 \in A_t$ such that $\mu(x_0) = t_1 < t$. Therefore, $\mu_t \subseteq \mu_{t_1}$ and $x_0 \notin \mu_t$.

Also,
$$x_0 \in \mu_{t_1} \subseteq \mu_{t_i} \text{ for all } t_i \leqslant t_1.$$
Hence
$$\mu_{t_1} \subseteq \cap_{t_i < t, t_i \in \text{Im}(\mu)} \mu_{t_i}.$$
However, $\cap_{t_i \leq t_1, t_i \in \text{Im}(\mu)} \mu_{t_i} = \cap_{t_i < t, t_i \in \text{Im}(\mu)} \mu_{t_i}$. Thus
$$\mu_t \subset \cap_{t_i < t, t_i \in \text{Im}(\mu)} \mu_{t_i}.$$
Conversely, suppose that μ does not have the sup property. Then there exists a subset A of G such that
$$\mu(x) \neq t_1 \text{ for all } x \in A,$$
where $t_1 = \vee\{\mu(x) \mid x \in A\}$. As above,
$$\mu_{t_1} \subseteq \cap_{t_i < t, t_i \in \text{Im}(\mu)} \mu_{t_i}.$$
Let
$$\text{Im}_A\mu = \{t_i \in \text{Im}(\mu) \mid \mu(x) = t_i \text{ for some } x \in A\}.$$
Then
$$t_1 = \vee\{t_i \mid t_i \in \text{Im}_A\mu\} \leqslant \vee\{t_i | t_i < t_1, t_i \in \text{Im}(\mu)\} \leqslant t_1.$$
Therefore, if $x \in \cap_{t_i < t_1, t_i \in \text{Im}(\mu)} \mu_{t_i}$, then $\mu(x) \leqslant t_i$ for all $t_i < t_1$ and $t_i \in \text{Im}\mu$. Thus
$$\mu(x) \leqslant \vee\{t_i | t_i < t_1, t_i \in \text{Im}(\mu)\} = t_1.$$
Hence $x \in \mu_{t_1}$. Consequently,
$$\mu_{t_1} = \cap_{t_i < t_1, t_i \in \text{Im}(\mu)} \mu_{t_i}.$$
That is, it is not the case that $\mu_t \subset \cap_{t_i < t, t_i \in \text{Im}(\mu)} \mu_{t_i}$. $\qquad\square$

Theorem 9.7.5. *Let f be a homomorphism from G into a group G' and let μ be a fuzzy subgroup of G with the sup property. Then $f(\mu)$ is a fuzzy subgroup of G' with the sup property.*

Proof. By Theorem 1.2.10, $f(\mu)$ is a fuzzy subgroup of G'. Let A be a subset of G'. If $f^{-1}(A) = \phi$, then the result is obvious. Suppse $f^{-1}(A) \neq \phi$. Then
$$\vee\{f(\mu)(y) \mid y \in A\} = \vee\{\vee\{\mu(x) \mid x \in f^{-1}(y) \mid y \in A\}\} = \vee\{\mu(x) \mid x \in f^{-1}(y)\}.$$

Since μ has sup property, there exists $x_0 \in f^{-1}(A)$ such that $\vee\{\mu(x) | x \in f^{-1}(A) = \mu(x_0)$. Hence $y_0 = f(x_0) \in A$ and $f(\mu)(y_0) = \vee\{\mu(x) \mid x \in f^{-1}(y_0)\}$. However, $x_0 \in f^{-1}(y_0) \subseteq f^{-1}(A)$. Therefore,
$$\vee\{\mu(x) \mid x \in f^{-1}(A)\} \geqslant \vee\{\mu(x) \mid x \in f^{-1}(y_0)\}.$$
Thus
$$\mu(x_0) \geqslant \vee\{\mu(x) \mid x \in f^{-1}(y_0)\} \geqslant \mu(x_0)$$
since $x \in f^{-1}(y_0)$.
Hence
$$\vee\mu(x) \mid x \in f^{-1}(y_0)\} = \mu(x_0).$$
Thus
$$\vee\{f(\mu)(y) \mid y \in A\} = \mu(x_0) = \vee\{\mu(x) \mid x \in f^{-1}(y_0)\} = f(\mu)(y_0).$$
Therefore, $f(\mu)$ has sup property. $\qquad\square$

Theorem 9.7.6. *Let f be a homomorphism from G into a group G' and let ν be a fuzzy subgroup with sup property of G'. Then the preimage $f^{-1}(\nu)$ is a fuzzy subgroup of G with the sup property.*

Proof. That $f^{-1}(\nu)$ is a fuzzy subgroup of G follows from Theorem 1.2.11. Let A be a subset of G. Then
$$\vee\{f^{-1}(\nu)(x) \mid x \in A\} = \vee\{\nu(f(x)) \mid x \in A\} = \vee\{\nu(y) \mid y \in f(A)\}.$$

Since ν has sup property, there exists $y_0 \in f(A)$ such that
$$\vee\{\nu(y) \mid y \in f(A)\} = \nu(y_0).$$
Hence
$$\vee\{f^{-1}(\nu)(x) \mid x \in A\} = \nu(y_0).$$
Since $y_0 \in f(A)$, there exists $x_0 \in A$ such that $y_0 \in f(x_0)$. Thus
$$f^{-1}(\nu)(x_0) = \nu(f(x_0)) = \nu(y_0).$$
Therefore, $f^{-1}(\nu)$ has the sup property. $\qquad\square$

The notion of the set product of fuzzy subsets was introduced in [20] as an extension of the notion of the product of complexes of classical group theory, and established some basic results to justify its appropriateness. The product of complexes plays an important role in the development of various aspects of group theory. The key to the use of this notion is that the product of two subgroups of a group is a subgroup if and only if the supbgroups commute. This fact leads to the important result that if the product of subgroups is a subgroup, then it is the least subgroup containing their union. In view of their importance in fuzzy group theory, we recall here some known results. Mashinchi and Zahedi [23, 24] have already obtained some of these results in more generality.

Lemma 9.7.7. *Let μ and ν be fuzzy subgroups of G. Then, $\mu \subseteq \mu \circ \nu$ if and only if $\mu(e) \leqslant \nu(e)$.*

Proof. Let $z \in G$. Then
$$\mu \circ \nu(z) = \vee\{\mu(x) \wedge \nu(y) \mid z = xy, x, y \in G\}$$
$$\geqslant \mu(z) \wedge \nu(e)$$
$$\geqslant \mu(z) \wedge \mu(e) \text{ if } \mu(e) \leqslant \nu(e)$$
$$= \mu(z).$$

Conversely,
$$\mu \circ \nu(e) = \vee\{\mu(x) \wedge \nu(x^{-1}) \mid x \in G\}$$
$$\leqslant \mu(e) \wedge \nu(e).$$
Thus if $\mu(e) > \nu(e)$, then $\mu \circ \nu(e) = \nu(e) < \mu(e)$ and so $\mu \nsubseteq \mu \circ \nu$.   □

Lemma 9.7.8. *Let μ and ν be fuzzy subgroups of G. Then $\mu \subseteq \mu \circ \nu$ and $\nu \subseteq \mu \circ \nu$ if and only if $\mu(e) = \nu(e)$.*

Proof. The proof follows by Lemma 9.7.7.   □

The next three results are restatements of Theorems 1.2.8, 1.2.9 and 1.3.1 respectively.

Theorem 9.7.9. *Let μ be a fuzzy subset of G. Then μ is a fuzzy subgroup of G if and only if*
(1) $\mu \circ \mu = \mu$,
(2) $\mu(x) = \mu(x^{-1})$ *for $x \in G$.*

Theorem 9.7.10. *Let μ and ν be fuzzy subgroups of G. Then $\mu \circ \nu = \nu \circ \mu$ if either μ or ν is normal in G.*

Theorem 9.7.11. *Let μ and ν be fuzzy subgroups of G. Then $\mu \circ \nu$ is a fuzzy subgroup of G if and only if $\mu \circ \nu = \nu \circ \mu$.*

The above theorem is established in a more general setting by Mashinchi and Zahedi in [24]. A similar generalization can be found in [[33], Theorem 5.1.10, p. 114].

Theorem 9.7.12. *Let μ and ν be fuzzy subgroups of G such that $\mu(e) = \nu(e)$. If $\mu \circ \nu$ is a fuzzy subgroup of G, then $\mu \circ \nu$ is generated by μ and ν.*

Proof. Suppose that $\mu \circ \nu$ is a fuzzy subgroup of G. Then we show that $\mu \circ \nu$ is the smallest fuzzy subgroup of G containing μ and ν. By Lemma 9.7.8, we see that $\mu \subseteq \mu \circ \nu$ and $\nu \subseteq \mu \circ \nu$.

Let θ be any fuzzy subgroup of G containing both μ and ν. Let $z \in G$. Then
$\mu \circ \nu(z) = \vee\{(\mu(x) \wedge \nu(y) | z = xy, x, y \in G\} \leqslant \vee\{\theta(x) \wedge \theta(y) \mid z = xy, x, y \in G\}$
$= \theta \circ \theta(z)$ since $\mu(x) \leqslant \theta(x)$ and $\nu(y) \leqslant \theta(y)$.

Since θ is a fuzzy subgroup of G, $\theta \circ \theta = \theta$ by Theorem 9.7.9. Hence $\mu \circ \nu \subseteq \theta$. Consequently, $\mu \circ \nu$ is a fuzzy subgroup generated by μ and ν.   □

Lemma 9.7.13. *Let μ and η be fuzzy subgroups of G. Then $(\mu \circ \eta)_t^{>} = \mu_t^{>} \mu_t^{>}$ for all $t \in [0, 1]$.*

Proof. Let $z \in G$. Then
$$z \in (\mu \circ \nu)_t^> \Leftrightarrow \mu \circ \nu(z) > t$$
$$\Leftrightarrow \vee\{\mu(x) \wedge \eta(y) \mid z = xy, xy, \in G\} > t$$
$$\Leftrightarrow \mu(x_0) \wedge \eta(y_0) > t \text{ for some } x_0, y_0 \in G, \ z = x_0 y_0$$
$$\Leftrightarrow x_0 \in \mu_t^>, \ y_0 \in \mu_t^> \text{ for some } x_0, y_0 \in G, \ z = x_0 y_0$$
$$\Leftrightarrow z \in \mu_t^> \mu_t^>. \qquad \square$$

9.8 Lattices of Fuzzy Subgroups

We now consider lattices of fuzzy subgroups. In this section, we continue to examine special types of lattices of fuzzy subgroups of a given group. It is shown in [6] that the class L_t of all fuzzy subgroups with tip "t" of a group G is a complete sublattice of the lattice $L = L(G)$ of all its fuzzy subgroups. On the other hand, it is well known in classical group theory that the set of all normal subgroups of a group forms a modular sublattice of the lattice of its subgroups. It is this result which motivates the following discussion.

Let $t \in [0, 1]$. We denote by L_{n_t}, the class of all normal fuzzy subgroups of a group G with tip "t." As usual, we denote by L the lattice of all fuzzy subgroups of G.

The proof of the following result is similar to that of previous results such as Theorem 1.3.3.

Theorem 9.8.1. *Let μ be a fuzzy subgroup of G. Then the following statements are equivalent:*

(1) *μ is normal.*
(2) *$\mu_r^>$ is a normal subgroup of G for all $r \in [0, t)$, $t = \mu(e)$.*
(3) *$\mu_r^>$ is a normal subgroup of G for all $r \in \operatorname{Im}\mu \backslash \{t\}$.*
(4) *Every nonempty strong level subset $\mu_t^>$ is a normal subgroup of G.*

Theorem 9.8.2. *L_{n_t} is a sublattice of L_t and therefore of L.*

Proof. Let $\mu, \nu \in L_{n_t}$. Then by Theorems 9.7.11 and 9.7.12, $\mu \circ \nu$ is the least fuzzy subgroup containing both μ and ν. Thus $\mu \vee \nu = \mu \circ \nu$.

Clearly, $\mu \circ \nu(e) = t$. In order to show that $\mu \vee \nu \in L_{n_t}$, it is sufficient to show that $\mu \circ \nu$ is normal. By Lemma 9.7.13,
$$(\mu \circ \nu)_r^> = (\mu)_r^> \cdot (\nu)_r^> \text{ for all } r \in [0, t).$$
By Theorem 9.8.1, $(\mu)_r^>$ and $(\nu)_r^>$ are normal subgroups of G for each $r \in [0, t)$. Therefore, their product $(\mu)_r^> \cdot (\nu)_r^>$ is a normal subgroup of G. Hence $(\mu \circ \nu)_r^>$ is a normal subgroup of G for all $r \in [0, t)$. By Theorem 9.8.1, it follows that $\mu \circ \nu$ is a normal fuzzy subgroup of G. The intersection $\mu \cap \nu$ is a normal fuzzy subgroup of G by Theorem 1.4.6. In fact, $\mu \cap \nu$ is the largest fuzzy subgroup contained in μ and ν. Therefore, $\mu \wedge \nu = \mu \cap \nu \in L_{n_t}$. $\qquad \square$

In order to construct the lattice of normal fuzzy subgroup with sup property, we begin with the following result.

Theorem 9.8.3. *Let μ and ν be fuzzy subgroups of G. If μ and ν have the sup property, then $\mu \cap \nu$ is a fuzzy subgroup of G with the sup property.*

Proof. It has been previously shown that $\mu \cap \nu$ is a fuzzy subgroup of G. Let A be a subset of G. Then
$$\vee \{\mu \cap \nu(z) \mid z \in A\} = \vee \{\mu(z) \wedge \nu(z) \mid z \in A\}$$
$$= (\vee \{\mu(z) \mid z \in X_A\}) \vee (\vee \{\nu(z) \mid z \in Y_A\}),$$
where
$$X_A = \{z \in A \mid \mu(z) \leqslant \nu(z)\},$$
$$Y_A = \{z \in A \mid \nu(z) \leqslant \mu(z)\}.$$
Since μ and ν have the sup property, there exist $x_0 \in X_A$ and $y_0 \in Y_A$ such that
$$\vee \{\mu(z) \mid z \in X_A\} = \mu(x_0) \text{ and } \vee \{\nu(z) \mid z \in Y_A\} = \nu(y_0).$$
Therefore,
$$\vee \{\mu \cap \nu(z) \mid z \in A\} = (\mu(x_0) \vee \nu(y_0).$$
Suppose that $\mu(x_0) \geqslant \eta(y_0)$. Then
$$\vee \{\mu \cap \nu(z) \mid z \in A\} = \mu(x_0).$$
Also,
$$\mu \cap \nu(x_0) = \mu(x_0) \wedge \eta(x_0) = \mu(x_0)$$
since $x_0 \in X_A$. Hence $\mu \cap \nu$ attains its supremum at x_0 and $x_0 \in X_A \subseteq A$. Suppose that $\mu(x_0) \leqslant \nu(y_0)$. Then, as in the previous case, it can be shown that
$$\vee \{\mu \cap \nu(z) \mid z \in A\} = \nu(y_0) = \mu \cap \nu(y_0)$$
since $y_0 \in Y_A \subseteq A$. Therefore, $\mu \cap \nu$ has the sup property. $\qquad \square$

Theorem 9.8.4. *Let μ and ν be fuzzy subgroups of G. If μ and ν have the sup property, then $\mu \circ \nu$ has the sup property.*

Proof. Let $A \subseteq G$. Then
$$\vee \{\mu \circ \nu(z) \mid z \in A\} = \vee \{\vee \{\mu(x) \wedge \mu(y) \mid z = xy, x, y \in G \mid z \in A\}\}.$$
Define the subsets X and Y of $G \times G$ as follows:
$$X = \{(x, y) \in G \times G \mid \mu(x) \leqslant \nu(y), z = xy \text{ and } z \in A\},$$
$$Y = \{(x, y) \in G \times G \mid \nu(y) \leqslant \mu(x), z = xy \text{ and } z \in A\}.$$
Then
$$\vee \{\mu \circ \nu(z) \mid z \in A\}$$
$$= (\vee \{\mu(x) \wedge \nu(y) \mid (x, y) \in X\}) \vee (\vee \{\mu(x) \wedge \nu(y) \mid (x, y) \in Y\}\})$$
$$= (\mu(x) \mid (x, y) \in X\}) \vee (\vee \{\nu(y) \mid (x, y) \in Y\}).$$
Define the subsets X_l and Y_r of G as follows:
$$X_l = \{x \in G \mid (x, y) \in X\},$$
$$Y_r = \{y \in G \mid (x, y) \in Y\}.$$
Then
$$\vee \{\mu \circ \nu(z) \mid z \in A\} = (\vee \{\mu(x) \mid x \in X_l\}) \vee (\vee \{\nu(y) \mid y \in Y_r\}).$$
Since μ and ν have the sup property, there exists $x_0 \in X_l$ and $y_0 \in Y_r$ such that

$$\vee\{\mu(x) \mid x \in X_l\} = \mu(x_0) \text{ and } \vee\{\nu(y) \mid y \in Y_r\} = \nu(y_0).$$

Therefore,
$$\vee\{\mu \circ \nu(z) \mid z \in A\} = \mu(x_0) \vee \nu(y_0).$$

Consider the cases $\mu(x_0) \geq \nu(y_0)$ and $\mu(x_0) \leqslant \nu(y_0)$.

Suppose $\mu(x_0) \geqslant \nu(y_0)$. Then
$$\vee\{\mu \circ \nu(z) \mid z \in A\} = \mu(x_0).$$

Since $x_0 \in X_l$, there exists $y_0' \in G$ such that $(x_0, y_0') \in X$. Hence there exists $z_0 \in A$ such that $z_0 = x_0 y_0'$ and $\mu(x_0) \leqslant \nu(y_0')$.

Choose such a y_0' and consider $z_0 = x_0 y_0' \in A$. We show that
$$\vee\{\mu \circ \nu(z) \mid z \in A\} = \mu \circ \nu(z_0).$$

Define subsets
$$X^{z_0} = \{(x,y) \in X \mid z_0 = xy\},$$
$$Y^{z_0} = \{(x,y) \in Y \mid z_0 = xy\},$$

and
$$X_l^{z_0} = \{x \in G \mid (x,y) \in X^{z_0}\},$$
$$Y_r^{z_0} = \{y \in G \mid (x,y) \in Y^{z_0}\}.$$

Then
$$\mu \circ \nu(z_0)$$
$$= \vee\{\mu(x) \wedge \nu(y) \mid z_0 = xy, x, y \in G\}$$
$$= (\vee\{\mu(x) \wedge \nu(y) \mid (x,y) \in X^{z_0}\}) \vee (\vee\{\mu(x) \wedge \nu(y) \mid (x,y) \in Y^{z_0}\})$$
$$= (\vee\{\mu(x) \mid x \in X_l^{z_0}\}) \vee (\vee\{\nu(y) \mid y \in Y_r^{z_0}\}.$$

Since $X_l^{z_0} \subseteq X_l$ and $x_0 \in X_l^{z_0}$, it follows that
$$\mu(x_0) \leqslant \vee\{\mu(x) \mid x \in X_l^{z_0}\} \leqslant \vee\{\mu(x) \mid x \in X_l\} = \mu(x_0).$$

Hence
$$\vee\{\mu(x) \mid x \in X_l^{z_0}\} = \vee\{\mu(x) \mid x \in X_l\} = \mu(x_0).$$

Since $Y_r^{z_0} \subseteq Y_r$, we also have
$$\vee\{\nu(y) \mid y \in Y_r^{z_0}\} \leqslant \vee\{\nu(y) \mid y \in Y_r\} = \nu(y_0).$$

Thus
$$\vee\{\nu(y) \mid y \in Y_r^{z_0}\} \leqslant \nu(y_0) \leqslant \mu(x_0).$$

Therefore,
$$\mu \circ \nu(z_0) = \mu(x_0) \vee (\vee\{\nu(y) \mid y \in Y_r^{z_0}\}) = \mu(x_0).$$

Thus $\vee\{\mu \circ \nu(z) \mid z \in A\} = \mu \circ \nu(z_0)$. Hence $\mu \circ \nu$ attains its supremum at $z_0 \in A$.

Suppose $\mu(x_0) \leqslant \nu(y_0)$. Then
$$\vee\{\mu \circ \nu(z) \mid z \in A\} = \nu(y_0).$$

Here it follows that
$$\vee\{\mu \circ \nu(z) \mid z \in A\} = \mu \circ \nu(z_0),$$

where $z_0 \in A$ such that $z_0 = x_0' y_0$ and $\nu(y_0) \leqslant \mu(x_0')$. Hence in both cases, $\mu \circ \nu$ attains its supremum at some element of A. Therefore, $\mu \circ \nu$ has the sup property. $\qquad\square$

Let L_{ns_t} denote the set of all normal fuzzy subgroups of G with the sup property and with tip "t."

Theorem 9.8.5. *L_{ns_t} is a sublattice of L_{n_t}.*

Proof. Let $\mu, \nu \in L_{ns_t}$. Then $\mu, \nu \in L_{n_t}$. Therefore, as in Theorem 9.8.2,
$$\mu \vee \nu = \mu \circ \nu.$$
Now by Theorem 9.8.4, $\mu \circ \nu$ has the sup property. Thus $\mu \vee \nu \in L_{ns_t}$. Now $\mu \wedge \nu = \mu \cap \nu$. Therefore, by Theorem 9.8.3, $\mu \cap \nu$ has the sup property. Hence, $\mu \wedge \nu \in L_{ns_t}$. □

Theorem 9.8.6. L_{ns_t} *is a modular sublattice of* L_t.

Proof. The proof follows in a similar manner as that of Theorem 3.10 of [6].
□

Corollary 9.8.7. L_{ns_t} *is a modular lattice.*

Proof. L_{ns_t} is modular since a sublattice of a modular lattice is modular. □

References

1. N. Ajmal, Set product and fuzzy subgroups, Proceedings of IFSA, World Congress, Belgium, 1991, 3 - 7.
2. N. Ajmal, Homomorphism of fuzzy subgroups, correspondence theorem and fuzzy quotient group, *Fuzzy Sets and Systems* 61 (1994) 329-339.
3. N. Ajmal, The lattice of fuzzy normal subgroups is modular, *Inform. Sci.* 83 (1995) 199-209.
4. N. Ajmal, Fuzzy groups with sup property, *Inform. Sci.* 93 (1996) 247-264.
5. N. Ajmal and S. Kumar, Lattice of subalgebras in the category of fuzzy groups, *J. Fuzzy Math.* 10 (2002) 359-369.
6. N. Ajmal and K.V. Thomas, The lattices of fuzzy subgroups and fuzzy normal subgroups, *Inform. Sci.* 76 (1994) 1-11.
7. N. Ajmal and K. V. Thomas, Fuzzy lattices, *Inform. Sci.* 79 (1994) 271 - 291.
8. N. Ajmal and K.V. Thomas, A complete study of the lattices of fuzzy congruences and fuzzy normal subgroups, *Inform. Sci.* 82 (1995) 197-218.
9. N. Ajmal and K. V. Thomas, The join of fuzzy algebraic substructures of a group and their lattices, *Fuzzy Sets and Systems* 99 (1998) 213-224.
10. N. Ajmal and K. V. Thomas, A new blue print for fuzzification: application to lattices of fuzzy congruences, *J. Fuzzy Math.* 7 (1999) 499-512.
11. N. Ajmal and K. V. Thomas, Lattice of subalgebras in the category of fuzzy groups, *J. Fuzzy Math.* 10 (2002) 359-369.
12. D - G Chen and W - X Gu, Product structure of the fuzzy factor groups, *Fuzzy Sets and Systems* 60 (1993) 229 - 232.
13. P. S. Das, Fuzzy group and level subgroups, *J. Math. Anal. Appl.* 84 (1981) 264-269.
14. V. N. Dixit, R. Kumar and N. Ajmal, Level subgroups and union of fuzzy subgroups, *Fuzzy Sets and Systems* 37 (1990) 359-371.
15. T. Head, A metatheorem for deriving fuzzy theorems from crisp versions, *Fuzzy Sets and Systems* 73 (1995) 349-358.
16. T. Head, Erratum to "A metatheorem for deriving fuzzy theorems from crisp versions," *Fuzzy Sets and Systems* 79 (1996) 277-278.

17. T. Head, Embedding lattices of fuzzy subgroups into lattices of crisp subgroups, Proceedings, Biennial Conference of the North American Fuzzy Information Proceedings Society NAFIPS' 1996,184-186.

18. A. Jain, Tom Head's join structure of fuzzy subgroups, *Fuzzy Sets and Systems* 125 (2002) 191-200.

19. N. Kuroki, Fuzzy Congruences and fuzzy normal subgroups, *Inform. Sci.* 60 (1992) 247-259.

20. W. J. Liu, Fuzzy invariant subgroups and fuzzy ideals, *Fuzzy Sets and Systems* 21 (1982) 133-139.

21. B. B. Makamba, Direct product and isomorphism of fuzzy subgroups, *Inform. Sci.* 65 (1992) 33-43.

22. B. B. Makamba and V. Murali, Normality and congruence in fuzzy subgroups, *Inform. Sci.* 59 (1992) 121-129.

23. M. Mashinchi and M.M. Zahedi, Lattice structure of fuzzy subgroups, *Bull. Iranian Math. Soc.* 18 (2) (1992) 17-29.

24. M. Mashinchi and M.M. Zahedi, On the product of T-fuzzy subgroups, Ann. Univ. Sci. Budapest. Sect. Comput., 12 (1991) 167-171.

25. N. P. Mukherjee and P. Bhattacharya, Fuzzy normal subgroups and fuzzy cosets, *Inform. Sci.* 34 (1984) 225 - 239.

26. V. Murali, Fuzzy equivalence relations, *Fuzzy Sets and Systems* 30 (1989) 155-163.

27. V. Murali, Fuzzy congruence relations, *Fuzzy Sets and Systems* 41 (1991) 359-369.

28. V. Murali, Lattice of fuzzy subalgebras and closure systems in I, *Fuzzy Sets and Systems* 41 (1991) 101-111.

29. A. Rosenfeld, Fuzzy groups, *J. Math. Anal. Appl.* 35 (1971) 512-517.

30. N. Sultana and N. Ajmal, Generated fuzzy subgroup: a modification, *Fuzzy Sets and Systems* 107 (1999) 241-243.

31. W. M. Wu, Normal fuzzy subgroups, *Fuzzy Math.*, I(1) (1981) 21-30 (in Chinese).

32. W. M. Wu, Fuzzy congruences and normal fuzzy subgroups, *Math. Appl.*, 1 (3) (1988) 9-20.

33. Yu Yandong, J. N. Mordeson, S - C Cheng, *Elements of L-algebra*, Lecture Notes Vol. 1, Creighton University, Omaha, Nebraska, 1994.

34. L. Zadeh, Fuzzy sets, *Inform. and Control* 8 (1965) 338-353.

10

Membership Functions From Similarity Relations

Let G be a group and μ a fuzzy subgroup of G. Then μ can be thought to be the membership function of a fuzzy subgroup of G. In this chapter, we sometimes refer to μ in this way. We show that μ satisfies the equation $\mu(x) = \sigma(e, x)$, where σ is a similarity relation on G which is invariant under left-translation. We also show that under certain natural assumptions the elements x of G can be represented as permutations P_x of a suitable universe Ω such that $\mu(x)$ equals the proportion of elements in Ω which are fixed by P_x. These results provide a deeper insight on the relationship of the group operation to the membership values $\mu(x)$.

The membership values of a fuzzy subset of the universe X classify the elements of X into disjoint subsets $X_t = \{x \in X \mid \mu(x) = t\}$. In many cases, the actual value taken by the membership function μ is not significant in that one can replace the membership values t by $f(t)$, where f is a strictly increasing function from $[0, 1]$ onto $[0, 1]$, and not change the basic properties of μ. This is still the case when there is a binary operation on X.

Since any group can be viewed as a group of permutations of a suitable universe Ω (which can be taken to be G itself) with the identity element e representing the identity transformation on Ω, it is natural to define μ by comparing x with e in some way. We show that this is possible by considering similarity relations on Ω (or on G). The consideration of similarity relations is motivated by the fact that they satisfy a "min-transitivity" property that is similar to (1) of Definition 1.2.3. We then show that, under a natural assumption, the membership values $\mu(x)$ of a fuzzy subgroup can be viewed as the proportion of the elements of a suitable universe Ω which are the fixed-points of P_x, when every $x \in G$ is represented as a permutation P_x of Ω.

For the sake of completeness and ease of reading, we occasionally restate and/or reprove results of previous chapters. When doing so, we try to give a different approach and proof. The first three sections are mainly from [7].

10.1 Similarity Relations and Membership Functions

Let Ω be a set and σ be a fuzzy subset of $\Omega \times \Omega$. Then σ is called a **similarity relation** on Ω if the following properties properties hold:

(S1) Reflexivity: $\sigma(x, x) = 1$ for all $x \in \Omega$.

(S2) Symmetry: $\sigma(x, y) = \sigma(y, x)$ for all $x, y \in \Omega$.

(S3) Min-transitivity: $\sigma(x, z) \geq \sigma(x, y) \wedge \sigma(y, z)$ for all $x, y, z \in \Omega$.

Let σ' be a similarity relation on Ω and G a group of permutations of Ω onto itself. Then it follows easily that σ defined on G by Expression 10.1.1 below satisfies (S1)-(S3), and hence is a similarity relation on G. This construction is similar to that for defining "distance" between two x and y of Ω into $\mathbb{R} : \|x - y\| = \vee\{|x(w) - y(w)|; \ w \in \Omega\}$. A general metric $d(x(w), y(w))$ on Ω can be used here rather than the absolute value $|x(w) - y(w)|$. Since the notion of "similarity" is opposite to that of "distance", the infimum is used in Expression 10.1.1 instead of the supremum. Let σ be the fuzzy subset of $G \times G$ defined as follows: $\forall x, y \in G$,

$$\sigma(x, y) = \wedge\{\sigma'(x(w), y(w)) \mid w \in \Omega\}. \tag{10.1.1}$$

The min-transitivity property (S3) is verified as follows: $\forall x, y, z \in G$,

$$\begin{aligned}
\sigma(x, z) &= \wedge\{\sigma'(x(w), z(w)) \mid w \in \Omega\} \\
&\geq \wedge\{\sigma'(x(w), y(w)) \wedge \sigma'(y(w), z(w)) \mid w \in \Omega\} \\
&\geq (\wedge\{\sigma'(x(w), y(w)) \mid w \in \Omega\}) \wedge (\wedge\{\sigma'(y(w), z(w)) \mid w \in \Omega\}) \\
&= \sigma(x, y) \wedge \sigma(y, z).
\end{aligned}$$

Let σ be a fuzzy subset of G. Then σ is said to be **right-invariant** if

$$\sigma(x, y) = \sigma(xz, yz) \quad \text{for all } x, y, z \text{ in } G. \tag{10.1.2}$$

For all $z \in G$, $\{z(w) \mid w \in \Omega\} = \Omega$. Hence it follows that σ defined by Expression (10.1.1) is right-invariant. The notion of *left-invariance* is defined in a similar way. However, σ as defined in Expression 10.1.1 may not be left-invariant. We say that σ is **translation invariant** if it is both left and right-invariant.

Theorem 10.1.2 below shows that the converse of Expression 10.1.1 also holds. That is, each right-invariant similarity relation on a group is obtained in that way. First, we give an example to illustrate the construction in Expression 10.1.1. We make use of the following algorithm to determine the transitive closure of a fuzzy relation ρ on a set X.

1. $\rho' = \rho \cup (\rho \circ \rho)$.
2. If $\rho' \neq \rho$, set $\rho = \rho'$ and go to step 1.
3. Stop ρ' is the transitive max-min closure of ρ.

Example 10.1.1. *Let* $\Omega = \{1, 2, 3, 4\}$. *Define the functions* e, a, b, c *of* Ω *onto itself as follows:*

$e(1) = 1, e(2) = 2, e(3) = 3, e(4) = 4,$
$a(1) = 1, a(2) = 2, a(3) = 4, a(4) = 3,$
$b(1) = 2, b(2) = 1, b(3) = 3, b(4) = 4,$
$c(1) = 2, c(2) = 1, c(3) = 4, c(4) = 3.$

Then e, a, b, c are one-to-one functions of Ω onto itself. Under composition of functions, they form the Klein 4-group $G = \{e, a, b, c\}$ with $ab = ba = c$ and $a^2 = b^2 = c^2 = e$. Define the fuzzy relation σ' on Ω as follows:

σ'	1	2	3	4
1	1.0	0.2	0	0
2	0.2	1.0	0	0
3	0	0	1.0	0.3
4	0	0	0.3	1.0

Clearly, σ' is reflexive and symmetric. A simple application of the algorithm preceding the example shows that σ' is its transitive closure. Hence σ' is a similarity relation on Ω. We now consider the corresponding similarity relation σ on G obtained by Expression 10.1.1. The values $\sigma'(i, j)$ and $\sigma'(j, i)$ for $i \in \{1, 2\}$ and $j \in \{3, 4\}$ do not have any influence here on the definition of σ. Now $\sigma((x, y)) = \wedge\{\sigma'(x(w), y(w)) \mid w \in \Omega\}$. Thus

$$\sigma((a, b)) = \sigma'(a(1), b(1)) \wedge \sigma'(a(2), b(2)) \wedge \sigma'(a(3), b(3)) \wedge \sigma'(a(4), b(4))$$
$$= \sigma'((1, 2)) \wedge \sigma'((2, 1)) \wedge \sigma'((4, 3)) \wedge \sigma'((3, 4))$$
$$= .2 \wedge .2 \wedge .3 \wedge .3$$
$$= .2$$

The remaining entries in the table given below are determined in a similar manner.

σ	e	a	b	c
e	1.0	0.3	0.2	0.2
a	0.3	1.0	0.2	0.2
b	0.2	0.2	1.0	0.3
c	0.2	0.2	0.3	1.0

Theorem 10.1.2. *Let σ be a similarity relation on a group G. If σ is right-invariant, then there is a universe Ω and a similarity relation σ' on Ω such that G is isomorphic to a group of permutations of Ω onto itself and σ is obtained from σ' as in Expression 10.1.1.*

Proof. Let $\Omega = G$ and let every $x \in G$ correspond to the left translations $L_x : G \to G$, where $L_x(z) = xz \; \forall z \in G$. The group product xy corresponds to the function composition $L_x \circ L_y$ (where L_y is applied first) and the group inverse x^{-1} corresponds to the inverse transformation L_x^{-1} where $x, y \in G$. Define the fuzzy subset σ' of $G \times G$ by $\forall(x, y) \in G \times G, \sigma'(x, y) = \sigma(x, y)$. Since $\sigma'(L_x(z), L_y(z)) = \sigma'(xz, yz) = \sigma(xz, yz) = \sigma(x, y)$, the right-hand side of Expression 10.1.1 equals $\sigma(x, y)$. Hence we regain σ from σ' as desired. $\square$

We note that if G is a group permutations on a given set Ω, then it may not be easy or even always possible to obtain a suitable σ' on Ω which will give rise to the given σ via Expression 10.1.1.

The next theorem shows the relationship between a right-invariant similarity relation on a group G and a fuzzy subgroup μ of G.

Theorem 10.1.3. *Let μ be a fuzzy subgroup of G such that $\mu(e) = 1$. Then there exists a right-invariant similarity relation σ on G such that $\mu(x) = \sigma(e, x) \; \forall x \in G$. Conversely, let σ be a right-invariant similarity relation on G. Then there exists a fuzzy group μ of G such that $\mu(x) = \sigma(e, x) \forall x \in G$.*

Proof. Suppose that μ is a fuzzy subgroup of G such that $\mu(e) = 1$. Define the fuzzy subset σ of $G \times G$ by $\sigma(x, y) = \mu(xy^{-1})$ for all $x, y \in G$. Then $\sigma(x, x) = \mu(xx^{-1}) = \mu(e) = 1$. Hence (S1) holds. Since $\mu(z) = \mu(z^{-1}) \; \forall z \in G$, (S2) holds. For all $x, y, z \in G$,

$$\sigma(x, z) = \mu(xz^{-1}) = \mu(xy^{-1}yz^{-1}) \geq \mu(xy^{-1}) \wedge \mu(yz^{-1}) = \sigma(x, y) \wedge \sigma(y, z).$$

Thus (S3) holds.

Conversely, suppose σ is a right-invariant similarity relation on G. Since e has the highest membership value, it is natural to have a higher value of $\mu(x)$ if x is more similar to e according to σ. Define the fuzzy subset μ of G by $\mu(x) = \sigma(e, x) \forall x \in G$. Since σ is right-invariant, we have

$$\begin{aligned}
\mu(x) = \sigma(e, x) &= \sigma(ex^{-1}, xx^{-1}) = \sigma(x^{-1}, e) \\
&= \sigma(e, x^{-1}) \quad \text{by (S2)} \\
&= \mu(x^{-1}).
\end{aligned}$$

Now $\mu(xy) = \sigma(e, xy) \geq \sigma(e, y) \wedge \sigma(y, xy) = \sigma(e, y) \wedge \sigma(e, x) = \mu(y) \wedge \mu(x)$. Thus μ is a fuzzy subgroup of G such that $\mu(e) = \sigma(e, e) = 1$. $\square$

Clearly, Theorem 10.1.3 holds if we replace the right-invariance of σ by left-invariance. In that case, given the fuzzy subgroup μ of G, we define the fuzzy subset σ of $G \times G$ by $\sigma(x, y) = \mu(x^{-1}y)$ for all $x, y \in G$. Hence $\sigma(zx, zy) = \mu(x^{-1}z^{-1}zy) = \mu(x^{-1}y) = \sigma(x, y)$ for all $z \in G$. Suppose that σ is left-invariant. Define μ as before. Then $\forall x, y \in G$,

$$\begin{aligned}
\mu(xy) = \sigma(e, xy) &\geq \sigma(e, x) \wedge \sigma(x, xy) \\
&= \sigma(e, x) \wedge \sigma(e, y) \quad \text{by left-invariance} \\
&= \mu(x) \wedge \mu(y).
\end{aligned}$$

Corollary 10.1.4. *Let μ be a fuzzy subgroup of G and let σ be the similarity relation on G defined by $\sigma(x, y) = \mu(xy^{-1})$ for all $x, y \in G$. Then σ is both left and right-invariant if and only if $\mu(xy) = \mu(yx)$ for all $x, y \in G$.*

Proof. Suppose $\mu(xy) = \mu(yx)$ for all $x, y \in G$. Then $\mu(x^{-1}y) = \mu((x^{-1}y)^{-1}) = \mu(y^{-1}x) = \mu(xy^{-1})$ for all $x, y \in G$. Thus it follows from the proof of Theorem 10.1.2 and the remarks following it that σ is both left and right-invariant. Conversely, suppose σ is left and right-invariant. Then it follows that for all $x, y \in G$,

$$
\begin{aligned}
\mu(xy) &= \mu((xy)^{-1}) = \sigma(e, (xy)^{-1}) = \sigma(e, y^{-1}x^{-1}) \\
&= \sigma(y, x^{-1}) \quad \text{by left-invariance} \\
&= \sigma(yx, e) \quad \text{by right-invariance} \\
&= \sigma(e, yx) \\
&= \mu(yx).
\end{aligned}
$$

$\square$

We recall that a fuzzy subgroup μ of a group G is called commutative if $\mu(xy) = \mu(yx)$ for all $x, y \in G$. The equality $\mu(xy) = \mu(yx)$ holds for certain pairs of elements x and y of G, namely those $x, y \in G$ such that $\mu(x) \neq \mu(y)$. This can be seen from the following result.

Lemma 10.1.5. *Let μ be a fuzzy subgroup of G. Let $x, y \in G$. If $\mu(x) \neq \mu(y)$, then $\mu(xy) = \mu(yx) = \mu(x) \wedge \mu(y)$.*

Proof. Suppose $\mu(x) > \mu(y)$. Then $\mu(y) = \mu(x^{-1}xy) \geq \mu(x^{-1}) \wedge \mu(xy) \geq \mu(x) \wedge \mu(x) \wedge \mu(y) = \mu(y)$. Thus $\mu(x) \wedge \mu(xy) = \mu(y)$. Since $\mu(x) > \mu(y), \mu(xy) = \mu(y) = \mu(x) \wedge \mu(y)$. Similarly, $\mu(yx) = \mu(x) \wedge \mu(y)$. $\square$

The following result follows from Theorem 1.3.3.

Lemma 10.1.6. *Let μ be a fuzzy subgroup of G. Then $\mu(xy) = \mu(yx)$ for all $x, y \in G$ if and only if μ_t is a normal subgroup of G $\forall t \in [0, \mu(0)]$.*

Suppose that μ is commutative. Then the similarity relation σ defined by $\sigma(x, y) = \mu(xy^{-1})$ $\forall x, y \in G$ is such that σ is a fuzzy subgroup of the group $G \times G$. To see this, note that $\sigma(x, y) = \mu(xy^{-1}) = \mu(y^{-1}x) = \mu((y^{-1}x)^{-1}) = \mu(x^{-1}y) = \sigma(x^{-1}, y^{-1})$ and that (x^{-1}, y^{-1}) is the inverse of (x, y) in $G \times G \forall x, y \in G$. Moreover,

$$
\begin{aligned}
\sigma((x, y)(x', y')) &= \sigma(xx', yy') = \mu(xx'(yy')^{-1}) = \mu(xx'(y')^{-1}y^{-1}) \\
&= \mu(y^{-1}xx'(y')^{-1}) \geq \mu(y^{-1}x) \wedge \mu(x'(y')^{-1}) \\
&= \mu(xy^{-1}) \wedge \mu(x'(y')^{-1}) = \sigma(x, y) \wedge \sigma(x', y').
\end{aligned}
$$

The converse is true as can be seen by the following argument. If the condition $\sigma(x, y) = \mu(xy^{-1}) \forall x, y \in G$ implies σ is a fuzzy subgroup of $G \times G$, then $\mu(xy) = \mu(yx)$ for all $x, y \in G$. This follows since $\mu(xy) = \sigma(x, y^{-1}) = \sigma(x^{-1}, y) \forall x, y \in G$ by (G1) applied to $G \times G$ since (x^{-1}, y) is the inverse of (x, y^{-1}). Finally, $\sigma(x^{-1}, y) = \mu(x^{-1}y^{-1}) = \mu((yx)^{-1}) = \mu(yx) \forall x, y \in G$.

Given a fuzzy subgroup μ of G, we can construct a new fuzzy subgroup $\bar{\mu}$ of G that satisfies $\bar{\mu}(xy) = \bar{\mu}(yx)$ for all $x, y \in G$ by using the observations made above. We first construct a similarity relation $\bar{\sigma}$ from σ which is both left and right invariant as follows: $\forall x, y \in G$,

$$\bar{\sigma}(x, y) = \wedge\{\sigma(zx, zy) \mid z \in G\}. \tag{10.1.3}$$

It follows easily that $\bar{\sigma}$ satisfies (S1)-(S3). We then define the fuzzy subset $\bar{\mu}$ of G by $\forall x \in G$, $\bar{\mu}(x) = \bar{\sigma}(e, x) = \wedge\{\sigma(z, zx) \mid z \in G\} = \wedge\{\sigma(e, zxz^{-1}) \mid z \in G\} = \wedge\{\mu(zxz^{-1}) \mid z \in G\}$. It follows that $\bar{\mu}$ is the largest fuzzy subgroup of G such that $\bar{\mu} \subseteq \mu$ and $\bar{\mu}$ is commutative. The following theorem follows from these observations.

Theorem 10.1.7. *Let μ be a fuzzy subgroup of G. Then $\bar{\mu}$ is the largest fuzzy subgroup of G such that $\bar{\mu} \subseteq \mu$ and $\bar{\mu}$ is commutative, where $\bar{\mu}$ is defined by $\bar{\mu}(x) = \wedge\{\mu(zxz^{-1}) \mid z \in G\} \; \forall x \in G$.*

The following theorem follows from the observations made above and is easily proved directly as well from the fact that $\sigma((x, y)(x', y')) = \sigma(xx', yy') \geq \sigma(xx', yx') \wedge \sigma(yx', yy') = \sigma(x, y) \wedge \sigma(x', y') \; \forall x, y, x', y' \in G$.

Theorem 10.1.8. *Suppose σ is a translation invariant similarity relation on G. Then σ is a fuzzy subgroup of $G \times G$ if and only if μ is a fuzzy subgroup of G such that $\mu(x) = \sigma(e, x)$ and μ is commutative.*

If the similarity relation σ defined by $\sigma(x, y) = \mu(xy^{-1}) \; \forall x, y \in G$ is translation invariant, then there are two ways of embedding G and μ into $G \times G$ and σ. The mapping $x \to (e, x)$ preserves both the group operation and the membership values since $\mu(x) = \sigma(e, x) \; \forall x \in G$. The same is true for the mapping $x \to (x, e)$. On the other hand, the mapping $x \to (x, x)$ preserves only the group operation, but not the membership values.

10.2 Level Subgroups, Cosets, and Equivalence Classes

Let σ be a fuzzy relation on a set Ω. It follows that σ is a similarity relation on Ω if and only if $\sigma_t = \{(x, y) \mid \sigma(x, y) \geq t\}$ is an equivalence relation on Ω $\forall$ $t \in [0, 1]$.

Lemma 10.2.1. *Let μ be a fuzzy subgroup of G and let σ be the similarity relation on G defined by $\sigma(x, y) = \mu(xy^{-1})$ (or, $\sigma(x, y) = \mu(x^{-1}y)$) for all $x, y \in G$. Let $[y]_t = \{x \mid \sigma(y, x) \geq t\} \forall y \in G$ and $\forall t \in [0, 1]$. Then $\mu_t = [e]_t$. Moreover, $[y]_t$ equals the right-coset (resp., left-coset) of y with respect to the subgroup μ_t.*

Proof. We have

$$[y]_t = \{x \mid \sigma(y, x) \geq t\} = \{x \mid \sigma(x, y) \geq t\}$$
$$= \{x \mid \sigma(e, xy^{-1}) \geq t\} \quad (\text{resp.}, = \{x \mid \sigma(e, y^{-1}x) \geq 1\})$$
$$= \{x \mid xy^{-1} \in \mu_t\} \quad (\text{resp.}, = \{x \mid y^{-1}x \in \mu_t\})$$
$$= \mu_t y \qquad (\text{resp.}, = y\mu_t).$$

In particular, $[e]_t = \mu_t$.   □

Example 10.2.2. *Consider the similarity relation σ in Example 10.1.1 on $G = \{e, a, b, c\}$ is $\mu(e) = 1$, $\mu(a) = .3$, $\mu(b) = \mu(c) = .2$, where $\mu(x) = \sigma(e, x) \forall x \in G$. There are three distinct level subgroups, namely,*

$\mu_1 = \{e\} = \mu_t \quad$ *for $.3 < t \leq 1$,*
$\mu_{.3} = \{e, a\} = \mu_t \quad$ *for $.2 < t \leq .3$*
$\mu_{.2} = G = \mu_t \quad$ *for $0 \leq t \leq .2$.*

For $t = 1$, there are 4 right-cosets $\{e\}$, $\{a\}$, $\{b\}$, and $\{c\}$ of the level-subgroup μ_1, which correspond to the equivalence-classes of

$$\sigma_1 = \{(e, e), (a, a), (b, b), (c, c)\}.$$

For $t = .3$, there are 2 right-cosets $\{e, a\}$ and $\{b, c\}$ of $\mu_{.3}$, which correspond to the equivalence classes of

$$\sigma_{.3} = \sigma_1 \cup \{(e, a), (a, e), (b, c), (c, b)\}.$$

Finally, for $t = .2$, there is only 1 right-coset $\{e, a, b, c\} = G$ of $\mu_{.2}$, which corresponds to the equivalence class of $\sigma_{.2} = G \times G$.

An important consequence of Lemma 10.2.1 is that each equivalence class determined by σ_t has the same size. As shown in the next theorem, this is all that can be said about σ as a fuzzy similarity relation on a finite group. In particular, the values taken by σ do not have any theoretical significance (cf., Theorem 10.3.1).

Theorem 10.2.3. *Suppose σ is a fuzzy similarity relation on a finite set G. Then there is a binary operation on G such that G is a group and a fuzzy subgroup μ of G such that $\sigma(x, y) = \mu(xy^{-1})$ (or, $= \mu(x^{-1}y)$) for all $x, y \in G$ if and only if the equivalence classes determined by the crisp equivalence relation $\sigma_t = \{(x, y) \mid \sigma(x, y) \geq t\}$ have the same size for all $t \in [0, 1]$.*

Proof. Suppose the equivalence classes determined by σ_t are of the same size for all $t \in [0, 1]$. If $\sigma(x, y) = 1$ for all $x, y \in G$, then let $\mu(x) = 1$ for all $x \in G$ and any group operation of G, e.g., we can let G be a cyclic group of order $|G|$. Now assume that $|G| = N$ and $0 \leq t_1 < t_2 < \cdots < t_n = 1$ are the distinct values of σ, where $n \geq 2$. Let s_j be the size of an equivalence class for $\sigma_{t_j}, j = 1, ..., n$. Since $\sigma_{t_{j-1}} \supset \sigma_{t_j}$ as crisp equivalence relations, each equivalence class in $\sigma_{t_{j-1}}$ is a disjoint union of equivalence classes of

$\sigma_{t_j}, j = 2, ..., n$. It follows that $s_n|s_{n-1}|s_{n-2}|...|s_2|s_1 \ (= N)$. Let $N_j = N/s_j$, the number of equivalence classes in σ_{t_j}. Clearly, $(1 =) \ N_1|N_2|...|N_{n-1}|N_n$ $(\le N)$.

We first consider the cyclic group H generated by an element b of order N with the membership function (fuzzy subgroup) ν defined on H as follows: $\forall h \in H$,

$\qquad \nu(h) = t_n \qquad$ for $h = b^{kN_n}$, $k = 1, 2, ..., s_n$

$\qquad \nu(h) = t_{n-1} \quad$ for $h = b^{kN_{n-1}}$, $k = 1, 2, ..., s_{n-1}$ except for k being a multiple of N_n/N_{n-1}

$\qquad \nu(h) = t_{n-2} \quad$ for $h = b^{kN_{n-2}}$, $k = 1, 2, ..., s_{n-2}$ except for k being a multiple of N_{n-1}/N_{n-2}

$\qquad \cdots$

$\qquad \cdots$

$\qquad \nu(h) = t_1 \qquad$ for $h = b^{kN_1}$, $k = 1, 2, ..., s_1$ except for k being a multiple of N_2/N_1.

Clearly, the level subgroup ν_{t_j} has the size s_j for every j. By Lemma 10.2.1, it suffices to identify the elements of G with those H in a one-to-one and onto fashion so that for each $t = t_j$, the equivalence class structures of σ_t coincide with that of the similarity relation obtained from $\nu(h)$ for $h \in H$. $\square$

The following example shows that there may be more than one way to make the identification alluded to in the proof of the previous theorem.

Example 10.2.4. *Let* $n = 3$, $N = 18 = s_1$. *Let* $s_2 = 9$ *and* $s_3 = 3$. *Then* $N_1 = 1$, $N_2 = 2$, $N_3 = 6$, *and the following membership values on the cyclic group* H *of order 18 are as follows:* $\nu(e) = \nu(b^6) = \nu(b^{12}) = t_3$, $\nu(b^2) = \nu(b^8) = \nu(b^{10}) = \nu(b^{14}) = \nu(b^{16}) = t_2$, *and the remaining* $\nu(b^j)$'s *equal* t_1. *The illustration below shows the equivalence classes of the similarity relation* σ *such that* $\sigma(x, y) = \nu(xy^{-1}) \forall x, y \in H$. *For an arbitrary set* G *of cardinality 18 and a similarity relation on* G *which takes 3 distinct values* $t_1 < t_2 < t_3 = 1$ *such that the equivalence classes for the corresponding* σ_{t_j} *have the equal size* s_j *for each* $j = 1, 2, 3$, *there is a one-to-one function from* G *onto* H *to match the equivalence class structures in many ways. Any such function* f *determines a fuzzy subgroup* μ *of* G *such that* $\mu(z) = \nu(f(z)) \forall z \in G$ *and moreover* $\sigma(z, z') = \mu(z(z')^{-1}) \forall z, z' \in G$. *We have the following equivalence classes.*

t_3 *-equivalence classes:*

$\{b, b^7, b^{13}\}, \{b^3, b^9, b^{15}\}, \{b^5, b^{11}, b^{17}\}, \{e, b^6, b^{12}\}, \{b^2, b^8, b^{14}\}, \{b^4, b^{10}, b^{16}\}$

t_2-*equivalence classes:*

$\{b, b^7, b^{13}\} \cup \{b^3, b^9, b^{15}\} \cup \{b^5, b^{11}, b^{17}\}, \{e, b^6, b^{12}\} \cup \{b^2, b^8, b^{14}\} \cup \{b^4, b^{10}, b^{16}\}$

t_1-*equivalence classes:*

$\{b, b^7, b^{13}\} \cup \{b^3, b^9, b^{15}\} \cup \{b^5, b^{11}, b^{17}\} \cup \{e, b^6, b^{12}\} \cup \{b^2, b^8, b^{14}\} \cup \{b^4, b^{10}, b^{16}\}$.

Corollary 10.2.5. *Let μ be a fuzzy subset of a finite set G with the membership values $0 \leq t_1 < t_2 < ... < t_n = 1$. Then there exists a group operation on G such that μ is a fuzzy group of G if and only if $|\mu_{t_j}|$ divides $|\mu_{t_{j-1}}|$ for $j = 2, ..., n$.*

Proof. If G is a group, then $|\mu_{t_j}|$ divides $|\mu_{t_{j-1}}|$ since μ_{t_j} is a subgroup of $\mu_{t_{j-1}}, j = 2, ..., n$. Suppose $|\mu_{t_j}|$ divides $|\mu_{t_{j-1}}|$ for $j = 2, ..., n$. We decompose G into a series of disjoint subsets of equal size as shown in the above illustration. If $t_1 > 0$, then we decompose the set $G \backslash \mu_{t_1}$ arbitrarily into disjoint subsets each of size $|\mu_{t_1}|$. If $t_1 = 0$, then $G = \mu_{t_1}$. We regard this decomposition as the equivalence class decomposition for a crisp equivalence relation σ_{t_1}. Next we decompose $\mu_{t_1} \backslash \mu_{t_2}$ into disjoint subsets of equal size $|\mu_{t_2}|$, and also each of the other subsets of size $|\mu_{t_1}|$ obtained in the previous step are decomposed arbitrarily into disjoint subsets of size $|\mu_{t_2}|$. This decomposition is taken to be the equivalence class decomposition for a crisp equivalence relation $R_{t_2} \subset R_{t_1}$. The process is continued until t_n, defining the equivalence class decomposition for the relation R_{t_n}. We now define the similarity relation σ on G by $\sigma(x, y) = \vee \{t_j \mid (x, y) \in R_{t_j}\} \forall x, y \in G$. An application of Theorem 10.2.3 yields the desired result. $\qquad \square$

We now consider how the behavior of a fuzzy subset ν of a universe X compares with the result of Theorem 10.2.3. We must first associate a suitable similarity relation ρ on X with ν. Define the fuzzy subset ρ of $X \times X$ as follows: $\forall x, y \in X$,

$$\rho(x, y) = \begin{cases} 1 & \text{if } x = y \\ \nu(x) \wedge \nu(y) & \text{otherwise.} \end{cases} \qquad (10.2.1)$$

It follows easily that conditions $(S1) - (S3)$ hold for ρ. If we assume that ν is normalized, that is, $\nu(z) = 1$ for some $z \in X$, then we can recover ν from ρ simply by $\nu(x) = \rho(z, x)$. In general, we cannot recover ν from ρ as defined by the above Expression 10.2.1. In particular, $\vee \{\rho(z, x) \mid z \in X\}$ may not equal $\nu(x)$. Let $E_t = \{(x, y) \mid \rho(x, y) \geq t\}$ denote the associated crisp equivalence relation. The equivalence classes $[[y]]_t$ with respect to E_t are singleton sets if $\nu(y) < t$, except possibly for one equivalence class which equals the level set ν_t if $\nu(y) \geq t$. This is illustrated in the Example 10.2.6.

Example 10.2.6. *Let $X = \{a, b, c, d\}$ and let ν be the fuzzy subset of X defined as follows: $\nu(a) = 1$, $\nu(b) = 0.7$, $\nu(c) = 0.3$, and $\nu(d) = 0.2$. . The similarity relation ρ, the level sets ν_t, and the equivalence classes $[[y]]_t$ are given as follows:*

ρ	a	b	c	d
a	1.0	0.7	0.3	0.2
b	0.7	1.0	0.3	0.2
c	0.3	0.3	1.0	0.2
d	0.2	0.2	0.2	1.0

$$\nu_1 = \{a\} = \nu_t, \qquad for\ 1.0 > t > 0.7$$
$$\nu_{.7} = \{a, b\} = \nu_t, \qquad for\ 0.7 > t > 0.3$$
$$\nu_{.3} = \{a, b, c\} = \nu_t, \quad for\ 0.3 > t > 0.2$$
$$\nu_{.2} = X = \nu_t, \qquad for\ 0.2 > t \geq 0.0$$

For $0.7 < t \leq 1.0 : \{a\} = \nu_t, \{b\}, \{c\}, \{d\}$; *4 classes*
For $0.3 < t \leq 0.7 : \{a, b\} = \nu_t, \{c\}, \{d\}$; *3 classes*
For $0.2 < t \leq 0.3 : \{a, b, c\} = \nu_t, \{d\}$; *2 classes*
For $0.0 \leq t \leq 0.3 : \{a, b, c, d\} = \nu_t.$ *1 class*

All equivalence classes $[[y]]_t$ are singleton sets except possibly for one which equals the level set ν_t.

Suppose G is an Abelian group. Then unlike σ, ρ may not always be a fuzzy subgroup of $G \times G$. Clearly, $\rho(x^{-1}, x^{-1}) = 1 = \rho(x, x)$ and for $x \neq y$, $\rho(x^{-1}, y^{-1}) = \mu(x^{-1}) \wedge \mu(y^{-1}) = \mu(x) \wedge \mu(y) = \rho(x, y)$. However, consider the membership function $\mu(e) = 1.0$, $\mu(a) = 0.3$, $\mu(b) = 0.2 = \mu(c)$ obtained from σ in Example 10.1.1, and apply the construction for ρ to μ. We have $\rho(b, b) = 1$ and $\rho(e, a) = \mu(e) \wedge \mu(a) = 0.3$, but $\rho((e, a)(b, b)) = \rho(eb, c) = \mu(b) \wedge \mu(c) = 0.2 < \rho(e, a) \wedge \rho(b, b)$. It is not surprising that ρ fails to be a fuzzy subgroup of $G \times G$ since Expression 10.2.1 does not make use of the group property of G.

Theorem 10.2.7. *Let G be an Abelian group and let μ be a fuzzy subgroup of G. Then σ defined by $\sigma(x, y) = \mu(xy^{-1})$ $\forall x, y \in G$ is the smallest similarity relation containing ρ as defined in the Expression 10.2.1 with ν replaced by μ and such that σ is a fuzzy subgroup of $G \times G$.*

Proof. Suppose that σ is a similarity relation on G such that $\sigma \supseteq \rho$, where ρ is a similarity relation on G such that σ is a fuzzy subgroup of $G \times G$. Then $\sigma(x, y) = \sigma((xy^{-1}, e)(y, y)) \geq \sigma(xy^{-1}, e) \wedge \sigma(y, y) \geq \rho(xy^{-1}, e) \wedge \rho(y, y) = \mu(xy^{-1}) \wedge \mu(e)$ and $1 = \mu(xy^{-1})$. Since $\sigma(x, y) = \mu(xy^{-1})$ $\forall x, y \in G$, σ is a fuzzy subgroup of $G \times G$. $\square$

10.3 Representation of Membership Functions

In the previous sections, we considered similarity relations on a group G, their invariant properties, and their relationship to fuzzy subgroups of G. We now consider to what extent the membership function μ can represent realistic properties of the group elements. As seen from Theorem 10.2.3 and Corollary 10.2.5 for a finite group, the values t_j taken by μ have no direct role except for separating the elements into disjoint equivalence classes for the associated similarity relation σ. If we regard each $x \in G$ as a 1-1 onto function P_x of a set Ω onto itself and we attempt to define μ in a direct way by looking at the properties of P_x, then it is not always easy to satisfy the property that

$\mu(xy) \geq \mu(x) \wedge \mu(y)$. For example, suppose Ω is finite. Then define μ in terms of the set of fixed points of P_x as follows: $\forall x \in \Omega$,

$$\mu(x) = |\phi(x)|/|\Omega|, \text{ where } \phi : \Omega \rightarrow \text{ P}(\Omega) \text{ maps } \phi(x) = \{w \in \Omega \mid P_x(w) = w\}.$$
$$\tag{10.3.1}$$

In general, we can only say that $\mu(xy) \geq 0 \vee (\mu(x) + \mu(y) - 1)$, which is weaker than the property that $\mu(xy) \geq \mu(x) \wedge \mu(y)$, even though the properties $\mu(e) = 1$ and $\mu(x) = \mu(x^{-1})$ hold, where $x, y \in G$. For the mappings b and c in Example 10.1.1, we have $\mu(c) = 0$ and $\mu(a) = 2/4 = \mu(b)$ and so μ satisfies the above Expression 10.3.1, but $\mu(ab) \not\geq \mu(a) \wedge \mu(b)$.

However, if we assume that for all $x, y \in G$ either $\phi(x)$ contains $\phi(y)$ or $\phi(y)$ contains $\phi(x)$, then the property $\mu(xy) \geq \mu(x) \wedge \mu(y)$ holds since $\phi(xy) \supseteq \phi(x) \cap \phi(y)$. We show below that, under some natural assumptions, there is a representation of the elements of G as one-to-one functions of a suitable universe Ω onto itself such that the property $\phi(x) \supseteq \phi(y)$ or $\phi(y) \supseteq \phi(x)$ holds for all $x, y \in G$. Note that the similarity relation σ associated with μ given by Expression 10.3.1 is such that $\sigma(x, y)$ equals $|\{w \mid P_x(w) = P_y(w)\}|/|\Omega|$, which is the proportion of the elements in Ω, where the mappings associated with x and y agree.

Suppose $\mu(x) = 1$ for some $x \neq e$. Then $P_x = P_e$, the identity mapping on Ω. Thus the association $x \rightarrow P_x$ is not a group isomorphism (from G to the group of one-to-one functions of Ω onto itself). Henceforth, we assume that μ takes at least two distinct values. We say μ satisfies the **identity property** if $\mu(x) < 1 = \mu(e)$ for all $x \neq e$.

Theorem 10.3.1. *Suppose μ is a fuzzy subgroup of a finite group G such that $\mu(e) = 1$. Let $0 \leq t_1 < t_2 < ... < t_n = 1$ be the membership values of μ with $n \geq 2$. Then there exists a representation of G involving permutations P_x on a suitable universe Ω such that Expression 10.3.1 holds if and only if each t_j is a rational number and $\mu(xy) = \mu(yx)$ for all x and y. The permutations P_x are distinct if μ has the identity property.*

The only-if part in Theorem 10.3.1 follows easily since given the mappings P_x, $\phi(yxy^{-1}) = \{P_y(w) \mid w \in \phi(x)\}$ and so $|\phi(x)| = |\phi(yxy^{-1})|$ for all $x, y \in G$. Thus $\mu(x) = \mu(yxy^{-1})$ and so $\mu(xy) = \mu(yx)$ for all $x, y \in G$. By Expression 10.3.1, each t_j or $\mu(x)$ is a rational number.

Before we prove the if-part of the theorem, we need the following lemma.

Lemma 10.3.2. *Suppose μ is a fuzzy subgroup of G and μ_t is a normal subgroup of G for some $t \in \mu(G)$. Consider the quotient group G/μ_t. Then μ' is a fuzzy subgroup of G/μ_t with the identity property, where μ' is defined on the cosets of G/μ_t as follows: $\forall [x] \in G/\mu_t$,*

$$\mu'([x]) = \begin{cases} 1 & \text{if } x \in \mu_t, \\ \mu(x) & \text{otherwise.} \end{cases} \tag{10.3.2}$$

Moreover, if G is commutative so is μ'.

Proof. We first show that μ' is well-defined. Suppose $[x] = [x']$ (or equivalently, $x' = xz$ for some $z \in \mu_t$). Then $\mu'([x]) = \mu'([x'])$. If both x and $x' \in \mu_t$, then there is nothing to prove. If both x and x' don't belong to μ_t, then $\mu(x') \geq \mu(x) \wedge \mu(z) = \mu(x)$. Similarly from $x = x'z^{-1}$, where $z^{-1} \in \mu_t$, we have $\mu(x) \geq \mu(x')$ and hence $\mu(x) = \mu(x')$. That $\mu'([x]) = \mu'([x]^{-1}) \forall x \in G$ follows immediately. To prove that $\mu'([x][y]) \geq \mu'([x]) \wedge \mu'([y])$, we only need to consider the case $\mu'([x][y]) = \mu'([xy]) < 1$. In that case, $xy \notin \mu_t$ and hence $yx = yxyy^{-1} \notin \mu_t$ and $\mu'([xy]) = \mu(xy)$. If both x and y do not belong to μ_t, then $\mu'([xy]) = \mu(xy) \geq \mu(x) \wedge (y) = \mu'([x]) \wedge \mu'([y])$ and we have the desired result. If $x \in \mu_t$, then $y \notin \mu_t$ and hence $\mu'([x]) \wedge \mu'([y]) = 1 \wedge \mu(y) = \mu(y)$. Also, since μ_t is a normal subgroup of G, $[xy] = xy\mu_t = \mu_t xy = \mu_t y = y\mu_t = [y]$. Thus $\mu(xy) = \mu(y)$ and so $\mu'([x][y]) = \mu'([x]) \wedge \mu'([y])$. A similar argument applies if $x \notin \mu_t$ and $y \in \mu_t$.

That μ' is commutative follows from the fact that $xy \in \mu_t$ if and only if $yx \in \mu_t$. That μ' satisfies the identity property is immediate. $\square$

Proof of Theorem 10.3.1 We may assume without loss of generality that μ has the identity property. Otherwise, let $G' = G/\mu_1$ and let $[x]$ denote the coset of x with respect to μ_1. By Lemma 10.3.2, $\mu'([x]) = \mu(x)$ and so μ' is a fuzzy subgroup of G', where μ' is commutative and has the identity property. Also, $\text{Im}(\mu') = \text{Im}(\mu)$. Assume that there is a suitable representation for the elements of G' as in Theorem 10.3.1, where $P_{[x]}$ is the mapping associated with $[x]$. We immediately obtain a desired representation of G by lifting the representation of G' via the mapping $x \to [x]$ from G to G', i.e., we associate $P_{[x]}$ with x.

We prove the theorem by induction on the number $n \geq 2$ of distinct values of μ. Suppose that $n = 2$. If $t_1 = 0$, then we let $\Omega = G$ and let P_x be the left-translation $P_x(z) = L_x(z) = xz$ from Ω onto Ω. This gives $\phi(x) = \emptyset$ for $x \neq e$ and $\phi(e) = \Omega$. Thus Expression 10.3.1 is satisfied and clearly $P_{xy} = P_x \circ P_y$. The P_x's are distinct since $|G| \geq 2$. If $t_1 = p/q > 0$, we take Ω to be q copies of the set G. We define P_x as follows. For each of the first $q - p$ copies of G, we define P_x to be the same as L_x and for each of the remaining p copies of G, we define P_x to be the identity map. Once again the mappings P_x satisfy expression (10.3.1), are distinct, and $P_{xy} = P_x \circ P_y$.

Now assume that the theorem is true for $n = m$ (≥ 2). We prove that it is true for $n = m + 1$. Consider the following two fuzzy subsets of G defined as follows: $\forall x \in G$

$$\mu_1(x) = \begin{cases} 1 & \text{if } x = e, \\ t_{n-2} & \text{if } \mu(x) = t_{n-1}, \\ \mu(x) & \text{otherwise;} \end{cases} \tag{10.3.3}$$

$$\mu_2(x) = \begin{cases} 1 & \text{if } x = e, \\ 1 & \text{if } \mu(x) = t_{n-1}, \\ \mu(x) & \text{otherwise.} \end{cases} \tag{10.3.4}$$

Both μ_1 and μ_2 take only m distinct values $\{t_1, t_2, ..., t_{n-2}, 1\}$. Since $(\mu_i)_t = \{x \mid \mu_i(x) \geq t\}$ is clearly a subgroup of G for all $t \geq 0$ and $i = 1, 2$, both μ_1 and μ_2 are fuzzy subgroups of G. Also, both μ_1 and μ_2 are commutative, but only μ_1 has the identity property. By the induction hypothesis, for each $i = 1, 2$, we have a representation of μ_i as 1-1 functions $P_x^{(i)}$ of a finite domain Ω_i onto itself such that $\mu_i(x) = |\phi_i(x)|/|\Omega_i|$, where $\phi_i(x)$ is the set of fixed points of $P_x^{(i)}$ in $\Omega_i, i = 1, 2$. Moreover, the functions $P_x^{(i)}$ are distinct. Let N_1 and N_2 be integers such that $N_1 t_{n-2}|\Omega_1| + N_2|\Omega_2| = t_{n-1}(N_1|\Omega_1| + N_2|\Omega_2|)$. Since $t_{n-2} < t_{n-1} < 1$, this is always possible. Let Ω consist of N_1 copies of Ω_1 and N_2 copies of Ω_2. We copy the functions $P_x^{(i)}$ on each copy of Ω_i. For all $x \in G$, we then have $|\phi(x)| = N_1|\phi_1(x)| + N_2|\phi_2(x)|$. In particular, if $\mu(x) = t_j$, $j \leq n - 2$, then $|\phi(x)| = N_1 t_j|\Omega_1| + N_2 t_j|\Omega_2| = t_j|\Omega|$. Clearly, $\mu(e) = 1$. Finally, if $\mu(x) = t_{n-1}$, then $|\phi(x)| = N_1 t_{n-2}|\Omega_1| + N_2|\Omega_2| = t_{n-1}|\Omega|$, where $x \in G$.

The following example illustrates the construction in the previous theorem.

Example 10.3.3. *We illustrate the construction in the proof of Theorem 10.3.1 using the group and fuzzy subgroup considered in Example 10.2.2, namely, $G = \{e, a, b, c\}$, where $c = ab = ba$ and $a^2 = b^2 = e$, and the membership function given by $\mu(e) = 1$, $\mu(a) = 0.3$, and $\mu(b) = 0.2 = \mu(c)$. The left translations $L_x(z) = xz$ on G are as follows:*
$$L_e(e) = e, L_e(a) = a, L_e(b) = b, L_e(c) = c,$$
$$L_a(e) = a, L_a(a) = e, L_a(b) = c, L_a(c) = b,$$
$$L_b(e) = b, L_b(a) = c, L_b(b) = e, L_b(c) = a,$$
$$L_c(e) = c, L_c(a) = b, L_c(b) = a, L_c(c) = e.$$

The representations of G based on the fuzzy subgroups μ_1 and μ_2 respectively, given by Expression 10.3.3 can be determined. The final representation is obtained by choosing $N_1 = 7$ copies of (i) below and $N_2 = 2$ copies of (ii) below. This gives $|\Omega| = 7(4 \times 4 + 4) + 2(2 \times 4 + 2) = 160$ and $|\phi(b)| = 2 \times 10 + 7 \times 4 = 48$. The proportion of fixed-points of P_a equals $48/160 = 0.3 = \mu(a)$, as desired. Similarly, the proportion of fixed-points of both P_b and P_c equal 0.2.

We have 4 copies of the following:
$$P_e(e) = e, P_e(a) = a, P_e(b) = b, P_e(c) = c,$$
$$P_a(e) = a, P_a(a) = e, P_a(b) = c, P_a(c) = b,$$
$$P_b(e) = b, P_b(a) = c, P_b(b) = e, P_b(c) = a,$$
$$P_c(e) = c, P_c(a) = b, P_c(b) = a, P_c(c) = e.$$

We have 1 copy of the following:
$$P_x(z) = z \; \forall x, z \in \{e, a, b, c\}.$$
$$(i)$$
We have 4 copies of the following:
$$P_e([e]) = [e], P_e([b]) = [b], P_a([e]) = [e], P_a([b]) = [b], P_b([e]) = [b], P_b([b]) =$$
$[e], P_c([e]) = [b], P_c([b]) = [e]$, *where* $[e] = [a], [b] = [c]$.

We have 1 copy of the following:
$P_e([e]) = [e], P_e([b]) = [b], P_a([e]) = [e], P_a([b]) = [b], P_b([e]) = [e], P_b([b]) = [b], P_c([e]) = [e], P_c([b]) = [b],$ *where* $[e] = [a], [b] = [c]$.

(*ii*)

(*i*) *gives the representation of the group* $G = \{e, a, b, c\}$ *for the membership function* $\mu_1(e) = 1$ *and* $\mu_1(a) = \mu_1(b) = \mu_1(c) = 0.2$. (*ii*) *gives the representation of* G, *based on that of the quotient group* $G/(\mu_2).3 = \{e, a, b, c\}/\{e, a\}$, *where* $\mu_2(e) = 1 = \mu_2(a)$ *and* $\mu_2(b) = 0.2 = \mu_2(c)$.

10.4 Fuzzy Subgroups Based on Group Properties

In this section, an algebraic approach to the construction of fuzzy subgroups is discussed [16]. Let $L = (L, \vee, \wedge, 0, 1)$ be a complete lattice. An **L-fuzzy subset** of a set X is a function of X into L. The approach in this section is the construction of L-fuzzy subgroups μ of a group G so that for each $x \in G$, the lattice element $\mu(x)$ signifies the membership grade of x and $\mu(x)$ is determined by the extent to which x satisfies some algebraic property.

Definition 10.4.1. *A **complete Heyting algebra** H is a complete lattice* $(H, \vee, \wedge, 0, 1)$ *such that* $\forall a \in H$ *and* $\forall B \subseteq H, \vee\{a \wedge b \mid b \in B\} = a \wedge (\vee B)$.

Let $I = [0, 1]$. Then $I = (I, \vee, \wedge, 0, 1)$ and $\mathcal{P}(S) = (\mathcal{P}(S), \cup, \cap, \emptyset, S)$ are complete Heyting algebras, where S is a nonempty set and $\mathcal{P}(S)$ is the power set of S. We adopt the usual convention that $\vee\emptyset = 0$ and $\wedge\emptyset = 1$ for I and $\vee\emptyset = \emptyset$ and $\wedge\emptyset = S$ for $\mathcal{P}(S)$.

Definition 10.4.2. *Let H be a complete Heyting algebra and let G be a group. An H-fuzzy subset μ of G is called an H-**fuzzy subgroup** of G if the following conditions hold:*

(1) $\mu(e) = 1$, where e is the identity of G,
(2) $\mu(xy) \geq \mu(x) \wedge \mu(y)$ for all $x, y \in G$,
(3) $\mu(x^{-1}) \geq \mu(x)$ for all $x \in G$.

Example 10.4.3. *Let G be an (additive) Abelian group and let $nG = \{nx \mid x \in G\}$, where n is a positive integer. Let m be a positive integer. Let μ be the I-fuzzy subset of G defined as follows:* $\forall x \in G, \mu(x) = \vee\{1 - 2^{-k} \mid k \in N$ *and* $x \in m^k G\}$. *Then* $\mu(0) = 1$. *Note that μ measures the membership grade of x in μ by the degree to which x is divisible by m. The higher the power of m dividing x, the greater the degree of membership of x. We now show that μ is an I-fuzzy subgroup of G. Let $x, y \in G$. Suppose $x \in m^p G$ and $y \in m^q G$. then $x + y \in m^{p \wedge q} G$. Thus $\mu(x + y) \geq 1 - 2^{-p \wedge q} = (1 - 2^{-p}) \wedge (1 - 2^{-q})$. Hence $\mu(x + y) \geq \vee\{(1 - 2^{-p}) \wedge (1 - 2^{-q}) \mid p, q \in N\} = \mu(x) \wedge \mu(y)$. Since $m^k G$ is a group for all positive integers m and for all $k \in N$, it follows that $x \in m^k G$ if and only if $x^{-1} \in m^k G$. Thus it follows that $\mu(x) = \mu(x^{-1})$. Hence μ is an I-fuzzy subgroup of G.*

Example 10.4.4. *Let G be an (additive) Abelian group and let m be an integer. Define the $\mathcal{P}(\mathbb{N})$-fuzzy subset μ of G as follows: $\forall x \in G, \mu(x) = \{k \in \mathbb{N} \mid x \in m^k G\}$. Then $\mu(0) = \mathbb{N}$. Let $x, y \in G$. Since $x \in m^k G$ if and only if $-x \in m^k G, \mu(x) = \mu(-x)$. Now $x \in m^k G$ and $y \in m^j G$ imply $x + y \in m^{k \wedge j} G$. Thus $\mu(x + y) \supseteq \mu(x) \cap \mu(y)$. Hence it follows that μ is a $\mathcal{P}(\mathbb{N})$-fuzzy subgroup of G.*

Example 10.4.5. *Let G be an (additive) Abelian group and let $\mathbb{Z}^+$ denote the positive integers. Define the $\mathcal{P}(\mathbb{Z}^+)$-fuzzy subset μ of G as follows: $\forall x \in G$, $\mu(x) = \{n \in \mathbb{Z}^+ \mid x \in nG\}$. Then $\mu(0) = \mathbb{Z}^+$. Let $x, y \in G$. Then $x \in nG$ if and only if $-x \in nG$. Thus $\mu(x) = \mu(-x)$. Now $x \in nG$ and $y \in mG$ imply $x + y \in (n \wedge m)G$. Hence $\mu(x + y) \supseteq \mu(x) \cap \mu(y)$. Thus it follows that μ is an $\mathcal{P}(\mathbb{Z}^+)$-fuzzy subgroup of G. It is clear that G is divisible if and only if μ is a constant on G.*

Example 10.4.6. *Let P denote the set of all primes in $\mathbb{Z}$. Define the $\mathcal{P}(P)$-fuzzy subset μ of $\mathbb{Z}$ as follows: $\forall n \in \mathbb{Z}, \mu(n) = \{p \in P \mid p \text{ divides } n\}$. Then $\mu(0) = P$. Let $n, m \in \mathbb{Z}$. Then p divides n if and only if p divides $-n$. Thus $\mu(n) = \mu(-n)$. Also, p divides n and p divides m imply p divides $n + m$. Hence $\mu(n + m) \supseteq \mu(n) \cap \mu(m)$. Thus μ is a $\mathcal{P}(P)$-fuzzy subgroup of $\mathbb{Z}$. (In fact, μ is a fuzzy subring of $\mathbb{Z}$.) We can think of μ as measuring each $n \neq \pm 1$ by the extent to which n is a composite.*

Example 10.4.7. *Let $K[x]$ denote the ring of polynomials over the field K. Define the $\mathcal{P}(K)$-fuzzy subset μ of $K[x]$ as follows: $\forall p(x) \in K[x], \mu(p(x)) = \{k \in K \mid p(k) = 0\}$. Then $\mu(0) = K$. Let $p(x), q(x) \in K[x]$. Then $p(k) = 0$ if and only if $-p(k) = 0$. Thus $\mu(p(x)) = 0 = \mu(-p(x))$. Suppose $p(k) = 0$ and $q(k) = 0$. Then $p(k) + q(k) = 0$. Hence $\mu(p(x) + q(x)) \supseteq \mu(p(x)) \cap \mu(q(x))$. Thus μ is a $\mathcal{P}(K)$-fuzzy subgroup of $K[x]$. (In fact, μ is a $\mathcal{P}(K)$-fuzzy subring and $\mathcal{P}(K)$-fuzzy subspace of $K[x]$.) We can think of μ as measuring the polynomial by the set of its roots in K. If $p(x)$ is a constant polynomial, but not 0, then $\mu(p(x)) = \emptyset$, while $\mu(p(x)) = K$ if $p(x)) = 0$.*

Example 10.4.8. *Let K be a field and $Mat_n[K]$ denote the set of all $n \times n$ matrices over K, where $n \in \mathbb{N}$ with $n > 1$. Let K^n denote the set of all ordered n-tuples with entries from K. Define the $\mathcal{P}(K^n)$-fuzzy subset μ of $Mat_n[K]$ as follows: $\forall A \in Mat_n[K], \mu(A) = \{x \in K^n \mid Ax = kx \text{ for some } k \in K\}$. Then $\mu(\oslash) = K^n$, where $\oslash$ denotes the zero matrix. Let $A, B \in Mat_n[K]$. Then $Ax = kx$ if and only if $(-A)x = (-k)x$. Thus $\mu(A) = \mu(-A)$. Now $Ax = k_1 x$ and $Bx = k_2 x$ imply $(A + B)x = (k_1 + k_2)x$. Hence $\mu(A + B) \supseteq \mu(A) \cap \mu(B)$. Thus μ is a $\mathcal{P}(K^n)$-fuzzy subgroup of $Mat_n[K]$. (In fact, μ is a $\mathcal{P}(K^n)$-fuzzy subring and a $\mathcal{P}(K^n)$-fuzzy subspace of $Mat_n[K]$.) We can think of μ as measuring A by its set of eigenvectors.*

Example 10.4.9. *Let G and H be groups and let $Hom(G, H)$ denote the set of all homomorphisms of G into H. Let S be a subgroup of H. Define the $\mathcal{P}(Hom(G, H))$-fuzzy subset μ of G as follows: $\forall x \in G, \mu(x) = \{f \in$*

$Hom(G, H) \mid f(x) \in S\}$. Then $\mu(e) = Hom(G, H)$. Let $x, y \in G$. Then $f(x) \in S$ if and only if $f(-x) \in S$. Thus $\mu(x) = \mu(-x)$. Now $f(x), f(y) \in S$ implies $f(xy) = f(x)f(y) \in S$. Hence $\mu(xy) \supseteq \mu(x) \cap \mu(y)$. Then μ is a $\mathcal{P}(Hom(G, H))$-fuzzy subgroup of G.

Example 10.4.10. Let G be a group and H be (additive) Abelian group. Let $Hom(G, H)$ denote the set of all homomorphisms of G into H. Let S be a subgroup of H. Now $Hom(G, H)$ is a an Abelian group under the operation of pointwise addition of homomorphisms. Define the $\mathcal{P}(G)$-fuzzy subset μ of $Hom(G, H)$ as follows: $\forall f \in Hom(G, H)$, $\mu(f) = \{x \in G \mid f(x) \in S\}$. Then $\mu(\oslash) = S$, where $\oslash(x) = 0 \; \forall x \in G$. Let $f, g \in Hom(G, H)$. Then $f(x) \in S$ if and only if $(-f)(x) = -f(x) \in S$. Thus $\mu(f) = \mu(-f)$. Now $f(x), g(x) \in S$ imply $(f + g)(x) = f(x) + g(x) \in S$. Hence $\mu(f + g) \supseteq \mu(f) \cap \mu(g)$. Thus μ is a $\mathcal{P}(G)$-fuzzy subgrouip of $Hom(G, H)$. If S consists of just the zero element, then μ can be thought of measuring f by the size of its kernel.

Example 10.4.11. Let X be a set and S_X be the symmetric group on X, i.e., the group of all permutations of X. Define the $\mathcal{P}(X)$-fuzzy subset μ of S_X as follows: $\forall f \in S_X$, $\mu(f) = \{x \in X \mid f(x) = x\}$. Then $\mu(i) = X$, where i is the identity function on X. Let $f, g \in S_X$. Then $f(x) = x$ if and only if $f^{-1}(x) = x$. Thus $\mu(f) = \mu(f^{-1})$. Now $f(x) = x$ and $g(x) = x$ imply $(g \circ f)(x) = g(f(x)) = g(x) = x$. Hence $\mu(g \circ f) \supseteq \mu(g) \cap \mu(f)$. Thus μ is a $\mathcal{P}(X)$-fuzzy subgroup of S_X. We can think of μ as assigning higher degrees of membership to those permutations that fix larger subsets of X.

Example 10.4.12. Let $X = \{1, 2, ..., n\}$ be the vertices of a regular n-gon. Let D_n denote the dihedral group of all rigid motions of the n-gon that leave the n-gon coincident with itself. Define the $\mathcal{P}(X)$-fuzzy subset μ of D_n as follows: $\forall f \in D_n$, $\mu(f) = \{x \in X \mid f(x) = x\}$. Then $\mu(i) = X$, where i denotes the identity of D_n. Let $f, g \in D_n$. Then $f(x) = x$ if and only if $f^{-1}(x) = x$. Thus $\mu(f) = \mu(f^{-1})$. Now $f(x) = x$ and $g(x) = x$ imply $(g \circ f)(x) = x$. Hence $\mu(g \circ f) \supseteq \mu(g) \cap \mu(g)$. Thus μ is $\mathcal{P}(X)$-fuzzy subgroup of D_n. It is natural to call μ the fuzzy dihedral subgroup D_n.

Example 10.4.13. Let H be a group and let G be a group acting on H. That is, there is a homomorphism $f : G \rightarrow Aut(H)$, where $Aut(H)$ denotes the group of all automorphisms of H. Denote $f(x)(y)$ by y^x, where $x \in G$ and $y \in H$. Define the $P(G)$-fuzzy subset μ of H as follows: $\forall y \in H$, $\mu(y) = \{x \in G \mid y^x = y\}$. Then $\mu(e) = G$. Let $y \in H$. Then $f(x)(y) = y$ if and only if $f(x)(y^{-1}) = y^{-1}$. Thus $\mu(y) = \mu(y^{-1})$. Let $y_1, y_2 \in H$. Then $f(x)(y_1) = y_1$ and $f(x)(y_2) = y_2$ imply $f(x)(y_1y_2) = f(x)(y_1)f(x)(y_2) = y_1y_2$. Hence $\mu(y_1y_2) \supseteq \mu(y_1) \cap \mu(y_2)$. Thus μ is a $\mathcal{P}G)$-fuzzy subgroup of H. We can think of μ as measuring the degree of membership of an element h by the size of its isotropic subgroup. Special cases of this can also be considered. For example, the action of a group on itself by right translations or the action of a group on a normal subgroup by conjugation. It is natural to call this class of fuzzy subgroups as the class of isotropic fuzzy subgroups.

Example 10.4.14. *Let H be a group and let G be a group acting on H. Let S be a subgroup of G. Define the $\mathcal{P}(G)$-fuzzy subset μ of H as follows: $\forall y \in H$, $\mu(y) = \{x \in G \mid y^x \in S\}$. Then $\mu(e) = G$. Let $y \in H$. Then $f(x)(y) \in S$ if and only if $f(x)(y^{-1}) \in S$. Thus $\mu(y) = \mu(y^{-1})$. Let $y_1, y_2 \in H$. Then $f(x)(y_1) \in S$ and $f(x)(y_2) \in S$ imply $f(x)(y_1 y_2) = f(x)(y_1) f(x)(y_2) \in S$. Hence $\mu(y_1 y_2) \supseteq \mu(y_1) \cap \mu(y_2)$. Thus μ is a $\mathcal{P}(G)$-fuzzy subgroup of H.*

Example 10.4.15. *Let G be a group. For all $x \in G$, let $C(x)$ denote the centralizer of x in G. Define the $\mathcal{P}(G)$-fuzzy subset μ of G as follows: $\forall x \in G$, $\mu(x) = C(x) \cap C(x^{-1})$. Then $\mu(e) = G$. Let $x, y \in G$. Since $C(x) \cap C(x^{-1}) = C(x^{-1}) \cap C(x)$, $\mu(x) = \mu(x^{-1})$. Since $\forall z \in G$, $zx = xz$, $zx^{-1} = x^{-1}z$, $zy = yz$, $zy^{-1} = y^{-1}z$ imply $(xy)z = z(xy)$ and $(xy)^{-1}z = z(xy)^{-1}$, we have that $C(xy) \cap C((xy)^{-1}) \supseteq C(x) \cap C(x^{-1}) \cap C(y) \cap C(y^{-1})$. Hence $\mu(xy) \supseteq \mu(x) \cap \mu(y)$. Thus μ is a $\mathcal{P}(G)$-fuzzy subgroup of G. We can think of μ as measuring every element of G by its contribution toward making G Abelian.*

Example 10.4.16. *Let G be a group. Define a $\mathcal{P}(G)$-fuzzy subset μ of G as follows: $\forall x \in G$, $\mu(x) = \{z \in G \mid x \in \langle z \rangle\}$. Then $\mu(e) = G$. Let $x, y \in G$. Since $x \in \langle z \rangle$ if and only if $x^{-1} \in \langle z \rangle$, $\mu(x) = \mu(x^{-1})$. Since $x, y \in \langle z \rangle$ imply $xy \in \langle z \rangle$, $\mu(xy) \supseteq \mu(x) \cap \mu(y)$. Thus μ is a $\mathcal{P}(G)$-fuzzy subgroup of G. If G is abelian, then μ measures x by the set of all elements of G of which x is a multiple.*

Example 10.4.17. *Let R be a ring and let M_1 and M_2 be R-modules. Let X denote the set $Hom(M_1, M_2)$ of group homomorphisms of M_1 into M_2. Then X is a group, where the operation is point-wise addition. Define a $\mathcal{P}(X)$-fuzzy subset μ of X as follows: $\forall f \in X$, $\mu(f) = \{r \in R \mid rf(x) = f(rx)$ for all $x \in M_1\}$. Then μ is a $\mathcal{P}(R)$-fuzzy subgroup of X. Note that, $\mu(\oslash) = R$, where $\oslash(x) = 0$, $\forall x \in M_1$. We can think of μ as grading a group homomorphism from M_1 into M_2 by how close it comes to being an R-module homomorphism.*

Example 10.4.18. *Let M be an R-module, where R is a ring. Let $Aut(M)$ be the group of all group automorphisms of M, where only the group structure of M is considered. For all $f \in Aut(M)$, let $M_f = \{r \in R \mid rf(x) = f(rx)$ for all $x \in M\}$. Define the $\mathcal{P}(R)$-fuzzy subset μ of $Aut(M)$ as follows: $\forall f \in Aut(M)$, $\mu(f) = M_f \cap M_{f^{-1}}$. Then μ is a $\mathcal{P}(R)$-fuzzy subgroup of $Aut(M)$. Note that $\mu(i) = R$, where $i(x) = x$ $\forall x \in M$.*

10.5 Applications

The results in this section are mainly from [2]. Some of the results are also from [4]. The closure property of a fuzzy subgroup μ of a group $(G, \cdot)$ can be obtained by the inequality $\mu(x \cdot y) \geq T(\mu(x), \mu(y))$, where T is a t-norm. In Rosenfeld's original definition, T was the function 'minimum'. However, any t-norm provides a meaningful generalization of the closure property. We investigate two classes of fuzzy subgroups. The fuzzy subgroups in one class are called in [2] subgroup generated while those in the other are called function generated. Every fuzzy subgroup in these classes satisfies the above inequality with T given by $T(a, b) = (a + b - 1) \vee 0$. It turns out that every fuzzy subgroup in either class is isomorphic to one in the other. We show that a fuzzy subgroup satisfies the above inequality with $T = \wedge$ if and only if it is subgroup generated of a very special type. We then apply these notions to some abstract pattern recognition problems and to coding theory.

Fuzzy subgroups of a group were first defined by Rosenfeld [18]. Subsequently his definition was generalized by Negoita and Ralescu [17] and by Anthony and Sherwood [1]. This section studies the structure of two classes of the more general fuzzy subgroups. That structure is used to characterize Rosenfeld's original fuzzy subgroups.

Whenever $*$ is introduced as a group operation it is usually suppressed and juxtaposition is used.

Definition 10.5.1. *A function* $T : [0, 1] \times [0, 1] \to [0, 1]$ *is called a t-**norm** if the following properties hold for all* x, y, z *in* $[0, 1]$,
 (1) $T(x, 1) = x$,
 (2) $T(x, y) \leq T(z, y)$ *if* $x \leq z$,
 (3) $T(x, y) = T(y, x)$,
 (4) $T(x, T(y, z)) = T(T(x, y), z)$.

Some t-norms that are frequently encountered in the literature are T_m, *Prod*, and $\wedge$, where $T_m(x, y) = \vee\{x + y - 1, 0\}$ and $Prod(x, y) = xy$ for all $x, y \in G$.

Definition 10.5.2. *Let* $(G, *)$ *be a group. A function* $\mu : G \to [0, 1]$ *is called a **fuzzy subgroup** of* G *with respect to a t-norm* T *if* $\forall \, x, y \in G$,
 (1) $\mu(x, y) \geq T(\mu(x), \mu(y))$;
 (2) $\mu(x^{-1}) = \mu(x)$;
 (3) $\mu(e) = 1$, *where* e *is the identity of* G.

A t-norm T_1 is said to be **stronger** than a t-norm T_2 if and only if $T_1(x, y) \geq T_2(x, y)$ for all $x, y \in [0, 1]$. Clearly, if μ is a fuzzy subgroup with respect to a t-norm T, then μ is a fuzzy subgroup with respect to any 'weaker' t-norm. Since $\wedge$ is the strongest of all t-norms (see e.g. [19]), any fuzzy subgroup with respect to $\wedge$ is a fuzzy subgroup with respect to any other t-norm.

The following lemma provides some justification for the inclusion of condition (3) in the definition of a fuzzy subgroup.

Lemma 10.5.3. *Let* $(G, *)$ *be a group and* μ *a fuzzy subset of* G *such that the following conditions hold:*
 (1) $\mu(xy) \geq \mu(x) \wedge \mu(y)$,
 (2) $\mu(x^{-1}) = \mu(x)$,

(3) $\mu(e) > 0$.

Define the function η of G into $[0,1]$ by for all $x \in G, \eta(x) = \mu(x)/\mu(e)$. Then η is a fuzzy subgroup of G with respect to $\wedge$ such that $\eta(e) = 1$.

Proof. For all $x \in G$,

$$\begin{aligned}
\eta(e) &= \mu(e)/\mu(e) = \mu(xx^{-1})/\mu(e) \\
&\geq \mu(x) \wedge \mu(x^{-1})/\mu(e) \\
&= \mu(x) \wedge \mu(x)/\mu(e) = \mu(x)/\mu(e) = \eta(x).
\end{aligned}$$

Hence it follows that $\eta(e) = 1$ and that $0 \leq \eta(x) \leq 1$ for all $x \in G$. Now

$$\begin{aligned}
\eta(xy) &= \mu(xy)/\mu(e) \geq \mu(x) \wedge \mu(y)/\mu(e) \\
&= \mu(x)/\mu(e) \wedge \mu(y)/\mu(e) \\
&= \eta(x) \wedge \eta(y).
\end{aligned}$$

Moreover, $\eta(x^{-1}) = \mu(x^{-1})/\mu(e) = \mu(x)/\mu(e) = \eta(x)$. Therefore η is a fuzzy subgroup of G with respect to $\wedge$. $\qquad\square$

Definition 10.5.4. *Let G_1 and G_2 be groups and let μ_1 and μ_2 be fuzzy subgroups of G_1 and G_2, respectively, with respect to a t-norm T. The fuzzy subgroups μ_1 and μ_2 are called **isomorphic** if there an isomorphism f of G_1 onto G_2 such that $\mu_1 = \mu_2 \circ f$.*

In the next result, we can think that the value of the fuzzy subgroup at a particular point x will be found in a randomly selected subgroup. This yields a particular way of generating fuzzy subgroups.

We first review some basic definitions. Let X be a set and let $\mathcal{A}$ be a collection of subsets of X. Then $\mathcal{A}$ is called an **algebra of sets** if (1) $A \cup B \in \mathcal{A}$ whenever $A, B \in \mathcal{A}$ and (2) the complement of A, cA, is in $\mathcal{A}$ whenever A is in $\mathcal{A}$. An algebra $\mathcal{A}$ of sets is called a σ-**algebra** if $\emptyset \in \mathcal{A}$ and if every union of a countable collection of sets in $\mathcal{A}$ is again in $\mathcal{A}$. A function $P : \mathcal{A} \to \mathbb{R}$ is called a **probability measure** if $P(A) > 0 \forall A \in \mathcal{A} \setminus \{\emptyset\}$, $P(\Omega) = 1$, and $P(\cup_{i=1}^{\infty} A_i) = \Sigma_{i=1}^{\infty} P(A_i)$ for any countable union of disjoint sets $A_i, i = 1, 2, \dots$. The triple $(\Omega, \mathcal{A}, P)$ is called a **probability space**.

Theorem 10.5.5. *Let $(G, *)$ be a group and let $\mathcal{S}$ be the set of all subgroups of G. For all $x \in G$, let $S_x = \{S \in \mathcal{S} \mid x \in S\}$ and let $\mathbb{S} = \{S_x \mid x \in G\}$. Let $\mathbb{A}$ be any σ-algebra on $\mathcal{S}$ which contains the σ-algebra generated by $\mathbb{S}$ and let m be a probability measure on $(\mathcal{S}, \mathbb{A})$. Then $\mu : G \to [0,1]$ defined by $\mu(x) = m(S_x)$ for all $x \in G$ is a fuzzy subgroup of G with respect to T_m. A fuzzy subgroup obtained in this manner is called **subgroup generated**.*

Proof. Let $x, y \in G$. Suppose $S \in S_x \cap S_y$. Then S is a subgroup of G containing both x and y. Thus $xy \in S$. Hence $S \in S_{xy}$. Therefore, $S_{xy} \supseteq S_x \cap S_y$. Now

$$\mu(xy) = m(S_{xy}) \geq m(S_x \cap S_y) = m(S_x) + m(S_y) - m(S_x \cup S_y)$$
$$\geq \mu(x) + \mu(y) - 1.$$

Since $\mu(xy) \geq 0$, it follows that

$$\mu(xy) \geq (\mu(x) + \mu(y) - 1) \vee 0 = T_m(\mu(x), \mu(y)).$$

Clearly, $S_{x^{-1}} = S_x$. Thus $\mu(x^{-1}) = m(S_{x^{-1}}) = m(S_x) = \mu(x)$. Moreover, $S_e = S$. Hence $\mu(e) = \mu(S) = 1$. By Definition 10.5.2, μ is a fuzzy subgroup of G with respect to T_m. $\qquad\square$

In the next result, we can think of a point which travels in some random fashion through a group and we compute the probability of finding the point in a particular subgroup.

Theorem 10.5.6. *Let $(\mathbb{G}, +)$ be a group and let H be a fixed subgroup of $\mathbb{G}$. Let $(\Omega, \mathcal{A}, P)$ be a probability space and $(\mathcal{G}, \oplus)$ be a group of functions mapping Ω into $\mathbb{G}$ with $\oplus$ defined by point-wise addition in the range space. Assume that for all $f \in \mathcal{G}$, $G_f = \{\omega \in \Omega \mid f(\omega) \in H\}$ is an element of $\mathcal{A}$. Then $\nu : \mathcal{G} \to [0,1]$ defined by $\nu(f) = P(G_t)$ for all $f \in \mathcal{G}$ is a fuzzy subgroup of $\mathcal{G}$ with respect to T_m. A fuzzy subgroup obtained in this manner is called **function generated**.*

Proof. Let $f, g \in \mathcal{G}$. Suppose $\omega \in G_f \cap G_g$. Then $f(\omega) \in H$ and $g(\omega) \in H$. Since H is a subgroup of $\mathbb{G}$, $f(\omega) + g(\omega) = (f \oplus g)(w) \in H$. Thus $w \in G_{f \oplus g}$. Therefore, $G_{f \oplus g} \supset G_f \cap G_g$. Now

$$\nu(f \oplus g) = P(G_{f \oplus g}) \geq P(G_f \cap G_g) = P(G_f) + P(G_g) - P(G_f \cup G_g)$$
$$\geq \nu(f) + \nu(g) - 1.$$

Since $\nu(f \oplus g) \geq 0$, it follows that

$$\nu(f \oplus g) \geq (\nu(g) + \nu(g) - 1) \vee 0 = T_m(\nu(f), \nu(g)).$$

Note that $G_{\ominus f} = G_f$. Hence $\nu(\ominus f) = P(G_{\ominus f}) = P(G_f) = \nu(f)$. The identity in $\mathcal{G}$ is the function $0 : \Omega \to \mathbb{G}$ defined by $0(\omega) = 0$ for all $\omega \in \Omega$. Now $G_0 = \Omega$ and so $\nu(0) = P(\Omega) = 1$. Therefore, ν is a fuzzy subgroup of $\mathcal{G}$ with respect to T_m. $\qquad\square$

We next establish a basic equivalence between the notions of subgroup generated and function generated. Fuzzy subgroups with respect to the t-norm $\wedge$ are then characterized in terms of these concepts.

Theorem 10.5.7. *Every function generated fuzzy subgroup is subgroup generated.*

Proof. Let ν be a function generated fuzzy subgroup with $(\mathbb{G}, +)$, H, $(\mathcal{G}, \oplus)$ and $(\Omega, \mathcal{A}, P)$ satisfying the properties given in Theorem 10.5.6. Let $\mathcal{S}$ be the family of all subgroups of $\mathcal{G}$. For all $\omega \in \Omega$, let $S_\omega = \{f \in \mathcal{G} \mid f(\omega) \in H\}$. It follows easily that S_ω is a subgroup of $\mathcal{G}$ for all $\omega \in \Omega$. Let $\sigma : \Omega \to \mathcal{S}$ be defined by $\sigma(\omega) = S_\omega$ for all $\omega \in \Omega$. Let $\mathbb{A}$ and μ be the σ-algebra and measure induced on $\mathcal{S}$ by σ and the probability space $(\Omega, \mathcal{A}, P)$, that is, a collection A of subgroups of $\mathcal{G}$ is measurable if and only if $\sigma^{-1}(A) = \{\omega \in \Omega \mid S_\omega \in A\} \in \mathcal{A}$ and $m(A) = P(\sigma^{-1}(A))$. Consider subsets of $\mathcal{S}$ of the form $S_f = \{S \in \mathcal{S} \mid f \in S\}$. Let G_f be the subset of Ω described in Theorem 10.5.6. It follows that $\sigma^{-1}(S_f) = G_f \in \mathcal{A}$. Hence $S_f \in \mathbb{A}$ and $m(S_f) = P(G_f)$. Now $(\mathcal{G}, \oplus)$ and $(\mathcal{S}, \mathbb{A}, m)$ satisfy the descriptions of $(G, \cdot)$ and $(\mathcal{S}, \mathbb{A}, m)$ given in Theorem 10.5.5. Define the fuzzy subset μ of $\mathcal{G}$ by $\mu(f) = m(S_f)$ for all $f \in \mathcal{G}$. Then μ is a subgroup generated fuzzy subgroup of $\mathcal{G}$ by Theorem 10.5.5. However, $\mu(f) = m(S_f) = P(G_f) = \nu(f)$ for every $f \in \mathcal{G}$. Therefore, $\mu = \nu$ and so ν is a subgroup generated. $\qquad\square$

Theorem 10.5.8. *Every subgroup generated fuzzy subgroup is isomorphic to a function generated fuzzy subgroup.*

Proof. Let μ be a subgroup generated fuzzy subgroup with $(G, \cdot)$ and $(\mathcal{S}, \mathbb{A}, m)$ satisfying the properties given in Theorem 10.5.5. For every subgroup S of G, let $G_s = G$. Let $\mathbb{G} = \prod_{s \in \mathcal{S}} G_S$ and $H = \prod_{S \in \mathcal{S}} S$. It follows easily that $\mathbb{G}$ is a group with H a subgroup of $\mathbb{G}$, where the operation $+$ is inherited coordinate-wise from the operation $\cdot$ in G. For all $x \in G$, let $\phi_x : \mathcal{S} \to \mathbb{G}$ be defined by $\phi_x(S) = \psi_{x,S}$ for all $S \in \mathcal{S}$, where

$$\psi_{x,S}(S^*) = \begin{cases} e \text{ if } S^* \neq S, \\ x \text{ if } S^* = S. \end{cases}$$

Let $\mathcal{G} = \{\phi_x \mid x \in G\}$. Define an operation $\oplus$ on $\mathcal{G}$ by point-wise addition in $\mathbb{G}$, that is, $(\phi_x \oplus \phi_y)(S) = \psi_{x,S} + \psi_{y,S} = \psi_{xy,S} = \phi_{xy}(S)$ for every $S \in \mathcal{S}$. Now $(\mathcal{G}, \oplus)$ is a group and the function $\xi : G \to \mathcal{G}$ defined by $\xi(x) = \phi_x$ for every $x \in G$ is an isomorphism. Consider the sets $G_{\phi_x} = \{S \in \mathcal{S} \mid \phi_x(S) \in H\}$. Let S_x be the subset of $\mathcal{S}$ described in Theorem 10.5.5. Suppose $S \in G_{\phi_x}$. Then $\phi_x(S) = \psi_{x,S} \in H$. Hence $\psi_{x,S}(S) = x \in S$. Therefore, $S \in S_x$. Conversely, if $S \in S_x$, then $x \in S$ and $\psi_{x,S} = \psi_{x,S}(S) = x \in S$. Hence $S \in S_x$. Therefore, $G_{\phi_x} = S_x \in \mathbb{A}$. Now $(\mathbb{G}, +)$, H, $(\mathcal{S}, \mathbb{A}, m)$ and $(\mathcal{G}, \oplus)$ satisfy the descriptions of $(\mathbb{G}, +)$, H, $(\Omega, \mathcal{A}, P)$ and $(\mathcal{G}, \oplus)$ given in Theorem 10.5.6. Thus by Theorem 10.5.6, $\nu : \mathcal{G} \to [0, 1]$ defined by $\nu(\phi_x) = m(G_{\phi_x})$ for all $\phi_x \in \mathcal{G}$ is a function generated fuzzy subgroup of $\mathcal{G}$. Moreover, $\mu(x) = m(S_x) = m(G_{\phi_x}) = \nu(\phi_x) = \nu \circ \xi(x)$. Therefore, μ and ν isomorphic by Definition 10.5.4. $\qquad\square$

The preceding two theorems show that the notions of function generated and subgroup generated are essentially equivalent. These ideas provide some basic intuition concerning the meaning of the values of certain fuzzy subgroups. An obvious problem is to decide which fuzzy subgroups are generated in this way. The next few results provide a partial solution to this problem which will include all the fuzzy subgroups originally introduced by Rosenfeld in [18]. First, we review some definitions. Let A be a set of real numbers and consider a countable collection $\{I_n\}$ of open intervals which cover A, i.e., $A \subseteq \cup_n I_n$. Let $l(I_n)$ denote the length of the interval I_n. Define the **outer measure** $m^*(A)$ of A to be the infimum of all sums of the lengths of a cover, i.e., $m^*(A) = \wedge\{\Sigma l(I_n) | A \subseteq \cup I_n\}$. A set E is said to be **measurable** if for every set A, $m^*(A) = m^*(A \cap E) + m^*(A \cap cA)$. Let E be a measurable set. Define the **Lebesgue measure** $m(E)$ to be the outer measure of E. Then m is the set function obtained by restricting the set function m^* to the family $\mathfrak{M}$ of measurable sets.

Theorem 10.5.9. *Every fuzzy subgroup ν of G with respect to $\wedge$ is subgroup generated.*

Proof. Let $\mathcal{S}$ be the collection of all subgroups of G and let $\eta : [0,1] \to \mathcal{S}$ be defined by $\eta(t) = \nu_t$ for all t in $[0,1]$. Let $\mathbb{A}$ and m be the σ-algebra and measure induced on $\mathcal{S}$ by the function η, using Lebesgue measure P, on $[0,1]$. That is, a subset A of $\mathcal{S}$ is measurable if and only if $\eta^{-1}(A)$ is Lebesgue measurable and then $m(A) = P(\eta^{-1}(A))$. Let $x \in G$ and consider the set $S_x = \{S \in \mathcal{S} \mid x \in S\}$. For every t in $[0, \nu(x)]$, $\nu(x) \geq t$ and so $x \in \nu_t$. Moreover, if $t > \nu(x)$, then $x \notin \nu_t$. Therefore, $\nu_t \in S_x$ if and only if $t \in [0, \nu(x)]$. Hence $\eta^{-1}(S_x) = [0, \nu(x)]$ which is Lebesgue measurable. Now $(G, \cdot)$ and $(\mathcal{S}, \mathbb{A}, m)$ satisfy the conditions of Theorem 10.5.5. Hence $\mu : G \to [0,1]$ defined by $\mu(x) = m(S_x)$ for every x in G is a subgroup generated fuzzy subgroup. However, $\nu(x) = P[0, \nu(x)] = P(\eta^{-1}(S_x)) = m(S_x) = \mu(x)$ for every x in G. Therefore, $\nu = \mu$ and ν is subgroup generated. $\square$

Theorem 10.5.10. *Let μ be a subgroup generated fuzzy subgroup with $(G, \cdot)$ and $(\mathcal{S}, \mathbb{A}, m)$ and the sets S_x, for $x \in G$, as described in Theorem 10.5.5. If there exists $\mathcal{S}^* \in \mathbb{A}$ which is linearly ordered by set inclusion such that $m(\mathcal{S}^*) = 1$, then μ is a fuzzy subgroup with respect to $\wedge$.*

Proof. Let $x, y \in G$. Since $\mathcal{S}^*$ is linearly ordered, either $S_x \cap \mathcal{S}^* \subseteq S_y \cap \mathcal{S}^*$ or $S_y \cap \mathcal{S}^* \subseteq S_x \cap \mathcal{S}^*$. We may assume without loss of generality that $S_x \cap \mathcal{S}^*$. Suppose that $S \in S_x \cap \mathcal{S}^*$. Then $S \in S_y \cap \mathcal{S}^*$. Hence both x and y are in S. Since S is a group, $xy \in S$ and so $S \in S_{xy}$. Therefore, $S_x \cap \mathcal{S}^* \subseteq S_{xy}$. Now $m(S_x \cap \mathcal{S}^*) = m(S_x)$. Also, $m(S_{xy}) \geq m(S_x \cap \mathcal{S}^*) = m(S_x) \geq m(S_x) \wedge m(S_y)$. Therefore, $\mu(xy) \geq \mu(x) \wedge \mu(y)$ and so μ is a fuzzy subgroup with respect to $\wedge$. $\square$

Theorems 10.5.9 and 10.5.10 combine to yield the following characterization of fuzzy subgroups with respect to $\wedge$.

Theorem 10.5.11. *A fuzzy subgroup is a fuzzy subgroup with respect to $\wedge$ if and only if it is subgroup generated and the generating family possesses a subfamily of measure one which is linearly ordered by set inclusion.*

We now present an **application** for function generated fuzzy subgroups. We first consider a generalized recognition problem.

Suppose that F is a device which receives a stream of discrete inputs and produces a stream of discrete outputs. We make the following assumptions about F and about knowledge of the input and output.

(1) F is deterministic and acts independently on each individual input. That is, a particular input produces the same output each time that it is provided to F. However, the output which is produced from any specific input is not known.

(2) There is complete knowledge of the outputs. That is, the output stream is observable.

(3) The input stream is not observable. The possible inputs are known and estimates can be obtained of their relative frequencies in a large segment of the input stream.

(4) The outputs have an algebraic character in the sense that they can be identified with the objects in a group. Thus there is a method of combining the outputs which has the ordinary properties of a group operation.

Let $\mathcal{I}$ denote the collection of inputs and let $\mathcal{O}$ denote the collection of outputs. If $T \in \mathcal{I}$ then $F(T) \in \mathcal{O}$. Thus F is identified with a function from $\mathcal{I}$ into $\mathcal{O}$. Suppose that f is a known function of $\mathcal{I}$ into $\mathcal{O}$. Moreover, suppose that some particular characteristic of F which we call "faithfullness" is associated with solvability for x of an equation in the output group of the form $x + f(T) = F(T)$, where $+$ is the group operation. If for some $T \in \mathcal{I}$ a solution for x can be found in a given subgroup H, then the output $F(T)$ is called $H - f$ **faithful** to the input T. For a sufficiently large finite segment of the output stream and for a given function f and subgroup H, We consider the problem of estimating the proportion of the outputs which are $H - f$ faithful to their respective inputs.

In order to translate this problem into the setting of fuzzy subgroups, certain identifications are necessary. The outputs have already been identified with a group $(\mathbb{G}, +)$. The inputs may be identified with a probability space $(\Omega, \mathcal{A}, P)$, where $\Omega = I$, $\mathcal{A}$ is the power set of Ω, and $P(T)$ is the known estimate of the relative frequency of T in the input stream for each $T \in \Omega$. If $(\mathcal{G}, \oplus)$ is the set of all functions from Ω into $\mathbb{G}$ with $\oplus$ defined by point-wise addition in the range space, then both F and f may be identified with elements of $\mathcal{G}$. The function f is known while F is not known. Also, H is a fixed subgroup of $\mathbb{G}$. By Theorem 10.5.6, the fuzzy subset ν of $\mathcal{G}$ defined by $\nu(g) = P\{T \in \Omega \mid g(T) \in H\}$ is a function generated fuzzy subgroup of $\mathcal{G}$ with respect to T_m. Now $\nu(F)$ can be estimated by observing throughput stream over some finite segment and computing the percentage of those outputs which are in H. Also, $\nu(\ominus f) = \nu(f)$ is a known quantity since the function f is known. An output, $F(T)$, is $H - f$ faithful to T if and only if $x + f(T) = F(T)$ has a solution for x which is in H. This occurs if and only if $x = F(T) - f(T) = (F \ominus f)(T) \in H$. Therefore, $\nu(F \ominus f)$ is the probability that $F(T)$ is $H - f$ faithful to T. The solution to the original problem may now be identified with $\nu(F \ominus f)$. This may be estimated using $\nu(f)$, an estimate of $\nu(F)$, and the properties of the fuzzy subgroup, ν, in the following way: Since $T_m(\nu(F), \nu(\ominus f)) = T_m(\nu(F), \nu(f)) = (\nu(F) + \nu(f) - 1) \vee 0 \geq \nu(F) + \nu(f) - 1$, we have

$$\nu(F \ominus f) \geq T_m(\nu(F), \nu(\ominus f)) \tag{10.5.1}$$
$$= T_m(\nu(F), \nu(f))$$
$$= (\nu(F) + \nu(f) - 1) \vee 0$$
$$\geq \nu(F) + \nu(f) - 1.$$

Similarly,

$$\nu(F) = \nu(F \ominus f + f) \tag{10.5.2}$$
$$\geq T_m(\nu(F \ominus f), \nu(f))$$
$$= (\nu(F \ominus f) + \nu(f) - 1) \vee 0$$
$$\geq \nu(F \ominus f) + \nu(f) - 1$$

and

$$\nu(f) = \nu(f \ominus F \oplus F) \tag{10.5.3}$$
$$\geq T_m(\nu(f \ominus F), \nu(F))$$
$$= T_m(\nu(F \ominus f), \nu(F))$$
$$= (\nu(F \ominus f) + \nu(F) - 1) \vee 0$$
$$\geq \nu(F \ominus f) + \nu(F) - 1$$

From (10.5.1) and (10.5.2), we obtain

$$\nu(F) - (1 - \nu(f)) \leq \nu(F \ominus f) \tag{10.5.4}$$
$$\leq \nu(F) + (1 - \nu(F)).$$

From (10.5.1) and (10.5.3), we obtain

$$\nu(f) - (1 - \nu(F)) \leq \nu(F \ominus f) \tag{10.5.5}$$
$$\leq \nu(f) + (1 - \nu(F)).$$

Thus we obtain the following estimate for the solution $\nu(F \ominus f)$:

$$|\nu(F \ominus f) - \nu(f) \wedge \nu(F)| \leq 1 - \nu(f) \vee \nu(F). \tag{10.5.6}$$

The estimate is close only when $\nu(f)$ or $\nu(F)$ is close to 1. However, if ν can be shown to be a fuzzy subgroup with respect to $\wedge$ the situation changes considerably.

Suppose ν is a fuzzy subgroup with respect to $\wedge$. In this case (10.5.1), (10.5.2) and (10.5.3) become, respectively,

$$\nu(F \ominus f) \geq \nu(F) \wedge \nu(f), \tag{10.5.1'}$$

$$\nu(F) \geq \nu(F \ominus f) \wedge \nu(f), \tag{10.5.2'}$$

$$\nu(f) \geq \nu(F \ominus f) \wedge \nu(F). \tag{10.5.3'}$$

Now if $\nu(f) > \nu(F)$, then from (10.5.1') we have that $\nu(F \ominus f) \geq \nu(F)$ and from (10.5.2') we have that $\nu(F) \geq \nu(F \ominus f)$, i.e., $\nu(F \ominus f) = \nu(F)$. Similarly, if $\nu(f) < \nu(F)$, then from (10.5.1') we conclude that $\nu(F \ominus f) \geq \nu(f)$ and from (10.5.3') we conclude that $\nu(f) \geq \nu(F \ominus f)$. Hence $\nu(F \ominus f) = \nu(f)$. Therefore, if $\nu(f) \neq \nu(F)$, then

$$\nu(F \ominus f) = \nu(f) \wedge \nu(F)$$

and so we know the solution exactly. Finally, if $\nu(F) = \nu(f)$ the best one can say is

$$\nu(F) = \nu(f) \leq \nu(F \ominus f) \leq 1.$$

We now consider a more specific recognition problem. Let n be a natural number. An $n \times n$ array of the integers $1, 2, 3, ..., n^2$ is called a **pattern.** Suppose that F is a machine which accepts input patterns and produces output patterns. Each pattern may be identified with a transformation in S_{n^2}, the permutation group of n^2 objects, in the following way.

$$
\begin{array}{cccc}
k_1 & k_2 & \dots & k_n \\
k_{n+1} & k_{n+2} & \dots & k_{2n} \\
\dots & \dots & \dots & \dots \\
k_{n^2-n} & k_{n^2-n+1} & \dots & k_{n^2}
\end{array}
\Leftrightarrow
\begin{pmatrix}
1 & 2 & 3 & \dots & n^2 \\
k_1 & k_2 & k_3 & \dots & k_{n^2}
\end{pmatrix}
$$

Thus F is identifiable with a function from S_{n^2} into S_{n^2}. An output pattern is called **recognizable** if it is a composition of translations and rotations of the input. There is a subgroup H of S_{n^2} such that an output pattern $F(T)$ is recognizable if and only if there exists a transformation T^* in H such that $T^* \circ T = F(T)$. Suppose that estimates of the relative frequency of patterns in the input stream can be obtained. Let $\Omega = S_{n^2}$, $\mathcal{A}$ be the power set of S_{n^2} and P be a probability measure on S_{n^2} obtained from the estimates of the relative frequency of input patterns. Now $(\Omega, \mathcal{A}, P) = (S_{n^2}, \mathcal{A}, P)$ and $(\mathbb{G}, +) = (S_{n^2}, \circ)$ with $(\mathcal{G}, \oplus)$ and H defined appropriately. Let $f_1 : S_{n^2} \to S_{n^2}$ be defined by $f_1(T) = T$ for every $T \in S_{n^2}$. Then the output pattern $F(T)$ is recognizable if and only if the equation $x \circ f_1(T) = F(T)$ has a solution for x in H. This is the definition of $F(T)$ being $H - f_1$ faithful to T. Note that

$$\nu(f_1) = P\{T \in S_{n^2} | \ f_1(T) = T \in H\} = P(H).$$

From the discussion of the generalized recognition problem, the probability that the output is recognizable ($H - f_1$ faithful) is $\nu(F \ominus f_1)$ which may be estimated using the inequality

$$|\nu(F \ominus f_1) - P(H) \wedge \nu(F)| \leq 1 = P(H) \vee \nu(F).$$

Once again, in the event ν can be shown to be a fuzzy subgroup with respect to $\wedge$ we obtain

$$\nu(F \ominus f) = P(H) \wedge \nu(F)$$

if $P(H) \neq \nu(F)$; otherwise $P(H) = \nu(F) \leq \nu(F \ominus f) \leq 1$. It should be remembered that $P(H)$ is known and $\nu(F)$ can be estimated by the percentage of outputs which are in H.

We now consider the standard problem concerning the transmission of strings of 0's and 1's across a symmetric binary channel with noise. Let $B = \{0, 1\}$ and B^n denote the set of all binary n-tuples, $n \geq 2$. Then B^n is a group under componentwise addition modulo 2. Let $C \subseteq B^n$ denote the set of all codewords. Then C is a subgroup of B^n. We make the following identifications: $C = H = \mathcal{I} = \Omega$ and $B^n = G = \mathcal{O}$, where $H, G, \mathcal{I}, \Omega$, and $\mathcal{O}$ are as described above. In this situation, f is known, $f(T)$ is unknown, F is unknown, and $F(T)$ is known, where $T \in H$. We let f be the identity map since in the ideal situation their is no noise and so the output equals the input. We recall that $f(T)$ is observable and $\nu(F)$ can be estimated. Since f is the identity map, $\nu(f) = 1$. Thus $\nu(F \ominus f) = \nu(F)$ by equality (10.5.1). Thus $F(T)$ is $H - f$ faithful.

We now consider a more general situation. Assume that $1 \geq \nu(f) > \nu(F)$. This is a reasonable assumption since f represents the ideal situation while F represents the real world situation. Then by inequality (10.5.1'), $|\nu(F \ominus f) - \nu(f) \wedge \nu(f)| \leq 1 - \nu(f) \vee \nu(F)$ and so $|\nu(F \ominus f) - \nu(F)| \leq 1 - \nu(f)$. Now assume that $\forall a, b \in \text{Im}(\nu), a \neq b, 1 - \nu(f) < |a - b|$. Then $|\nu(F \ominus f) - \nu(F)| = 0$. Once again, $F(T)$ is $H - f$ faithful. Consider the fuzzy coset ν_f, where $\nu_f(g) = \nu(g \ominus f) \forall g \in \mathcal{G}$. Then $\nu_f(F) = \nu(F)$. Also, $g \in (\nu_f)_a \Leftrightarrow \nu_f(g) \geq a \Leftrightarrow \nu(g \ominus f) \geq a \Leftrightarrow g \ominus f \in \nu_a \Leftrightarrow g \in f \oplus \nu_a$.

We now note a structure result for the group $(\mathcal{G}, \oplus)$ and the fuzzy subgroup ν of $\mathcal{G}$. For all $f \in \mathcal{G}, (g \oplus g)(T) = g(T) \oplus g(T) = 0 \ \forall T \in \mathcal{I}$. Thus $2\mathcal{G} = \{\theta\}$, where $\theta(T) = 0 \ \forall T \in \mathcal{I}$. Thus $\mathcal{G} = \oplus_{g \in \mathcal{G}} \langle g \rangle$. For all $g \in \mathcal{G}$, define the fuzzy subset $\nu^{(g)}$ of $\mathcal{G}$ as follows: $\nu^{(g)}(g) = \nu(g)$ if $g \in \nu^*$ and $\nu^{(g)}(h) = 0$ if $h \in \mathcal{G} \backslash \{\theta, g\}$. Then $\nu^{(g)}$ is a fuzzy subgroup of $\mathcal{G}$. Hence $\nu = \oplus_{g \in S} \nu^{(g)}$ for some subset S of ν^*, where $|S| = |\nu^* : \mathbb{Z}_2|$ by [[12], Theorem 2.3, p. 96].

We now discuss some other applications of fuzzy group theory. The work in [11] is concerned with the classification of knowledges when they are endowed with some algebraic structure. By using the quotient group of symmetric knowledges, an algebraic method is given in [11] to classify them. Also the anti-fuzzy subgroup construction is used to classify knowledges.

In [8], fuzzy points are regarded as data and fuzzy objects are constructed from the set of given data on an arbitrary group. Using the method of least squares, optimal fuzzy subgroups are defined for the set of data and it is shown that one of them is obtained as a fuzzy subgroup by a set of some modified data.

In [20], a decomposition of an L-valued set (L a lattice) gives a family of characteristic functions which can be considered as a binary block-code.

Conditions are given under which an arbitrary block-code corresponds to an L-valued fuzzy set. An explicit description of the Hamming distance, as well as of any code distance, is also given, all in lattice-theoretic terms. A necessary and sufficient condition is given for a linear code to correspond to an L-valued fuzzy set. In such a case the lattice has to be Boolean.

References

1. J. M. Anthony and H. Sherwood, Fuzzy groups redefined, *J. Math. Anal. Appl.* 69 (1979) 124-130.
2. J. M. Anthony and H. Sherwood, A characterization of fuzzy subgroups, *Fuzzy Sets and Systems* 7 (1982) 297-305.
3. S. Burris and H. P. Sankappanavar, *A Course in Universal Algebra*, Springer-Verlag, New York, 1981
4. S-C Cheng and J. N. Mordeson, Applications of fuzzy algebra in automata theory and coding theory, Fifth IEEE International Conference on Fuzzy Systems, Proceedings vol. 1, 1996, 125-129.
5. P. S. Das, Fuzzy groups and level subgroups, *J. Math Anal. Appl.* 84 (1981) 264-269.
6. G. J. Klir, T.A. Folger, *Fuzzy Sets, Uncertainty, and Information*, 2nd ed., Prentice-Hall, Englewood Cliffs, NJ, 1994.
7. S. Kundu, Membership functions for a fuzzy group from similarity relations, *Fuzzy Sets and Systems* 101 (1999) 391-402.
8. T. Kuraoka and N-Y Suzuki, Optimal fuzzy objects for the set of given data in the case of the group theory, *Inform. Sci.* 92 (1996) 197-210.
9. R. Lowen, Convex fuzzy sets, *Fuzzy Sets and Systems* 19 (1980) 291-310.
10. D. S. Malik, J, N. Mordeson, and P. S. Nair, Fuzzy generators and fuzzy direct sums of abelian groups, *Fuzzy Sets and Systems* 50 (1992) 193-199.
11. M. Mashinchi and M. Mukaidono, Algebraic knowledge classification, *J. Fuzzy Math.* 2 (1994) 233-247.
12. J. N. Mordeson, Fuzzy subfields of finite fields, *Fuzzy Sets and Systems* 52 (1992) 93-96.
13. J. N. Mordeson, Bases of fuzzy vector spaces, *Inf. Sci.* 67 (1993) 87-92.
14. J. N. Mordeson, Bases of fuzzy algebraic substructures, *Fuzzy Sets and Systems* 62 (1994) 185-191.
15. G. C. Muganda and M. Garzon, On the structure of fuzzy groups, In P. P. Wang, editor, *Advances in Fuzzy Theory and Technology*, Vol. I, Book Wrights, Durham North Carolina, 1993, 23-42.
16. G. C. Muganda, Fuzzy algebras based on algebraic properties, *J. Fuzzy Math.* 6 (1998) 649-658.
17. C. V. Negoita and D. A. Ralescu, *Applications of Fuzzy Sets to Systems Analysis*, Wiley, New York, 1975, 54-59.
18. A. Rosenfeld, Fuzzy groups, *J. Math. Anal. Appl.* 35 (1971) 512-517.
19. B. Schweizer and A. Sklar, Statistical metric spaces, *Pacific J. Math.* 10 (1960) 313-334.
20. B. Seselja, A. Tepavcevic, and G. Vojvodic, *L*-fuzzy sets and codes, *Fuzzy Sets and Systems* 53 (1993) 217-222.

21. F-G Shi, L-fuzzy relations and L-fuzzy subgroups, *J. Fuzzy Math.* 8(2000) 491-499.
22. Z. Wang, Y. Yu and F. Dai, On T-congruence L-relations on groups and rings, *Fuzzy Sets and Systems* 119 (2001) 393-407.

Index

Index of Symbols

Printing: Krips bv, Meppel
Binding: Stürtz, Würzburg